Understanding
Semiconductor
Devices

The Oxford Series in Electrical and Computer Engineering

Adel S. Sedra, Series Editor, Electrical Engineering
Michael R. Lightner, Series Editor, Computer Engineering

Allen and Holberg, *CMOS Analog Circuit Design*
Bobrow, *Elementary Linear Circuit Analysis, 2nd Ed.*
Bobrow, *Fundamentals of Electrical Engineering, 2nd Ed.*
Campbell, *The Science and Engineering of Microelectronic Fabrication*
Chen, *Analog & Digital Control System Design*
Chen, *Linear System Theory and Design, 3rd Ed.*
Chen, *System and Signal Analysis, 2nd Ed.*
Comer, *Digital Logic and State Machine Design, 3rd Ed.*
Cooper and McGillem, *Probabilistic Methods of Signal and System Analysis, 3rd Ed.*
Dimitrijev, *Understanding Semiconductor Devices*
Franco, *Electric Circuits Fundamentals*
Fortney, *Principles of Electronics: Analog & Digital*
Granzow, *Digital Transmission Lines*
Guru and Hiziroğlu, *Electric Machinery & Transformers, 2nd Ed.*
Hoole and Hoole, *A Modern Short Course In Engineering Electromagnetics*
Jones, *Introduction to Optical Fiber Communication Systems*
Krein, *Elements of Power Electronics*
Kuo, *Digital Control Systems, 3rd Ed.*
Lathi, *Modern Digital and Analog Communications Systems, 3rd Ed.*
Martin, *Digital Integrated Circuit Design*
McGillem and Cooper, *Continuous and Discrete Signal and System Analysis, 3rd Ed.*
Miner, *Lines and Electromagnetic Fields for Engineers*
Roberts and Sedra, *SPICE, 2nd Ed.*
Roulston, *An Introduction to the Physics of Semiconductor Devices*
Sadiku, *Elements of Electromagnetics, 2nd Ed.*
Santina, Stubberud, and Hostetter, *Digital Control System Design, 2nd Ed.*
Schaumann and Van Valkenburg, *Design of Analog Filters*
Schwarz, *Electromagnetics for Engineers*
Schwarz and Oldham, *Electrical Engineering: An Introduction, 2nd Ed.*
Sedra and Smith, *Microelectronic Circuits, 4th Ed.*
Stefani, Savant, Shahian, and Hostetter, *Design of Feedback Control Systems, 3rd Ed.*
Van Valkenburg, *Analog Filter Design*
Warner and Grung, *Semiconductor Device Electronics*
Wolovich, *Automatic Control Systems*
Yariv, *Optical Electronics in Modern Communications, 5th Ed.*

Understanding Semiconductor Devices

Sima Dimitrijev
Griffith University

New York • Oxford
OXFORD UNIVERSITY PRESS
2000

Oxford University Press

Oxford   New York
Athens   Auckland   Bangkok   Bogotá   Buenos Aires   Calcutta
Cape Town   Chennai   Dar es Salaam   Delhi   Florence   Hong Kong   Istanbul
Karachi   Kuala Lumpur   Madrid   Melbourne   Mexico City   Mumbai
Nairobi   Paris   São Paulo   Singapore   Taipei   Tokyo   Toronto   Warsaw

and associated companies in
Berlin   Ibadan

Copyright © 2000 by Oxford University Press, Inc.

Published by Oxford University Press, Inc.,
198 Madison Avenue, New York, New York, 10016
http://www.oup-usa.org

Oxford is a registered trademark of Oxford University Press

Library of Congress Cataloging-in-Publication Data

Dimitrijev, Sima, 1958–
    Understanding semiconductor devices / by Sima Dimitrijev.
        p.   cm. — (The Oxford series in electrical and computer engineering)
    Includes bibliographical references.
    ISBN 0-19-513186-X (cloth)
    1. Semiconductors.   2. Integrated circuits–Design and
construction.   I. Title.   II. Series.
    TK7871.85 .D547   1999   621.3815'2—dc21   99-35429
                                                          CIP

Printing (last digit): 9 8 7 6 5 4 3 2

Printed in the United States of America
on acid-free paper

To Milan and Mirjana

CONTENTS

PART 2: ADVANCED TOPICS

PREFACE

Electronics is a rapidly developing discipline, placing two seemingly contradictory demands on electronic-engineering education: (1) to provide knowledge that will remain valuable as new technologies and concepts are developed and (2) to prepare the graduate for a quick and productive start in the real world of electronics. To achieve the first goal, there is no alternative but to ensure the depth of fundamentals in the undergraduate curriculum. Understanding semiconductor devices, as the building blocks of electronic circuits, is clearly of fundamental importance. The challenge is to present the semiconductor devices in a way that most efficiently reconciles the above-stated demands on contemporary education. In response to this challenge, this book is written to achieve the following aims.

1. MEETING THE STUDENTS' NEED TO UNDERSTAND

Understanding of principles, effects, and phenomena enables engineers to design and create useful things. The necessary threshold of understanding is achieved when the natural human curiosity is awoken. This is because an interplay of self-learning and creation begins beyond that point. This process never ends, lasting throughout an entire engineering career; we can refer to this as *life-long learning*. The following features of the book address the issue of understanding from the students' perspective:

(a) intuitive explanations of the underlying scientific concepts,

(b) explained fundamental equations, as opposed to using mathematics to explain physical phenomena,

(c) explained energy-band diagrams to provide rigorous and powerful presentations, which are easy to understand through an analogy with balls on a solid surface and bubbles in water, and

(d) intriguing review questions and a comprehensive set of worked-out examples and problems.

2. MOTIVATING APPROACH AND FLEXIBLE TEXT ORGANIZATION

Overloaded curriculum and *lack of motivation* to deal with "dry" equations and theories are two important issues that have to be properly addressed to ensure the necessary depth of fundamentals is actually achieved. As the professors design courses to maximize the efficiency in the given circumstances, the students need a flexible textbook that can be used without strictly prescribed prerequisite knowledge. The organization of this book provides a vertical (electronics-to-physics) hierarchy and a lateral (in terms of contents) flexibility:

(a) The *electronics-to-physics* approach motivates the students to learn the underlying scientific concepts and mathematical models by providing appropriate context and a continuous progression from common sense toward more abstract concepts.

(b) **Part I** covers the fundamental material that every electronic-engineering student should be introduced to:

Chapter 1: Resistors: Introduction to Semiconductors
Chapter 2: Capacitors: Reverse-Biased P–N Junction and MOS Structure
Chapter 3: Diodes: Forward-Biased P–N Junction and Metal–Semiconductor Contact
Chapter 4: Basics of Transistor Applications (Controlled Switch and Amplifier)
Chapter 5: MOSFET (Principles, Technology, Models, and SPICE Parameters)
Chapter 6: BJT (Principles, Technology, Models, and SPICE Parameters)

Some of that material would be introduced in prerequisite courses, in which case there is no need to formally cover it, although the students will find it quite useful as a link to the relevant material presented elsewhere. In particular, the starting points for considerations of different devices/group of devices (e.g. Sections 1.1, 2.1, 3.1.1, and Chapter 4) are fairly easy and there is no problem in formally omitting them; however, many students will find them quite helpful.

Part II consists of six advanced topics:
Chapter 7: Advanced and Specific IC Devices and Technologies
Chapter 8: Photonic Devices
Chapter 9: Microwave FETs and Diodes
Chapter 10: Power Devices
Chapter 11: Semiconductor Device Reliability
Chapter 12: Quantum Mechanics

This enables the professors to appropriately insert selected topics to suit the specific aims and coverage of their course.

(c) There is no need for a solid-state physics prerequisite, which helps with the increasingly overloaded electronics-engineering curriculum. However, it is assumed that the students will be familiar with concepts of general physics, essential mathematics (including derivatives and integrals), and the basics of network analysis.

3. PROVIDING CAD "NUTS AND BOLTS"

SPICE simulator is widely used for computer-aided design of electronic circuits. It is relatively easy to get the SPICE "engine" going and to obtain some results. The situation changes, though, when the analysis involves the characteristics of the circuit elements, and the user has to start playing with the numerous device parameters. Although the manuals and circuit books typically list the device parameters, they do not explain them; it is assumed that the students and engineers know those parameters. The SPICE contents of this book provide a unique reference and pedagogical tools:

(a) tabular presentations of the equations used as device models,
(b) hierarchically classified and explained SPICE parameters,
(c) case by case descriptions of parameter measurement techniques, and
(d) SPICE-based exercises that illustrate the effects and the meaning of individual device parameters (*Computer Exercises Manual* is provided on the CD enclosed).

4. LINKING THE THEORETICAL PRINCIPLES TO REAL-LIFE ISSUES

As stated earlier, the students need to acquire both knowledge of fundamentals (for a long-lasting successful career) and pragmatic skills (for a quick start in the dynamically developing world of electronics). Although it is generally accepted that the theoretical principles and the real-life issues should not be presented as separate blocks, their integration is still seen as the greatest teaching challenge. The following points illustrate the level of integration achieved in this book:

(a) The device descriptions develop from application concepts (basic circuits). The underlying scientific concepts are not introduced as inventory knowledge, but are "discovered" *as needed*. The introductions of particular device effects are explained by directly linked descriptions of what is happening on an atomic scale.
(b) Direct links are established between the fundamental equations and pragmatic SPICE models.
(c) Practically important issues such as reliability and device parameter measurements are presented with language and approaches that are directly related to the descriptions of fundamental device electronics and technology.

The following supporting material is provided on the CD, enclosed with the book: (1) *Computer Exercises Manual* and (2) *Interactive MATLAB Animations*. The SPICE-based computer exercises relate to practically important CAD "nuts and bolts." On the other hand, the interactive MATLAB animations are designed to enable a quicker and deeper introduction and understanding of the underlying theoretical concepts. In addition to the interactive feature, they also utilize motion and color to directly support the explanations and graphs from many sections of the textbook. Further information and supporting material will be placed on *http://www.gu.edu.au/school/mee/PPages/Sima/*.

ACKNOWLEDGMENTS

A number of people have contributed to the development of this book in different ways. I wish to sincerely thank the anonymous reviewers who provided me with an irreplaceable guide during different stages of the book development. The following colleagues read different versions of the manuscript and gave me extremely valuable feedback and support: Professor Hong Koo Kim, University of Pittsburgh; Professor John F. Wager, Oregon State University; and Professor Anthony J. Walton, University of Edinburgh.

I am indebted to my students whose feedback was a valuable guide in my attempts to find easy and motivating explanations. I wish to specifically thank Mr. Aaron Harwood, who read the first chapters of the first draft and provided me with valuable comments, and in particular for his help on related software projects.

I truly appreciate the specific and quite important roles of three people: Professor H. Barry Harrison, Griffith University; Professor Ninoslav Stojadinovic, University of Nis; and Professor Kenneth F. Galloway, Vanderbilt University. Indirectly, they profoundly influenced the development of this book.

I am particularly grateful to Mr. Peter Gordon of Oxford University Press for his demonstrated vision, valuable advice, and continuing motivation. I believe it takes both *vision* and *persistence* to complete a project like this, and Peter has provided me with the much-needed intellectual support. Other people at Oxford University Press helped me with a number of important aspects, in particular, Ms. Jasmine Urmeneta as Assistant Editor, Ms. Karen Shapiro as Senior Project Editor, and Ms. Arline Keithe as Copyeditor.

Finally, I wish to acknowledge the role of my wife, Vesna, and to thank her for the understanding and the support I received during the time spent on this book.

Sima Dimitrijev

PART 1

THE FUNDAMENTALS

Chapter *1*

RESISTORS
Introduction to Semiconductors

Resistors are very simple devices, and it is easy to explain their basic characteristics and integrated-circuit implementation using simple models. However, the text of this chapter extends well beyond the simplest models needed to explain resistors. The aim is to introduce some very important, although apparently abstract concepts, using the simplicity of resistors as a sound reference. The solid-state physics concepts introduced in this chapter include drift current, diffusion, carrier mobility, carrier concentration, and energy-band model. This chapter also introduces some fundamental microelectronic-device processing steps, namely photolithography and doping by diffusion.

1.1 THE BASICS: RESISTOR STRUCTURE AND DRIFT CURRENT

1.1.1 Integral Form of Ohm's Law

If an electric–potential difference is established between the terminals of a resistor, for example, by connecting the resistor to a voltage source (battery) as shown in Fig. 1.1, a limited current will flow through the resistor. How large a current, I, will flow through the resistor depends on (1) how large the potential difference $V = \varphi_1 - \varphi_0$ is, and (2) how large the resistance, R, of the resistor is. An increase in the potential difference, or in other words the voltage applied across the resistor, increases the current. An increase in the resistance, however, reduces the current. These dependencies are expressed by Ohm's law:

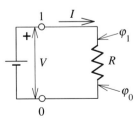

Figure 1.1 Current I induced by voltage V is limited by resistor R.

3

$$I = \frac{V}{R}.$$ (1.1)

1.1.2 Structure of Integrated-Circuit Resistors

As mentioned earlier, the aim of this chapter is to introduce some fundamental semiconductor phenomena using the easily understood concepts related to resistors. To this end, only integrated-circuit resistors are considered in this chapter. The basic integrated-circuit concepts are introduced in Appendix A.

Resistors in integrated circuits are made of a resistive body surrounded by an insulating medium, and contacted at the ends by conductive stripes (usually thin metal layers), as shown in Fig. 1.2. The resistance of the resistor depends on the resistive property of the body, called resistivity (ρ), and its dimensions in the following way: (1) an increase in the length of the resistor, L, increases its resistance as the carriers making the current have to travel a longer distance and (2) an increase in the cross-sectional area, $A = x_j W$, decreases the resistance as more current carriers can flow in parallel. Therefore, the resistance can be expressed as

$$R = \rho \frac{L}{x_j W}$$ (1.2)

Note that the curved ends of the resistive body are neglected and the resistor shape is considered as a rectangular prism with dimensions L, W, and x_j.

The reciprocal value of the resistivity, called conductivity, is more frequently used to characterize the resistive/conductive properties of semiconductor layers. The conductivity is denoted by σ, thus

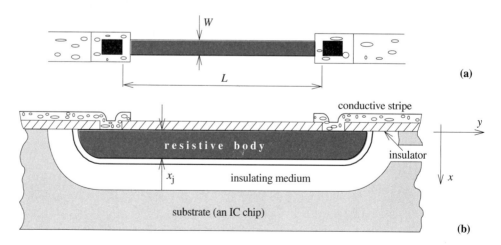

Figure 1.2 Integrated-circuit resistor: (a) top view and (b) cross section.

$$\sigma = \frac{1}{\rho} \tag{1.3}$$

Different resistors in an integrated circuit can have different and independent lengths L and widths W, although there is a minimum dimension that is technologically achievable. The value of the ratio L/W is used to adjust the resistor values while designing integrated circuits. The conductivity (σ) and the thickness of the resistive layer (x_j) are technological parameters, and they have the same values for all the resistors made with one type of resistive layer. There will be only a few different types of layers (with different conductivities and thicknesses) available for making resistors. The term $\rho/x_j = 1/(\sigma x_j)$, which appears in Eq. (1.2), is frequently expressed as one variable called sheet resistance, R_S:

$$R_S = \frac{1}{\sigma x_j} = \frac{\rho}{x_j} \tag{1.4}$$

The sheet resistance (R_S) is a quantity that can be measured more easily than the conductivity σ and the layer thickness x_j. If the resistance of a test resistor with known L and W is measured, the sheet resistance of the associated layer can easily be calculated from the following equation:

$$R = R_S \frac{L}{W} \tag{1.5}$$

The above equation is obviously obtained after replacement of ρ/x_j in Eq. (1.2) by R_S. The unit of the sheet resistance is essentially Ω, however, it is typically expressed as $\Omega/\square$, indicating that it represents the resistance of a squared resistor ($L = W$).

▆ **Example 1.1 Using Sheet Resistance**

Design a resistor of 3.5 $k\Omega$ using a layer with sheet resistance of 200 $\Omega/\square$. The minimum dimension achievable by the particular technology is 1 μm.

Solution: From Eq. (1.5) we find

$$\frac{L}{W} = \frac{R}{R_S} = \frac{3500}{200} = 17.5$$

Any combination of L and W that gives the ratio of $L/W = 17.5$ will provide the needed resistor of 3.5 $k\Omega$. However, a rule in microelectronics states that the performance of the microelectronic circuits is maximized if the device dimensions are minimized. Because of that, the optimum solution is $W = 1$ μm, and $L = 17.5$ μm.

▪ **Example 1.2 Selecting Sheet Resistance**

A resistor of $0.5\,k\Omega$ is needed. There are four layers available with the following sheet resistances: $R_{S1} = 5\,\Omega/\square$, $R_{S2} = 200\,\Omega/\square$, $R_{S3} = 1.5\,k\Omega/\square$, $R_{S4} = 4\,k\Omega/\square$. Which one of the four layers would you use?

Solution: The value of the resistor is closest to the sheet resistance of the second layer ($200\,\Omega/\square$), therefore the smallest resistor (in terms of area $L \times W$) can be obtained with this layer. It is easy to find that $L/W = 2.5$.

1.1.3 Differential and Integral Physical Quantities

The difference and relationship between the resistance R and resistivity ρ needs more careful consideration. We can assign a resistivity value to any point in the resistive body. If the resistive body is a homogeneous one, the resistivity will be the same throughout the body, but if it is not the resistivity will have different values at different points. Typically, the resistivity is not uniform in integrated-circuit resistors, but is smallest at the surface, increasing inside the material. Equivalently, the conductivity is highest at the surface, reducing inside the material. This is illustrated in Fig. 1.3 for the resistor shown in Fig. 1.2. Because of this property, the conductivity and the resistivity are called local or *differential* quantities. Differential quantities are needed to describe the internal structure (sometimes

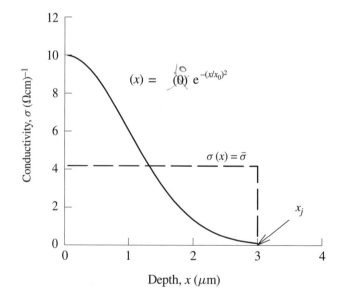

Figure 1.3 A typical variation of conductivity from the surface into the bulk of an integrated circuit resistor (solid line) and uniform-conductivity approximation (dashed line); in the example, $\sigma(0) = 10\ (\Omega\,\text{cm})^{-1}$ and $x_0 = \sqrt{2}\ \mu\text{m}$

called microstructure) of the semiconductor devices. Information on the conductivity at different points of the resistive body, however, cannot be used to directly determine the value to which the resistor will limit the terminal current (I) when a certain terminal voltage (V) is applied. That is why it is necessary to determine the overall, or *integral* resistance R, after which Ohm's law (Eq. 1.1) can simply be applied. Therefore, the integral quantities are needed to describe terminal or integral characteristics of the semiconductor devices.

The equations that relate differential and integral quantities to each other have to be carefully used. The relationships between the resistance R as an integral quantity and the conductivity σ as a differential quantity (Eqs. 1.2, 1.4, and 1.5) appear very simple, but this is because they are simplified to account for only the simplest case of uniform (or homogeneous) resistive bodies. It is these simplifications that are frequently overlooked and consequently mistakes are made. It will always be necessary to make some simplifications, or approximations, when modeling the real world in order to be able to understand it. Moreover, we will always be trying to make as many approximations as possible because it makes our understanding more efficient, but it is important to keep in mind the limitations that are introduced by the approximations made.

In the case of the resistor illustrated in Figs. 1.2 and 1.3, we see that Eq. (1.4) cannot be directly used to calculate the sheet resistance R_S. To be able to use Eq. (1.4), it is necessary to find a single value for the conductivity that will provide the most suitable uniform-conductivity approximation of the real conductivity dependence. Typically, the average value of the conductivity is used as the best uniform-conductivity approximation, as shown in Fig. 1.3 by the dashed lines. The average conductivity, $\overline{\sigma}$, can be found by equating the rectangular area $\overline{\sigma}x_j$ defined by the dashed lines in Fig. 1.3 with the area enclosed between the real conductivity curve and x and y axes, which is calculated as

$$\int_0^{x_j} \sigma(x)\, dx \approx \int_0^{\infty} \sigma(x)\, dx \tag{1.6}$$

Therefore,

$$\overline{\sigma} \approx \frac{1}{x_j} \int_0^{\infty} \sigma(x)\, dx \tag{1.7}$$

The average value of the conductivity determined in this way can now be used to calculate the sheet resistance by Eq. (1.4): $R_S = 1/(\overline{\sigma}x_j)$.

The above-described approach of calculating the sheet resistance enables a satisfactory estimation of the overall resistance, and its result can be used to determine the terminal voltage and current using Ohm's law. This agreement is due to the fact that the terminal current is an integral quantity, integrating any possible current flow through the resistive body in a way similar to what Eq. (1.7) does with the conductivity. It is not difficult to imagine that a larger portion of the terminal (integral) current I will flow close to the surface, due to the higher conductivity, than through the lower part of the resistor where the conductivity is low, and that all these larger and smaller current streams integrate into the terminal current I at the resistor terminals. However, the use of this approach of averaging the conductivity to find the terminal current does not provide any information on how the current flow is distributed inside the resistive body.

■ Example 1.3 **Average Conductivity**

(a) Find the average conductivity for the layer shown in Fig. 1.3.

(b) Design a 100-$k\Omega$ resistor using this layer, if it is known that the minimum dimension achievable by the particular technology is 1 μm.

Solution:

(a) Putting the conductivity function as given in Fig. 1.3 into the integral of Eq. (1.7) one obtains

$$\overline{\sigma} = \frac{10}{x_j} \int_0^\infty e^{-x^2/2} \, dx$$

where $x_j = 3.0 \, \mu m$. It is known from mathematics that the solution of the above integral (Laplace integral) is

$$\int_0^\infty e^{-x^2/2} \, dx = \sqrt{\pi/2} \, \mu m$$

The solution of the integral is expressed in micrometers as x and $x_0 = \sqrt{2}$ are given in micrometers. Therefore, the average conductance is

$$\overline{\sigma} = \frac{10 \, (\Omega \, cm)^{-1}}{3 \, \mu m} \sqrt{\pi/2} \, \mu m = 4.18 \, (\Omega \, cm)^{-1}$$

(b) Calculate first the sheet resistance:

$$R_S = \frac{1}{\overline{\sigma} x_j} = \frac{1}{4.18 \, (\Omega \, cm)^{-1} \times 3 \times 10^{-4} \, cm} = 797.4 \, \Omega/\square$$

Thus, the number of squares needed is

$$\frac{L}{W} = \frac{R}{R_S} = \frac{100,000}{797.4} = 125$$

As $W = 1 \, \mu m$, the length of the resistor needs to be $L = 125 \, \mu m$.

1.1.4 Drift Current: Differential Form of Ohm's Law

The approach of averaging the conductivity of semiconductor layers, described in the previous section, enables someone to design resistors without knowing much of what is happening inside the resistor. Can you imagine, however, the feeling of the designer if it happens that a proudly designed resistor does not show a linear current-to-voltage (I–V) characteristic but rather a saturating-type curve? There must have been a limitation the designer was not aware of! Was that a new effect? Could it be exploited for something useful? Was the effect reproducible? The answers to all these questions lie in the knowledge of what is happening inside the resistor. This section will not provide answers to all these questions, however, it will provide an insight into how the distribution of the carrier current is related to the conductivity.

To help ourselves, consider the simplest (even though rarely realistic) case of a homogeneous resistive body with length L and cross-sectional area A (can be $A = Wx_j$ as in Fig. 1.2). Knowing that it is the differential quantities that enable us to describe the internal structure of the semiconductor devices, let us transform the quantities from Ohm's law (Eq. 1.1), which are all integral quantities, into their differential counterparts. We already know that for this simple case we can easily transform the resistance R into the conductivity σ using Eqs. (1.4) and (1.5). The differential counterpart for the terminal current I is the current density, denoted by j. The current density in a homogeneous resistive body has the same value at any point of the resistor cross section, which is simply calculated as

$$j = \frac{I}{A} \tag{1.8}$$

The differential counterpart of the terminal voltage is electric field, denoted by E. If the electric field has a constant value in the considered domain, as is the case for our example, than the electric field can be calculated as

$$E = \frac{V}{L} \tag{1.9}$$

Replacing I, V, and R in the integral form of Ohm's law (Eq. 1.1) by j, E, and σ as given in Eqs. (1.4), (1.5), (1.8), and (1.9), the differential form of Ohm's law is obtained:

$$j = \sigma E \tag{1.10}$$

The differential form of Ohm's law can also be applied in the case of nonhomogeneous resistive bodies when the current density is not uniformly distributed over the resistor cross section, because of which it is said to be more general than the integral form.

The differential form of Ohm's law as given by Eq. (1.10) is applicable in the case in which the current flows in one direction only, which is basically the direction of the electric field. In the example of our resistor (Fig. 1.2), this one-dimensional form of Ohm's law can be applied to the central part of the resistor where the current flow is parallel to the surface and an axis, say y axis, can be placed along the current path. The one-dimensional form of Ohm's law, however, is not useful if we want to study the corner effects, as two dimensions are needed to express the circle-like current paths in these regions. That is why in the most general case the current density is considered to be a vector quantity, expressing not only the value of the current density at a certain point of the resistive body, but also the direction of the current flow. Note that the current density can in general take any direction in space, unlike the terminal current, which can flow only in one of two possible directions, and therefore is successfully expressed by making the current value positive or negative depending on the direction. The electric current takes no other path but the one traced by the electric-field lines, which means that the direction of the current density vector and the electric field vector are the same. Expressing the current density and the electric field as vectors, $\mathbf{j}$ and $\mathbf{E}$, the most general form of Ohm's law is obtained:

$$\mathbf{j} = \sigma \mathbf{E} \tag{1.11}$$

It is important to remind ourselves that the electric field E is essentially related to electric potential φ, the electric field being equal to the negative value of the slope of electric potential change along a direction in space (say y axis): $E \approx -\Delta\varphi/\Delta y$, or in a strict mathematical form:

$$E = -d\varphi/dy \tag{1.12}$$

In the specific case of $E = \text{const}$, $d\varphi/dy = \Delta\varphi/\Delta y = -(\varphi_1 - \varphi_0)/L = -V/L$, which reduces Eq. (1.12) to Eq. (1.9). In the three-dimensional case, Eq. (1.12) is generalized as

$$\mathbf{E} = -\frac{\partial\varphi}{\partial x}\mathbf{x}_u - \frac{\partial\varphi}{\partial y}\mathbf{y}_u - \frac{\partial\varphi}{\partial z}\mathbf{z}_u \tag{1.13}$$

where $\mathbf{x}_u$, $\mathbf{y}_u$, and $\mathbf{z}_u$ are the unit vectors in the x-, y-, and z-directions, respectively, while ∂s denote partial derivatives [they are used to denote that a derivation is with respect to one variable while others are treated as constants for that particular derivative; if there is one only variable, the partial derivative reduces to the ordinary derivative, as in Eq. (1.12)]. Using Eqs. (1.12) and (1.13), the one-dimensional (along y axis) and three-dimensional drift current can be expressed as

$$j = -\sigma\frac{d\varphi}{dy}, \quad \mathbf{j} = -\sigma\left(\frac{\partial}{\partial x}\mathbf{x}_u + \frac{\partial}{\partial y}\mathbf{y}_u + \frac{\partial}{\partial z}\mathbf{z}_u\right)\varphi = -\sigma\nabla\varphi \tag{1.14}$$

The importance of Eq. (1.11), or equivalently Eq. (1.14), lies in the fact that it models one of the most important transport mechanisms appearing in semiconductor devices, which is the transport induced by an electric field, or equivalently an electric potential difference. If an electric potential difference is created across a medium having mobile charged particles, a nonequilibrium situation is created, and the particles will start flowing as a reaction aimed at diminishing the potential difference. This means that if the carriers are negatively charged, for example, electrons, they will flow toward the higher potential in order to find the positive charges creating the potential difference and neutralize them.

> The electric current induced by an electric potential difference, or equivalently an electric field, is called *drift current*.

It is important to know that the drift current is by no means the unique current mechanism. Other types of forces that have nothing to do with the electric potential difference or the electric field can produce an electric current; two examples are the currents induced by carrier-concentration difference (diffusion current) and by temperature difference (thermal current). The diffusion current is explained in Section 1.3.

The drift-current equation (1.11) can be used to determine the current density, but only if the values for the conductivity and the electric field are known. That is why we still cannot handle the problem of the corner effects, as we need to know precisely how the electric field is bent there. To handle this problem an additional fundamental equation, introduced in Chapter 2, is needed. In this chapter we will limit ourselves to the conductivity σ, which is as important as the electric field for calculating the drift current. A further analysis will be done in order to understand how the conductivity of a semiconductor can be influenced externally, which is the tool used to create all the semiconductor devices, not only the resistors.

There are basically two things that influence the conductivity: (1) the concentration of the carriers available to contribute toward the electrical current when an electric field is applied and (2) the mobility of these carriers. With the concentration of the carriers, which is the number of carriers per unit volume, it is clear that a higher carrier concentration means a higher current for the same electric field, which further means that the conductivity is higher. The second factor, the mobility, accounts for the effect that different carriers, or the same carriers in different conditions, do not flow equally easily. To illustrate this effect, consider the ions in a metal lattice: these are charged particles with a finite concentration, but they would not make any current if an electric field is applied as they are completely immobile, their mobility is equal to zero.

The conductivity is proportional to the carrier concentration and the carrier mobility:

$$\sigma \propto \text{(carrier concentration)} \times \text{(carrier mobility)}. \tag{1.15}$$

The concepts of the mobility and the carrier concentration are analyzed in more detail in Sections 1.4 and 1.5, after some more fundamental concepts of the current carriers are introduced in Section 1.2, and basic processing used to technologically control the current carriers is described in Section 1.3.

■ **Example 1.4 Current Density Versus Terminal Current**

Calculate the maximum current density and the terminal current for the resistor designed in Example 1.3 if a voltage of 5 V is applied to the resistor terminals. Neglect the corner effects.

Solution: Neglecting the corner effects, the one-dimensional form of Ohm's law [Eq. 1.10] can be used to calculate the current density. In that case the electric field is uniform and equal to $E = V/L = 5/(125 \times 10^{-6}) = 40,000\,V/m = 40\,kV/m$. The maximum current is obtained for the maximum conductivity, which is $\sigma_{max} = \sigma(0) = 10\,(\Omega\,cm)^{-1}$ as can be seen from Fig. 1.3. Therefore, $j_{max} = \sigma_{max}E = 4 \times 10^7\,A/m^2$.

The terminal current can be obtained by integrating the current density as distributed from the surface to x_j (or to ∞, which is the same as no current flows between x_j and ∞), and multiplying by the resistor width:

$$I = W \int_0^\infty j(x)\,dx = W\frac{V}{L}\sigma(0) \int_0^\infty e^{-(x/x_0)^2}\,dx$$

The integral that appears in the above equation is already discussed in Example 1.3. Using the values for all the quantities in SI units, one obtains

$$I = 10^{-6} \frac{5}{125 \times 10^{-6}} 1000 \sqrt{\pi/2} = 5 \times 10^{-5} \, \text{A} = 50 \, \mu\text{A}$$

Note that the integral form of the Ohm's law gives the same result for the terminal current ($I = V/R = 5 \, \text{V}/100 \, \text{k}\Omega = 50 \, \mu\text{A}$), but it cannot be used to find the maximum current density inside the resistor body.

1.2 INSIGHT INTO CONDUCTIVITY INGREDIENTS: CHEMICAL-BOND MODEL

The conductivity of silicon, the most frequently used semiconductor, can be changed to be as low as $5 \times 10^{-2} \, (\Omega \, \text{cm})^{-1}$ and as high as $5 \times 10^{5} \, (\Omega \, \text{cm})^{-1}$, which is a difference of seven orders of magnitude. There are not many physical quantities that can be changed in such a wide range. It is difficult to change the size of a 1-cm object to 10^{7} cm, which is 100 km, or slow down a process that takes 1 s to 10^{7} s, which is nearly 4 months, or heat up a furnace from 20°C to 20×10^{7}°C. The possibility of changing the conductivity of semiconductor materials in such a wide range enables the creation of very useful microelectronic devices. This section provides an insight into the conductivity ingredients.

1.2.1 Semiconductor Crystals

The most frequently used semiconductor material in microelectronics is silicon. Silicon is located in the fourth column of the periodic table (as shown in Table 1.1), which means it has four electrons in the outer shell of eight electron places. Therefore, the silicon atoms are prone to either give away the four electrons from the outer shell (called valence electrons) or accept an additional four electrons in order to achieve the stability of a completely filled outer shell. Silicon as a semiconductor material is a crystal following a tetrahedral pattern in which each silicon atom couples one valence electron with a valence electron from each of four neighboring atoms. These electron pairs are shared between the neighboring silicon atoms, by which the stability of the completely filled outer shell is "felt" by either taking all four electrons from the neighboring atoms to fill the shell of eight electron places or giving the four valence electrons to the neighboring atoms.

The shared electron pairs hold the silicon atoms together in a crystal; they are called *covalent bonds*.

Three-dimensional presentation of the silicon tetrahedral structure is given in Appendix B. The three-dimensional presentation of a semiconductor crystal is important for a number of crystal-related concepts. However, the basic electronic properties can be explained using the simplified two–dimensional representation of Fig. 1.4. This representation shows silicon positively charged cores (four charge units) with the four valence electrons forming covalent bonds in a simple quadratic structure.

TABLE 1.1 Semiconductor-Related Elements in the
Periodic Table (with Atomic Number and Atomic Weight)

III	IV	V
⊖ ⊖ $(+3)$ ⊖ ⊖	⊖ ⊖ $(+4)$ ⊖ ⊖	⊖ ⊖ $(+5)$ ⊖ ⊖ ⊖
5　　　　　　B	6　　　　　　C	7　　　　　　N
Boron	Carbon	Nitrogen
10.82	12.01	14.008
13　　　　　Al	14　　　　　Si	15　　　　　P
Aluminum	Silicon	Phosphorus
26.97	28.09	31.02
31　　　　　Ga	32　　　　　Ge	33　　　　　As
Gallium	Germanium	Arsenic
69.72	72.60	74.91
49　　　　　In	50　　　　　Sn	51　　　　　Sb
Indium	Tin	Antimony
114.8	118.7	121.8

An important thing to note from Fig. 1.4 is that all the electrons are bound through covalent bonds only at the temperature of 0 K. There is always a certain amount of broken covalent bonds at temperatures higher then 0 K. This happens when heat energy absorbed by a silicon atom is released through breakage of a covalent bond and release of a free electron, which carries the energy away. The energy needed to break a covalent bond in a silicon crystal is 1.1 eV at room temperature and it is slightly different at different temperatures. Obviously, there will be more broken covalent bonds at higher temperatures as the silicon atoms possess more thermal energy, which eventually destroys covalent bonds. When a silicon atom releases an electron, it becomes positively charged with a hole in its bond structure, as illustrated in Fig. 1.4b.

Germanium is another element from the fourth column of the periodic table (Table 1.1) that forms a semiconductor crystal similar to the way silicon does. There are, however, definite and important differences between the two materials, which eventually make the silicon preferred material for a majority of microelectronic devices, regardless of the initial (back to the 1950s) popularity of germanium. Semiconductor materials can also be created as compound crystals consisting of atoms from the third and the fifth columns of the periodic table (called III–V materials), or even the second and the sixth columns (called II–VI materials). The most successful III–V material in view of microelectronic application, which is the second most important semiconductor material, is gallium arsenide (GaAs). Although the detailed structure of GaAs is different from the silicon structure, the general semiconductor concepts are the same. As silicon is by far the most used semiconductor material, these concepts are introduced using the example of silicon in this book. The differences in properties of GaAs material will be explained when specific GaAs devices are considered, in particular in Sections 8.1.4 and 9.1.

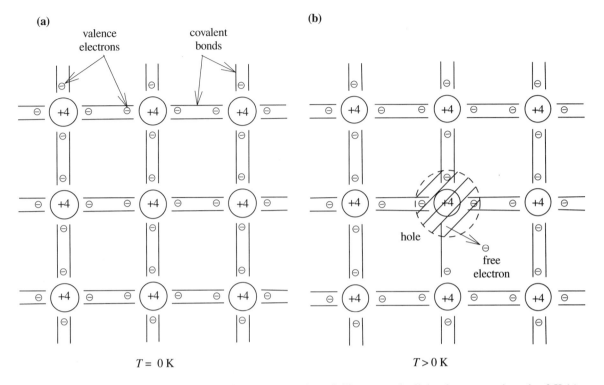

(a)

valence
electrons

covalent
bonds

$T = 0$ K

(b)

hole

free
electron

$T > 0$ K

Figure 1.4 Two-dimensional representation of silicon crystal; all the electrons are bound at 0 K (a), while there are broken bonds at temperatures > 0 K, creating free electron–hole pairs (b).

1.2.2 Two Types of Current Carriers in Semiconductors

The electron released from a covalent bond broken by thermal energy (Fig. 1.4) is a mobile charged particle, which therefore contributes to the current flow when an electric field is applied. If there are more free electrons in the crystal, a higher current will flow for the same applied electric field. The number of these electrons per unit volume, i.e., the *concentration* of free electrons, directly influences the conductivity of a semiconductor layer, as shown in Eq. (1.15).

In addition, it is necessary to consider what happens with the hole in the bond structure of the silicon atom that released the free electron (Fig. 1.4). This silicon atom is definitely unstable with the hole in its bond structure, and it will "use" any opportunity to "steal" an electron from a neighboring atom to rebuild its four covalent bonds, making itself "comfortable" again. If there is an electric field applied in the crystal, as shown in Fig. 1.5, the field would help this atom to take an electron over from a neighboring atom, leaving the neighboring atom with a hole in its bond structure. We can say that the hole has moved to the neighboring atom, as illustrated in Fig. 1.5. Obviously, there is no reason why this new silicon atom should stay with the hole for much longer than the first atom; it can equally

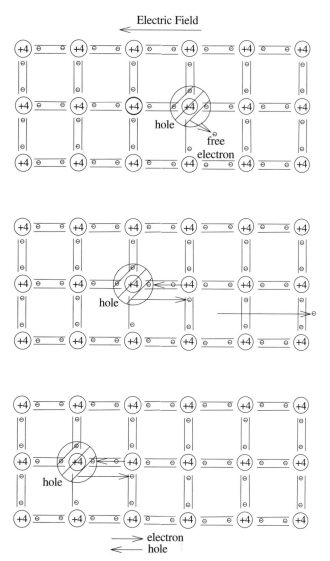

Figure 1.5 Model of hole as mobile carrier of positive charge.

well use the field's help to take an electron over from its neighbor in the direction of the field direction. As a consequence, the hole has moved one step further in the direction of the field direction, a process illustrated again in Fig. 1.5. As this process continues, we get an impression of the hole as a positive charge carrier moving in the crystal along the electric field lines.

Although this motion of the hole is associated with electron transitions from atom to atom, it does appear as an additional carrier to the free electron.

To simplify our presentation of charge transport mechanisms in semiconductors, it is not necessary to constantly keep in mind the details of the process of hole motion as described above; we can simply treat the hole as a mobile carrier of positive charge. The hole is considered a positively charged particle (or quasiparticle).

Having established that there are two types of current carriers in semiconductors, free electrons and holes, the conductivity can now be expressed more precisely than in Eq. (1.15):

$$\sigma = qn\mu_n + qp\mu_p \qquad (1.16)$$

In Eq. (1.16), n and p are concentrations of the free electrons and holes, respectively, μ_n and μ_p are the free-electron and hole mobilities, respectively, and q is the electronic charge ($q = 1.6 \times 10^{-19}$ C). Obviously, $qn\mu_n$ is the electron's contribution to the conductivity, while $qp\mu_p$ is the contribution of the second type of current carriers, the holes.

1.2.3 Intrinsic Semiconductors

A semiconductor crystal with only the native atoms is called *an intrinsic semiconductor*. Even though it is not possible in reality to obtain an ideally pure semiconductor, the intrinsic-semiconductor model is very useful in explaining semiconductor properties. An important characteristic of intrinsic semiconductors is that the concentration of free electrons n is always equal to the concentration of holes p. This is due to the fact that the free electrons and holes are created in pairs in the process of covalent-bond breakage (Fig. 1.4). The concentration of free electrons and holes in an intrinsic semiconductor is denoted by n_i, and is called the *intrinsic carrier concentration*. Therefore, in an intrinsic semiconductor

$$n = p = n_i \qquad (1.17)$$

The intrinsic carrier concentration n_i, as well as carrier mobilities in an intrinsic semiconductor μ_n and μ_p, are constant for a given semiconductor at a given temperature. Their values for silicon[1] and gallium arsenide at room temperature are given in Table 1.2.

The intrinsic carrier concentration n_i and the carrier mobilities μ_n and μ_p are, however, temperature dependent. As previously explained, an increase in the temperature means that the crystal–lattice atoms possess more thermal energy; because of this more covalent bonds are destroyed and, therefore, the concentration of free electrons and holes (n_i) is higher.

TABLE 1.2 Intrinsic Concentration and Carrier Mobilities for Si and GaAs at 300 K

	n_i [cm^{-3}]	μ_n [cm^2/(V s)]	μ_p [cm^2/(V s)]
Si	1.02×10^{10}	1450	500
GaAs	2.1×10^{6}	8500	400

[1]*Source:* M.A. Green, "Intrinsic concentration, effective densities of states, and effective mass in silicon." *J. Appl. Phys.*, **67**, 2944–2954, 1990.

Electron and hole mobilities also depend on temperature, but in a more complicated way. It is obvious that semiconductor resistors can be used as temperature sensors. We definitely need to know more of solid-state physics before we can introduce more details about the dependencies of carrier mobility and intrinsic carrier concentration on temperature. More detailed considerations of these effects are provided in Sections 1.4 and 1.5. At this stage, it should be pointed out that it is not only the temperature that can destroy the covalent bonds, thereby creating free electrons and holes, but other types of energy can do the job as well. An important example is *light*. If the surface of a semiconductor is illuminated, the absorbed photons of the light can transfer their energy to the electrons in the crystal, a process in which the covalent bonds can be destroyed and pairs of free electrons and holes can be generated. This is the effect that makes the use of semiconductor-based devices as light detectors and solar cells possible.

1.2.4 Doped Semiconductors

To create useful microelectronic devices, the conductivity is changed through an increase in carrier concentration by technological means. Semiconductors with technologically changed concentrations of free electrons and/or holes are called *doped semiconductors*. In the process of semiconductor doping, some silicon atoms are replaced by different types of atoms, which are called *doping atoms* or *impurity atoms*. In a real silicon crystal many silicon-atom places are occupied by impurity atoms. Useful properties are obtained when some silicon-atom places are taken by atoms from the third or the fifth column of the periodic table (Table 1.1). In the process of semiconductor doping, many more silicon atoms than would normally happen are intentionally replaced by impurity atoms. This section considers the effects of the introduction of impurity atoms from the third and the fifth columns of the periodic table into the silicon crystal lattice.

N-Type Doping

Consider first the case in which a silicon atom is replaced by an atom from the fifth column of the periodic table, say phosphorus, as shown in Fig. 1.6a. The phosphorus atom has five electrons in the outer shell. To replace a silicon atom in the lattice, it will use four electrons to form the four covalent bonds with the four neighboring silicon atoms, as shown in Fig. 1.6a. The fifth electron does not fit into this structure; it would not be able to find a comfortable place around the parent atom, and with a little help from the thermal energy it would leave the phosphorus atom "looking for a better place." In other words, the fifth phosphorus electron is easily liberated by the thermal energy, by which it becomes a free electron. This electron is not different from the free electrons produced by covalent-bond breakage in its ability to contribute to the current flow when an electric field is applied. Therefore, a replacement of a silicon atom by a phosphorus atom produces a free electron, increasing the concentration of free electrons n. This process is called *N-type doping*. The elements from the fifth group of the periodic table that can produce free electrons when inserted into the silicon crystal lattice are phosphorus, arsenic, and antimony. These doping elements are called *donors* (they donate electrons), their concentration in the silicon crystal being denoted by N_D.

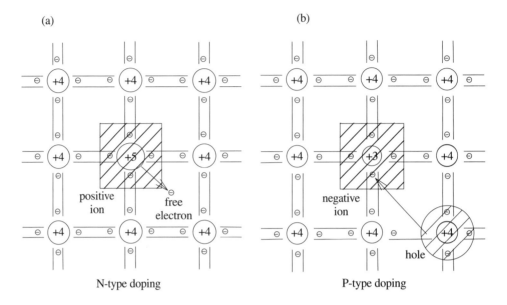

Figure 1.6 Effects of N-type (a) and P-type (b) doping.

As almost every donor atom produces a free electron at room temperature, through the process described above, the concentration of the doping-induced free electrons is $n \approx N_D$.

N-type doping, however, does not produce free electrons only—generation of every free electron in turn creates a positive ion. There is always the same number of positive and negative charges in a crystal, to preserve its overall electroneutrality. When a phosphorus atom liberates its fifth electron, it remains positively charged, as illustrated in Fig. 1.6a. There is an important difference between this positive charge and the positive charge created as a broken covalent bond (Fig. 1.4b). The positive charge of Fig. 1.4b is due to a missing electron, or a hole in the bond structure of a silicon atom. There is not a hole in the bond structure that phosphorus (or another fifth-group element) creates with the neighboring silicon atoms; the structure is stable as all four bonds are satisfied.

The positive charge created due to the absence of the fifth (extra) phosphorus electron will neither be neutralized nor moved to a neighboring atom. This charge appears as a fixed positive ion.

This positive-ion charge cannot contribute to a current flow because it is immobile. The positive-ion charge cannot be forgotten, however, as it creates important effects in semiconductor devices by its charge and the associated electric field. To distinguish from holes and electrons, squares are used as symbols for the immobile ion charge.

TABLE 1.3 Charges in N-Type Semiconductors

Mobile Charges
  1. Thermally generated holes (minority carriers)
  2. Thermally generated electrons (negligible)
  3. Doping-induced electrons ($\approx N_D$)

Fixed Charge
  4. Doping-induced positive ions ($\approx N_D$)

concentration of electrons $\gg$ concentration of holes
net charge $= p - n + N_D = 0$

Table 1.3 summarizes the types of charges that appear in an N-type semiconductor. There are both holes and electrons as mobile charges, however, electrons are coming from two independent sources: thermal generation and doping atoms. As shown in Table 1.2, the thermal generation produces about 10^{10} electrons/cm^3 in the intrinsic silicon. Doping can, however, introduce as many as 10^{21} electrons/cm^3. In fact, the lowest amount of electrons that can be introduced by doping in a controllable way is not much smaller than 10^{14} electrons/cm^3. These numbers show that the concentration of thermally generated electrons is negligible when compared to the concentration of doping-induced electrons in an N-type semiconductor. Frequently, we can also neglect the concentration of thermally generated holes compared to the concentration of doping-induced electrons. If the conductivity of N-type-doped silicon is considered, the contribution of the thermally generated electrons and holes would be orders of magnitude smaller than the contribution of the doping-induced electrons; because of this Eq. (1.16) can be simplified as

$$\sigma \approx q \mu_n N_D. \qquad \gg p \qquad (1.18)$$

Sometimes holes and electrons, unlike the above example, produce different effects. In this case, we still can neglect the concentration of the thermally generated electrons compared to the doping-induced electrons, but the concentration of holes cannot be neglected as the effect they produce does not merge with the electron-based effect in that case. It is important to keep in mind that there are some holes in an N-type semiconductor, and it will be important to calculate their concentration. The holes in an N-type semiconductor are called *minority carriers* as opposed to electrons, which are called *majority carriers*.

P-Type Doping

P-type doping is obtained when a number of silicon atoms are replaced by atoms from the third column of the periodic table, basically boron atoms. Boron atoms have three electrons in the outer shell, which are taken to create three covalent bonds with the neighboring silicon atoms.

As an additional electron per boron atom is needed to satisfy the fourth covalent bond, the boron atoms will capture thermally generated free electrons (generated by breakage of covalent bonds of silicon atoms), creating in effect excess holes as positive mobile charge (Fig. 1.6b). The boron atoms that accept electrons to complete their covalent-bond structure with silicon atoms become negatively charged ions—immobile charged atoms analogous to the positively charged donor atoms in the case of N-type doping.

Boron atoms, or in general P-type doping atoms, are called *acceptors*. Similar to the case of N-type doping, almost all boron (acceptor) atoms are ionized at room temperature, which means they have taken electrons creating mobile holes. The concentration of doping-induced holes is, therefore, very close to the concentration of the doping acceptor atoms N_A, as indicated in Table 1.4, which lists the charges in P-type semiconductors. Again, the concentration of the doping-induced mobile charge (holes in this case) is much higher than the concentrations of thermally generated mobile charge, thus $\sigma \approx q\mu_p N_A$.

1.2.5 Electron and Hole Generation and Recombination[2] in Thermal Equilibrium

Although we already know that the concentration of majority carriers is approximately equal to the concentration of doping atoms, a deeper consideration of generation and recombination processes is needed to be able to determine the concentration of minority carriers. It was stated above that the minority carriers sometimes produce different effects than the majority carriers; because of this they cannot always be neglected.

Generation of free electrons and holes is a process in which bound electrons are given enough energy to (1) liberate themselves from silicon atoms creating electron–hole pairs (Fig. 1.4), or (2) liberate themselves from donor-type atoms creating free electrons and fixed positive charge (Fig. 1.6a), or (3) liberate themselves from silicon atoms to provide only the fourth bond of acceptor-type atoms, creating mobile holes and negative fixed charge (Fig. 1.6b). The doping-assisted generation of electrons/holes is obviously limited to the level of doping—no more electrons/holes can be generated once all the doping atoms are ionized. **What does, however, limit the process of thermal generation of electron–hole**

TABLE 1.4 Charges in P-Type Semiconductors

Mobile Charges
1. Thermally generated electrons (minority carriers)
2. Thermally generated holes (negligible)
3. Doping-induced holes ($\approx N_A$)

Fixed Charge
4. Doping-induced negative ions ($\approx N_A$)

concentration of holes $\gg$ concentration of electrons
net charge $= p - n - N_A = 0$

---

[2]A more detailed description and modeling of recombination–generation processes is presented in Sections 8.1 and 8.2.

pairs due to breakage of covalent bonds? Would such created electrons and holes accumulate in time to very high concentrations?

A free electron that carries energy taken from the crystal lattice can easily return this energy to the lattice and bond itself again when it finds a silicon atom with a hole in its bond structure. This process, called *recombination*, results in annihilation of free electron–hole pairs. Obviously, if the concentration of free electrons is higher, it is more likely that a hole will be "met" by an electron and recombined. Similarly, an increase in the hole concentration also increases the recombination rate, i.e., the concentration of electron–hole pairs recombined per unit time. It is the recombination rate that balances the generation rate, limiting the concentration of free electrons and holes to a certain level. If the generation rate is increased, by an increase in the temperature for example, the resulting increase in concentration of free electrons and holes will automatically make the recombination events more probable, therefore the recombination rate is automatically increased to the level of the generation rate.

In thermal equilibrium, the generation rate is always equal to the recombination rate.

These rates are, however, different at different temperatures—at 0 K they are both equal to 0 as is the concentration of free electrons and holes; with an increase in the temperature, the recombination and generation rates are increased, which results in increased concentrations of free electrons and holes.

Doping influences the recombination and generation rates as well. Take as an example an N-type semiconductor in which the concentration of free electrons is increased by doping-induced electrons. We have concluded above that an increased concentration means it is more likely that a hole will be "met" by an electron and recombined. Consequently, the concentration of holes in an N-type semiconductor is smaller than in the intrinsic semiconductor, as the holes are recombined not only by the thermally generated electrons but also, and much more, by the doping-induced electrons. Therefore, with an increase in the concentration of electrons, the concentration of holes is reduced. This dependence can be expressed as

$$p = C\frac{1}{n} \qquad (1.19)$$

where the proportionality factor C is a temperature-dependent constant. This equation is written for N-type semiconductors, but it can also be applied to P-type semiconductors as well as intrinsic semiconductors. It can be rewritten as $n = C/p$ for P-type semiconductors, but it is essentially the same equation. The constant C can be determined if Eq. (1.19) is applied to the case of intrinsic semiconductor. As $n = p = n_i$ in that case, we see that the constant C is given as

$$C = n_i^2 \qquad (1.20)$$

Using the above-determined value for the constant C, Eq.(1.19) becomes

$$\boxed{np = n_i^2} \qquad (1.21)$$

Equation (1.21) enables calculation of the concentration of minority carriers in doped semiconductors.

∎ **Example 1.5 Minority-Carrier Concentration**

Calculate the concentration of minority carriers at room temperature if the concentration of donor atoms in N-type silicon is $N_D = 10^{16}$ cm^{-3}.

Solution: As $n \approx N_D$ in the N-type semiconductor, the concentration of holes (minority carriers) can be determined using Eq. (1.21):

$$p = n_i^2/N_D$$

The intrinsic carrier concentration can be found in Table 1.2. Thus, $p = 1.02^2 \times 10^{20}/10^{16} = 1.04 \times 10^4$ cm^{-3}.

∎ **Example 1.6 Doping, Minority Carriers, and Conductivity**

Determine the conductivity of the material considered in Example 1.5, assuming the values for carrier mobilities as given in Table 1.2.

Solution: The concentration of holes is 12 orders of magnitude smaller than the concentration of electrons, therefore, the hole contribution to the conductivity can be neglected and the approximate Eq. (1.18) used: $\sigma = q\mu_n N_D = 1.6 \times 10^{-19} \times 1450 \times 10^{16} = 2.32$ $(\Omega\,\text{cm})^{-1}$.

∎ **Example 1.7 Doping Compensation**

In a silicon crystal, $N_D = 10^{17}$ cm^{-3} and $N_A = 10^{16}$ cm^{-3}. Find the resistivity of the crystal at room temperature, if the electron mobility is $\mu_n = 770$ cm^2/Vs.

Solution: In this case we have both the N-type and P-type doping. As $N_D > N_A$, the semiconductor effectively appears as N-type. However, the holes from the P-type doping will recombine electrons, reducing the electron concentration to $n = N_D - N_A = 9 \times 10^{16}$ cm^{-3}.

The resistivity is, therefore, obtained as

$$\rho = 1/q\mu_n n = 0.09\ \Omega\,\text{cm}$$

■ **Example 1.8 Light and Resistivity**

P-type silicon has resistivity of 0.5 Ω cm. Assuming constant temperature of 300 K, find

(a) the hole and electron concentrations;

(b) the maximum change in resistivity caused by a flash of light, if the light creates 2×10^{16} additional electron–hole pairs per cm^3.

Solution:

(a) In a P-type semiconductor the conductivity is $\sigma = q\mu_p N_A$. The resistivity is

$$\rho = \frac{1}{q\mu_p N_A}$$

The concentration of acceptor ions, and therefore holes, is then $N_A = p = 1/(q\mu_p\rho) = 2.5 \times 10^{16}$ cm^{-3}. The concentration of electrons is found as

$$n = \frac{n_i^2}{p} = 4.2 \times 10^3 \text{ cm}^{-3}$$

(b) The flash of light produces excess electrons and holes, therefore reducing the resistivity. When the light is removed, the excess electrons and holes will gradually recombine with each other, increasing the resistivity to its original value, i.e., the equilibrium value. To find the maximum change in the resistivity, we need to determine the resistivity of the specimen when the light is on. In that case the concentration of holes is $p = 2.5 \times 10^{16} + 2 \times 10^{16} = 4.5 \times 10^{16}$ cm^{-3}. The concentration of electrons is $n = 2 \times 10^{16}$ cm^{-3}, as generated by the light, and cannot be neglected when compared to the concentration of holes. The conductivity is calculated as

$$\sigma = q\mu_p p + q\mu_n n = 8.24 \ (\Omega \text{ cm})^{-1}$$

The corresponding resistivity is $\rho = 1/\sigma = 0.12$ Ω cm. The maximum difference is, therefore, $\Delta\rho_{max} = 0.5 - 0.12 = 0.38$ Ω cm.

1.3 MAKING A SEMICONDUCTOR RESISTOR: LITHOGRAPHY AND DIFFUSION

This section introduces diffusion as one of the possible techniques for doping of semiconductors, as well as lithography as a technique used to define the lateral dimensions of the doped layers. The section concludes with diffusion modeling. Although diffusion modeling is important in understanding diffusion processes used for semiconductor doping, it is even more important in understanding the diffusion as a fundamental transport mechanism playing an essential role in the operation of some semiconductor devices. In particular, two fundamental equations of semiconductor-device physics (the diffusion-current equation and the continuity equation) are introduced.

1.3.1 The Concept of Diffusion

The essence of diffusion is random thermal motion of the diffusing particles (these can be electrons, holes, doping atoms such as boron or phosphorus, molecules in the air, or smoke particles). Imagine that thousands of smoke particles are produced near the wall of a house (Fig. 1.7). If there is a window in the wall, some particles will pass through the window. In their random motion, half of the outside particles move toward the window, as well as half of the inside particles, with the same chance of passing through it. Having more particles outside the window, more particles will be passing through the window from the outside, entering the house, compared to those going in the opposite direction.

The diffusion of particles creates an effective particle current toward the points of lower *particle concentration* (number of particles per unit volume). There is no effective particle current when the concentration of particles is uniform, as in this case equal numbers of particles move either way. The current of particles (charged or uncharged) produced by a difference in the particle concentration is called *diffusion current*. The force behind this current has nothing to do with gravity or with electric-field forces; these are separate forces that can produce a current of particles on their own.[3] The force behind the diffusion current is random thermal motion.

The diffusion is not limited to gases, it happens in liquids and solids as well, although a very high temperature is typically needed for diffusion of atoms to be clearly observed in solids. If a semiconductor crystal is heated, the diffusion of doping atoms can be achieved in a way similar to the diffusion of the smoke particles into the house. Figure 1.8 is analogous

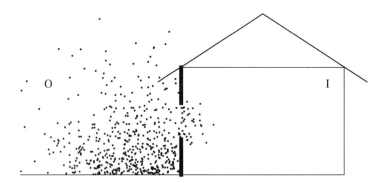

Figure 1.7 The concept of diffusion.

---

[3]Electric-field-induced current of charged particles (like electrons and holes), called drift current, was considered in Section 1.1.

impurity atoms

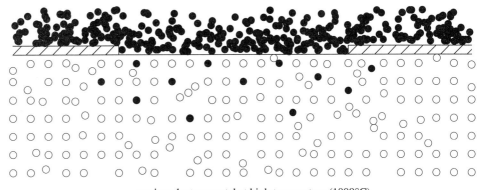

semiconductor crystal at high temperature (1000°C)

Figure 1.8 Doping of a semiconductor by diffusion.

to Fig. 1.7, except that it shows diffusion of doping atoms into a piece of semiconductor through a window in a diffusion-protective "wall."

The semiconductor crystal is heated to about 1000°C, so that a sufficient number of semiconductor atoms are released from their crystal-lattice positions. The semiconductor atoms that leave their crystal-lattice positions are called *interstitials*, whereas the empty positions left behind are called *vacancies*. The doping atoms that are provided at the surface of the semiconductor can diffuse into the semiconductor at this high temperature, and place themselves into the created vacancies, taking therefore crystal-lattice positions. When the semiconductor is cooled down to room temperature, the thermal motion of semiconductor and doping atoms becomes insignificant, therefore, the doping atoms will stay "frozen" in their positions. The doped semiconductor layer created in this way expands only to a certain depth and appears roughly under the provided window in the diffusion-protective "wall." Using diffusion at a high temperature, a doped semiconductor layer having a desired depth, length, and width can be created.

1.3.2 Making the Micropatterns: Lithography

Figure 1.8 illustrates the importance of the protective "wall" with its window in the process of selective doping of semiconductors. Typically, the "wall" is a layer of SiO_2 (a thin glass layer), although Si_3N_4 as well as some other materials are sometimes used. If the semiconductor in question is silicon, the SiO_2 layer at the surface can be created by thermal oxidation of the silicon. Alternatively, it can be deposited at the surface. These processes are described in sections 2.5.2 and 6.2.2, respectively. This section introduces lithography as a process used to enable selective etching of thin films deposited or grown onto the semiconductor surface. Etching of the window in the SiO_2 layer, to enable selective diffusion, is only one example of the need for selective etching of layers. Another example is etching of a metal layer to define metal lines that interconnect the devices created in the semiconductor. The lithography is basically a set of processes that enables the patterns of an integrated circuit (IC) to be created.

Figure 1.9 illustrates the set of processes referred to as lithography. The starting material in this example is an N-type silicon wafer with a thin film of silicon dioxide grown at the surface. To etch a window in the silicon dioxide film, the areas of the oxide that are not to be etched need to be protected. To this end, a special film, referred to as *photoresist*, is deposited onto the oxide surface.

The photoresist has to fulfill a double role: (1) it should be light sensitive to enable the transfer of the desired pattern by way of exposure/nonexposure to the light, and (2) it should be resistant to chemicals used to etch the underlying layer. Generally, there are two types of photoresists: negative and positive. The exposed areas of the positive photoresist are washed away by a developing chemical, while the unexposed areas remain at the surface as shown in Fig. 1.9d . In the case of the negative photoresist, the unexposed areas are removed during the developing. The photoresist is originally in a liquid form, which is

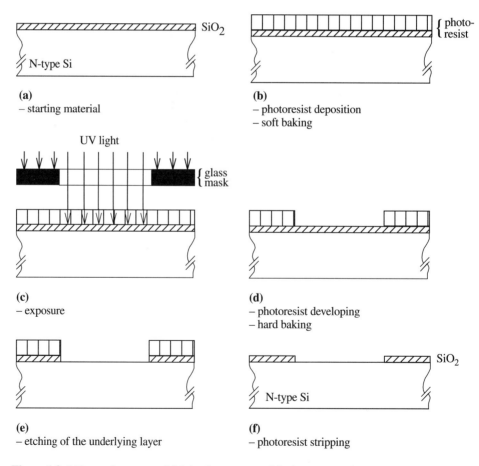

(a)
– starting material

(b)
– photoresist deposition
– soft baking

(c)
– exposure

(d)
– photoresist developing
– hard baking

(e)
– etching of the underlying layer

(f)
– photoresist stripping

Figure 1.9 Lithography—a set of fabrication steps used for layer patterning.

spun onto the substrate surface to create a thin film. The deposited photoresist is soft baked (80–100°C) before the exposure, to evaporate the solvents, and hard baked (120–150°C) after the developing, to improve the adhesion to the underlying layer.

The photoresist is being exposed through a glass mask by an unltraviolet (UV) light (Fig. 1.9c). The pattern that is to be transferred onto the substrate is provided on the glass mask in the form of clear and opaque fields. The mask itself is made by computer-controlled exposure of a thin film deposited on the glass mask and sensitive to either UV light or an electron beam; the clear and opaque fields are obtained after film development. Figure 1.9c illustrates so-called *contact lithography*, which requires a 1× image on the mask. *Projection lithography* is typically now used, where the mask is positioned above the substrate. When the light passes through the mask, it is focused onto the substrate, which results in image reduction. This has several advantages: (1) the image on the mask can be much larger than the image on the substrate, (2) the substrate can be exposed part by part (chip by chip), which enables better focusing, and (3) the mask is not contaminated due to contact with the photoresist.

In principle, it is possible to expose an electron-beam resist deposited directly on the substrate by a computer-controlled electron beam, in which case there would be no need for the glass mask. Although this may seem advantageous, there is a practical problem. There can be as many as 10^{10} picture elements in todays ICs, which are exposed simultaneously using the glass mask. It would be absurdly slow to use an electron beam to directly write these 10^{10} picture elements, and of course repeat the process for every IC. As there is a limit in terms of the finest pattern that can be obtained by the UV light (the limit imposed by the wavelength of the UV light), the research of deep-submicron and nanometer structures is carried out using direct electron-beam writing.

Once the photoresist is developed and hard baked, the underlying layer can be etched (Fig. 1.9e). This etching can be wet (etching by an appropriate chemical solution) or dry (plasma etching). After the etching is completed, the photoresist is removed to obtain the desired structure as shown in Fig. 1.9f.

1.3.3 Making an IC Resistor

Combining the lithography and the diffusion, semiconductor resistors can be made. The process of making a resistor is illustrated in Fig. 1.10. In this example, the initial material is N-type silicon with a window in the overlying oxide layer, prepared as illustrated previously in Fig. 1.9. Let us assume that the concentration of donor atoms in the N-type substrate is $N_D = 10^{16}$ cm^{-3}, as illustrated in Fig. 1.11 by the dashed line. The concentration of the donors is uniform throughout the substrate as this doping is performed while the crystal substrate is being grown. To create a P-type resistive body at the surface (refer also to Fig. 1.2), boron diffusion is performed. The boron diffusion will create a nonuniform doping profile as most of the boron atoms incorporated by the diffusion in the silicon will remain at the surface; going deeper into the substrate, lower concentrations of the boron atoms will be found (Fig. 1.11). If a higher concentration of boron atoms is achieved at the surface (say $N_A = 10^{18}$ cm^{-3}), the acceptor-type doping atoms will prevail and the surface of the silicon will appear as P-type. Figure 1.11 shows that there is a point at which $N_A = N_D$. This point is called the *P–N junction*—on one side of the junction the semiconductor is a

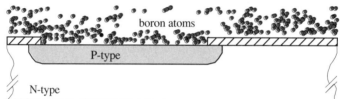

(a)
– Boron diffusion

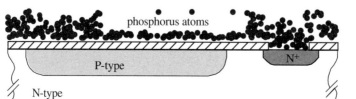

(b)
– oxide removal
– oxide deposition
– photolithography
  (N⁺-diff. windows)
– phosph. diffusion

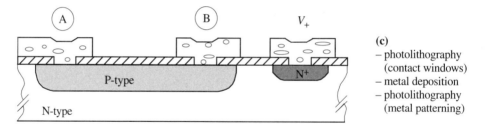

(c)
– photolithography
  (contact windows)
– metal deposition
– photolithography
  (metal patterning)

Figure 1.10 Process steps used in fabrication of an IC resistor.

P type while on the other side of the junction it is an N type. The P–N junction formed by the diffusion of boron into the N-type substrate is illustrated in Fig. 1.10a.

The resistor body is a P type, therefore the holes will be the current carriers in the resistor. For the resistor to function properly, it is necessary to confine the current flow within the P-type resistor body, i.e., not to let any current leak into the N-type substrate. This can be achieved if the N-type substrate is connected to a positive voltage that will repel any hole trying to enter the N-type substrate. It is known that metals create good ohmic contacts to P-type silicon, therefore, connecting the resistor ends to metal terminals is straightforward. In contrast, good contacts to N-type silicon can be achieved only with high concentrations of donor atoms ($N_D > 10^{19}$ cm^{-3}). Because of this, it is necessary to provide a highly doped N-type region (labeled as N⁺) for the contact with the metal terminal where the positive voltage is to be connected. To this end, the oxide mask (the oxide layer with the window that enabled the P-type layer to be formed by boron diffusion) has to be stripped, and a new oxide layer grown to create a diffusion mask with a window over the area where the N⁺ region is to be formed. This obviously involves an

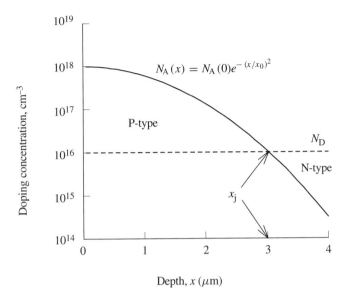

Figure 1.11 Diffusion of acceptors into an N-type substrate creates a P–N junction: P-type semiconductor between the surface and x_j, and an N-type semiconductor from x_j into the bulk of the substrate.

oxide deposition or growth and a lithography to open the window. After this, diffusion of phosphorus is performed to create the N^+ region needed for the contact, as shown in Fig. 1.10b.

Finally, the resistor terminals and the metal stripe for the positive voltage (V_+) have to be made. The areas where the metal should contact the silicon are defined as contact windows (another lithography process) in a freshly grown oxide layer. The oxide is a good insulator, so it will electrically isolate the metal layer from the silicon in all other areas. After the metal is deposited it appears all over the chip, and has to be removed from the areas where it is not wanted to create the desired metal stripes. Again, a lithography process is needed to enable selective etching of the metal layer, as shown in Fig. 1.10c. At this stage, the resistor is electrically functional.

To conclude, the lateral dimensions of the diffused resistors are controlled by lithography as they are defined by the dimensions on the glass mask. Different resistors in integrated circuits can have different and independent lengths and widths, as discussed in section 1.1.2. It is important to note that the actual length and width of the resistor body are larger than the dimensions on the glass mask due to the effect of *lateral diffusion*. Once the doping atoms are introduced into the semiconductor substrate, they diffuse laterally as well as vertically. The lateral distance between the edge of the window in the oxide mask and the P–N junction (x_{j-lat}) is experimentally estimated to be 80% of the junction depth x_j in the vertical direction:

$$x_{j-lat} = 0.8\, x_j \tag{1.22}$$

The other properties of the diffused layers, the conductivity and the depth, are controlled by the parameters of the diffusion process—temperature and time. The conductivity and the junction depth appear as technological parameters—they have the same values for all the resistors made with one type of diffusion layer, as discussed in section 1.1.2. If the diffusion is performed at a higher temperature and for a longer time, it will enable the doping atoms to penetrate deeper into the semiconductor, which results in a higher conductivity of the diffused layer. More detailed understanding of the diffusion process and precise models are necessary to be able to precisely determine the temperature and the time needed, in other words to design the diffusion process. Diffusion modeling is considered in the next section.

1.3.4 Modeling the Diffusion: Diffusion-Current Equation, Continuity Equation, and Doping Profiles

In the first two parts of this section, the diffusion phenomena are considered in more detail to introduce two fundamental equations, the diffusion-current equation and the continuity equation, which are generally valid and will be used in device modeling as well. The third part provides the equations needed to design a diffusion process as used for semiconductor doping.

The Diffusion-Current Equation

As illustrated in Figs. 1.7 and 1.12, a difference in concentration of particles performing random thermal motion is the only driving force behind the diffusion current. It is useful to express this understanding of the diffusion current by a mathematical equation. It should be noted that the current considered here is not necessarily electric current, i.e., current of charged particles. Uncharged particles can create diffusion current as well. The unit for the current of uncharged particles is not ampere ($A = C/s$), but simply $1/s$ expressing the number of particles per unit time. The current density (which is current per unit area) as a differential quantity is more general and more convenient to work with than the overall, i.e., integral current, as discussed in Section 1.1. The diffusion current density will be denoted by J_{diff}, its unit being $s^{-1}m^{-2}$.

As concluded above, the current density of the particles flowing from outside into the room ($J_{diff\rightarrow}$) is proportional to the concentration of particles available outside the window (N_O). The current density of the particles flowing in the opposite direction ($J_{diff\leftarrow}$) is proportional to the concentration of particles inside the window (N_I). The effective current density is equal to the difference of those two currents:

$$J_{diff} = J_{diff\rightarrow} - J_{diff\leftarrow} \propto N_O - N_I \equiv -\Delta N \qquad (1.23)$$

Note that the minus sign in front of ΔN indicates that the concentration N is decreasing along the x-axis; it is defined as $\Delta N \equiv N(x + \Delta x) - N(x)$.

It is not the difference in concentration alone that is important for the diffusion current—equally important is the distance at which this difference appears. The same difference in the concentrations of the particles appearing on each side of the window, ΔN, would lead to a smaller current density if the thickness of the window (or the wall) Δx was larger. Therefore,

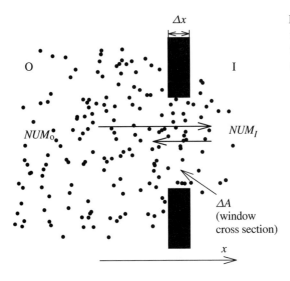

Figure 1.12 Diffusion current: particles flow through the window in both directions, but more from left to right as there are more particles on the left-hand side.

the current density is proportional to the difference in concentration ΔN appearing across a distance Δx:

$$J_{\text{diff}} \propto -\frac{\Delta N}{\Delta x} \tag{1.24}$$

The ratio $\Delta N/\Delta x$ basically expresses change in the concentration per unit length, or concentration gradient. A steeper concentration gradient will produce a larger diffusion current.

If a proportionality factor, denoted by D, is introduced into Eq. (1.24), and the finite differences Δ replaced by their infinitesimal counterparts ∂, the final form of the diffusion-current equation is obtained:

$$J_{\text{diff}} = -D\frac{\partial N}{\partial x} \tag{1.25}$$

The concentration gradient is given in Eq. (1.25) by the partial derivative $\partial N/\partial x$ in order to stress that the derivative expresses the change of concentration along the x-axis, and has nothing to do with any other dependence of the concentration, such as the time dependence. Equation (1.25) is the one-dimensional form of the diffusion-current equation. It can be expanded to include all the three space dimensions, in which case the current density is expressed as a vector:

$$\mathbf{J}_{\text{diff}} = -D\left(\frac{\partial N}{\partial x}\mathbf{x}_u + \frac{\partial N}{\partial y}\mathbf{y}_u + \frac{\partial N}{\partial z}\mathbf{z}_u\right) = -D\nabla N \tag{1.26}$$

The random thermal motion of particles is more pronounced at higher temperatures, therefore, the diffusion current is expected to be larger. It appears that our mathematical model of the diffusion current does not take into account the influence of the temperature on diffusion. The dependence of the random thermal motion of particles on temperature, however, is different in different materials and for different particles (different in gases and solids for that matter), so it is not possible to establish a general temperature dependence of the diffusion current. To deal with this problem, the temperature dependence of the diffusion current is taken into account by the proportionality factor D, termed the *diffusion coefficient*, which has to be determined for any individual material and any kind of diffusing particles. It is possible to establish temperature dependencies of the diffusion coefficient for different materials and diffusing particles, as will be shown later for the case of diffusion of doping atoms into semiconductor substrates (the last part of this section), as well as for the case of electron and hole diffusion in semiconductors (Chapter 2).

The Continuity Equation[4]

The diffusion process changes the concentration of particles with the time. Even if the concentration of the particles at the surface of the substrate is kept constant, the concentration profile changes as particles diffuse deeper into the substrate. The concentration of diffusion particles is, therefore, a function of two variables: $N(x, t)$. One more equation is needed to account for the time dependence of the concentration profile.

Let us consider how the particle concentration at the window in Fig. 1.13 changes in time. To do so, consider the change in the number of particles confined in the little box at the window during a small time interval Δt. This change is equal to the difference between the number of particles that enter the box (NUM_{in}) and the number of particles that come out of the box (NUM_{out}) in the time interval Δt:

$$\frac{NUM_{in} - NUM_{out}}{\Delta t} \equiv \frac{\Delta NUM}{\Delta t} \tag{1.27}$$

As $NUM_{in} = NUM_{O\rightarrow box} + NUM_{I\rightarrow box}$ and $NUM_{out} = NUM_{box\rightarrow O} + NUM_{box\rightarrow I}$, we can write that

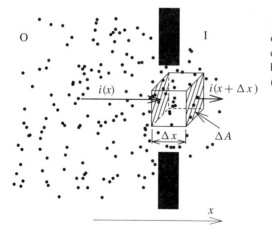

Figure 1.13 If the particle current flowing out of a box is different from the particle current flowing into the box, the number of particles in the box changes in time ($\partial N/\partial t \neq 0$).

---

[4]A more general form of the continuity equation, which includes recombination and generation terms, is presented in sections 8.1.3 and 8.2.3.

$$\frac{\Delta NUM}{\Delta t} = \frac{NUM_{O \to box} - NUM_{box \to O}}{\Delta t} - \frac{NUM_{box \to I} - NUM_{I \to box}}{\Delta t} \qquad (1.28)$$

The first term on the right-hand side of Eq. (1.28) expresses the effective number of particles entering the box from outside per unit time, which is the current of particles flowing into the box from outside (O). This current is denoted as $i(x)$ in Fig. 1.13. The current is not denoted as $i_{diff}(x)$ because this consideration is generally applicable, not only when the current is due to diffusion. Similarly, the second term on the right-hand side of Eq. (1.28) represents the current flowing out of the box into the room (I), labeled in Fig. 1.13 as $i(x + \Delta x)$. Therefore

$$\frac{\Delta NUM}{\Delta t} = i(x) - i(x + \Delta x) \equiv -\Delta i(x) \qquad (1.29)$$

Dividing Eq. (1.29) by the area of the sides of the small (elementary) box ΔA, the change in current density is obtained:

$$\frac{\Delta i(x)}{\Delta A} \equiv \Delta J(x) = -\frac{\Delta NUM/\Delta A}{\Delta t} \qquad (1.30)$$

To transfer the *number of particles NUM* into the *concentration N*, the number of particles has to be divided by a volume, in this example the volume of the elementary box $\Delta A \Delta x$. This can be achieved if Eq. (1.30) is divided by Δx:

$$\frac{\Delta J}{\Delta x} = -\frac{\Delta NUM/\Delta A \Delta x}{\Delta t} \equiv -\frac{\Delta N}{\Delta t} \qquad (1.31)$$

Finally, taking infinitesimal values, a differential equation called the *continuity equation* is obtained:

$$\frac{\partial J}{\partial x} = -\frac{\partial N}{\partial t} \qquad (1.32)$$

If there is no drift or any other current but the diffusion current, then the current density can be denoted as J_{diff}:

$$\frac{\partial J_{diff}}{\partial x} = -\frac{\partial N}{\partial t} \qquad (1.33)$$

The continuity equation is of general validity. It indicates that the concentration of particles in a considered element of volume can change only when the currents of particles flowing into and out of the volume are different. The concentration of particles *that are not being generated or annihilated* cannot change if there is no particle current or if the particle current is uniform.

Fick's Equation and Its Solutions: Doping Profiles

To model the diffusion of doping atoms in a semiconductor, we do not need to determine the actual value of the diffusion current density J_{diff}; what is needed is the doping-atom concentration $N(x, t)$. If the current density J_{diff} is eliminated from the diffusion-current equation (Eq. 1.25) and the continuity equation (Eq. 1.33), the so-called Fick's diffusion equation is obtained:

$$\frac{\partial N(x,t)}{\partial t} = D\frac{\partial^2 N(x,t)}{\partial x^2} \tag{1.34}$$

Solving the Fick's equation with appropriate boundary and initial conditions, the function $N(x,t)$ describing the time-dependent distribution of the doping atoms in the semiconductor can be obtained.

The distribution of the doping atoms (doping profile) can, obviously, be controlled by the time of the diffusion. The diffusion process also depends on temperature, through the diffusion coefficient D. In the case of diffusion of doping atoms, the diffusion coefficient exponentially depends on the temperature:

$$D = D_0 e^{-E_A/kT} \tag{1.35}$$

where T is the absolute temperature (temperature expressed in K), k is the Boltzmann constant, and the parameters E_A and D_0 are the activation energy and frequency factor, respectively. The parameters E_A and D_0 have to be determined for any individual semiconductor material and for individual doping species.

How strongly the diffusion coefficient depends on the temperature can be illustrated by considering the ratio of the diffusion coefficient at 1000°C and room temperature. For the case of $E_A = 3.5$ eV, which is a typical value, this ratio is $D(T = 1273 \text{ K})/D(T = 300 \text{ K}) = 8.5 \times 10^{44}$. This means that if an hour is needed to obtain a doped layer in the silicon at $T = 1000$°C , the same process would take 8.5×10^{44} at room temperature. This time expressed in years is 10^{41}, or it is 10^{39} centuries or 10^{38} millennia. It is certainly correct to say that there is no diffusion in the silicon at room temperature.

Integrated circuit diffusions are often done in two steps, the predeposition diffusion and the drive-in diffusion. The first is a constant-source diffusion, which means that a constant doping concentration N_0 is maintained at the surface of the semiconductor substrate. The second is a redistribution of the doping atoms in the semiconductor by heating the substrate while not providing any additional doping atoms to the surface of the substrate.

If the predeposition is carried out in an atmosphere sufficiently rich in doping atoms, then the surface doping concentration N_0 is at the solid-solubility limit (the maximum concentration of doping atoms that the host semiconductor crystal can accommodate without a serious distortion of the crystalline structure). The solid-solubility limit depends on the temperature, however, in silicon it is approximately equal to 4×10^{20} cm^{-3} for boron, 8×10^{20} cm^{-3} for phosphorus, 1.5×10^{21} cm^{-3} for arsenic, and 4×10^{19} cm^{-3} for antimony. The time dependence of the doping concentration in the semiconductor, $N(x,t)$, is given by the solution of Fick's diffusion equation for appropriate boundary conditions. In the case of constant-source diffusion, the following initial and boundary conditions will apply:

$$N(x,0) = 0; \quad N(0,t) = N_0; \quad N(\infty,t) = 0 \tag{1.36}$$

The solution of Eq. (1.34) that satisfies the boundary and initial conditions Eq. (1.36) is given by

$$N(x, t) = N_0 erfc \frac{x}{2\sqrt{Dt}} \tag{1.37}$$

the $erfc$ being the complementary error function defined by

$$erfc(z) = \frac{2}{\sqrt{\pi}} \int_z^\infty e^{-\alpha^2} \, d\alpha \tag{1.38}$$

This function is plotted in Fig. 1.14, while the normalized doping profiles for three different predeposition times/temperatures (more precisely Dt products) are plotted in Fig. 1.15a. Note that the doping profiles depend on the time–diffusion coefficient product Dt, where the diffusion coefficient depends exponentially on the temperature, as given by Eq. (1.35). A typical predeposition temperature is 950°C and a typical time is 30 min. The relatively low temperatures and times result in shallow doping profiles, which serve as the initial condition for the following drive-in diffusion.

The drive-in is done in an atmosphere containing oxygen or nitrogen, but not dopants. Because of that, the following so-called limited source boundary conditions will apply:

$$\int_0^\infty N(x, t) \, dx = \Phi; \quad N(\infty, t) = 0 \tag{1.39}$$

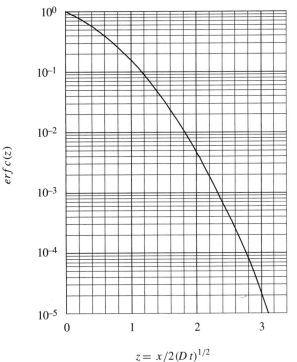

$$z = x/2(Dt)^{1/2}$$

Figure 1.14 The $erfc$ function.

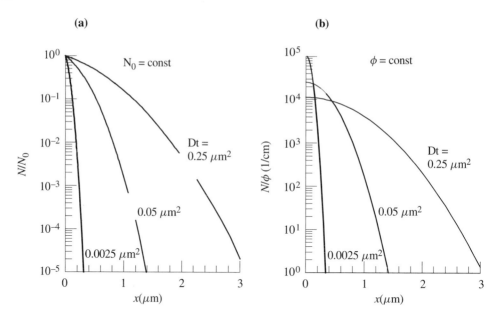

Figure 1.15 Doping profiles: (a) predeposition (constant source); (b) drive-in (limited source).

where Φ is called the dose of doping atoms that is incorporated into the semiconductor during the predeposition.

> The dose expresses how many doping atoms per unit area are in the semiconductor, regardless of their depth in the semiconductor, and is given in $1/m^2$.

The concentration of the doping atoms, $N(x, t)$, expressing their in-depth distribution in the semiconductor is integrated along the x-axis to express the overall number of doping atoms per unit area, that is the dose Φ (Eq. 1.39). The dose does not change during the drive-in diffusion [$\Phi \neq \Phi(t)$], as there are no new doping atoms that are incorporated into the semiconductor, but the already existing ones are being redistributed diffusing deeper into the semiconductor. The solution of Eq. (1.34) that satisfies boundary conditions (1.39) is given by

$$N(x, t) = \frac{\Phi}{\sqrt{\pi \, Dt}} e^{-\frac{x^2}{4Dt}} \tag{1.40}$$

The doping profiles obtained after drive-in diffusion are illustrated in Fig. 1.15b for three different time–diffusion coefficient products. Note that Eq. (1.40) is of the same form as those used in Example 1.3 and Fig. 1.11.

If the predeposition profile given by Eq. (1.37) is integrated to obtain Φ,

$$\Phi = N_0 \int_0^\infty erfc \frac{x}{2\sqrt{Dt}} \, dx \tag{1.41}$$

and the obtained result used in Eq. (1.40), the following equation is obtained as a model for the two-step (predeposition and drive-in) diffusion process:

$$N(x) = \underbrace{\frac{2N_0}{\pi}\sqrt{\frac{D_1 t_1}{D_2 t_2}}}_{N_s}\, e^{-\frac{x^2}{4D_2 t_2}} \tag{1.42}$$

In Eq. (1.42), D_1 and t_1 refer to the predeposition, D_2 and t_2 refer to the drive-in diffusion, and $N_s = N(0)$ obviously expresses the doping concentration at the surface after the drive-in.

■ **Example 1.9 Diffusion Coefficient**

Calculate the diffusion coefficient for boron at 1000°C and 1100°C using the following values for the frequency factor and the activation energy: $D_0 = 0.76\,\mathrm{cm^2/s}$, $E_A = 3.46$ eV. Comment on the results. The Boltzmann constant is $k = 8.62 \times 10^{-5}$ eV/K.

Solution: The diffusion coefficient can be calculated using Eq. (1.35). Note that the temperature should be expressed in K, therefore, $T_1 = 1000 + 273.15 = 1273.15$ K and $T_2 = 1100 + 273.15 = 1373.15$ K, respectively. The results are $D_1 = 1.54 \times 10^{-14}$ cm^2/s and $D_2 = 1.53 \times 10^{-13}$ cm^2/s for T_1 and T_2, respectively. Increase in the temperature from 1000°C to 1100°C (10%) increases the diffusion coefficient 10 times.

■ **Example 1.10 Constant-Source Diffusion**

A constant-source boron diffusion is carried out at 1050°C into an N-type silicon wafer with uniform background doping N_B, where $N_B = 10^{16}$ cm^{-3}. The surface concentration is maintained at 4×10^{20} cm^{-3}, which is the solid solubility limit. A junction depth of 1 μm is required. What should the diffusion time be?

Solution: Calculate first the diffusion coefficient at $T = 1050$°C. Repeating the procedure described in Example 1.9, we find that $D = 5.1 \times 10^{-14}$ cm^2/s.

At the P–N junction, the boron concentration $N(x)$ is equal to the substrate concentration N_B. Referring to Eq. (1.37) we obtain

$$N_B = N(x_j) = N_0 erfc(z), \quad z = x_j/(2\sqrt{Dt})$$

It is easy to find that $erfc(z) = N_B/N_0 = 10^{16}/4 \times 10^{20} = 2.5 \times 10^{-5}$. From Fig. 1.14, we can find that $erfc(z) = 2.5 \times 10^{-5}$ for $z = 3.0$. Using $z = x_j/(2\sqrt{Dt})$, the time required is obtained as

$$t = x_j^2/(4z^2 D) = 5447 \text{ s} = 1.51 \text{ hours}$$

■ **Example 1.11   Drive-in Diffusion**

How will the junction depth from Example 1.10 change if a drive-in diffusion is performed after the constant-source diffusion. The time and temperature of the drive-in diffusion are the same as the time and the temperature of the constant-source diffusion. What is the surface doping concentration after the drive-in diffusion?

Solution: The doping profile after a drive-in, which follows a constant-source diffusion, is given by Eq. (1.42). At the P–N junction, $(x = x_j)$, the boron concentration $N(x)$ is equal to the substrate concentration N_B (refer to Fig. 1.11):

$$N_B = N(x_j) = N_s e^{-[x_j^2/(4D_2 t_2)]}$$

From the above equation we calculate x_j as

$$x_j = \sqrt{4Dt \ \ln[2N_0/(\pi N_B)]} = 1.1 \times 10^{-4} \text{ cm} = 1.1 \ \mu\text{m}$$

The surface concentration after the constant-source diffusion and the drive-in is given by

$$N_s = 2N_0/\pi = 2.55 \times 10^{20} \text{ cm}^{-3}$$

During the drive-in, the junction depth is increased while the surface concentration is reduced.

■ **Example 1.12   Designing Two-Step Diffusion Process**

The doping concentration of the N-type substrate in the example of Fig. 1.10 is $N_B = 10^{16}$ cm^{-3}. Determine times and temperatures of the predeposition and drive-in diffusions of boron so to achieve the following parameters of the resistor P-type body: the surface concentration $N_s = 2 \times 10^{18}$ cm^{-3}, and the P–N junction depth $x_j = 2 \ \mu$m. Assume that the surface concentration during the predeposition is equal to the solid-solubility limit, $N_0 = 4 \times 10^{20}$ cm^{-3}. The activation energy and the frequency factor for the boron diffusion are $E_A = 3.46$ eV and $D_0 = 0.76$ cm^2/s, respectively.

Solution: The doping profile after a two-step boron diffusion (predeposition and drive-in) is described by Eq. (1.42). At the P–N junction,

$$N_B = N(x_j) = N_s e^{-[x_j^2/(4D_2 t_2)]}$$

The product of a drive-in time t_2 and diffusion coefficient D_2 that is necessary to achieve the desired x_j and N_s can be calculated using the above condition:

$$D_2 t_2 = \frac{x_j^2}{4} \frac{1}{\ln(N_s/N_B)} = 0.189 \ \mu\text{m}^2 = 1.89 \times 10^{-9} \text{ cm}^2$$

Theoretically, any combination of t_2 and D_2 that gives the above-calculated product is a correct solution. However, the time t_2 and the diffusion temperature T_2, which determines the value of D_2 through Eq. (1.35), should have practically meaningful values. The times should neither be too long nor too short, and the temperature should not be higher than 1200°C. If we assume the drive-in time of 1 h, that is $t_2 = 3600$ s, the diffusion coefficient should be $D_2 = 1.89 \times 10^{-9}/3600 = 5.25 \times 10^{-13}$ cm^2/s. Using Eq. (1.35), the drive-in temperature is calculated as

$$T_2 = \frac{E_A}{k} \frac{1}{\ln(D_0/D_2)} = 1433.5 \text{ K} = 1160°C$$

This is an acceptable temperature, therefore, $t_2 = 3600$ s and $T_2 = 1160°C$ can be used as a practically acceptable set of values.

The predeposition time t_1 and temperature T_1 (i.e., diffusion coefficient D_1) are determined from the N_s part of Eq. (1.42):

$$D_1 t_1 = \left(\frac{\pi N_s}{2N_0}\right)^2 D_2 t_2 = 1.17 \times 10^{-13} \text{ cm}^2$$

If the predeposition temperature is set to $T_1 = 900°C$, then $D_1 = 1.05 \times 10^{-15}$ cm^2/s, and $t_1 = 1.17 \times 10^{-13}/1.05 \times 10^{-15} = 111$ s. Therefore, $T_1 = 900°C$ and $t_1 = 111$ s is a possible set of values for the predeposition temperature and time.

1.4 CARRIER MOBILITY

Doping atoms introduce free electrons and holes into semiconductors, as described in Sections 1.2 and 1.3. The free electrons and holes are mobile charged particles that make the electric current in semiconductors, and are therefore referred to as the *current carriers*. Mobility of the carriers directly determines the conductivity of semiconductors, as described in Sections 1.1 and 1.2. This section provides an insight into the concept of mobility, as well as a description of the most important mobility-related phenomena.

1.4.1 Effective Mass, Thermal Velocity, and Drift Velocity

Section 1.2.2 introduces a model for the holes as positively charged particles (or quasi-particles) in order to simplify our considerations of hole-related transport phenomena in semiconductors. Thinking of electrons as negatively charged particles is probably nothing more than a suitable model as well; it depends on what we believe the particles are. Although it is generally incorrect to consider the electrons and holes as minute spherical particles, such a model is used to describe many electric-current-related phenomena. The motion of molecules in gases (such as the air) is frequently used analogously to describe the motion of electrons and holes in a semiconductor crystal, with the apparent difference that the electrons and holes are charged particles. The free electrons and holes are frequently referred to as electron/hole gas.

Undoubtedly, it is a good idea to use simplified models as long as they help us to comprehend and describe abstract things such as the electric current. It should never be forgotten, however, that simplified models are only toys, unable to express the details and latent properties of the real things. With this in mind, the model of electron/hole gas as a gas of charged spherical particles performing random thermal motion will be used when describing most of the transport-related phenomena in semiconductors. Limitations of the model will be mentioned whenever it is necessary.

Two important concepts related to moving particles are the particle mass and velocity. The mass of an isolated electron (electron in vacuum) is a constant, $m_0 = 9.1 \times 10^{-31}$ kg. However, the electrons in semiconductors are different due to interactions with the semiconductor atoms. To make the concept of carrier gas applicable, the mass of the free electrons and holes has to be empirically adjusted for any semiconductor material. The mass of electrons and holes, used in the carrier-gas model, is referred to as *effective mass* and is labeled by m^*. Typically, the holes appear heavier than electrons, while $m^* < m_0$ for both types of current carriers in most of the known semiconductors.

The particle momentum ($\mathbf{p}$) combines the two important concepts, the mass m^* and the velocity $\mathbf{v}$ as:

$$\mathbf{p} = m^*\mathbf{v} \quad \text{general case}$$
$$p_x = m^*v_x \quad \text{one-dimensional case} \tag{1.43}$$

The carrier-gas model assumes that the kinetic energy of any single carrier is given by

$$E_{\text{kin}} = \begin{cases} \frac{m^*|\mathbf{v}|^2}{2} = \frac{|\mathbf{p}|^2}{2m^*} & \text{general case} \\ \frac{m^*v_x^2}{2} = \frac{p_x^2}{2m^*} & \text{one-dimensional case} \end{cases} \tag{1.44}$$

The plots given in Fig. 1.16 illustrate this dependence. The particle momentum p_x is related to another quantity, k_x, which is the x-coordinate of so-called wave vector $\mathbf{k}$. The wave vector is shown in Fig. 1.16 because it will be used later when the importance of this simple graph will become clear.

Like the molecules in the air, the current carriers in semiconductors possess kinetic energy even when no electric field is applied. In that case, the kinetic energy of the carriers is related to the crystal temperature T:

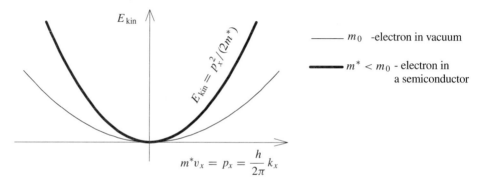

Figure 1.16　E–k dependence of a free electron.

$$E_{\text{kin}} = \frac{m^* v_{\text{th}}^2}{2} = \begin{cases} \frac{3}{2}kT & \text{three-dimensional case} \\ \frac{1}{2}kT & \text{one-dimensional case} \end{cases} \tag{1.45}$$

where k is the Boltzmann constant, T is the absolute temperature, and v_{th} is the *thermal velocity*. The thermal motion of carriers is essentially random due to collisions with each other like the molecules in the air, and more importantly due to collisions with the imperfections of the crystal lattice (doping atoms, thermal vibrations of crystal-lattice atoms, etc.). The collisions of electrons and holes are referred to as *carrier scattering*.

Carriers that move randomly due to the thermal energy that they possess, do not make any electric current as they do not effectively move from one side of the semiconductor to the other. Figure 1.17-Ia illustrates an electron that returns to the initial position after a number of scattering events, therefore performing no effective motion in any particular direction in the crystal. It is said that the *drift velocity* of the carriers is equal to zero.

If an electric field is applied to the electron gas, the electric-field force will deviate slightly from the electron paths between collisions, producing an effective shift of the electrons in the direction opposite to the direction of the electric field, as illustrated in Fig. 1.17-Ib. The effective shift of carriers per unit time is expressed as drift velocity. Figure 1.17-Ic illustrates that an increase in the electric field increases the drift velocity of the carriers. The graphs from rows II and III in Fig. 1.17 illustrate the influences of the effective carrier mass and scattering, respectively, which are considered later.

Let us now imagine a bar of semiconductor that has a cross-sectional area A and good contacts at the ends. Applying a voltage at the ends of the sample, an electric field is produced within the sample that forces the carriers, say electrons, to move at a drift velocity of v_{d}. If we count how many electrons per unit time t are collected at the positive contact, we will find that this number is equal to the number of electrons that are not further than $v_{\text{d}} t$ from the contact at the beginning of our counting ($v_{\text{d}} t$ is the distance that electrons travel in time t when moving with speed v_{d}). In other words, this number is equal to the number of electrons confined within the volume $v_{\text{d}} t A$. If there are n electrons per unit volume, that is the concentration of electrons is n, there will be $n v_{\text{d}} t A$ electrons that reach the positive contact during time t. The number of electrons that are collected per unit time t, which is $n v_{\text{d}} A$, multiplied by the charge that every electron carries ($-q$), is the electric current flowing through the terminal, I. Therefore,

$$I = -q n v_{\text{d}} A \tag{1.46}$$

The current I (expressed in A, which is C/s) turns into current density j (expressed in A/m^2) when divided by the area A:

$$j = \begin{cases} -q n v_{\text{d}} & \text{electrons} \\ q p v_{\text{d}} & \text{holes} \end{cases} \tag{1.47}$$

Equation (1.47) provides the relationship between the electric current density and the drift velocity of carriers. It is not difficult to understand this equation: if there are more carriers per unit volume (n higher) that move faster (v_{d} larger), the current density j will be

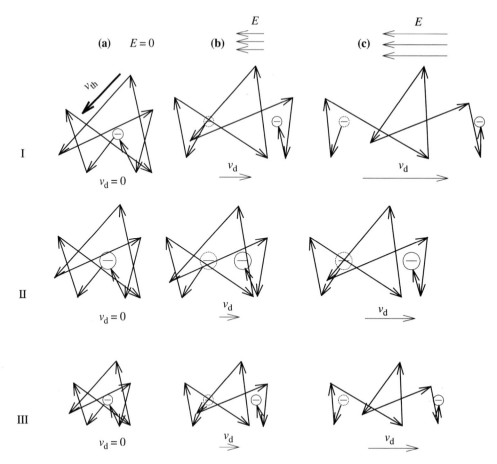

Figure 1.17 The concept of drift velocity (I) and its dependence on carrier effective mass (II) and free path (III). The first column (a) expresses the case of no electric field applied and the second (b) and the third (c) columns illustrate the cases when smaller and larger electric fields are applied, respectively.

proportionally larger. The factor q is there to convert particle current density expressed in $s^{-1}m^{-2}$ into electric current density expressed in $C\ s^{-1}\ m^{-2}$, that is $A\ m^{-2}$.

As illustrated in Fig. 1.17, the drift velocity depends on the electric field applied. Figure 1.18 shows measured values of the drift velocities of electrons and holes versus electric field in silicon. The expected linear relationship between the drift velocity and the electric field is observed only for small electric fields, that is small drift velocities. As the electric field is increased, the drift velocity tends to saturate at about $0.1\ \mu m/ps$ in the case of Si crystal. Let us think of what is going to happen with the electron trajectory in Fig 1.17-I if we continue to increase the electric field E. We can see that there is a limit to the increase of the drift velocity, as there is a limit of how much the scatter-like electron trajectory can be stretched in the direction of the electric field. This can account for the effect of drift velocity saturation in silicon.

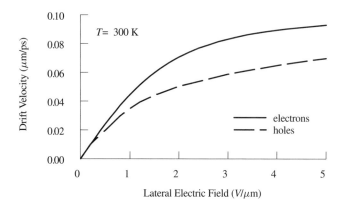

Figure 1.18 Drift velocity vs electric field in Si.

1.4.2 Mobility

Let us consider the semiconductor bar having good contacts at the ends again. If electric field E is applied, and the conductivity of the semiconductor is σ, then the current density is obtained from the differential form of Ohm's law [Eq. (1.10)]. If the semiconductor is N type, the conductivity is given by Eq. (1.18), where the concentration of donor atoms N_D is approximately equal to the concentration of electrons n. Therefore, Ohm's law expresses the current density in the following way:

$$j = \begin{cases} q\mu_n n E & \text{electrons} \\ q\mu_p p E & \text{holes} \end{cases} \tag{1.48}$$

Ohm's law is not based on consideration of a deep solid-state physics, it simply expresses an observation that the current density depends linearly on the electric field applied. Lacking a detailed physical background, it has to involve a proportionality constant, which is the mobility μ_n in the above equation, that needs to be determined experimentally. In the previous section, a current equation is derived [Eq. (1.47)] from a microscopic consideration of the flow of the current carriers. If these two equations are to give the same current densities for given carrier concentrations n and electric fields E, then the mobility has to be linked to the drift velocity as

$$v_d = \begin{cases} -\mu_n E & \text{electrons} \\ \mu_p E & \text{holes} \end{cases} \tag{1.49}$$

Equation (1.49) reveals that Ohm's law implicitly assumes a linear relationship between the drift velocity and the electric field. The experimental results given in Fig. 1.18 show that this assumption is valid only at small electric fields. This clearly shows an important limitation of Ohm's law when applied to semiconductor devices. The mobility has to be adjusted in order to preserve the validity of Eq. (1.48). This complicates the concept of mobility due to the fact that it can be considered as a constant only at low electric fields, while it becomes an electric-field–dependent parameter at high electric fields.

Equation (1.49) is basically the definition of mobility.

> The carrier mobility is *the proportionality coefficient in the dependence of drift velocity on the applied field*. The unit for mobility is given by the velocity unit over the electric field unit, which is $(m/s)/(V/m) = m^2/Vs$.

Although it is difficult to provide a more descriptive explanation of the mobility, it is easy to distinguish between a mobile and an immobile particle. If particles are immobile (like the ionized doping atoms in Fig.1.6), then no electric field can produce a nonzero drift velocity, that is a motion of these particles. In the case of mobile particles, however, Eq. (1.49) suggests that the same electric field can produce larger or smaller drift velocity. This shows that thinking of particles as being mobile or immobile is not precise enough, as the mobile particles can exhibit different mobilities.

What can make the current carriers different in terms of their mobilities? Figure 1.17-II illustrates one factor, which is the effective mass. Remember that the carriers, electrons for example, have different effective masses in different semiconductor materials; also holes generally have different effective mass compared to electrons. It is more difficult for the electric field to move a heavier electron (compare case II to case I in Fig. 1.17), which means the electric field is able to drive it to a shorter distance for the same time, therefore the drift velocity is smaller. If the same field produces a smaller drift velocity, that means that the mobility is smaller, according to Eq. (1.49).

Another factor that influences the carrier mobility is the degree of carrier scattering by the imperfections of the crystal lattice (as mentioned above, those are doping atoms, thermal vibrations of the crystal-lattice atoms, etc.). More frequent scattering means that the average distance between two scattering events (free path) is smaller, as illustrated in Fig. 1.17-IIIa. When the free path is smaller, the electric field is able to drive the carriers to a shorter distance, which means the drift velocity/mobility is smaller, as can be seen when case III is compared to case I in Fig. 1.17. Carrier scattering and its influence on mobility are considered in more detail in the next section.

1.4.3 Scattering: Dependence of Mobility on Temperature and Doping Concentration

The semiconductor atoms are basically tied to their crystal-lattice positions, however, they vibrate randomly around the central positions due to the thermal energy that is delivered to the crystal. The radius of the atom vibrations increases as the temperature increases, while they do not vibrate at all at 0 K (the vibration radius is equal to zero). The random thermal vibrations of a semiconductor atom are confined within a sphere, which is seen as a particle (or quasiparticle) called a *phonon*. The radius of the sphere increases with increases in the temperature, increasing the possibility that electrons and holes collide with the atom, an event referred to as *phonon scattering*. There would not be any phonon scattering at 0 K, and its frequency increases as the temperature is increased.

The doping atoms are another important source of electron and hole scattering. The doping atoms are ionized particles (donors positively and acceptors negatively, as shown in Fig. 1.6), therefore they repel or attract electrons or holes that are appearing in their vicinity, consequently changing the direction of their motion. This is referred to as *Coulomb scattering*. Coulomb scattering is more pronounced at lower temperatures, as the thermal velocity of the carriers is smaller and they stay in electric contact with the ionized doping atoms for a longer time, making the scattering more efficient.

As the temperature has opposite effects in the cases of phonon and Coulomb scattering, the dependence of mobility on temperature is not a straightforward function. If the doping concentration is high enough, causing Coulomb scattering to dominate over phonon scattering, the mobility continuously increases with temperature as Coulomb scattering is being weakened. Such a situation appears in silicon for a doping level of 10^{19} cm^{-3} and temperatures up to 500 K (Fig. 1.19). As the doping concentration is reduced, phonon scattering dominates at higher temperatures, which means that the increase in mobility at lower temperatures (due to weakened Coulomb scattering) is followed by a decrease in mobility at higher temperatures (due to strengthened phonon scattering). Figure 1.19 shows such a behavior of the mobility, observed for low and medium doping levels (10^{15} and 10^{17} cm^{-3}, respectively).

Figures 1.20 and 1.21 provide dependencies of electron and hole mobilities on doping concentration at different temperatures for silicon and gallium arsenide, respectively.

1.4.4 Mobility versus Diffusion Coefficient: Haynes–Shockley Experiment and Einstein Relationship

The discussions in this section have so far been concentrated on carrier mobility as an important factor in determining the level of *drift current* [Eq. (1.48)]. Section 1.3 described the *diffusion current*, given by Eq. (1.25). To apply Eq. (1.25) to the case of electron or hole current, the particle concentration N can be replaced by n or p and multiplied by q to convert the particle current density (expressed as number of particles per second and unit area) into electric current [expressed in C/[sm^2], that is A/m^2]:

$$j_{\text{diff}} = \begin{cases} q D_n \frac{\partial n}{\partial x} & \text{electrons} \\ -q D_p \frac{\partial p}{\partial x} & \text{holes} \end{cases} \qquad (1.50)$$

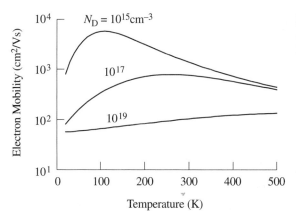

Figure 1.19 Temperature dependence of mobility for three different doping levels in Si.

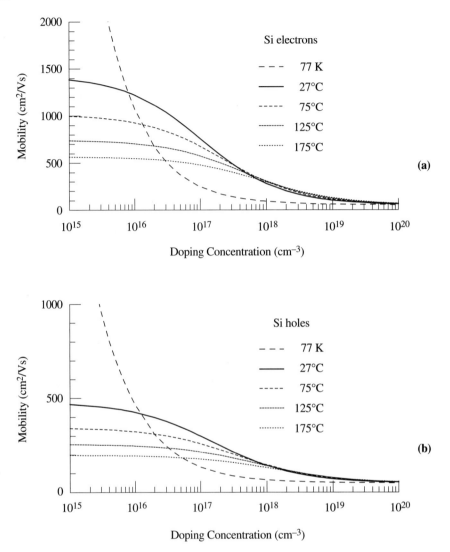

Figure 1.20 Low-field electron (a) and hole (b) mobility in Si.

The drift and diffusion are two independent current mechanisms. This is nicely demonstrated by the Haynes–Shockley experiment, illustrated in Fig. 1.22. To monitor the process of drift and diffusion, a pulse of minority carriers is generated in a very narrow region ($\Delta x \rightarrow 0$) of a semiconductor bar for a very short time interval ($\Delta t \rightarrow 0$). In the case shown in Fig. 1.22 this is achieved by a flash of light that illuminates a narrow region at the end of an N-type silicon bar at time $t = 0$. The light generates both electrons and holes, however, the direction of the electric field applied is such that the electrons are quickly collected by the positive contact, while the holes have to travel through the silicon bar to

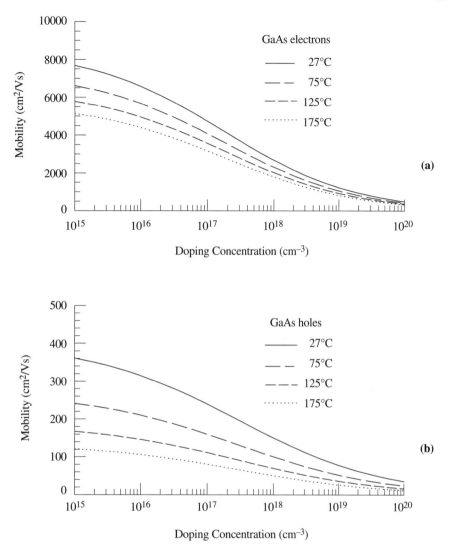

Figure 1.21 Low-field electron (a) and hole (b) mobility in GaAs.

be collected by the negatively biased contact. The drift velocity of holes is $v_d = \mu_p E$, and if the diffusion did not exist, a short current pulse would be detected by the ammeter as the holes are collected by the negative contact after the time interval equal to L/v_d. However, as the holes diffuse in either direction, the hole distribution widens with time, as illustrated in Fig. 1.22. Analogously to Eq. (1.40), the hole distribution can be expressed by the following form of the Gauss distribution:

$$p = p_{max} e^{-\frac{(x-x_{max})^2}{4D_p t}} \tag{1.51}$$

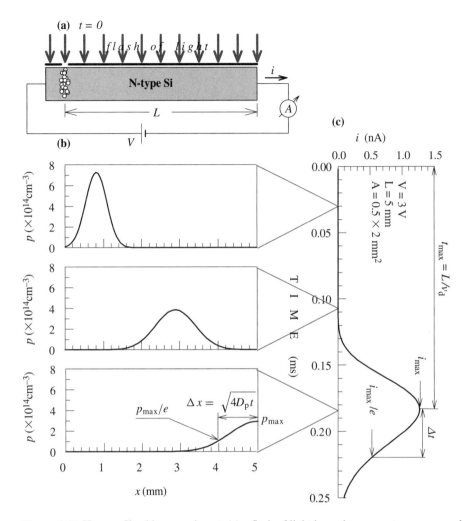

Figure 1.22 Haynes–Shockley experiment: (a) a flash of light is used to generate a narrow pulse of minority carriers (holes), (b) the motion of the hole peak illustrates the hole drift, while the widening of the hole distribution illustrates the hole diffusion, and (c) the measured current at the end of the semiconductor can be used to calculate μ_p and D_p.

The position of the distribution peak ($x = x_{max}$) shifts as the holes drift along the electric field: $x_{max} = v_d t$.

Both the minority-carrier mobility and diffusion coefficient can be extracted from the measured time dependence of the current. Figure 1.22c illustrates that the maximum of the current coincides with the arrival of the peak of the hole distribution at the negative contact. The drift velocity can be calculated from the measured time between the current maximum and the flash of light as $v_d = L/t_{max}$. The mobility is then calculated using Eq. (1.49):

$$\mu_p = \frac{v_d}{E} = \frac{L^2}{t_{max} V} \tag{1.52}$$

The diffusion coefficient can be determined from the width of the current pulse. Equation (1.51) shows that $p = p_{max}/e$ for $(x - x_{max})^2 = 4D_p t$, that is $\Delta x = x_{max} - x = \sqrt{4D_p t}$. The current that corresponds to this point of the hole distribution is i_{max}/e. As it takes the time of $\Delta t = \Delta x / v_d$ for the holes to travel the distance of $\Delta x = \sqrt{4D_p t} = \sqrt{4D_p(t_{max} + \Delta t)}$, the following equation can be written:

$$\Delta t = \frac{\Delta x}{v_d} = \frac{\sqrt{4D_p(t_{max} + \Delta t)}}{v_d} \tag{1.53}$$

Using $v_d = L/t_{max}$, the diffusion coefficient can be expressed in terms of the measured values of t_{max} and Δt:

$$D_p = \frac{\Delta t^2 (L/t_{max})^2}{4(t_{max} + \Delta t)} \tag{1.54}$$

The diffusion coefficient $D_{n,p}$ in Eq. (1.50) plays a role analogous to the role that mobility $\mu_{n,p}$ plays in Eq. (1.48), as it determines the level of the diffusion current that is produced by a certain concentration gradient. Even though the two current mechanisms are completely independent of each other, it somehow seems that a higher mobility of a carrier gas may be related to a higher diffusion coefficient.

Take the case of a low-mobility gas due to heavy effective mass of the carriers. If there is a concentration gradient, a diffusion current would exist. As particles would move very slowly in this case, it would take a significant time for the particles from the higher concentration part to be transported to the lower concentration parts, which means that the diffusion current is small. This example indicates that the mobility $\mu_{n,p}$ is proportional to the diffusion coefficient $D_{n,p}$.

Consider the effect of the temperature as both the mobility and the diffusion coefficient are found to be temperature dependent. The influence of the temperature on the mobility is basically through scattering, more pronounced scattering leading to reduced free path, and consequently reduced mobility. Reduction in the free path would proportionally reduce the diffusion coefficient, implying again that the mobility is proportional to the diffusion coefficient. The temperature, however, is essentially important for the process of diffusion. Imagine an intrinsic semiconductor (Coulomb scattering negligible) at a very low temperature. The thermal velocity of the carriers would be very small, but this does not stop the electric field from driving the carriers very efficiently, producing a significant drift current, which means the mobility is quite high. When the thermal velocity of the carriers is small, however, the diffusion process goes very slowly as the only driving force behind the diffusion is the random thermal motion. If the temperature is increased, the random thermal motion as well as the diffusion process become more pronounced, which means that the diffusion coefficient is increased.

The above considerations imply that the diffusion coefficient $D_{n,p}$ of a carrier gas is proportional to the mobility $\mu_{n,p}$ of the carriers as well as to the thermal energy kT (or thermal voltage kT/q):

$$D_{n,p} = \frac{kT}{q}\mu_{n,p} \tag{1.55}$$

The above relationship is known as the Einstein relation. It does not hold for very high carrier concentrations (so-called degenerate gases) when the carriers start competing for vacant positions during their motion, which influences the mobility.

■ **Example 1.13 Resistance and Mobility**

2-cm-long silicon piece, with cross-sectional area of 0.1 cm^2, is used to measure the electron mobility. What is the electron mobility if 90-Ω resistance is measured, and the doping level is known to be $N_D = 10^{15}$ cm^{-3}? Neglect any contact resistance.

Solution: Using Eqs. (1.2), (1.3), and (1.18), and replacing $x_j W$ by A as a more appropriate symbol for the cross-sectional area, it is found that

$$\mu_n = \frac{L}{A\,q\,N_D\,R} = 1390 \text{ cm}^2/Vs$$

■ **Example 1.14 TCR**

The temperature coefficient of a resistor, TCR, is defined as

$$TCR = \frac{1}{R}\frac{dR}{dT} \times 100 \ (\%/°C)$$

It expresses the percentage change of the resistance for any °C, or K, change in the temperature. Find the TCR (at room temperature) of a resistor made of N-type silicon. If the resistance of the resistor is 1 kΩ at 27°C, estimate the resistance at 75°C. The temperature dependence of electron mobility is approximated by

$$\mu_n = \text{ const } T^{-3/2} \ (T \text{ in K})$$

Solution: The resistance of a resistor depends on the resistivity ρ, resistor length L, and resistor cross-sectional area $x_j W$ [Eq. (1.2)], where the resistivity is the only temperature-dependent parameter. Replacing the resistance R in the above definition of TCR by Eq. (1.2), the TCR is obtained as

$$TCR = \frac{1}{\rho}\frac{d\rho}{dT} \times 100$$

The resistivity of N-type silicon is given by

$$\rho = 1/(q\mu_n N_D)$$

where, obviously, the mobility is the only temperature-dependent parameter. Using the given temperature dependence of mobility, the resistivity is expressed by

$$\rho = \frac{1}{q N_{\mathrm{D}} \mathrm{const}} T^{3/2}$$

Finding that

$$\frac{d\rho}{dT} = \frac{3}{2} \frac{1}{q N_{\mathrm{D}} \mathrm{const}} T^{1/2}$$

the TCR is obtained as

$$TCR = \frac{3}{2T} \times 100$$

Therefore,

$$TCR = \frac{3}{2 \times 300} \times 100 = 0.5\%/\mathrm{K} = 0.5\%/°\mathrm{C}$$

When the temperature changes from 27°C to 75°C, the resistance changes by

$$\Delta R \approx \frac{TCR}{100} R \Delta T = 0.005 \times 1\,\mathrm{k}\Omega \times (75°\mathrm{C} - 27°\mathrm{C}) = 0.24\,\mathrm{k}\Omega$$

The resistance at 75°C is, therefore, estimated at 1.24 kΩ.

1.5 ENERGY-BAND MODEL

It has been possible to introduce the most important resistor-related phenomena using the chemical-bond model alone. Nevertheless, it is a very simple model that does not clearly express the energy state of the current carriers; because of this it would fail to account for many important effects and, for that matter, would fail to explain the basic principle of operation in some devices. It is necessary to go a step deeper into solid-state physics and introduce what is referred to as the *energy-band model*. The energy-band model is commonly used to explain semiconductor devices. It is a very powerful tool for describing and comprehending the phenomena in semiconductor devices, but only when clearly understood.

This last section of the chapter devoted to the simplest device, the resistor, introduces the energy-band model. It is not intended to further explain the resistor characteristics, but rather to use a sound understanding of the semiconductor resistor properties to introduce the essentials of the energy-band model.

1.5.1 Energy Bands: Quantum Mechanics Background

As mentioned in the previous section, the model of electrons as minute negatively charged particles is not precise enough. We are not going to say now that our simple model used in the carrier-mobility section was completely wrong, and that electrons cannot be considered as negatively charged particles. Even when Davisson and Germer experimentally proved in 1927 that the electrons exhibit wave properties (diffraction), they did not say that the electrons were not particles. The particle properties of electrons are undeniable as it is proved that an electron carries a unit of charge ($-q = -1.6 \times 10^{-19}$ C) and it can be released from an atom, or captured by an atom only as a whole—there is no such thing as a half or a quarter of an electron. Equally well, however, it is proved that the electrons exhibit wave properties, like diffraction and interference. The electron microscopes quite successfully exploit the wave properties of electrons. This wave-particle duality is a general concept, and there is a general relationship between the momentum of a particle $m_0 v$ and the wavelength λ:

$$\lambda = \frac{h}{m_0 v} \tag{1.56}$$

In this relationship, known as the de Broglie postulate, $h = 6.626 \times 10^{-34}$ Js is the Planck constant.

The concept of electrons is usually introduced through Bohr's planetary model of atoms, which was used to introduce the chemical-bond model in Section 1.2. The version of the planetary model of silicon used in Section 1.2 is shown in more detail in Fig. 1.23a. Ten of the silicon 14 electrons are much closer to the core than the remaining four electrons. It is these four electrons that determine the chemical properties of silicon. To simplify our considerations, the 10 inner electrons are grouped together with the 14 positively charged protons into a larger "core" labeled as "+4" to indicate the net charge.

There are profound reasons for singling these four of the 14 silicon electrons. It will become clear toward the end of this section that these reasons are related to the fact that every single electron in the universe has different properties: it either belongs to a different atom, or it has different energy, or it has different angular momentum, or it has different magnetic moment, or, finally, it has different spin direction. The possible energy levels, the angular momentum, the magnetic moment, and the spin direction define possible *electron states*.

It is a very specific characteristic of the electrons that no more than one electron can occupy a single electron state. This is known as the *Pauli exclusion principle*.

In addition, because of the wave properties, the electrons cannot have any value of energy, angular momentum, or magnetic moment. In terms of the planetary model, the radii of electron orbits cannot take any value. The electrons on different orbits can be stable only as standing waves, which happens when the orbital circumference is an integral number (n) of the wavelength (λ):

$$2\pi r = n\lambda \tag{1.57}$$

Substituting Eq. (1.56) into Eq. (1.57) yields the following equation for the possible values of the orbital radii:

$$r = n\frac{h}{2\pi m v} \quad (n = 1, 2, \ldots) \tag{1.58}$$

The fact that the electron radius, energy, etc. can take only specific discrete values is known as *quantization*.

There are four *quantum numbers* describing the packing of electrons in the atoms: the principal (n), angular, magnetic, and spin quantum numbers. The electrons at the first orbit ($n = 1$) are spherically symmetrical,[5] and they cannot have different angular and magnetic quantum numbers, however, they can have different spin directions. As a result, there can be only two electrons at the first orbit, a fact expressed symbolically as $1s^2$. The second orbit can also have two spherically symmetrical electrons, which is expressed as $2s^2$. However, it can also have electrons with x-, y-, or z-directional symmetry, which means six additional electrons: three space times two spin directions. This is symbolically expressed as $2p^6$. When completely filled, the first two orbits take 10 electrons, which make the inner 10 electrons of the silicon atom. The third orbit ($n = 3$), illustrated in Fig. 1.23, also starts with s^2 and p^6 shells. This makes eight electron states, however, only four states are actually taken by electrons. The four electrons at the third orbit are referred to as *valence electrons*.

As mentioned earlier, the planetary model (Fig. 1.23a) does not clearly show the energy level of the electrons. A more precise atom model is shown in Fig. 1.23b. The electric field of the positively charged core creates the funnel-like shape of the electric potential, and therefore the potential energy. This shape of the potential energy surface (function), the cross section of which is shown in Fig. 1.23b, is frequently referred to as *potential well*. It is said that the electrons of the atom are confined inside its potential well. As the electrons are not stationary, they possess kinetic energy $m_0 v^2 / 2$, which adds to the potential energy. The cross section of the total energy surface is also shown in Fig. 1.23b. Note that the potential energy is negative, while the kinetic energy is positive.

The concept of the energy well (or funnel) helps to visualize the need for having electrons at different and quantized energy levels. As the electrons can be stable only at their energy levels as standing waves, the electrons can appear only at the energy levels that correspond to energy-funnel widths related to integer multiples of the half-wavelength. The width of the energy funnel at the energy levels corresponding to the third orbit of silicon is about 0.20 nm, which corresponds to three electron half-wavelengths.[6] There is an energy difference between the $3s^2$ and $3p^6$ electrons due to the fact that there is a slight difference in the wavelengths of the "s" and "p" electrons.

Let us move now from the situation of an isolated atom to the case of a semiconductor crystal, where the atoms are packed closely to each other. Fig. 1.23c illustrates three atoms from a one-dimensional cross section of a piece of silicon crystal. As the atoms are close to each other, the shape of the potential is changed. Figure 1.23c illustrates that the maximum of the potential energy walls is now below the zero level (the reference level corresponding to the energy of a free electron, the so-called vacuum level). As a result, the valence electrons are not strictly confined inside the potential wells of the individual atoms, although they are still confined inside the crystal as

---

[5]A deeper insight into the concepts of electron symmetry and "shape" in general goes through the electron wave function, which is described in Chapter 12.

[6]The energy levels of the electrons at the second orbit (double half-wavelength) and the first orbit (a single half-wavelength) are much lower, about -100 and -1800 eV, respectively (these levels are not shown in Fig. 1.23b).

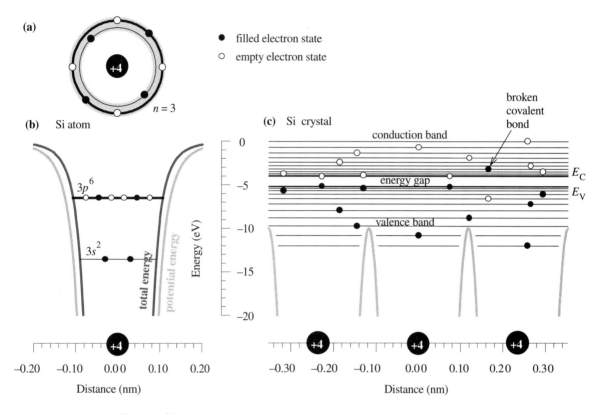

(a)

● filled electron state
○ empty electron state

$n = 3$

(b) Si atom

$3p^6$

$3s^2$

total energy

potential energy

Energy (eV)

Distance (nm)

(c) Si crystal

broken covalent bond

conduction band

energy gap

E_C

E_V

valence band

Distance (nm)

Figure 1.23 The electron orbits (a) and energy levels (b) in a single atom are analogous to energy bands in a crystal (c).

their total energy is still negative. The fact that the valence electrons are not confined inside the potential wells of the atoms means that the valence electrons interact with each other, and this is what enables the covalent bonds (Section 1.2) that link the atoms together.

According to the Pauli exclusion principle, there cannot be two identical electrons, that is two electrons at a single electron state. The "s" and "p" electrons cannot retain the $3s^2$ and $3p^6$ levels of the individual atoms, as they are not confined inside the potential wells of the individual atoms, and are not different simply because they belong to different atoms. As the electron states cannot disappear when the atoms are packed into a crystal, and they cannot be identical either, the interaction between the valence electrons leads to splitting of the electron states. The eight electron states available at the third orbit of every silicon atom have to split into $8N$ states in a crystal having N silicon atoms. As there is a large number of atoms in a piece of semiconductor, the gaps between the energy levels of different electron states are very small. A set of such energy levels, which are placed closely to each other, is referred to as an *energy band*. A remarkable energy gap can appear between two different energy bands, as illustrated in Fig. 1.23c.

A typical situation with the semiconductor materials is that the valence-electron levels of individual atoms produce two energy bands, called *valence* and *conduction* bands, that are separated by an *energy gap* (Fig. 1.23c). The electrons that form the covalent bonds (refer to Fig. 1.4) are energetically placed in the valence band. When a covalent bond is broken due to some thermal energy delivered to a valence electron, this electron jumps up into the conduction band, leaving behind a hole, as illustrated in Fig. 1.23c (energy-band model) and Fig. 1.4b (chemical-bond model).

The electrons in the conduction band are mobile particles as there are many free states that the electrons can move on to. Analogously, the holes in the valence band are mobile particles as well. The electrons in the valence band, which form the covalent bonds, are considered immobile particles.

The energy gap determines the basic material properties. Table 1.5 gives the values of the energy gap E_g at room temperature for different materials. It is the value of the energy gap that determines whether a certain material appears as conductor, insulator, or semiconductor. Figure 1.24a shows that there is no energy gap in metals (the conduction and the valence bands are merged). This makes all the electrons in the band conductive, as they all have access to a number of empty states to move to. Consequently, metals appear as good current conductors even at temperatures very close to 0 K. The energy gap in semiconductors (Fig. 1.24b) is relatively small, and some electrons have enough energy to jump up into the conduction band at room temperature. The electrons in the conduction band are mobile particles as they have plenty of empty states to move to, contributing to electric-current conduction. The electrons jumping into the conduction band, in the process of covalent-bond breakage, leave behind a hole that is a particle capable of conducting electric current as well. The energy gap in the insulators is relatively large (Fig. 1.24c), and the electrons do not obtain enough thermal energy at room temperature to jump over the gap. As there are neither electrons in the conduction band nor holes in the valence band, insulators do not conduct electric current. Referring to Table 1.5, silicon, gallium arsenide and germanium are classified as semiconductors, while silicon dioxide, silicon nitride, and carbon appear as insulators.

The conductivity of intrinsic semiconductors is typically very low, as there are not many free electrons and holes to conduct the current. The conductivity of semiconductors, however, can be increased to the level of good conductors by doping, as explained earlier. The effects of N-type and P-type doping are illustrated by the chemical-bond model in Fig. 1.6. An energy-band model presentation of the effects of N-type and P-type doping

TABLE 1.5 Energy Gap Values for Different Materials at 300 K

Material	E_g (eV)
Silicon	1.12
Gallium arsenide	1.42
Germanium	0.66
Silicon dioxide (SiO_2)	9
Silicon nitride (Si_3N_4)	5
Carbon	5.47

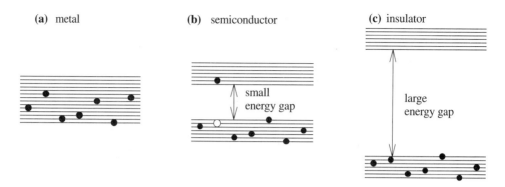

Figure 1.24 Energy band diagram for a metal (a), a semiconductor (b), and an insulator (c).

is given in Fig. 1.25. It shows that the donor atom (an atom from the Group V of the periodic table) introduces an energy level into the energy gap of the crystal. This level is close to the bottom of the conduction band and accommodates the fifth valence electron of the donor atom. This electron would remain at its level in the energy gap at very low temperature, but at room temperature it obtains enough energy to jump into the conduction band, becoming a free electron. It leaves behind a positively charged donor atom, which is immobile. Remember that the doping-induced electrons by far exceed the thermally generated electrons (Table 1.3), even though only one doping-induced and one thermally generated electron are shown in Fig. 1.25a. The effect of P-type doping is similarly expressed in Fig. 1.25b. In this case, the acceptor atom introduces an energy level into the energy gap that is close to the top of the valence band. As a consequence, an electron from the valence band jumps onto this level, leaving behind a mobile hole and creating a negatively charged and immobile acceptor atom.

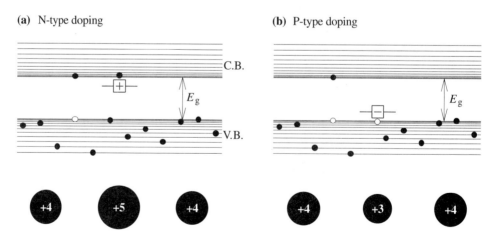

Figure 1.25 Effects of N-type (a) and P-type (b) doping in energy-band model presentation. C.B., conduction band; V.B., valence band.

1.5.2 Population of Energy Bands: Fermi–Dirac Distribution and Fermi Level

This section introduces a very important concept, which is the probability that an energy state is occupied by an electron (f). If the probability f is known, and there are N_C states per unit volume in the conduction band, then the concentration of free electrons can simply be calculated as

$$n = N_C f \tag{1.59}$$

It is reasonable to assume that the electrons in the conduction band are not uniformly distributed, but more electrons can be found at the bottom of the band than at higher energy levels. In general, higher energy levels are less likely to be populated. At the same time, it is important to consider what thermal energy is given, on average, to the particles. The same energy level in the conduction band is more likely to be populated at a higher temperature when the average thermal energy kT of the electrons is higher. If we further assume exponential-type dependence on energy and temperature [analogous to Eq. (1.35)], the probability that an energy state is occupied appears to be

$$f \propto \frac{1}{e^{\frac{E-E_F}{kT}}} \tag{1.60}$$

where E_F is a reference energy level. For $T = 0$ K, $1/e^{\infty} = 0$, which expresses the zero probability of finding an electron in the conduction band. As the temperature is increased, the value of f becomes larger than 0, and reduces with increase in energy (that is an increase in $E - E_F$).

The probability that an energy state is not occupied by an electron is $1 - f$, as the probability of the event of either having or not having an electron in that state has to be 1. The probability that an energy state in the valence band is not occupied by an electron is equal to the probability that the energy state is occupied by a hole. Therefore, the function $1 - f$ expresses the occupancy of energy levels in the valence band by holes. It appears that there is a high degree of symmetry in this problem—if the probability that an energy state is occupied by an electron is 0.8, then the probability that this state is occupied by a hole is 0.2, and vice versa. Obviously, the center of the symmetry is the energy at which the electron probability and the hole probability are equal, that is $f = 1 - f = 1/2$. If we take exactly this energy as the reference energy level E_F, then f should be equal to $1/2$ for $E - E_F = 0$. As $e^0 = 1$, an additional 1 should be added to the exponential term in the denominator of Eq. (1.60), which gives

$$f = \frac{1}{1 + e^{\frac{E-E_F}{kT}}} \tag{1.61}$$

The above equation is plotted in solid lines for three different temperatures in Fig. 1.26. The dashed lines are plots of the function $1 - f$, which is the probability for holes. It can be seen that the electron and hole probabilities are completely symmetrical, always giving 1 when added to each other. Equation (1.61) for f is known as the *Fermi–Dirac distribution function*, and it expresses the probability of having an electron with energy E. The reference energy level E_F is referred to as the *Fermi level*.

It can be noticed from Fig. 1.26 that, at 0 K the probability f is equal to 1 for $E < E_F$ and is equal to zero for $E > E_F$. If f is to express that at 0 K the probability of having an electron in the conduction band is equal to 0, while it is equal to 1 for the valence band, it is obvious that E_F has to be somewhere between the conduction and the valence band, therefore in the energy gap. This does not contradict the fact that there are no allowed energy states in the energy gap, as E_F is a reference energy level only. Despite the fact that $f = 0.5$ at $E = E_F$, there will be no electrons with $E = E_F$ as the electron concentration is obtained when the probability f is multiplied by the density of states [Eq. (1.59)], which is zero in the energy gap.

The logarithmic plots of f and $1 - f$, given in Fig. 1.26b by the solid and the dashed lines, respectively, more clearly express the electron probabilities for $E > E_F$ (toward the conduction band) and the hole probabilities for $E < E_F$ (toward the valence band). Again, the symmetry with respect to E_F is obvious. In an intrinsic semiconductor, the concentration of electrons is equal to the concentration of holes. Assuming that the density of states in the conduction band (N_C) and the valence band (N_V) are equal, the probabilities of finding electrons at the levels in the conduction band will have to be equal to the probabilities of finding holes at the levels in the valence band. This means that the top of the valence band E_V has to be as far from the reference energy E_F as the bottom of the conduction band E_C ($E_C - E_F = E_F - E_V$).

Fermi level E_F appears at the mid-gap in an intrinsic semiconductor.

In the intrinsic silicon, the top of the valence band is about 0.56 eV ($E_g/2$) below the Fermi level, while the bottom of the conduction band is about 0.56 eV above the Fermi level.

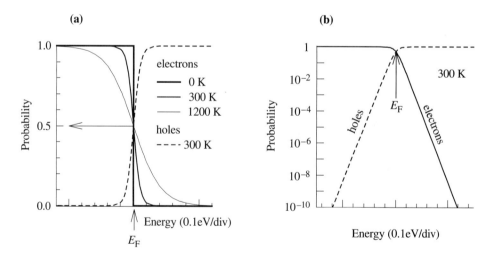

Figure 1.26 Fermi–Dirac distribution for electrons [solid lines, f as given by Eq. (1.61)] and holes (dashed lines, $1 - f$); the probability f is plotted on linear (a) and logarithmic axes (b).

Figure 1.26b shows that at those energies the probability of finding an electron and hole is less than 10^{-9}, which means that less than one state in every one billion is occupied by an electron in the conduction band and by a hole in the valence band. Yet, there are more than 10^{10} electrons and holes per cm^3 at room temperature. This is possible due to a high density of states close to the bottom of the conduction band (N_C) and to the top of the valence band (N_V). The density of states N_C and N_V are temperature-dependent parameters for a given semiconductor. Their room temperature values for the three most common semiconductors are given in Table 1.6.

Knowing the density of states at the bottom of the conduction band and the top of the valence band, and assuming that free electrons and holes are located very close to the bottom of the conduction band (E_C) and the top of the valence band (E_V), respectively, the concentration of electrons and holes can be related to the position of the Fermi level. Determine, first, the occupancies of the bottom of the conduction band and the top of the valence band by electrons and holes, respectively:

$$f(E_C) = \frac{1}{1 + e^{\frac{E_C - E_F}{kT}}} \approx e^{-\frac{E_C - E_F}{kT}} \tag{1.62}$$

$$f_h(E_V) \equiv 1 - f(E_V) = \frac{1}{1 + e^{\frac{E_F - E_V}{kT}}} \approx e^{-\frac{E_F - E_V}{kT}} \tag{1.63}$$

The approximations used in the above equations would not introduce any significant errors as long as $\exp[(E_C - E_F)/kT] \gg 1$ and $\exp[(E_F - E_V)/kT] \gg 1$, respectively. In other words, these approximations can be used in the regions where the probabilities are exponentially dependent on the energy, which are the straight-line regions on the semilogarithmic plots in Fig. 1.26b.

The exponential approximations cannot be used when the Fermi level is very close or within the conduction or valence band. In that case we are dealing with degenerate electron or hole gases.

Multiplying the occupancies of energy states, as given by Eqs. (1.62) and (1.63), by the density of states gives the concentration of electron and holes:

$$\boxed{n = N_C e^{-\frac{E_C - E_F}{kT}}} \tag{1.64}$$

TABLE 1.6 Effective Density of States at 300 K

	Effective Density of States at 300 K (cm^{-3})	
Semiconductor	Conduction Band $N_C = A_C T^{3/2}$	Valence Band $N_V = A_V T^{3/2}$
Silicon	2.86×10^{19}	3.10×10^{19}
Gallium arsenide	4.7×10^{17}	7.0×10^{18}
Germanium	1.0×10^{19}	6.0×10^{18}

$$p = N_V e^{-\frac{E_F - E_V}{kT}}$$ (1.65)

Equations (1.64) and (1.65) clearly account for the influence of temperature on the concentration of free electrons and holes. It is less obvious how these equations take into account the influence of doping level, which is even more important as doping is used to control the concentration of electrons and holes in semiconductors. As explained in Section 1.2, in an N-type semiconductor, the concentration of electrons is approximately equal to the donor-atom concentration ($n \approx N_D$), and similarly $p \approx N_A$ in P-type semiconductors. The fact that n is much higher in an N-type semiconductor than in the intrinsic semiconductor means that the occupancy of the states in the conduction band has to be higher. Figure 1.27b shows that this is possible if the Fermi level is closer to the conduction band. Indeed, if $E_C - E_F$ in Eq. (1.64) is smaller than in the intrinsic semiconductor, then the concentration n would exceed the intrinsic concentration. Analogously, the Fermi level moves toward the valence band to express a higher occupancy of the valence band by holes in a P-type semiconductor (Fig. 1.27c).

The position of the Fermi level in the equations for electron and hole concentrations expresses the doping type and level.

Using the fact that $n \approx N_D$ in N-type and $p \approx N_A$ in P-type semiconductors, the position of the Fermi level is obtained from Eqs. (1.64) and (1.65) as

$$E_F = E_C - kT \ \ln \frac{N_C}{N_D} \quad \text{N-type}$$ (1.66)

$$E_F = E_V + kT \ \ln \frac{N_V}{N_A} \quad \text{P-type}$$ (1.67)

The above equations are correct for doping levels that are lower than the density of states, that is $N_D < N_C$ and $N_A < N_V$. When $N_D = N_C$ the Fermi level approaches the bottom of the conduction band E_C, in which case the exponential approximation used in Eq. (1.62) is not good, but the complete Fermi–Dirac distribution has to be used; the electron gas becomes degenerate. Analogously, the hole gas becomes degenerate when N_A approaches the density of states N_V.

The semilogarithm plots of Fig. 1.27b show that the increase in the occupancy of the conduction-band levels by electrons in the N-type semiconductor is accompanied by a reduction in the occupancy of the valence-band levels by holes. This is due to the effect of electron and hole recombination discussed in section 1.2.5. Using Eqs. (1.64) and (1.65) we can now easily show that the product of electron and hole concentrations np is indeed a constant, thus independent of the doping level (position of the Fermi level):

$$np = N_C N_V e^{-\frac{E_C - E_V}{kT}}$$ (1.68)

Moreover, knowing that $np = n_i^2$ [Eq. (1.21)], we can now express the intrinsic carrier concentration n_i in terms of the energy gap of the material $E_g = E_C - E_V$:

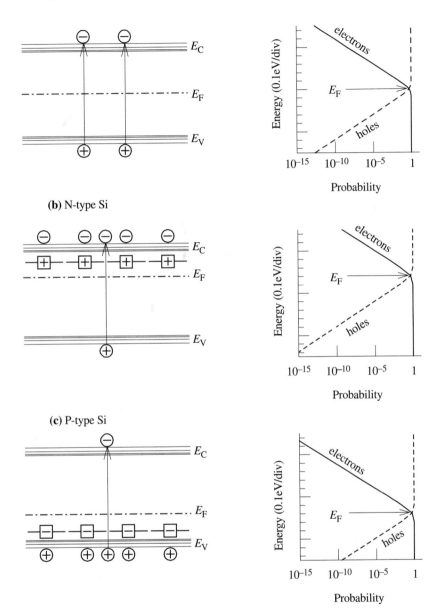

Figure 1.27 Position of the Fermi level expresses doping type and level: (a) intrinsic semiconductor, (b) N-type semiconductor, and (c) P-type semiconductor.

$$n_i = \sqrt{N_C N_V}\, e^{-\frac{E_g}{2kT}} \tag{1.69}$$

The above equation shows in its own way that the materials with large energy gap E_g appear as insulators (Fig. 1.24), as the intrinsic concentration of the carriers is fairly small.

1.5.3 Energy Bands with Applied Electric Field

The energy-band model is very useful for illustrating the electric potential inside a semiconductor structure. In fact, the bottom of the conduction band represents the potential energy of the electrons in the conduction band, and analogously, the top of the valence band represents the potential energy of the holes in the valence band. The potential energy E_{pot} and the electric potential φ are linked by the following fundamental relation:

$$\boxed{E_{pot} = -q\varphi} \tag{1.70}$$

where q is the unit of charge ($q = +1.6 \times 10^{-19}$ C). If the potential energy is expressed in electron volts, then the electric potential and the potential energy have the same numerical values with a different sign. This means that if a voltage of $V = \varphi_1 - \varphi_0 = -1$ V is applied across a semiconductor resistor, the difference between the potential energies of electrons and holes at the ends of the resistor will be 1 eV. The electric potential changes linearly from φ_0 to φ_1 inside the resistor body. Consequently, the potential energy (which means the bottom of the conduction band and the top of the valence band) linearly changes inside the resistor body as well. This situation is illustrated in Fig. 1.28.

Figure 1.28c gives the energy bands for the case in which an electric field is applied inside a semiconductor. As the electric field is equal to the negative gradient of the electric potential $(-d\varphi/dx)$, the slope of the energy bands $(dE_C/dx = dE_V/dx)$ directly expresses the electric field. This enables a very illustrative model for the drift current to be introduced.

> The conduction band can be considered as a vessel containing electrons that tend to roll down when the bottom of the conduction band is tilted.

> Analogously, the valence band can be considered as a vessel containing a liquid and bubbles (holes) in it. When the top of the valence band is tilted due to the appearance of an electric field, the holes tend to bubble up along the top of the valence band, moving effectively in the direction of the electric field applied.

This model is very useful in illustrating the principle of the operation of semiconductor devices.

Figure 1.28c shows the electron trajectory in even more detail, to illustrate the effects of electron scattering. The electrons move horizontally in the energy band between two collisions, which means that their total energy remains constant. This means the reduction in potential energy is compensated for by the kinetic energy that the electrons gain as they are accelerated by the electric field ($E_{kin} = p^2/2m^*$). During the collisions, electrons deliver their kinetic energy to the crystal lattice, which is consequently being heated.

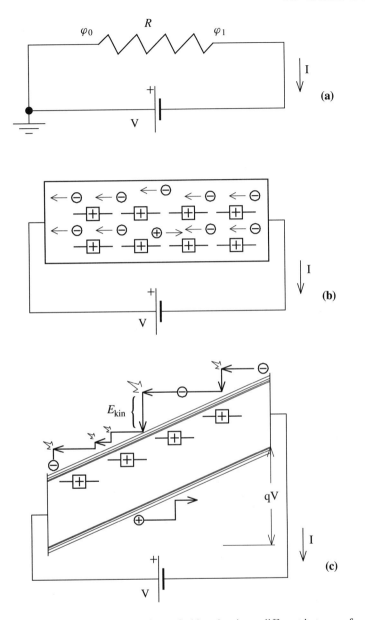

Figure 1.28 Three presentations of a biased resistor, different in terms of complexity and depth of insight: (a) resistor symbol, (b) chemical-bond model, and (c) energy-band model.

As the energy-band diagram of Fig. 1.28c shows the *potential energy* (E_C and E_V) in space, say in the x-direction, it can consequently be called an E–x type of energy-band diagram. This type of diagram is extensively used to explain different semiconductor devices. In some cases, however, it is important to visualize the *kinetic energy* of the carriers, which is related to the momentum $\mathbf{p}$ and the wave vector $\mathbf{k} = (2\pi/h)\mathbf{p}$. In that case, so-called E–k diagrams are used. Fig. 1.29 shows how the simplest, parabolic E–k diagrams are related to the E–x energy-band diagrams of the biased resistor.

The parabolic E–k diagrams reflect the carrier-gas model in which the electrons and holes are assigned effective masses and considered as free particles, so that their kinetic energy is given by $E_{kin} = p^2/2m^*$ (Section 1.4, Fig. 1.16). This model can account for the resistor current–voltage characteristic. However, it will prove inaccurate for some effects considered later. When this happens, more precise E–k diagrams will be introduced.

■ **Example 1.15   Fermi–Dirac and Maxwell–Boltzmann Distributions**

(a) Assuming that the Fermi level is at the mid-gap in the intrinsic silicon, calculate the probability of having an electron at the bottom of the conduction band ($E = E_C$) for three different temperatures: 0 K, 20°C, and 100°C.

(b) How are these probabilities related to the probabilities of finding a hole at $E = E_V$, which is the top of the valence band?

(c) How much will the above-obtained results change if the exponential-approximation (known also as Maxwell–Boltzmann distribution) is used?

Solution:

(a) To calculate these probabilities, the Fermi–Dirac distribution is used:

$$f(E_C) = 1/[1 + e^{(E_C - E_F)/kT}]$$

where $E_C - E_F = 0.56$ eV, which is half the value of the energy gap $E_g = 1.12$ eV. For the case of 0 K, $f(E_C) = 1/[1 + e^{\infty}] = 0$. For 20°C the following result is obtained:

$$f(E_C) = 1/[1 + e^{0.56/[8.62 \times 10^{-5}(20 + 273.15)]}] = 2.3744692 \times 10^{-10}$$

In a similar way, $f(E_C)$ is determined for 100°C, the result being 2.7476309×10^{-8}.

(b) The probability of finding a hole at $E = E_V$ is equal to the probability of finding an electron at E_C in the case of an intrinsic semiconductor.

(c) Using the exponential approximation to calculate $f(E_C)$ at 20°C one obtains

$$f(E_C) = e^{-(E_C - E_F)/kT} = 2.3744692 \times 10^{-10}$$

Comparing this result with the result obtained in (a), we see that there is no observable difference between the two calculations.

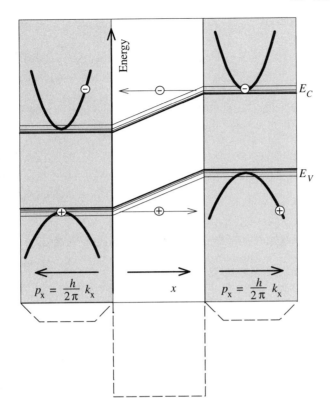

Figure 1.29 The relationship between E–k and E–x diagrams.

■ **Example 1.16 Doping and Fermi Level**

If the donor concentration in N-type silicon is $N_D = 10^{15}$ cm^{-3}, calculate the probability of finding an electron at $E = E_C$ and the probability of finding a hole at $E = E_V$ at 20°C. Compare these results to the results of Example 1.15.

Solution: As the doping shifts the Fermi level, it is important to determine first the position of the Fermi level with respect to E_C. Using Eq. (1.66) one obtains

$$E_C - E_F = kT \ln(N_C/N_D)$$
$$= 8.62 \times 10^{-5}(20 + 273.15) \ln(2.86 \times 10^{19}/10^{15}) = 0.259 \text{ eV}$$

When this value is used in the Fermi–Dirac distribution, the following result is obtained: $f(E_C) = 3.5 \times 10^{-5}$. The probability of finding a hole at $E = E_V$ is

$$f_h(E_V) = 1 - 1/[1 + e^{(E_V - E_F)/kT}] = 1/[1 + e^{(E_F - E_V)/kT}]$$

$E_F - E_V$ is determined in the following way:

$$E_F - E_V = E_C - E_V - (E_C - E_F) = E_g - (E_C - E_F) = 1.12 - 0.259 = 0.861 \text{ eV}$$

Using this result, the probability of finding a hole at $E = E_V$ is calculated as $f_h(E_V) = 1.6 \times 10^{-15}$.

Comparing these results to the results obtained in Example 1.15 for the case of the intrinsic silicon, we see that the probability of having electrons in the conduction band is much higher in the N-type semiconductor ($3.5 \times 10^{-5}/2.4 \times 10^{-10} = 145{,}833$ times). The probability of having holes in the valence band, however, is much smaller.

PROBLEMS

Section 1.1 The Basics: Resistor Structure and Drift Current

1.1 A diffusion layer with sheet resistance $R_S = 200 \ \Omega/\square$ is used for the resistor of Fig. 1.30. What is the resistance of this resistor?
- 200 Ω
- 400 Ω
- 5.8 kΩ
- 6.2 kΩ
- 7.4 kΩ
- 7.8 kΩ

1.2 A test resistor with length $L = 50 \ \mu$m and width $W = 5 \ \mu$m is used to measure the sheet resistance of a resistive layer. What is the sheet resistance if a test voltage of 1 V produces current of $I = 0.50$ mA?
- 50 $\Omega/\square$
- 100 $\Omega/\square$
- 200 $\Omega/\square$
- 300 $\Omega/\square$
- 400 $\Omega/\square$
- 450 $\Omega/\square$

1.3 A resistor of 50 Ω is to be made by standard bipolar integrated-circuit technology. The sheet resistance of a base-type diffusion layer is 200 $\Omega/\square$, while the sheet resistance of an emitter-type diffusion layer is 5 $\Omega/\square$. The minimum width of the diffusion lines is 5 μm. Which diffusion layer (emitter-type or base-type) would you use for this resistor and why? Determine the dimensions (length and width) of the resistor.

1.4 The voltage drop across a 200-Ω resistor in a bipolar integrated circuit is ≤ 5 V. Reliability considerations limit the average current density to $j_{max} = 10^9$ A/m^2. Design this resistor so that it can be implemented as a base-diffusion resistor, the sheet resistance and the junction depth of the base-diffusion layer being $R_S = 100 \ \Omega/\square$ and $x_j = 2 \ \mu$m, respectively.

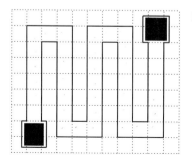

Figure 1.30 Top view of an IC resistor.

1.5 Repeat Problem 1.4 for the case in which the reliability constraint means that the current density should not exceed $j_{max} = 10^9$ A/m^2 at any point, and the nonuniform conductivity can be expressed as $\sigma(x) = \sigma(0) \exp[-(x/x_0)^2]$, where $x_0 = 1$ μm. Assume constant electric field $E = V/L$. It is also known that $\int_0^\infty \exp(-u^2/2)\, du = \sqrt{\pi/2}$. $\boxed{A}$

1.6 To measure the sheet resistance of a resistive layer, taking into account the parasitic series contact resistance, a test structure consisting of resistors with the same width and different lengths is provided. Measuring the resistances of the resistors with lengths $L_1 = 10$ μm and $L_2 = 30$ μm, the following values are obtained: $R_1 = 365$ Ω, and $R_2 = 1085$ Ω, respectively. If the width of the resistors is 5 μm, determine the sheet resistance and the contact resistance values. $\boxed{A}$

Section 1.2 Insight into Conductivity Ingredients: Chemical-Bond Model

1.7 A sample of N-type GaAs, doped with $N_D = 10^{16}$ cm^{-3}, is cooled to ≈ 0 K. What is the concentration of holes?
- 0
- 10,404 cm^{-3}
- 1.02×10^{10} cm^{-3}
- 4.41×10^{-6} cm^{-3}
- 2.1×10^6 cm^{-3}
- 10^{16} cm^{-3}

1.8 For the sample of Problem 1.7, what is the net charge concentration at 300 K?
- 0
- 10,404 cm^{-3}
- 1.02×10^{10} cm^{-3}
- 4.41×10^{-6} cm^{-3}
- 2.1×10^6 cm^{-3}
- 10^{16} cm^{-3}

1.9 P-type doped semiconductor layer has approximately uniform acceptor concentration $N_A = 5 \times 10^{16}$ cm^{-3} and thickness $x_j = 4$ μm. Calculate the sheet resistance, if the hole mobility is $\mu_p = 450$ cm^2/Vs.

1.10 Calculate the minority carrier concentration in a heavily doped N-type silicon having donor concentration $N_D = 10^{20}$ cm^{-3}.

1.11 The substrate concentration of an N-type Si wafer is $N_D = 10^{15}$ cm^{-3}. The wafer is doped with $N_A = 1.1 \times 10^{15}$, so that the very lightly doped P-type region is created at the surface. What is the concentration of electrons in the P-type region?

1.12 Repeat the calculations from Problem 1.11 for the case of Ge, having $n_i = 2.5 \times 10^{13}$ cm^{-3}. $\boxed{A}$

1.13 What is the tolerance $\Delta R/R$ of an N-type diffusion resistor if the tolerances of the resistor dimensions (including junction depth) are ± 0.3 μm and the tolerance of the doping density is 5%? $\boxed{A}$

Section 1.3 Making a Semiconductor Resistor: Lithography and Diffusion

1.14 The dimensions of a diffusion resistor on the photolithography mask are $W_m = 1$ μm and $L_m = 20$ μm. The resistor is to be created by boron diffusion to the junction depth of $x_j = 2$ μm. Assuming that the mask dimensions are perfectly transferred onto the masking oxide, what are the actual dimensions of the resistor?
- 1 μm; 20 μm
- 4.2 μm; 20 μm
- 2.6 μm; 21.6 μm
- 2.6 μm; 20 μm
- 4.2 μm; 23.2 μm
- 1 μm; 23.2 μm

1.15 Calculate the diffusion coefficient of phosphorus at $T = 1100°C$, knowing that $D_0 = 19.7 \text{ cm}^2/\text{s}$ and $E_A = 3.75 \text{ eV}$.

1.16 The concentration profile of drive-in diffusion is $N(x, t) = (\Phi/\sqrt{\pi Dt}) \exp[-x^2/(4Dt)]$. What deposition dose ($\Phi$) and drive-in time ($t$) are needed to obtain a layer with surface concentration $N_s = 5 \times 10^{16} \text{ cm}^{-3}$ and junction depth $x_j = 2 \text{ μm}$? The doping element is phosphorus, which is being diffused into a P-type wafer with uniform concentration $N_A = 10^{15} \text{ cm}^{-3}$ The drive-in temperature is $T = 1100°C$.

1.17 Calculate the sheet resistance of the layer designed in Problem 1.16. Assume constant electron mobility $\mu_n = 1250 \text{ cm}^2/\text{Vs}$. It is known that $\int_0^\infty \exp(-u^2/2)\, du = \sqrt{\pi/2}$.

1.18 What would the sheet resistance of the layer designed in Problem 1.16 be if the actual drive-in temperature was 1102°C (2° higher than the nominal 1100°C)? $\boxed{A}$

1.19 The temperature variation of a drive-in diffusion furnace is ±0.1%. Determine the junction depth tolerance if the nominal drive-in temperature is $T = 1050°C$. Express the result as a percentage. The doping element is

(a) phosphorus ($E_A = 3.75 \text{ eV}$),

(b) arsenic ($E_A = 3.90 \text{ eV}$). $\boxed{A}$

Section 1.4 Carrier Mobility

1.20 Identify

(a) drift-current,

(b) diffusion-current, and

(c) continuity equations, applied to electrons.

- $j = q D_n \frac{dn}{dx}$ • $j = \sigma \frac{dn}{dx}$ • $j = -q D_n \frac{d\varphi}{dx}$
- $j = q n \mu_n E$ • $\frac{\partial j}{\partial x} = q \frac{\partial n}{\partial t}$ • $j = q D_n n E$

1.21 An electric field $E = 1 \text{ V}/\mu\text{m}$ produces current density $j = 0.8 \times 10^9 \text{ A/m}^2$ through an N-type doped semiconductor ($N_D = 10^{17} \text{ cm}^{-3}$). What will the current density be if the electric field is increased five times, so that the electrons reach the velocity saturation $v_{sat} = 0.1 \text{ μm/ps}$?

- 0 • $1.6 \times 10^9 \text{ A/m}^2$ • $3.2 \times 10^9 \text{ A/m}^2$
- $0.8 \times 10^9 \text{ A/m}^2$ • $4.0 \times 10^9 \text{ A/m}^2$ • $8.0 \times 10^9 \text{ A/m}^2$

1.22 A bar of silicon 1 cm long, 0.5 cm wide, and 0.5 mm thick has a resistance of 190 Ω. The silicon has a uniform N-type doping concentration of 10^{15} cm^{-3}.

(a) Calculate the electron mobility.

(b) Find the drift velocity of the electrons when 10 V is applied to the ends of the bar of silicon.

10V / Length = 10V/cm = E?

(c) Find the corresponding electric field and relate it to the drift velocity and the electron mobility.

1.23 (a) Design a 1-kΩ diffused N-type resistor. The technology parameters are as follows: the concentration of donor impurities $N_D = 10^{17}$ cm^{-3}, the junction depth $x_j = 2$ μm, and the minimum diffusion width as constrained by the particular photolithography process and lateral diffusion is 4 μm.

(b) How long should the resistor be if P-type silicon with the same doping density is used instead of the N-type silicon?

(c) What will the resistance be if the concentration of donors is increased to $N_D = 10^{19}$ cm^{-3}? [A]

1.24 For the resistor designed in Problem 1.23 (a), find the resistance at 125°C?

1.25 The resistance of a P-type semiconductor layer is 1 kΩ. In what range will the resistance change if the operating temperature is 75 $\pm$ 10°C. The temperature dependence of hole mobility is given by $\mu_p = $ const $T^{-3/2}$.

1.26 For silicon doped with $N_D \approx 10^{16}$ cm^{-3}, the electron mobility at the liquid-nitrogen temperature ($T = 77$ K) is approximately the same as at the room temperature ($T = 27$°C): $\mu_n = 1250$ cm^2/Vs. Calculate the diffusion coefficient at these temperatures.

1.27 The donor distribution of a nonuniformly doped silicon sample can be approximated by $N(x) = N_s \exp(-x/x_0)$, where $x_0 = 1$ μm. As no current flows under the open-circuit condition, a built-in electric field is established, so that the drift current exactly compensates the diffusion current. Calculate the built-in electric field. [A]

Section 1.5 Energy-Band Model

1.28 Select an answer for each of the following questions:

(a) What is the value of the Fermi–Dirac function at E_F?

(b) What is the probability of finding a hole at E_F?

(c) What is the probability of finding an electron at E_V at room temperature?

(d) What is the probability of finding a hole at E_V at room temperature?

(e) What is the probability of finding an electron at E_C at 0 K?

(f) What is the probability of finding a hole at E_V at 0 K?

- 0   • 0.5   • 1
- 0.25   • 0.56   • anything between 0 and 1

1.29 The probability that an energy state in the conduction band is occupied by an electron is 0.001. Is this N-type, P-type, or intrinsic silicon?

1.30 Find the room temperature position of the Fermi level with respect to the top of the valence band for N-type silicon doped with $N_D = 10^{16}$ cm^{-3} donor atoms. Is it closer to the top of the valence band or to the bottom of the conduction band?

1.31 Calculate the electron concentration in a doped silicon if the Fermi level is as close to the top of the valence band as to the mid-gap. What is the hole concentration? Is this N-type or P-type silicon?

1.32 N_C and N_V equations given in Table 1.6 show that the effective density of states depends on temperature. The energy gap is also slightly temperature dependent; in the case of silicon, this dependence can be fitted by E_g [eV] $= 1.17 - 7.02 \times 10^{-4}T^2/(T$ [K] $+ 1108)$. Find the intrinsic carrier concentration of silicon at $T = 300°C$. Compare this value to the doping level of $N_D = 10^{16}$ cm^{-3}.

1.33 Repeat Problem 1.32 for 6H silicon carbide ($E_g = 3$ eV) to see the importance of the energy gap value for high-temperature applications. The room temperature densities of states are $N_C = N_V = 2.51 \times 10^{19}$ cm^{-3}, while the $E_g(T)$ dependence can be fitted by E_g [eV] $= 3.0 - 3.3 \times 10^{-4}$ (T [K] $- 300$).

1.34 N-type silicon is doped with $N_D = 5 \times 10^{15}$ cm^{-3}. Calculate the concentration of holes at $300°C$. [E_g [eV] $= 1.17 - 7.02 \times 10^{-4} T^2/(T$ [K] $+ 1108)$].

1.35 A bar of N-type semiconductor ($N_D = 5 \times 10^{17}$ cm^{-3}, $L = 1$ cm, $W \times T = 3 \times 0.5$ mm) is used as a resistor. Calculate and compare the sheet resistances at 25 and 700°C if the semiconductor material is:

(a) silicon ☐A

(b) 6H silicon carbide.

Assume constant mobilities $\mu_n = 400$ cm^2/Vs and $\mu_p = 200$ cm^2/Vs, and use the energy-gap and density-of-states data from Problems 1.32 and 1.33.

1.36 The bottom of the conduction band at one end of a resistor and the top of the valence band at the other end are at the same energy level. What current flows through the resistor if the resistance is 1.12 kΩ?

REVIEW QUESTIONS

R-1.1 What is the sheet resistance? In what unit is it expressed?

R-1.2 What is the difference between the integral and the differential forms of Ohm's law? What is the difference between differential and integral quantities?

R-1.3 Can Ohm's law be applied to the diffusion current? The drift current?

R-1.4 Is there any positive charge in N-type semiconductors? If yes, how many types of positive charge are there?

R-1.5 Which type of positive charge dominates in terms of the concentration in N-type semiconductors?

R-1.6 Does the dominant positive charge in N-type semiconductors contribute to the current flow?

R-1.7 Can the minority carriers be always neglected? Can they be neglected in the resistor models?

R-1.8 How is the concentration of the minority carriers calculated?

R-1.9 What force causes the diffusion current? Can charge-neutral particles make diffusion current? Drift current?

R-1.10 What is the difference between the dose and concentration of doping atoms and how are they related? What are the respective units?

R-1.11 Is the linear relationship between the current density and the electric field (Ohm's law) generally valid? What about the relationship between the current density and the drift velocity?

R-1.12 How is the effect of drift velocity saturation included in Ohm's law?

R-1.13 How does mobility of carriers change with temperature?

R-1.14 What is the basic meaning of energy bands and energy gap? Is there an energy gap in every material? In metals?

R-1.15 What are the two bands important for the properties of semiconductors and insulators, called?

R-1.16 **Where do the mobile electrons appear (conduction band or valence band)?**

R-1.17 Are there electrons in the valence band? Where do the mobile holes appear?

R-1.18 Do all electrons have the same energy?

R-1.19 What function models the electron distribution along the energy?

R-1.20 Is the Fermi–Dirac distribution f a general function? Can it be applied in the same form to metals where no energy gap exists?

R-1.21 How are f and the band properties of a material combined to express the carrier concentration?

R-1.22 What is the Fermi level?

R-1.23 What is the position of the Fermi level in the intrinsic semiconductor?

R-1.24 How is the doping expressed in the energy-band diagram?

R-1.25 What does it mean if the energy bands are tilted?

R-1.26 How do the electrons and holes behave in the region of tilted energy bands?

Chapter 2

CAPACITORS
Reverse-Biased P–N Junction and MOS Structure

Capacitors are frequently used in linear circuits. The need to understand integrated-circuit capacitors, however, by far exceeds the application in linear circuits. The capacitors are inherently present, as parasitic devices, in the structure of any diode or transistor, and it is these capacitors that typically determine high-frequency behavior of the circuit. Also, the principle of operation of field-effect transistors (the most frequently used semiconductor devices) is essentially based on the capacitance effect.

This chapter describes the two most commonly used types of capacitors, the P–N junction capacitor and MOS (metal–oxide–semiconductor) capacitor. Therefore, it introduces the basic concepts of two fundamental semiconductor structures, namely the P–N junction and the MOS structures. It also introduces an additional fundamental equation, the Poisson equation, which models the relationship between the electric potential and the charge density.

2.1 BASIC APPLICATIONS

2.1.1 Capacitance Effect

The capacitance effect is illustrated in Fig. 2.1. This effect is achieved when two conductive "plates" are placed close to each other, but are still physically separated by a dielectric. The dielectric suppresses any direct current (DC) flow through the capacitor. The electrons arriving from the negative terminal of the battery accumulate at the capacitor plate as they cannot cross over the dielectric to continue the flow toward the positive terminal of the battery. The electric field of these electrons, however, can penetrate through the dielectric repelling the electrons from the other capacitor plate inducing, in effect, positive charge accumulation at that plate. Therefore, if the voltage applied across the capacitor is varied, the amount of charge stored at the capacitor plates changes, which appears as current flow through the circuit. A stronger capacitance effect means that the same voltage change (ΔV_C) produces a larger change in the amount of charge at the capacitor plates (ΔQ_C). Using infinitesimal changes in the voltage and charge, the capacitance is defined as

$$C = \frac{dQ_C}{dV_C} \tag{2.1}$$

The strength of the capacitance effect depends on three factors: (1) the area of the capacitor plates, as larger plates accommodate more charge, (2) the dielectric thickness, as thinner dielectrics provide more effective penetration of the electric field produced by the charges at the plates, and (3) the degree to which the internal structure of the dielectric material permits the penetration of the electric field, which is called *dielectric permittivity*. The capacitance can, therefore, be calculated by

$$C = \varepsilon_d \frac{A}{t_d} \tag{2.2}$$

where ε_d is the dielectric permittivity, A is the area of the capacitor plates, and t_d is the dielectric thickness. The unit for the capacitance is the *farad* (F), while the dielectric permittivity is expressed in farads per meter (F/m).

2.1.2 High and Low Pass Filter

The current in the circuit of Fig. 2.1 flows in one direction to charge the capacitor and in the opposite direction when the capacitor is being discharged. This produces intervals of both positive and negative voltage drops across the resistor, as shown in Fig. 2.2a . If the value of the capacitor is large enough, very small changes of the voltage across the capacitor will induce large changes in the charge ($\Delta Q_C = C \Delta V_C$), which means large current flows through the circuit. Therefore, if the value of the capacitor is large enough, the changing voltage v_{in} is almost completely transferred onto the output as the voltage across the capacitor does not change very much, while large charging and discharging currents are induced producing the voltage drop across the output resistor. The constant voltage component V_{IN} cannot produce a current that flows in one direction only through the capacitor, therefore it does not produce any voltage drop across the resistor (the voltage drop V_{IN} appears across the capacitor). The constant voltage component V_{IN}, which shifts the input signal up is filtered out at the output. The circuit passes high-frequency signals and blocks any DC component in the signal, as well as any low-frequency signal.

The fact that a capacitor can accommodate large changes in the stored charge with very small voltage changes can also be used to create a low pass filter (Fig. 2.2b). The variable input voltage v_{in} produces variable current in the circuit, and consequently changes the charge in the capacitor, but it does not significantly change the voltage across the capacitor, the same effects as in the circuit of Fig. 2.2a. The difference here is that the capacitor is connected across the output, so it "absorbs" the voltage changes at the output, keeping the output voltage close to the constant level V_{IN}.

(a)

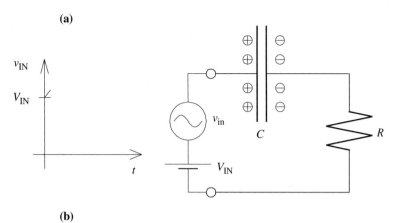

(b)

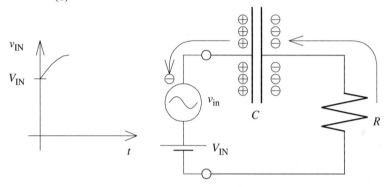

(c)

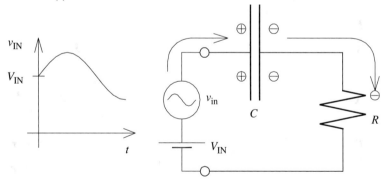

Figure 2.1 Change in the voltage across the capacitor produces change in the charge stored at the capacitor plates, and therefore produces current flow in the circuit.

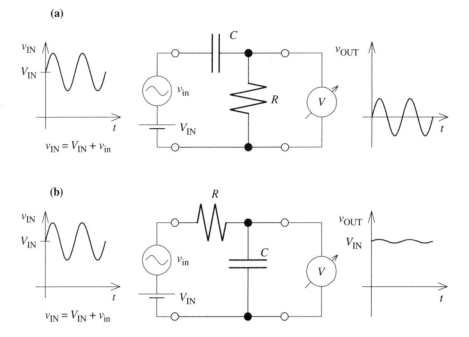

Figure 2.2 Capacitors are used to create both high-pass filters (a) and low-pass filters (b).

2.2 REVERSE-BIASED P–N JUNCTION

2.2.1 Integrated-Circuit Structure

A P–N junction can perform the function of a capacitor. The two conductive plates are the N-type region on one side and the P-type region on the other side. These two conductive plates are separated by a *depletion layer* (Fig. 2.3), which acts as the capacitor dielectric. The depletion layer is created by a recombination of electrons and holes, which meet each other in their random motion around the P–N junction. As a layer is being depleted of free electrons and holes around the P–N junction, excess fixed charges are left behind: positive donor ions at the N-type side and negative acceptor ions at the P-type side. These charges create a field at the P–N junction (*built-in field*) that consequently blocks any further transport of the electrons from the N-type side into the P-type side and vice versa. The depletion layer has a constant width when the steady state is achieved.

The P–N junction capacitor in integrated circuits has to be isolated from the other components built in the same substrate. Figure 2.3 illustrates that the principle of *reverse-biased P–N junction*, illustrated in Fig. 1.10 for the case of resistor isolation, can be used to isolate the capacitor as well. The N-type substrate in Fig. 2.3 is contacted to the highest potential in the circuit (V_+), which repels any hole trying to cross over the junction. In this way, the current flowing between terminals A and B of the capacitor is confined within the boundary of the P-type region, thus, it is isolated from any other device made in the N-substrate. It is true that the minority carriers (holes in the N-type and electrons in the P-type

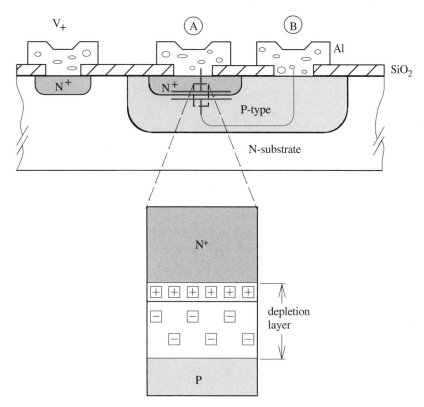

Figure 2.3 Structure of a P–N junction capacitor in integrated circuits.

region) easily pass through the junction, however, the current that they make is very small (frequently called leakage current) due to very small concentrations of the minority carriers.

It should be pointed out that the reverse-biased P–N junction used for isolation inevitably introduces an additional (parasitic) capacitance, linked between the *B* terminal and V_+. This parasitic capacitance has to be considered when designing a circuit with P–N junction capacitors.

2.2.2 Depletion-Layer Capacitance

Figure 2.4 shows a reverse-biased P–N junction used as a capacitor in the circuit of Fig. 2.1. It can be seen from Fig. 2.4b that the increase in the voltage applied increases the depletion-layer width as this voltage attracts electrons and holes toward the contacts. As a result, more positive donor ions on the N-type side, and negative acceptor ions on the P-type side, appear in the depletion layer. Fig. 2.4c illustrates that the decrease in the voltage causes reduction of the depletion-layer width, and consequently reduction of the depletion-layer charge. These changes in the capacitor charge, induced by the changes of the applied voltage, represent exactly the capacitance effect ($C = dQ_C/dV_C$). The current that charges and discharges the capacitor is also indicated in Fig. 2.4b and c.

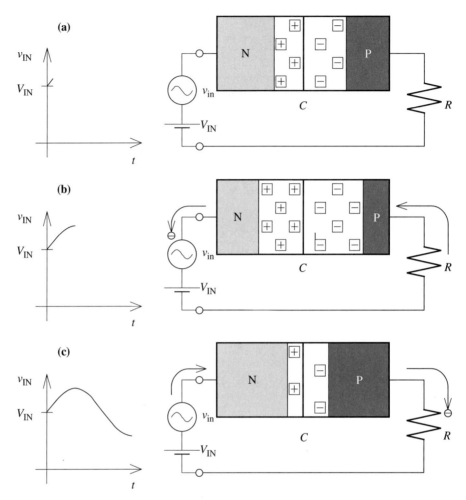

Figure 2.4 Reverse biased P–N junction used as a capacitor.

The capacitance of the reverse-biased P–N junction (also referred to as depletion-layer capacitance) is given by

$$C_d = \varepsilon_s \frac{A}{w_d} \qquad (2.3)$$

where ε_s is the silicon permittivity ($\varepsilon_s = 11.8 \times \varepsilon_0 = 11.8 \times 8.85 \times 10^{-12}$ F/m), A is the P–N junction area, and w_d is the average width of the depletion layer (depletion-layer width for zero-signal voltage).

The fact that the externally applied voltage changes the width of the depletion layer makes the P–N junction capacitor voltage dependent. If the DC component of the voltage, V_{IN}, in the circuit of Fig. 2.4 is increased, the average value of the depletion layer is also increased as the electrons and holes are attracted toward the contacts. In effect the capacitance is reduced [Eq. (2.3)]. The capacitance–voltage (C–V) dependence of a typical P–N junction is shown in Fig. 2.5. This voltage dependence of the capacitance is an additional feature of the P–N junction capacitor, which does not exist in discrete ceramic capacitors. The voltage dependence of the capacitance is very useful as it enables variable capacitors that are electrically controlled to be made.

2.2.3 Energy-Band Diagram

Presentation by the energy-band model provides much deeper insight into the electron and hole transport as it introduces the energy dimension. This section considers the effects associated with the reverse-biased P–N junction.

The energy-band diagrams of the N-type and P-type semiconductors are explained in Section 1.5. The diagrams given in Fig. 2.6a indicate the electrons in the conduction band (filled circles) and the holes in the valence band (open circles). The graphs illustrate that the electrons are mainly located near the bottom of the conduction band, while the holes mainly appear close to the top of the valence band. The probability of finding an electron (hole) at levels further away from the bottom of the conduction band (top of the valence band) decays exponentially according to the Fermi–Dirac distribution function (Section 1.5). Note that in the N-type and the P-type semiconductor, the Fermi level positions are different, expressing different concentrations of electrons and holes, respectively.

Figure 2.6b shows the energy-band diagram of an unbiased P–N junction along with a chemical-bond presentation illustrating electrons and holes in the N-type and P-type neutral regions as well as donor and acceptor ions in the depletion layer. The process of creation of the depletion layer due to electron–hole recombination, explained in Section 2.2.1, can also be described with the energy bands. When the N-type and P-type blocks of semiconductor are joined, a number of electrons from the N-type side that move randomly around will appear on the P-type side and a number of holes will appear on the N-type side. An excess electron appearing among the numerous holes on the P-type side will very soon meet a hole, a process in which they recombine with each other. The hole recombination leads to reduction of the hole concentration close to the P–N junction. This effect is reflected by a larger distance between the Fermi level and the top of the valence band at points

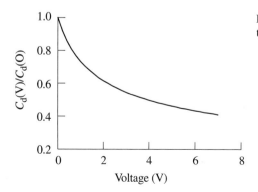

Figure 2.5 C–V curve for a typical P–N junction.

close to the P–N junction where the bands are bent. In a similar way holes and electrons recombine at the N-type side, resulting in reduction of the concentration of electrons close to the junction. This is reflected by a larger distance between the Fermi level and the bottom of the conduction band. When an equilibrium state is achieved, the occupancy of electron and hole states is stabilized throughout the system, which is expressed by a unique Fermi level position. This means that any electron–hole recombination is compensated for by an equal amount of electron–hole generation at any point in the P–N junction system.

For a system in thermal equilibrium, the Fermi level is constant throughout the system.

The energy-band diagram of Fig. 2.6b (unbiased P–N junction) is constructed in the following way:

1. The Fermi level line (dash–dot line in the figure) is drawn first. (The Fermi level is constant as the system is in thermal equilibrium.)
2. The conduction and valence bands are drawn for the N-type and P-type neutral regions (outside the depletion layer). The bands are placed appropriately with respect to the Fermi level. Neither the electron/hole concentration nor any other property of the N-type/P-type semiconductor is changed in the neutral regions (away of the junction). Therefore, the energy-band diagrams in the neutral N-type and P-type regions should be equivalent to those shown in Fig. 2.6a.
3. The conduction- and valence-band levels are joined by curved lines to complete the diagram.

Bent lines on the energy-band diagram express the existence of an electric field. The electric field is equal to the negative gradient of the electric potential $(-d\varphi/dx)$, which means it is proportional to the positive gradient of the potential energy $[(1/q) \, dE_{pot}/dx]$.

The bending of the energy bands produces a potential-energy difference between the N-type and P-type neutral regions, which is labeled $q V_{bi}$ in the graph. The corresponding difference in the electric potential, V_{bi}, is referred to as *built-in voltage*. The energy-band diagram provides a good insight into the value of the built-in voltage. It can be seen from Fig. 2.6b that $q V_{bi}$ is a bit smaller than the energy gap, which is 1.12 eV in silicon. If $q V_{bi}$ is found to be 0.8 eV, for example, the built-in voltage is consequently $V_{bi} = 0.8$ V. It will be shown in Example 2.1 that the built-in voltage is given as

$$V_{bi} = \frac{kT}{q} \ln \frac{N_D N_A}{n_i^2} \tag{2.4}$$

where N_D and N_A are the doping concentrations on the N-type and P-type side, respectively, while n_i is the intrinsic carrier concentration.

It is important to clarify that the appearance of the built-in voltage at the P–N junction by no means violates the second Kirchoff law, which states that the sum of voltage drops along any closed loop is equal to zero. Imagine that the semiconductor is a circle, half doped as P type and half as N type. There will be two P–N junctions with two equivalent built-in voltages, which cancel each other when added up to make the sum of the voltages along the closed loop. If the P–N junction is contacted by metal stripes, built-in voltages will appear

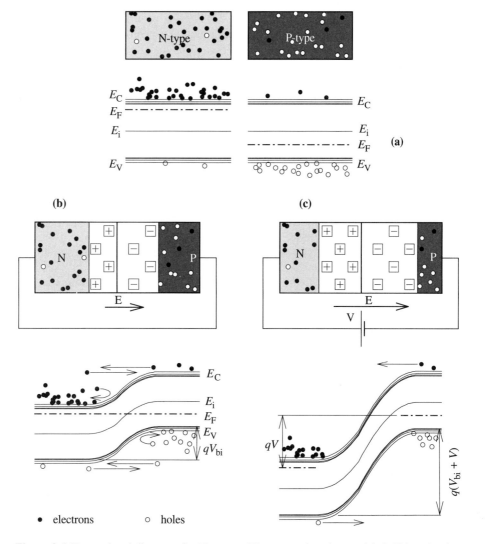

Figure 2.6 Energy band diagrams for N-type and P-type semiconductors (a), P–N junction in thermal equilibrium (b), and reverse-biased P–N junction (c).

at the contacts,[1] so that the sum of built-in voltages at the contacts and at the P–N junction is equal to zero.

The current flowing through the circuit in Fig. 2.6b is equal to zero. If there was any current the system would not be in thermal equilibrium. However, there are individual electrons and holes that do pass through the P–N junction. Figure 2.6b shows that most of

---

[1]In general, built-in voltage appears at any junction of materials that have different positions of the Fermi level.

the electrons in the N-type side are unable to move onto the P-type side, as they cannot overcome the energy barrier at the P–N junction. The electrons that do not have enough energy will hit the "wall" (bottom of the conduction band) if they move toward the P-type side.

Only the electrons possessing energy larger than the potential-energy barrier can make the transition to the P-type side, as shown in Fig. 2.6b.

On the other hand, any electron from the P-type region (minority carriers) can move into the N-type region, as there is nothing to stop them. The concentration of these electrons is very small (they are minority carriers), however, it is sufficient to produce transitions of electrons from the P-type to the N-type side, which exactly compensate for the transitions in the opposite direction. In average, the electron current through the junction is zero. The situation is analogous with the holes. The difference is that holes in the energy-band diagram are like bubbles in the water, which tend to bubble up. The slope of the energy bands at the P–N junction appears as a barrier for the holes on the P-type side (majority carriers).

When a reverse-biased voltage is applied to the P–N junction, the energy-band diagram is changed to express the applied voltage, which is illustrated in Fig. 2.6c. Because of the applied voltage, the P–N junction is not in thermal equilibrium any more, and the Fermi level is not constant throughout the system. The voltage applied to the P–N junction appears mainly across the depletion layer, expanding the depletion-layer width and increasing the electric field in the depletion layer (there cannot be a significant increase in the field in the neutral regions as it would move the mobile electrons and holes producing an unsustainable current flow). The voltage drop V is expressed in the band diagram as energy difference of $-qV$. Figure 2.6c shows that the initially leveled Fermi levels in the N-type and P-type side are now separated by qV to express the voltage drop across the depletion layer. The Fermi levels in this nonequilibrium case are referred to as quasi-Fermi levels.

The energy-band diagram of Fig. 2.6c (reverse-biased P–N junction) is therefore constructed in the following way:

1. Quasi-Fermi level lines (dash–dot line in the figure) separated by qV are drawn first. The lower quasi-Fermi level line is for the N-type region (left side in the figure), indicating that the N-type region is at higher electric potential ($E_{\text{pot}} = -q\varphi$).
2. The conduction and valence bands are drawn for the N-type and P-type neutral regions (outside the depletion layer). The bands are placed appropriately with respect to the Fermi level, indicating the band diagrams of the N-type and P-type semiconductors as in Fig. 2.6a.
3. The conduction and valence band levels are joined by curved lines to complete the diagram.

The slope of the energy bands in the depletion layer now appears larger compared to the case of Fig. 2.6b, which indicates the increased electric field in the depletion layer. Also, the barrier height between the N-type and P-type region is now higher (increased exactly by qV), indicating the increased voltage drop across the depletion layer.

The fact that the P–N junction is not in thermal equilibrium means that there must be a current flow that works to bring the system back in equilibrium.

Fig. 2.6c shows that the electrons from the P-type region (minority carriers) can roll down along the bottom of the conduction band as easily as in the case in which no bias is applied (Fig. 2.6b). Analogously, the holes as minority carriers in the N-type region bubble up along the top of the valence band, as in the case in which no bias is applied. The minority carrier current flow remains the same as in the case of thermal equilibrium at the junction. The difference in this case is that this current is not compensated for by the current of electrons from the N-type side and holes from the P-type side. The potential barrier is now too high and almost none of the electrons from the N-type region or the holes from the P-type region have enough kinetic energy to overcome it. As a result, the current of the minority carriers makes effective current flow through the junction. This is the current that drains the battery reducing the voltage V, and therefore reducing the qV gap between the quasi-Fermi levels. After some time (it may be very long), when the battery flattens out and the quasi-Fermi levels are leveled off, the P–N junction will be in equilibrium again. The situation in which the battery voltage is kept, or assumed, constant is referred to as *steady state*.

> The current flowing through the reverse-biased P–N junction does not increase with increase in the reverse-bias voltage.

The increased voltage increases the slope of the energy bands, but this does not produce any additional electrons (or holes) rolling down (bubbling up) along the slope of the bands. The minority-carrier current is not controlled by the voltage/field applied to the depletion layer but by the concentration of minority electrons and holes. The concentration of minority carriers is very small (usually more than 10 orders of magnitude smaller than the concentration of the majority carriers), which makes the minority-carrier current very small indeed. This enables the reverse-biased junction to be used as a capacitor.

■ **Example 2.1   Built-in Voltage Equation**

Express the built-in voltage V_{bi} of a P–N junction in terms of the doping concentrations N_D and N_A.

Solution: As can be seen from Fig. 2.6b, qV_{bi} appears as the difference between the mid-gap on the P-type side (E_{i-p}) and on the N-type side (E_{i-n}):

$$qV_{bi} = E_{i-p} - E_{i-n}$$

Let us express the mid-gap energies with respect to the constant Fermi level:

$$qV_{bi} = \underbrace{E_{i-p} - E_F}_{q\phi_{F-p}} + \underbrace{E_F - E_{i-n}}_{|q\phi_{F-n}|}$$

where $q\phi_{F-p}$ and $q\phi_{F-n}$ are known as *Fermi potentials*, expressing the doping level of P-type and N-type semiconductors, respectively. To find the relationship between $q\phi_{F-p}$ and the doping level N_A, we can apply Eq. (1.65) to the case of P-type doped ($p = N_A$) and intrinsic ($p = n_i$, $E_F = E_{i-p}$) semiconductor:

$$N_A = N_V e^{-\frac{E_F - E_V}{kT}}$$

$$n_i = N_V e^{-\frac{E_{i-p} - E_V}{kT}}$$

N_V and E_V will disappear if these two equations are divided:

$$\frac{N_A}{n_i} = e^{\frac{E_{i-p} - E_F}{kT}}$$

$$\underbrace{E_{i-p} - E_F}_{q\phi_{F-p}} = kT \ln \frac{N_A}{n_i}$$

Analogously, we can find that

$$\underbrace{E_F - E_{i-n}}_{-q\phi_{F-n}} = kT \ln \frac{N_D}{n_i}$$

Therefore,

$$q V_{bi} = q\phi_{F-p} + |q\phi_{F-n}| = kT \left(\ln \frac{N_A}{n_i} + \ln \frac{N_D}{n_i} \right)$$

$$V_{bi} = \frac{kT}{q} \ln \frac{N_A N_D}{n_i^2}$$

2.3 C–V DEPENDENCE OF THE REVERSE-BIASED P–N JUNCTION: SOLVING THE POISSON EQUATION

In Section 2.2.2, it was explained that the capacitance of a reverse-biased P–N junction depends on the reverse-biased voltage (Fig. 2.5). This dependence is due to the change in the depletion layer width w_d with a change in the reverse-biased voltage [Eq. (2.3)]. To derive a mathematical model for this effect, it is necessary to find the relationship between the depletion-layer width and the voltage applied. There is an equation that links the electric potential and the charge existing in a dielectric medium (like the positive donors and the negative acceptors in the depletion layer). This equation, called the Poisson equation, is introduced in this section. The Poisson equation is solved, and models for the C–V dependence of P–N junctions are derived for two illustrative (and extreme) cases, namely the abrupt and linear P–N junctions.

2.3.1 The Poisson Equation

Although the electrons and holes cannot flow through the capacitor dielectric, their electric field can penetrate through this dielectric. It appears very useful to establish an analogy between the flow of electrons or holes (current) and the penetration of the electric field through a dielectric. The electric field appearing in a conductor or semiconductor (where free current carriers exist) produces a current that is proportional to the field strength:

$$j = \sigma E \tag{2.5}$$

Equation (2.5) is exactly the differential form of Ohm's law [Eq. (1.10)] and it is repeated here for easier comparison. The electric field appearing in a dielectric cannot produce a flow of electrons or holes through the dielectric. There is, however, an analogous quantity to the current density j, which expresses the penetration of the electric field into the dielectric. This quantity is referred to as electric flux density, and is labeled by D. Analogous to the current density j, the electric flux density D is proportional to the strength of the electric field:

$$D = \varepsilon_d E \tag{2.6}$$

where the proportionality coefficient ε_d is the already mentioned dielectric permittivity. Equation (2.6) can be generalized for three dimensions in a way similar to Ohm's law [Eq. (1.11)], in which case D becomes a vector $\mathbf{D}$. Similar to the case of Ohm's law, Eq. (2.6) is not correct for very large electric fields E, as in that case the relationship between D and E is not linear any more. Unlike the situation with Ohm's law, however, deviations of the linear relationship between D and E are normally not observed in semiconductor devices.

To analyze changes in the electric flux density D, say along the x-axis, an equation analogous to the continuity equation for the electric current density [Eq. (1.32)] can be used. Although continuity equation [Eq. (1.32)] also involves time dependence of the current carrier concentration, we will limit ourselves to considerations of spatial changes only in this section. It is known that any variation of the electric field in time is coupled to a corresponding variation in the magnetic field, however, we do not need to involve the magnetic field to consider the depletion layer of our P–N junction. We consider here steady states, with no changes in time, thus the first derivatives with respect to time are zero. Therefore, the electric flux density equation, which is analogous to the continuity equation [Eq. (1.32)], is given by

$$\frac{\partial D}{\partial x} = 0 \tag{2.7}$$

The meaning of this equation is illustrated in Fig. 2.7a. It expresses the fact that the electric flux density does not change between two close points (x and $x + \Delta x$), as the number of electric-field lines that enter an imagined closed space [$D(x)$ in Fig. 2.7a] is equal to the number of electric-field lines that come out of this space [$D(x + \Delta x)$ in Fig. 2.7a]. This is correct as long as there is no source (or drain) for the electric-field lines in the considered space. This means that Eq. (2.7) cannot be applied when there are charges in the space that we want to consider, as the electric-field lines will originate from the positive charge centers and will terminate at the negative charge centers. The depletion layer of the P–N junction is exactly such a case: there are a lot of positive (donor ions) and negative (acceptor ions) charge centers in the depletion layer.

To generalize Eq. (2.7) to include the effect of these charges, let us consider the situation presented in Fig. 2.7b. It can be seen that the effect of one positive charge center appearing

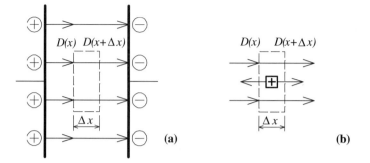

Figure 2.7 Relationship between electric flux density D and charge centers. (a) The number of electric-field lines entering a closed space (the dashed-line rectangle) is equal to the number of electric-field lines coming out of the closed space, as the electric-field lines neither originate nor terminate when no charge centers are present. (b) The presence of a charge center introduces a difference in the number of electric-field lines that enter and come out of the closed space.

in the imagined space (rectangle in the figure) is that an extra electric-field line comes out at $x + \Delta x$, while the effective number of lines at x is reduced by one as two lines with opposite directions cancel each other. This means that the electric flux density $D(x + \Delta x)$ is now larger than $D(x)$. Obviously, the difference between $D(x + \Delta x)$ and $D(x)$ is proportional to the number of charge centers appearing between x and $x + \Delta x$. In general, the electric flux density change along x (which is $\partial D/\partial x$) is directly proportional to the density of charge centers. The density of charge centers is denoted by ρ_{charge}, and is expressed in C/m^3. $\partial D/\partial x$ is zero only when ρ_{charge} is zero. Therefore, a more general form of Eq. (2.7) is

$$\frac{\partial D}{\partial x} = \rho_{charge} \tag{2.8}$$

Equation (2.8) is known as the differential form of the Gauss theorem. It can be further generalized to include all the three spatial dimensions, in which case it becomes

$$\frac{\partial \boldsymbol{D}}{\partial x} \boldsymbol{x}_u + \frac{\partial \boldsymbol{D}}{\partial y} \boldsymbol{y}_u + \frac{\partial \boldsymbol{D}}{\partial z} \boldsymbol{z}_u = \nabla \boldsymbol{D} = \rho_{charge} \tag{2.9}$$

If we assume a homogeneous dielectric in terms of its dielectric permittivity ε_d, and eliminate D from Eqs. (2.6) and (2.8), the following equation is obtained:

$$\frac{\partial E}{\partial x} = \frac{\rho_{charge}}{\varepsilon_d} \tag{2.10}$$

Finally, if the electric field E is expressed in terms of the electric potential φ, using the relationship $E = -\partial\varphi/\partial x$, the equation that links the electric potential to the charge density is obtained:

$$\frac{\partial^2 \varphi}{\partial x^2} = -\frac{\rho_{\text{charge}}}{\varepsilon_{\text{d}}} \qquad (2.11)$$

Equation (2.11) is the one-dimensional form of the Poisson equation. The three-dimensional form of this equation is as follows:

$$\frac{\partial^2 \varphi}{\partial x^2} + \frac{\partial^2 \varphi}{\partial y^2} + \frac{\partial^2 \varphi}{\partial z^2} = \nabla^2 \varphi = -\frac{\rho_{\text{charge}}}{\varepsilon_{\text{d}}} \qquad (2.12)$$

2.3.2 Solution for Abrupt P–N Junction

A P–N junction is considered abrupt if there are only donors on one side of the junction and only acceptors on the other side, as shown in Fig. 2.8a. Referring to the way the P–N junctions are produced (Section 1.3.3, Fig. 1.11), it becomes obvious that the abrupt junction is only an approximation. Nonetheless, it is a very useful approximation as, (1) it represents one extreme case (the other extreme case, the linear junction, is considered in the next section), and (2) capacitance versus voltage (C–V) dependence can be expressed in a simple form.

The relationship between the electric potential and the charge density at any single point in the depletion layer is given by the Poisson equation. Solving the Poisson equation we can find the relationship between the depletion layer width w_{d} and the voltage applied, and then using Eq. (2.3), the depletion layer capacitance C_{d} can be expressed in terms of the voltage applied.

Neglecting any edge effects at the perimeter of the P–N junction (mathematically, this is equivalent to the assumption of infinite junction along the y- and z-axes), the problem becomes purely one dimensional, so we do not need to use the partial derivatives. The Poisson equation can be expressed as

$$\frac{d^2 \varphi}{dx^2} = -\frac{\rho_{\text{charge}}}{\varepsilon_{\text{s}}} \qquad (2.13)$$

The symbol ε_{d} is replaced by ε_{s} to indicate that the dielectric permittivity of the semiconductor (index s) should be taken. For silicon, $\varepsilon_{\text{s}} = 11.8\varepsilon_0 = 11.8 \times 8.85 \times 10^{-12}$ F/m.

It can be seen from Fig. 2.8a that there are four different regions in terms of the charge appearing at and around the P–N junction:

1. The electroneutral N^+-type region in which the donor ions are compensated for by the electrons. The charge density in this region is $\rho_{\text{charge}} = 0$.
2. The depletion layer on the N^+-type side in which the donor ions appear as the only charge centers. There are N_{D} donor ions per cubic meter or cubic centimeter (this is the donor concentration, which is assumed to be uniform on the N^+-type side). Every donor ion carries one unit (q) of positive charge. To obtain the charge density ρ_{charge}, which expresses the amount of charge per unit volume (C/m^3), the number of donor ions per unit volume N_{D} is multiplied by the unit charge q ($q = 1.6 \times 10^{-19}$ C). Therefore, in this region $\rho_{\text{charge}} = qN_{\text{D}}$.

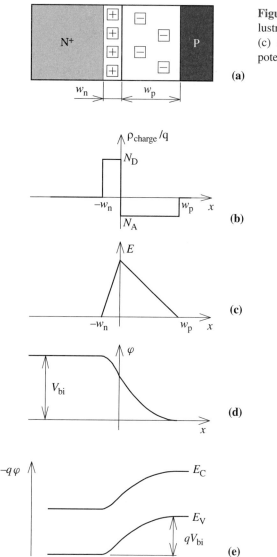

Figure 2.8 An abrupt P–N junction: (a) illustration, (b) charge density distribution, (c) electric-field distribution, (d) electric-potential distribution, and (e) energy bands.

3. The depletion layer on the P-type side in which the negative acceptor ions appear as the only charge centers. As there are N_A negative charge centers per unit volume, the charge density in this region is given as $\rho_{charge} = -qN_A$.

4. The electroneutral P-type region in which the acceptor ions are compensated for by the holes. The charge density in this region is $\rho_{charge} = 0$.

The charge density at and around the P–N junction is graphically presented in Fig. 2.8b. To solve the Poisson equation, it is useful to split it into four parts, corresponding to the four regions described above:

2.3 The Poisson Equation and C–V Dependence ■ 89

$$\frac{d^2\varphi}{dx^2} = \begin{cases} 0 & \text{for } x \leq -w_n \\ -\frac{qN_D}{\varepsilon_s} & \text{for } -w_n \leq x \leq 0 \\ \frac{qN_A}{\varepsilon_s} & \text{for } 0 \leq x \leq w_p \\ 0 & \text{for } x \geq w_p \end{cases} \tag{2.14}$$

If the left- and right-hand sides of Eq. (2.14) are integrated, the following is obtained:

$$\frac{d\varphi}{dx} = \begin{cases} C_1 & \text{for } x \leq -w_n \\ -\frac{qN_D}{\varepsilon_s}x + C_2 & \text{for } -w_n \leq x \leq 0 \\ \frac{qN_A}{\varepsilon_s}x + C_3 & \text{for } 0 \leq x \leq w_p \\ C_4 & \text{for } x \geq w_p \end{cases} \tag{2.15}$$

where C_1, C_2, C_3, and C_4 are integration constants. Remembering that the electric field is $E = -d\varphi/dx$, it can be seen that Eq. (2.15) gives the electric field multiplied by a minus sign. The electric field in the electroneutral regions ($x \leq -w_n$ and $x \geq w_p$) must be zero. If the field was not zero, the existing free electrons and holes would be moved by the field producing a current flow, which means that the P–N junction would not be in equilibrium. We wish to consider the P–N junction in equilibrium here. Thus, the constants $C_1 = 0$ and $C_4 = 0$. To obtain the other two integration constants, the boundary conditions at $x = -w_n$ and $x = w_p$ should be used. We already concluded that the field at $x = -w_n$ and $x = w_p$ (edges of the electroneutral regions) is zero. Applying Eq. (2.15) to the boundaries,

$$0 = \begin{cases} \frac{qN_D}{\varepsilon_s}w_n + C_2 \\ \frac{qN_A}{\varepsilon_s}w_p + C_3 \end{cases} \tag{2.16}$$

the constants are obtained as

$$C_2 = -\frac{qN_D}{\varepsilon_s}w_n$$

$$C_3 = -\frac{qN_A}{\varepsilon_s}w_p \tag{2.17}$$

After putting these constants into Eq. (2.15), it becomes:

$$\frac{d\varphi}{dx} = \begin{cases} 0 & \text{for } x \leq -w_n \\ -\frac{qN_D}{\varepsilon_s}(x + w_n) & \text{for } -w_n \leq x \leq 0 \\ \frac{qN_A}{\varepsilon_s}(x - w_p) & \text{for } 0 \leq x \leq w_p \\ 0 & \text{for } x \geq w_p \end{cases} \tag{2.18}$$

The above-described integration of the Poisson equation practically provides equations for the electric field at and around the P–N junction, as it is only necessary to multiply Eq. (2.18) by -1 to convert $d\varphi/dx$ into E:

$$E(x) = \begin{cases} 0 & \text{for } x \leq -w_n \\ \frac{qN_D}{\varepsilon_s}(x + w_n) & \text{for } -w_n \leq x \leq 0 \\ -\frac{qN_A}{\varepsilon_s}(x - w_p) & \text{for } 0 \leq x \leq w_p \\ 0 & \text{for } x \geq w_p \end{cases} \tag{2.19}$$

The electric field, as given by Eq. (2.19), is plotted in Fig. 2.8c. It can be seen that the maximum of the electric field appears right at the P–N junction ($x = 0$). Moreover, the maximum electric field $E_{max} = E(0)$ is expressed in two ways: (1) through N_D and w_n [the second line in Eq. (2.19)], and (2) through N_A and w_p [the third line in Eq. (2.19)]. This indicates that there is a relationship between N_D and w_n on the N$^+$-type side and N_A and w_p on the P-type side of the junction. This relationship can be found using the two equations for E_{max}:

$$E_{max} = E(0) = \frac{qN_D}{\varepsilon_s}w_n = \frac{qN_A}{\varepsilon_s}w_p \tag{2.20}$$

which gives

$$N_D w_n = N_A w_p \tag{2.21}$$

It is not hard to understand the meaning of Eq. (2.21). The number of donors (acceptors) N_D (N_A) per unit volume (expressed in m^{-3}), multiplied by the depletion layer width w_n (w_p) gives the number of donors (acceptors) per unit of junction area (expressed in m^{-2}). As the junction area is the same for the both donors and acceptors, Eq. (2.21) means that there is an equal number of donor and acceptor ions in the depletion layer of the P–N junction. We could not expect any other result, because the number of donor and acceptor ions in the depletion layer is equal to the number of electrons and holes recombined at the P–N junction in the process of depletion-layer creation, and the number of recombined electrons must be equal to the number of recombined holes (they recombine each other in pairs).

This means that if N_D is larger than N_A (as in the case shown in Fig. 2.8), depletion-layer width w_n is smaller than w_p. The depletion layer expands more on the side of the lower doped material.

Electric-potential distribution at and around the P–N junction can be obtained if Eq. (2.18) is integrated (the second integration of the Poisson equation; the first integration provided the electric field). The second integration gives

$$\varphi = \begin{cases} C_5 & \text{for } x \leq -w_n \\ -\frac{qN_D}{\varepsilon_s}\left(\frac{x^2}{2} + w_n x\right) + C_6 & \text{for } -w_n \leq x \leq 0 \\ \frac{qN_A}{\varepsilon_s}\left(\frac{x^2}{2} - w_p x\right) + C_7 & \text{for } 0 \leq x \leq w_p \\ C_8 & \text{for } x \geq w_p \end{cases} \tag{2.22}$$

As the field in the electroneutral regions ($x \leq -w_n$ and $x \geq w_p$) is zero, the electric potential is constant. Let us take the electric potential in the P-type region as the reference potential (the P-type side of the P–N junction is grounded). This means that $C_8 = 0$. As explained in

Section 2.2.3, the electric field created in the depletion layer results in a potential difference appearing across the P–N junction, referred to as the built-in voltage and denoted by V_{bi}. As the electric potential in the electroneutral N^+-type region is equal to V_{bi}, it is found that $C_5 = V_{bi}$. To find the other two constants, the following boundary conditions should be used: $\varphi(-w_n) = V_{bi}$, $\varphi(w_p) = 0$. Applying Eq. (2.22) to the boundaries $x = -w_n$ and $x = w_p$,

$$V_{bi} = \frac{q N_D}{\varepsilon_s} \frac{w_n^2}{2} + C_6$$

$$0 = -\frac{q N_A}{\varepsilon_s} \frac{w_p^2}{2} + C_7 \tag{2.23}$$

the constants are obtained as

$$C_6 = V_{bi} - \frac{q N_D}{\varepsilon_s} \frac{w_n^2}{2}$$

$$C_7 = \frac{q N_A}{\varepsilon_s} \frac{w_p^2}{2} \tag{2.24}$$

Putting the obtained constants into Eq. (2.22), the electric-potential distribution is obtained as

$$\varphi = \begin{cases} V_{bi} & \text{for } x \leq -w_n \\ V_{bi} - \frac{q N_D}{2\varepsilon_s}(x + w_n)^2 & \text{for } -w_n \leq x \leq 0 \\ \frac{q N_A}{2\varepsilon_s}(x - w_p)^2 & \text{for } 0 \leq x \leq w_p \\ 0 & \text{for } x \geq w_p \end{cases} \tag{2.25}$$

The electric potential, as given by Eq. (2.25), is plotted in Fig. 2.8d. The electric-potential distribution at and around the P–N junction can be used to plot the energy bands as well. The electric potential is related to the potential energy: $E_{pot}(x) = -q\varphi(x)$. Using this equation, and knowing that the bottom of the conduction band is separated from the top of the valence band by the energy-gap value (E_g), the energy bands are constructed as illustrated in Fig. 2.8e.

The earlier derived Eq. (2.21) provides a relationship between the depletion-layer width components (w_n and w_p), expressed in terms of presumably known doping concentrations N_D and N_A. The electric-potential distribution, given by Eq. (2.25), can be used to provide the second relationship between w_n and w_p. With two equations, the two unknown components of the depletion layer, namely w_n and w_p, can be found. To obtain the second relationship between w_n and w_p, observe that the electric potential at $x = 0$ is expressed in two ways: (1) through the parameters of the N^+-type side [the second line in Eq. (2.25)], and (2) through the parameters of the P-type side [the third line in Eq. (2.25)]. Therefore, using the fact that the electric potential at $x = 0$ is unique, a relationship between the parameters of the N^+-type and P-type sides can be established:

$$\varphi(0) = V_{bi} - \frac{q N_D}{2\varepsilon_s} w_n^2 = \frac{q N_A}{2\varepsilon_s} w_p^2 \tag{2.26}$$

Solving the system of two equations (2.21) and (2.26), w_n and w_p are obtained as

$$w_n = \sqrt{\frac{2\varepsilon_s V_{bi}}{q N_D[1 + (N_D/N_A)]}}$$ (2.27)

$$w_p = \sqrt{\frac{2\varepsilon_s V_{bi}}{q N_A[1 + (N_A/N_D)]}}$$ (2.28)

The total depletion-layer width is, obviously, given as

$$w_d = w_n + w_p$$ (2.29)

Equations (2.27), (2.28), and (2.29) can be used to calculate the components of the depletion-layer width, as well as the total depletion-layer width, for an abrupt P–N junction in thermal equilibrium. (The P–N junction is in thermal equilibrium when no voltage is applied at the terminals.) Our final aim, however, is to determine the dependence of the depletion-layer width (and thus depletion-layer capacitance) on the reverse-biased voltage applied to the terminals of the P–N junction capacitor. If the current of the reverse-biased P–N junction is neglected (the leakage current), the Poisson equation can be solved in a way similar to that described above. An important difference would be the boundary condition for the electric potential at $x = -w_n$. If a reverse-biased voltage V_R is applied to the N^+-type region, it should be added to the existing built-in voltage V_{bi} to obtain the correct boundary condition in this case. This is analogous to the transition from Fig. 2.6b to Fig. 2.6c. Therefore, in the final equations for w_n and w_p, $V_{bi} + V_R$ will appear instead of V_{bi} alone:

$$w_n = \sqrt{\frac{2\varepsilon_s(V_{bi} + V_R)}{q N_D[1 + (N_D/N_A)]}}$$ (2.30)

$$w_p = \sqrt{\frac{2\varepsilon_s(V_{bi} + V_R)}{q N_A[1 + (N_A/N_D)]}}$$ (2.31)

The depletion-layer capacitance, given by Eq. (2.3), can now be related to the applied reverse-biased voltage V_R through w_n and w_p:

$$C_d = A\frac{\varepsilon_s}{w_d} = A\frac{\varepsilon_s}{w_n + w_p}$$ (2.32)

It is important to note that in most practical cases it is either $N_D \gg N_A$ (the case in Fig. 2.8) or $N_A \gg N_D$. This is because either the P-type substrate has to be converted into the N-type layer at the surface by diffusing a much higher concentration of donors at the surface, or vice versa, to create a P–N junction. Accordingly, the equations for the depletion layer widths can be simplified as follows:

$$w_p \approx \sqrt{\frac{2\varepsilon_s(V_{bi}+V_R)}{qN_A}}$$
$$\left.\begin{array}{l} \\ w_n \ll w_p \\ \\ w_d \approx w_p \end{array}\right\} \quad \text{for } N_D \gg N_A \qquad (2.33)$$

$$w_n \approx \sqrt{\frac{2\varepsilon_s(V_{bi}+V_R)}{qN_D}}$$
$$\left.\begin{array}{l} \\ w_p \ll w_n \\ \\ w_d \approx w_n \end{array}\right\} \quad \text{for } N_A \gg N_D \qquad (2.34)$$

Using Eqs. (2.32), (2.33), and (2.34), the depletion-layer capacitance can be expressed in the following compact form:

$$C_d = C_d(0)\left(1 + \frac{V_R}{V_{bi}}\right)^{-1/2} \qquad (2.35)$$

where $C_d(0)$ is

$$C_d(0) = \begin{cases} \dfrac{A}{2}\sqrt{\dfrac{2\varepsilon_s q N_A}{V_{bi}}} & \text{for } N_D \gg N_A \\[2mm] \dfrac{A}{2}\sqrt{\dfrac{2\varepsilon_s q N_D}{V_{bi}}} & \text{for } N_A \gg N_D \end{cases} \qquad (2.36)$$

$C_d(0)$ can be considered as a parameter in the equation for the capacitance versus applied voltage (Eq. 2.35). Its physical meaning is obvious: it represents the P–N junction capacitance at zero bias ($V_R = 0$). It is referred to as *zero-bias junction capacitance*.

In conclusion, the depletion-layer capacitance of the abrupt P–N junction is shown to be proportional to $(1 + V_R/V_{bi})^{-1/2}$, where V_R is the reverse-bias voltage.

2.3.3 Solution for Linear P–N Junction

The linear P–N junction is the other extreme approximation, where the charge density changes from the most positive value (the donors on the N-type side) to the most negative value (the acceptors on the P-type side) in the smoothest possible way. The linear P–N junction is the one where the charge density in the depletion layer changes linearly, as illustrated in Fig. 2.9a and b.

To solve the Poisson equation in the depletion layer of the linear P–N junction, the charge density is expressed as

$$\rho_{charge} = -ax \qquad (2.37)$$

where a is the slope of the linear dependence. In this case, the Poisson equation (2.13) becomes

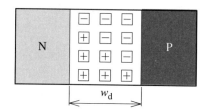

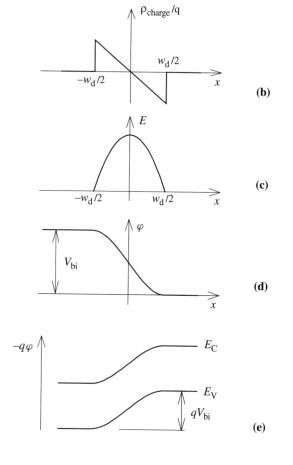

Figure 2.9 A linear P–N junction: (a) illustration, (b) charge-density distribution, (c) electric-field distribution, (d) electric-potential distribution, and (e) energy bands.

$$\frac{d^2\varphi}{dx^2} = \frac{a}{\varepsilon_s}x \qquad (2.38)$$

Of course, Eq. (2.38) is correct only for $-w_d/2 \leq x \leq w_d/2$ (the depletion layer), however, solving the Poisson equation in the depletion layer will be quite enough to find the depletion-layer width, provided the boundary conditions are properly established.

The first integration of Eq. (2.38) leads to

$$\frac{d\varphi}{dx} = \frac{a}{\varepsilon_s}\frac{x^2}{2} + C_a \qquad (2.39)$$

where C_a is the integration constant. The integration constant can be found from the condition in which the electric field, and therefore $d\varphi/dx$, is equal to zero at $x = -w_d/2$:

$$C_a = -\frac{a}{\varepsilon_s}\frac{w_d^2}{8} \qquad (2.40)$$

Using this value for the constant C_a, $d\varphi/dx$ becomes

$$\frac{d\varphi}{dx} = \frac{a}{2\varepsilon_s}\left[x^2 - \left(\frac{w_d}{2}\right)^2\right] \qquad (2.41)$$

while the electric field ($E = -d\varphi/dx$) is obtained as

$$E(x) = \frac{a}{2\varepsilon_s}\left[\left(\frac{w_d}{2}\right)^2 - x^2\right] \qquad (2.42)$$

The electric field, as given by Eq. (2.42), is plotted in Fig. 2.9c.

Integrating the left-hand and right-hand sides of Eq. (2.41) (the second integration of the Poisson equation) leads to an equation for the electric potential φ:

$$\varphi(x) = \frac{a}{2\varepsilon_s}\left[\frac{x^3}{3} - \left(\frac{w_d}{2}\right)^2 x\right] + C_b \qquad (2.43)$$

The integration constant C_b can be found from the boundary condition $\varphi(w_d/2) = 0$ (it is again assumed that the potential of the P-type side is the reference potential). Using this boundary condition, the constant is found as

$$C_b = \frac{a}{\varepsilon_s}\frac{w_d^3}{24} \qquad (2.44)$$

which means the electric-potential distribution in the depletion layer is given by

$$\varphi(x) = \frac{a}{2\varepsilon_s}\left[\frac{x^3}{3} - \left(\frac{w_d}{2}\right)^2 x + \frac{w_d^3}{12}\right] \qquad (2.45)$$

The electric potential, as given by Eq. (2.45), is plotted in Fig. 2.9d. The electric-potential function $\varphi(x)$ can be used to plot the energy bands as well. Using the fact that the energy bands follow the potential-energy function $E_{pot}(x) = -q\varphi(x)$, and that the bottom of the conduction band and the top of the valence band are separated by the energy gap E_g, the energy bands are constructed as in Fig. 2.9e.

The electric-potential distribution $\varphi(x)$ can also be used to determine the depletion-layer width. This can be achieved if the second boundary condition for the electric potential is employed, which is the electric potential at $x = -w_d/2$. As the electric potential at $x = w_d/2$ is taken as the reference potential [$\varphi(w_d/2) = 0$], the electric potential at the other side of the depletion layer ($x = -w_d/2$) must be equal to the built-in voltage V_{bi}. Therefore,

$$V_{bi} = \frac{a}{2\varepsilon_s}\left(-\frac{w_d^3}{24} + \frac{w_d^3}{8} + \frac{w_d^3}{12}\right) = \frac{aw_d^3}{12\varepsilon_s} \tag{2.46}$$

The depletion-layer width w_d is obtained from Eq. (2.46) as

$$w_d = \left(\frac{12\varepsilon_s V_{bi}}{a}\right)^{1/3} \tag{2.47}$$

This is the depletion-layer width of the linear P–N junction in thermal equilibrium (zero bias). The Poisson equation can similarly be solved for a reverse-biased linear P–N junction. An important difference will be the boundary condition at $x = -w_d/2$. In that case the reverse-bias voltage applied V_R has to be added to the existing built-in voltage V_{bi}, which means that the boundary condition becomes $\varphi(-w_d/2) = V_{bi} + V_R$. Accordingly, $V_{bi} + V_R$ will appear in the final equation for w_d instead of V_{bi} alone:

$$w_d = \left[\frac{12\varepsilon_s(V_{bi} + V_R)}{a}\right]^{1/3} \tag{2.48}$$

The depletion-layer capacitance of the linear P–N junction can now be related to the applied reverse-bias voltage V_R through w_d:

$$C_d = A\frac{\varepsilon_s}{w_d} = A\left[\frac{a\varepsilon_s^2}{12(V_{bi} + V_R)}\right]^{1/3} \tag{2.49}$$

The depletion layer capacitance C_d can be written in a more suitable form:

$$C_d = C_d(0)\left(1 + \frac{V_R}{V_{bi}}\right)^{-1/3} \tag{2.50}$$

where

$$C_d(0) = A\left(\frac{a\varepsilon_s^2}{12V_{bi}}\right)^{1/3} \tag{2.51}$$

$C_d(0)$ in Eq. (2.50) can be considered as a parameter. This parameter represents the zero-biased junction capacitance, analogously to the case of the abrupt P–N junction.

In conclusion, the depletion-layer capacitance of the linear P–N junction is shown to be proportional to $(1 + V_R/V_{bi})^{-1/3}$, where V_R is the reverse-bias voltage.

■ **Example 2.2 P–N Junction as a Varactor**

A P–N junction is to be used as a varactor (a voltage-dependent capacitor used as a tuning element in microwave circuits). The doping concentration in the P-type and N^+-type layers are known: $N_A = 10^{18}$ cm^{-3} and $N_D = 10^{20}$ cm^{-3}, respectively.

(a) Design the varactor (determine the needed junction area) so that the maximum capacitance is $C_{max} = 30$ pF.

(b) Calculate the varactor sensitivity (dC/dV_R) at $V_R = 5$ V.

Solution:

(a) The depletion-layer capacitance of a reverse-biased P–N junction is maximum when the reverse-bias voltage is zero. Any increase in the reverse-bias voltage increases the depletion-layer width, reducing the capacitance. Therefore, Eq. (2.36) can be used to calculate the maximum capacitance. We need to calculate V_{bi} first, which can be done using Eq. (2.4):

$$V_{bi} = \frac{kT}{q} \ln \frac{N_A N_D}{n_i^2} = 0.02585 \ln \frac{10^{18} \times 10^{20}}{(1.02 \times 10^{10})^2} = 1.07 \text{ V}$$

Using Eq. (2.36), the maximum capacitance per unit area is determined:

$$\frac{C_{max}}{A} = \frac{1}{2} \sqrt{\frac{2\varepsilon_s q N_A}{V_{bi}}}$$

$$= 0.5 \sqrt{\frac{2 \times 11.8 \times 8.85 \times 10^{-12} \times 1.6 \times 10^{-19} \times 10^{24}}{1.07}}$$

$$= 2.8 \text{ mF/m}^2$$

The needed junction area is then

$$A = \frac{30 \text{ pF}}{2.8 \text{ mF/m}^2} = 1.07 \times 10^{-4} \text{ cm}^2 = 10{,}700 \text{ } \mu\text{m}^2$$

(b) The capacitance versus applied reverse-bias voltage is given by Eq. (2.35), which can be rewritten as

$$C = C_{max}(1 + V_R/V_{bi})^{-1/2}$$

The first derivative of the capacitance with respect to V_R will provide the equation for the capacitor sensitivity:

$$dC/dV_R = -\frac{1}{2} \frac{C_{max}}{V_{bi}} (1 + V_R/V_{bi})^{-3/2}$$

At $V_R = 5$ V, the sensitivity is $dC/dV_R = -1.0$ pF/V.

■ **Example 2.3 Minimum P–N Junction Capacitance**

Calculate the minimum capacitance that can be achieved by a linear P–N junction capacitor when the reverse-bias voltage changes between 0 and 5 V, if the maximum capacitance is 2.5 pF. Assume that the built-in voltage is $V_{bi} = 0.8$ V.

Solution: The capacitance of the linear P–N junction is given by Eq. (2.50). The capacitance is maximum for $V_R = 0$ V, which means that the parameter $C_d(0) = 2.5$ pF. The reverse-bias voltage V_R reduces the capacitance, which means it is minimum for the largest V_R, which is $V_{R-max} = 5$ V in this example. Therefore, the minimum capacitance is calculated as

$$C_{min} = C_d(0)(1 + V_{R-max}/V_{bi})^{-1/3} = 2.5 \times (1 + 5/0.8)^{-1/3} = 1.3 \text{ pF}$$

2.4 SPICE PARAMETERS AND THEIR MEASUREMENT

The SPICE program is used worldwide as an essential computer aid for circuit design. There are many different versions available from different suppliers, however, they are typically built on a common set of device models, developed originally at University of California at Berkeley. The accompanying *Computer Exercises* manual provides a brief description of the SPICE program. Using the circuit simulators like SPICE, it is possible to predict the detailed performance of a circuit prior to its fabrication. A number of circuit modifications can be tested in this way in a time and cost-effective way. Accurate circuit simulation is possible, however, only by specifying accurate and meaningful values of the parameters used in the device models. Consequently, the model parameters should be extracted from corresponding measurement data (test devices are usually available to enable the collection of the experimental data required).

This section describes the SPICE mathematical model for the depletion-layer capacitance of a reverse-biased P–N junction, and uses this example to introduce the basic parameter measurement techniques.

2.4.1 SPICE Model for the Depletion-Layer Capacitance

The previous section has introduced the two extreme approximations of the P–N junction: the abrupt and the linear P–N junction. The theoretical equations for the dependence of the depletion-layer capacitance on the reverse-bias voltage have been derived for both approximations. It has been found that the capacitance is proportional to $(1 + V_R/V_{bi})^{-1/2}$ for the abrupt P–N junction and is proportional to $(1 + V_R/V_{bi})^{-1/3}$ for the linear P–N junction. It can be seen that the only difference between these two extreme cases is the power coefficient, which is $-1/2$ for the abrupt P–N junction and $-1/3$ for the linear P–N junction. The dependence of the depletion-layer capacitance C_d on the reverse-bias voltage V_R can, therefore, be expressed in the following compact form:

$$C_d = C_d(0) \left(1 + \frac{V_R}{V_{bi}} \right)^{-m} \tag{2.52}$$

where $m = 1/2$ for the abrupt P–N junction and $m = 1/3$ for the linear P–N junction. As the abrupt and the linear P–N junctions are the two extreme cases, it is obvious that Eq. (2.52) becomes applicable to any P–N junction when m is allowed to take any value between the two extremes ($1/2 \geq m \geq 1/3$). Thus, considering m as a parameter whose value has to be determined experimentally, Eq. (2.52) can be used for real P–N junctions that are neither abrupt nor linear. The parameter m is referred to as the *grading coefficient*.

In SPICE, there is no separate device called a reverse-biased P–N junction capacitor. The P–N junction is represented by a unique device, called a *diode*. The complete operation of the diode is considered in the next chapter. The capacitance effect of the P–N junction has nothing to do with the principle of diode operation; rather it appears as a parasitic effect. There is an equation for the depletion-layer capacitance of the P–N junction in the set of equations used in SPICE to model the diode. This equation is equivalent to Eq. (2.52). If a reverse-biased P–N junction is used as a capacitor, an appropriately biased diode can be used in SPICE to simulate the P–N junction capacitor. It is true that SPICE will use all the other equations existing in the diode model, not only Eq. (2.52). This is an advantage, as the leakage current will be included, and if it happens that the voltage appearing across the P–N junction is not appropriate, SPICE will give correct (but not expected) results helping notice a problem due to wrongly assumed biasing conditions.

In conclusion, Eq. (2.52) is used in SPICE to model the depletion-layer capacitance. $C_d(0)$, V_{bi}, and m are all considered as parameters of the model, in a sense that their values can be adjusted to achieve the best agreement between the model and the experimental characteristics of the P–N junction used. Table 2.1 provides a summary of the SPICE model for the reverse-biased P–N junction capacitor.

2.4.2 Parameter Measurement Techniques

Parameter measurement is a procedure used to determine the values of the model parameters that give the "best fit" between the model (the mathematical equation) and the experimental data.

Graphic Method

Consider the example of Fig. 2.10, in which a set of experimental data is shown by the symbols. It appears that the data express a linear relationship between y and x. The simplest linear equation $y = x$ does not provide a good model—it is obvious that the dashed line (which is the $y = x$ plot) does not fit the data appropriately. The most general form of the linear relationship is given as $y = a_1 x + a_0$, which means $a_1 = 1$ and $a_0 = 0$ in the case of the dashed line. Using different values for a_1 and a_0, a different straight line can be plotted, like the dotted line in Fig. 2.10. The dotted line may be a better approximation of the experimental data, but obviously, still not good enough. Instead of guessing a_1 and a_0 and plotting straight lines to check if the data are properly fitted, it may be more suitable to take a ruler and draw a straight line through the data, and then determine the a_1 and

TABLE 2.1 Summary of the SPICE Model of the Reverse-Biased P–N Junction Capacitor

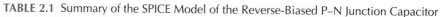

		Parameters	
Symbol	Usual SPICE Keyword	Parameter Name	Unit
$C_d(0)$	CJO	Zero-bias junction capacitance	F
V_{bi}	VJ	Built-in (junction) voltage	V
m	M	Grading coefficient	

Model

$$C_d = \text{CJO}\left(1 + \frac{V_R}{VJ}\right)^{-M}$$

C_d is the depletion layer capacitance and V_R is the reverse-biased voltage ($V_R = -V_D$)

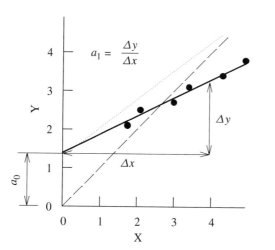

Figure 2.10 Fitting experimental data (symbols) by the linear $y = a_1 x + a_0$ relationship. The solid line shows the case for properly determined parameters a_1 and a_0; the dashed and the dotted lines show two cases for improper parameters a_1 and a_0.

a_0 for this supposedly "best" line. To determine a_1 and a_0 for a drawn line, the following procedure can be used:

- As $y = a_0$ for $x = 0$, a_0 is directly determined by reading or measuring the y value at the cross section between the straight line and the y-axis.
- a_1 represents the slope of the straight line, which can be determined from the y and x increments between any two points on the line.

The above-described procedure is referred to as the *graphic method* for parameter measurement. It may appear that the graphic method is limited to linearly related data. It should be noted, however, that a number of different relationships can be linearized by appropriate transformations. For example, the relationship $z = a \exp(bx)$ can be linearized if $\ln z$ is plotted vs x, instead of z vs x. It is obvious that $\ln z = \ln a + bx$ is equivalent to the linear relationship, where $y = \ln z$, $a_0 = \ln a$, and $a_1 = b$.

The graphic method is very illustrative—it is very convenient to check or analyze the fitting procedure on a graph. In its described form, however, it lacks a unique criterion for the best fit. Looking at a graph we may decide that the agreement between the model and the experimental data is acceptable, however, we cannot be sure whether the best possible fitting is achieved. The following text introduces the criterion for the best fit.

Linear Regression

The solid line in Fig. 2.10 provides the best fit because the differences between the experimental data and the line are the smallest. Denote the experimental data points by x_i and y_i, where $i = 1, ..., n$ is related to each data point. The difference between the model and the ith experimental point, along the y-axis, can be expressed as

$$(a_1 x_i + a_0) - y_i \tag{2.53}$$

To involve all the experimental data points, it may seem appropriate to add all the $(a_1 x_i + a_0) - y_i$ differences, however, this would not be a good approach, as fairly large positive and negative differences can cancel each other, producing a small net result, which wrongly implies that a good fit is achieved. To avoid this problem, the $(a_1 x_i + a_0) - y_i$ differences are squared, and then added to provide an indication about the overall fitting:

$$S = \sum_{i=1}^{n} [(a_1 x_i + a_0) - y_i]^2 \tag{2.54}$$

It is said that *the best fit is achieved when the sum of squared deviations of the model from the experimental data points is minimum*. In other words, the minimum of S is taken as the criterion for the best fit.

Having established the criterion for the best fit, we can check if a model with its parameters represents the best fit. However, it is even more useful if we knew how to directly determine the parameters that guarantee the best fit. Using the above-established criterion and employing some mathematics, this can be achieved. It is enough to express the first derivatives of S with respect to a_0 and a_1, and use the condition that these derivatives are equal to zero for the values of a_0 and a_1 that minimize S:

$$\frac{\partial S}{\partial a_0} = 2 \sum_{i=1}^{n} (a_1 x_i + a_0 - y_i) = 0$$

$$\frac{\partial S}{\partial a_1} = 2 \sum_{i=1}^{n} x_i (a_1 x_i + a_0 - y_i) = 0 \tag{2.55}$$

These conditions lead to the following system of two linear equations:

$$na_0 + \left(\sum_{i=1}^{n} x_i\right) a_1 = \sum_{i=1}^{n} y_i$$

$$\left(\sum_{i=1}^{n} x_i\right) a_0 + \left(\sum_{i=1}^{n} x_i^2\right) a_1 = \sum_{i=1}^{n} x_i y_i \qquad (2.56)$$

Solving this system of linear equations, the values of a_0 and a_1 that guarantee the minimum S are obtained.

This method of parameter measurement is referred to as the *linear regression*. Obviously, it is much more convenient to solve a system of linear equations and obtain the best parameters directly, than to guess parameters, plot the graph, and then guess the parameters again. It is still very useful to plot the graph with the parameters determined by the linear regression, to check for possible errors and to visualize the best fit. The regression method guarantees the best fit by the model chosen, but it cannot guarantee that the model chosen can provide an acceptable fit.

When there are more than two parameters, the linear regression method will require a system of more than two linear equations to be solved. There is a number of software packages that can help solve the system of linear equations involved in the linear regression analysis.

Note again that this method is not strictly limited to the case of linearly related experimental data as many relationships can be linearized. Frequently, however, it is difficult or impossible to find suitable transformations to linearize the model. The following text introduces a method that can be used in the most general case, for an arbitrary form of the model.

Curve Fitting

The procedures used to fit nonlinear models directly, without transforming the data and performing linear regression analysis, are referred to as the *curve fitting* or *nonlinear regression*. The curve fitting is also based on the criterion that the best fit is achieved when *the sum of squared differences between the model and the experimental data is minimum*.

The curve fitting process is *iterative*:

1. A set of *initial values* of the model parameters is specified.
2. The model predictions are calculated for all the existing experimental values of the independent variable(s), and the squared differences between the model predictions and the experimental data added to create *sum of squared differences*, or shorter *sum of squares*.
3. The parameter values are varied by an algorithm, and the sum of squared differences is calculated for this new set of parameters.
4. The new and the preceding sums of squared differences are compared:
   4.1 If the new sum of squared differences is smaller, steps 3 and 4 are repeated—this is referred to as the *iteration*. The condition in which every new sum is smaller than the previous one is known as *convergence*.

4.2 If the new sum of squared differences is larger, the parameters are varied using a different algorithm in step 3, and steps 3 and 4 are repeated. The condition in which every new sum is larger than the previous one is known as *divergence*, and it has to be prevented by adjusting the algorithm used in step 3. If the divergence does occur, than the iterative process has to be reinitiated with a set of initial parameters that is closer to the final solution.

4.3 If the differences between the residual sum of squares no longer decrease significantly, the iterative process is terminated. If the final sum of squares is acceptable (smaller than a preset value), the final parameter values are considered as acceptable. If the final sum of squares is unacceptable, the iterative process has to be reinitiated with a set of initial parameters that is closer to the final solution.

There are different possible algorithms that can be used to vary the parameters in the attempt to ensure convergence (and preferably fast convergence) of the iterative process. These algorithms are explained in books on numerical analysis. Moreover, there are a number of software packages that can be used for curve fitting without a detailed knowledge of the particular algorithms used. It is necessary, however, to carefully consider the problem of choosing the initial parameters, as this can make the difference between obtaining and not obtaining an acceptable fit.

The most significant problem with the described curve fitting process is that the sum of squares can have one or more local (or relative) minima in addition to the global (or absolute) minimum. This means that the first derivatives of the sum of squares with respect to the model parameters can be zero for a number of different sets of the model parameters. Although the sum of squares in some of its local minima (the first derivatives are equal to zero) can be quite large, the iterative process may end up at that minimum. This problem does not occur when the initial parameters are chosen close enough to the global minimum—in that case the iterative process ends up at the global minimum. Sometimes there is not a big difference in the sum of squares between the global and some of the local minima. In such a case, it is desirable to use the minimum, which is associated with the parameters that have the best physical justification. Again, this can be ensured if the initial parameters are chosen close to the desired minimum.

The choice of the initial parameters can be made in different ways. In some cases (very simple models with a unique minimum) a pure guess may prove sufficient. More frequently, however, it is necessary to apply more sophisticated techniques. These techniques are considered individually for any device model introduced. It should be noted here that these techniques generally involve the graphic and/or the linear regression method. The P–N junction capacitance model is considered in the next section.

2.4.3 Measurement of the Capacitance-Model Parameters

It is not possible to completely linearize the model for the reverse-biased P–N junction capacitance given by Eq. (2.52). Therefore, the graphic or the linear regression method cannot be directly applied. The situation is further complicated by the fact that the measured data contain an additional, parasitic capacitance component. The P–N junction capacitance can be measured in different ways, perhaps most suitable being by means of a bridge. The measurement frequency can be set low enough so that the parasitic series resistance becomes negligible compared to the impedance of the capacitor. However, the parasitic capacitance, caused mainly by pin capacitance, stray capacitance, and pad capacitance, cannot be avoided. Assuming that the parasitic capacitance C_p does not depend on the voltage applied, the measured capacitance can be expressed as

$$C_{meas} = C_d(0) \left(1 + \frac{V_R}{V_{bi}}\right)^{-m} + C_p \tag{2.57}$$

Although the parameter C_p in Eq. (2.57) is not needed as a SPICE parameter, it has to be extracted from the experimental data.

Curve fitting can be the most effective way of parameter measurement in this case, provided the initial parameter values are properly determined. There are four parameters in Eq. (2.57): $C_d(0)$, V_{bi}, m, and C_p. The built-in voltage V_{bi} depends on the doping levels in the P- and N-type regions, as given by Eq. (2.4). It is useful to estimate the likely extreme values of this parameter. The lowest value is obtained when the lowest doping levels are assumed—let this be $N_A N_D = 10^{15} \times 10^{16}$ cm^{-3} × cm^{-3}. With this V_{bi}(min) $= 0.02585 \ln[10^{31}/(1.02 \times 10^{10})^2] = 0.65$ V. The highest value is obtained when the highest doping levels are assumed, say $N_A N_D = 10^{20} \times 10^{21}$ cm^{-3} × cm^{-3}. With this, V_{bi}(max) $= 1.25$ V. If the value of $V_{bi} = 0.9$ V is assumed, the maximum error cannot be much bigger than 0.3 V, which is $0.3/0.9 \times 100 = 33\%$.

Assuming a constant value for the parameter V_{bi} enables linearization of the equation for the reverse-biased P–N junction capacitance in the following way:

$$\log(C_{meas} - C_p) = \log C_d(0) - m \log \left(1 + \frac{V_R}{V_{bi}}\right) \tag{2.58}$$

where $C_{meas} - C_p$ represents the reverse-biased P–N junction capacitance (the depletion-layer capacitance C_d). If the parasitic capacitance C_p was zero, and V_{bi} was correctly assumed, plotting the logarithm of the measured capacitances vs $\log(1 + V_R/V_{bi})$ would give a straight line. The parameters of the linear relationship $y = a_0 + a_1 x$ would be related to the parameters m and V_{bi}: $\log C_d(0) = a_0$, and $-m = a_1$. This would enable the measurement of m and $C_d(0)$ by the graphic or linear regression method. Although the value of C_p cannot be neglected, $C_p = 0$ can still be used as the initial value to enable the measurement of initial values of m and $C_d(0)$.

To illustrate the above-described technique, consider the example of experimental data given in Table 2.2. Assuming $V_{bi} = 0.9$ V, and $C_p = 0$, $\log(1 + V_R/V_{bi})$ and $\log(C_{meas} - C_p)$ can be calculated as shown in Table 2.2 as well. The $\log(C_{meas} - C_p)$ vs $\log(1 + V_R/V_{bi})$ graph (for $V_{bi} = 0.9$ V and $C_p = 0$) is shown in Fig. 2.11a by the squared symbols. The dotted line represents the best linear fit for these data. The parameters of the dotted line are found to be $a_1 = -0.25$ and $a_0 = 0.639$. This means that $m = -a_1 = 0.25$ and $C_d(0) = 10^{0.639} = 4.35$ pF. This completes the set of possible values of the four parameters. Using these values, the corresponding theoretical curve (dotted line) for C_{meas} vs V_R is compared to the experimental data (symbols) in Fig. 2.11b.

TABLE 2.2 Example of Experimental Data for the Capacitance-Model Parameter Measurement

V_R (V)	C_{meas} (pF)	$\log\left(1+\frac{V_R}{0.9}\right)$	$\log C_{meas}$	$\log (C_{meas} - 1.26 \text{ pF})$
0.0	4.45	0.0000	0.6484	0.5038
1.0	3.59	0.3245	0.5551	0.3686
2.0	3.21	0.5082	0.5065	0.2981
3.0	2.98	0.6368	0.4742	0.2379
4.0	2.82	0.7360	0.4502	0.1963
5.0	2.70	0.8166	0.4314	0.1625
6.0	2.61	0.8846	0.4166	0.1340
7.0	2.54	0.9434	0.4048	0.1094
8.0	2.47	0.9951	0.3927	0.0876
9.0	2.42	1.0414	0.3838	0.0682
10.0	2.37	1.0832	0.3747	0.0507

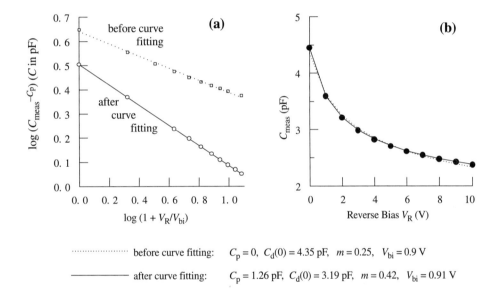

before curve fitting: $C_p = 0$, $C_d(0) = 4.35$ pF, $m = 0.25$, $V_{bi} = 0.9$ V

after curve fitting: $C_p = 1.26$ pF, $C_d(0) = 3.19$ pF, $m = 0.42$, $V_{bi} = 0.91$ V

Figure 2.11 (a, b) Measurement of the parameters of the capacitance model: (1) $V_{bi} = 0.9$ V and $C_p = 0$ is assumed, and linear regression is applied to estimate m and $C_d(0)$—the square symbols and the dotted line in (a); (2) curve fitting is performed using the estimated values as the initial parameters, which improves the fit (the solid lines) but more importantly provides parameters that are physically justified.

The fit between the dotted line and the symbols in Fig. 2.11b appears as quite reasonable. However, the value of the parameter m is questionable. It has been shown in Section 2.3 that m is the smallest for the extreme case of the linear junction, in which case it is $m = 0.33$. The value of $m = 0.25$ is not physically justified. The good fit is achieved due to the compensating effect of the assumption $C_p = 0$. We should not accept this situation. The value of C_p can be as high as 2 pF, which means that the measured capacitance can be more than twice as high as the real junction capacitance.

The above-described procedure can be repeated assuming, for example, $C_p = 2$ pF. If a better fit is achieved, it would indicate that the real C_p is closer to 2 pF than 0 pF. Making yet another guess and performing an additional iteration may improve the fit further. This is in essence the iterative process used for curve fitting. As mentioned above, there are a number of software packages that automatically perform the iterative process used in curve fitting. The initially extracted values of the four parameters can be attempted as the initial parameters for the curve fitting procedure. Using $C_p = 0$, $C_d(0) = 4.35$ pF, $m = 0.25$, and $V_{bi} = 0.9$ V as the initial values, the following final values are obtained by the curve fitting available in *SigmaPlot* scientific graphing software (Jandel Scientific): $C_p = 1.26$ pF, $C_d(0) = 3.19$ pF, $m = 0.42$, $V_{bi} = 0.91$ V. The solid lines in Fig. 2.11 represent the model with these parameters. It can be seen that the fitting is improved (the fact that the iterative process used for the curve fitting converged indicated that the best fit was achieved). More importantly, the parameter $m = 0.42$ is now within the extreme limits of 0.33 and 0.5; in addition, the value $C_p = 1.26$ pF appears as practically reasonable. Plotting $\log(C_{meas} - C_p)$ vs $\log(1 + V_R/V_{bi})$ (the solid line in Fig. 2.11a), a straight line is obtained, which confirms that the measurement procedure is successfully completed.

■ **Example 2.4 Linear Regression Analysis**

Using the data given in columns 3 and 5 of Table 2.2, apply the linear regression analysis to obtain the parameters m and $C_d(0)$ in Eq. 2.57.

Solution: To be able to apply the linear regression analysis, the linear form of Eq. (2.57) has to be used, as given by Eq. (2.58). Using the numbers from the third and fifth columns of Table 2.2 as the experimental x_i and y_i data, respectively, the system of two linear equations (2.56) can directly be applied. Transforming the parameters as $\log C_d(0) = a_0$ and $-m = a_1$, the system of two equations can be written as

$$11a_0 + 7.9698a_1 = 2.2108$$

$$7.9698a_0 + 6.8982a_1 = 1.1314$$

Solving the above system, it is found that $a_0 = 0.504$ and $a_1 = -0.42$. Therefore, $C_d(0) = 10^{a_0} = 10^{0.504} = 3.19$ pF and $m = -a_1 = 0.42$.

2.5 METAL–OXIDE–SEMICONDUCTOR (MOS) CAPACITOR AND THERMAL OXIDE

MOS capacitors are used in linear circuits, and even more frequently in digital circuits as the charge storage components in random-access memories (RAMs). Additionally, the most frequently used component in modern electronics, the metal–oxide–semiconductor field-effect transistor (MOSFET), is based on the MOS capacitor. Therefore, the importance of this section is not limited to capacitor applications, it also introduces many MOSFET-related concepts and parameters.[2]

2.5.1 Integrated-Circuit Structure

An MOS capacitor is created by using semiconductor substrate as one of the capacitor conductive plates, which is separated by a thin silicon dioxide layer (referred shortly to as oxide) from the other conductive plate, which is deposited onto the oxide layer (Fig. 2.12). Originally, a metal film (most usually aluminum) was used to create the second capacitor plate, which is directly reflected in the name *metal–oxide–semiconductor*. We will refer to this capacitor plate (electrode) as a *gate*, because that is the name used for the corresponding MOSFET electrode. In modern integrated circuits, the aluminum gates are replaced by silicon gates. In this case, a thin film of silicon is deposited onto the oxide and heavily doped to create the gate electrode. When the silicon is deposited onto the oxide, the structure of the silicon film is not monocrystalline but polycrystalline; because of this the electrode is frequently referred to as a *polysilicon gate* or even *polygate*. The name *metal–oxide–semiconductor (MOS)* is, however, retained irrespective of the material (metal or silicon) used for the gate electrode.

The capacitor active area (area of the thin oxide, referred also to as *gate oxide*) is usually surrounded by a thicker oxide called a *field oxide*, as shown in Fig. 2.12. This is to define the area of the semiconductor plate of the capacitor and minimize fringing-field effects, thus it serves as a kind of isolation for the capacitor. The field oxide is widely used as isolation technique in integrated circuits using MOSFETs. A contact is made to the semiconductor plate of the capacitor to enable access to this electrode from the surface of the IC. As Fig. 2.12 shows, the doping of the semiconductor substrate is usually increased (P^+ area in the figure) to provide better contact between the metal film and the semiconductor substrate. A P-type semiconductor substrate is used in the MOS

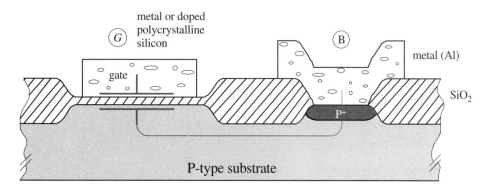

Figure 2.12 Cross section of a metal–oxide–semiconductor (MOS) capacitor.

---

[2]This section can be postponed until Chapter 5.

capacitor illustrated in Fig. 2.12. MOS capacitors can also be created using an N-type substrate.

In the case of a P–N junction capacitor, only reverse-bias voltage can be used. The presence of the oxide enables the application of both negative and positive voltages in the case of an MOS capacitor; because of this the capacitance–voltage dependence is much more complex. A typical C–V curve for an MOS capacitor on a P-type substrate is shown in Fig. 2.13.

Accumulation

When negative voltage is applied between the gate and the substrate (V_G), a number of holes from the P-type substrate are attracted to the semiconductor surface. This effect, where the surface concentration of the majority carriers (holes in this case) is increased, is called *accumulation*. In the accumulation mode, the positive charge appearing at the semiconductor plate is separated only by the gate oxide, which means that the capacitance is given as

$$C = C'_{ox} = \frac{\varepsilon_{ox}}{t_{ox}} A \tag{2.59}$$

where t_{ox} is the gate oxide thickness, ε_{ox} is the oxide permittivity ($\varepsilon_{ox} = 3.9 \times 8.85 \times 10^{-12}$ F/m), and A is the gate area. Frequently, it is much more convenient to express the capacitance per unit area, in which case the oxide capacitance (per unit area) is given by

$$C_{ox} = \frac{\varepsilon_{ox}}{t_{ox}} \tag{2.60}$$

where C_{ox} is in F/m². As long as the capacitor is in accumulation mode, the capacitance is equal to C_{ox}, which means that it does not depend on the voltage applied.

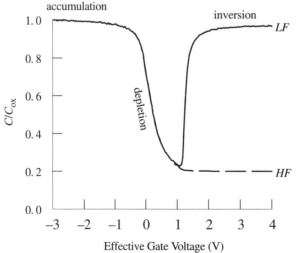

Figure 2.13 MOS capacitor C–V curves: solid line, low frequency (quasistatic); dashed line, high frequency.

Depletion

When a positive voltage is applied to the gate of the MOS capacitor, the holes are repelled from the surface, which creates a depletion layer at the surface. The capacitor conductive plates are separated by the oxide plus the depletion layer, which means that the capacitance is smaller than the oxide capacitance C_{ox}. The MOS capacitor is said to be in *depletion mode*. The depletion-layer width depends on the voltage applied (a higher gate voltage corresponds to a wider depletion layer as more holes are repelled from the surface). As a consequence, the capacitance depends on the voltage applied, where a higher gate voltage corresponds to a smaller capacitance.

Inversion

If the magnitude of the positive gate voltage is increased beyond a certain value, the electric field penetrating through the gate oxide becomes strong enough to attract electrons (the minority carriers) to the semiconductor surface. This is referred to as *inversion mode*. If the gate voltage is changed slowly (low-frequency signal applied to the gate), the electrons at the surface of the semiconductor follow the change in the positive gate charge. The depletion layer does not influence the capacitance effect, thus, the gate oxide represents the effective capacitor dielectric. This means that the capacitance is again equal to C_{ox} and independent on the gate voltage. The electrons coming to the surface of the semiconductor as a response to the gate voltage changes are thermally generated (electrons in the P-type substrate are the minority carriers existing due to the process of thermal generation). The process of the thermal generation of electrons is, however, very slow. Because of that, the electron density at the semiconductor surface does not change if a high-frequency (>100 Hz) voltage signal is applied to the gate. In this case, the electrons at the semiconductor surface are ineffective, as they do not respond to the high-frequency signal. The high-frequency capacitance of an MOS capacitor in the inversion mode remains smaller than C_{ox}, as determined by the sum of the oxide thickness and the depletion-layer width.

2.5.2 Thermal Oxide: Growth and Properties

The properties of silicon dioxide, used as the MOS capacitor dielectric, are fundamental to the success of silicon integrated-circuit technologies. If it were not for the properties of the silicon dioxide, the advantages of silicon-based technologies over other semiconductor materials would probably be lost. In fact, the silicon dioxide–silicon interface (or shortly oxide–silicon interface) is the only known interface that is good enough to enable operation of MOSFETs to industrial standards.

Basic Properties of the Oxide and the Oxide–Silicon Interface

The properties of the silicon dioxide films are summarized in Table 2.3. A two-dimensional chemical-bond model of the oxide–semiconductor interface is given in Fig. 2.14a. Although the oxide is not a crystal, the silicon and oxygen atoms are packed in an orderly manner: each silicon atom is bonded to four oxygen atoms and each oxygen atom is bonded to two silicon atoms. Cells formed by one silicon atom and the four surrounding oxygen atoms have tetrahedral shapes in reality (three dimensions) as explained in Table 2.3. The energy-band model of the oxide–silicon interface is shown in Fig. 2.14b. The energy gap of the oxide is about 9 eV. This value places the oxide among very good insulators. It is more than eight times larger than the energy gap of silicon (1.12 eV). This disproportion in the energy gaps of the oxide and the silicon means that there must be a discontinuity in the energy bands at the oxide–silicon interface. Figure 2.14b illustrates that the electrons in the silicon face a barrier of $q\phi_B = 3.2$ eV when they move toward the oxide silicon interface. As for the holes, their barrier is even higher: $9 - 3.2 - 1.12 \approx 4.7$ eV. These barriers are high enough to prevent any motion of electrons or holes from the silicon into the oxide in normal conditions.

Interface Traps and Oxide Charge

Practically, the oxide–silicon interface is not limited to one atomic layer, as the transition from the silicon to the oxide develops over a number of atomic layers. Although Fig. 2.14 does not express this fact, it does illustrate the most frequent defects appearing at the interface and in the oxide.

To begin with, the average distance between the oxygen atoms is larger than the average distance between the silicon atoms in the silicon, which means that some of the interface atoms from the silicon will inevitably miss oxygen atoms to create Si–O bonds. Atoms from the silicon that remain bonded only to three silicon atoms with the fourth bond unsaturated (trivalent interfacial silicon atoms) represent interface defects. The energy levels associated with the fourth unsaturated bond of the trivalent silicon atoms do not appear in the conduction or the valence band, but rather in the silicon energy gap. It is believed that every trivalent silicon atom introduces a pair of energy levels; one can be occupied by an electron (acceptor type) and the other can be occupied by a hole (donor type). Electrons and holes that appear on these levels cannot move freely as there is a relatively large distance between the neighboring interfacial trivalent silicon atoms (these levels are localized and isolated from each other). As these levels can effectively trap the mobile electrons and holes (from the conduction and valence bands, respectively), they are called *interface traps*. Impurity atoms and groups (such as H, OH, and N) can be bonded to the unsaturated bonds of the interfacial trivalent silicon atoms, which results in a shift of the corresponding energy levels into the conduction and the valence bands (defect B in Fig. 2.14). Although this process effectively neutralizes the interface traps, it is not possible to enforce such a saturation of all the interfacial trivalent silicon atoms, which means that the density of the interface traps can never be reduced to

TABLE 2.3 Properties of Thermally Grown SiO_2

Structure	Amorphous silica in which Si atoms are surrounded tetrahedrally by four O atoms: Si–O distances vary from 0.152 to 0.169 nm, Si–O–Si angles vary from 120° to 180°, and the O–Si–O angle is about 109.5°.
Dielectric constant	3.9
Dielectric strength	$\approx 10^7$ V/cm
Energy gap	≈ 9 eV
Specific resistivity	10^{12}–10^{16} Ω cm

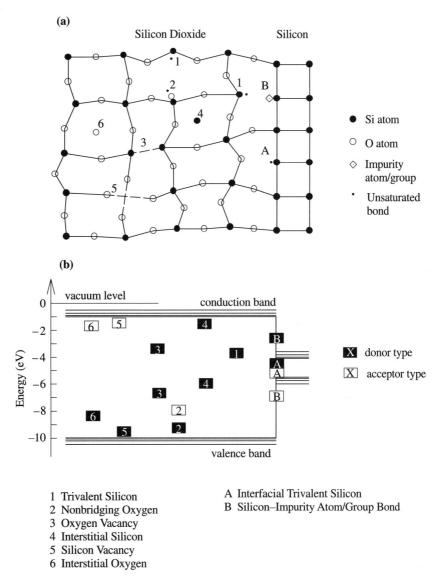

Figure 2.14 Illustration of the oxide–silicon interface and the associated defects: (a) a two-dimensional chemical-bond model and (b) the energy-band model.

zero. The interface trap density will be denoted by N_{it} to express the number of interface trap per unit area (in m^{-2}), or by $q N_{it}$ to express the associated charge per unit area (in $C\, m^{-2}$).

Trivalent silicon atoms can also appear in the oxide—these are silicon atoms bonded to three neighboring oxygen atoms with the fourth bond unsaturated (defect 1 in Fig. 2.14). There is also a number of other possible defects in the oxide: nonbridging oxygen, oxygen vacancy, interstitial silicon, silicon vacancy, and interstitial oxygen. All these defects are illustrated in Fig. 2.14 as well. The oxide defects introduce energy levels in the oxide energy gap, which can trap electrons and holes. The charge due to the trapped electrons and holes onto the oxide defects is referred to as the *oxide charge*. Although the oxide traps do not continuously exchange the electrons and holes with the silicon, the oxide charge does affect the electrons and holes in the silicon by its electric field. In general the oxide charge is usually positive and is mostly located close to the oxide–silicon interface. The density of oxide charge will be labeled by N_{oc} to express the number of charge centers per unit area (in m^{-2}) or by $q N_{oc}$ to express it in $C\ m^{-2}$.

Oxide Growth and Local Oxidation (LOCOS)

It can be imagined that the density of the interface traps and oxide charge is largely dependent on the processing conditions. Although it is possible to deposit the oxide film onto the silicon surface, such a process does not provide a good enough oxide–silicon interface to be used in MOS capacitors. The density of interface traps in this case may exceed the density of electrons–holes that can ever be attracted to the surface. This means that the interface traps would make the appearance of any significant amount of the free carriers at the silicon surface impossible.

A high-quality oxide–silicon interface can be achieved if the oxide is thermally grown on the silicon surface. When the silicon is exposed to oxygen or water vapor at high temperature (around 1000°C), silicon dioxide is created through the following reactions:

$$Si + O_2 \Rightarrow SiO_2 \qquad \text{Dry Oxidation}$$

$$Si + 2H_2O \Rightarrow SiO_2 + 2H_2 \qquad \text{Wet Oxidation} \qquad (2.61)$$

This process of thermal oxidation, when conducted in an ultrapure atmosphere and after sophisticated cleaning of the silicon surface, produces a high-quality oxide–silicon interface. The interface traps density is in the order of $10^{10}\ cm^{-2}$. This has proved sufficient for the integrated circuits that use gate oxides thicker than 5 nm. Continuing development of integrated-circuit technology, necessitating the development of sub-5 nm gate oxides, has led to research into the possibility of further improving the quality of the oxide–silicon interface. It appears that the solution for sub-5-nm gate dielectrics will be found in some kind of nitrogen-enriched silicon dioxides, as it has been demonstrated that the nitrogen neutralizes a number of defects in the oxide and at the oxide–silicon interface.

The oxidation reaction takes place at the interface, which means that after a layer of the oxide is created, the oxygen or the water molecules have to diffuse through this already created layer to interact with the silicon. As a consequence, the growth rate is slowed down as the oxide thickness is increased. The growth rate is very dependent on the oxidation temperature, and is also different for the dry (O_2) and the wet (H_2O) processes. An increase in the oxidation temperature remarkably increases the growth rate, which at any temperature is higher for wet processing.

Silicon nitride (Si_3N_4) can be used to protect parts of the silicon surface against oxidation, as the silicon nitride represents a very efficient barrier against the diffusion of oxygen and water molecules. Figure 2.15 illustrates the sequence of process steps used to locally oxidize the silicon surface in order to create the field oxide. First, the silicon nitride layer is deposited and patterned to open windows in the areas that are to be oxidized (Fig. 2.15a). Usually a very thin buffer silicon dioxide layer is placed between the silicon nitride and the silicon surface (not shown in the figure).

After that, thermal oxidation is performed to create the field oxide of desired thickness, as illustrated in Fig. 2.15b. This figure also illustrates the fact that approximately half of the created oxide grows at the expense of the silicon, while the other half builds up at the top of the original surface. This is because silicon atoms are used to create the oxide (silicon is consumed) and oxygen atoms are incorporated into the oxide film (oxide rises above the original silicon surface). The precise ratio between the depth of the consumed silicon and the total oxide thickness is 0.46. This is a useful property as thick isolation field oxide can be created without the need to have as large oxide-to-silicon steps, which is hard to reliably cover by thin metal films. In addition, some lateral oxidation occurs, which smooths the step, making it easier to cover by the metal film used for contacts and interconnections.

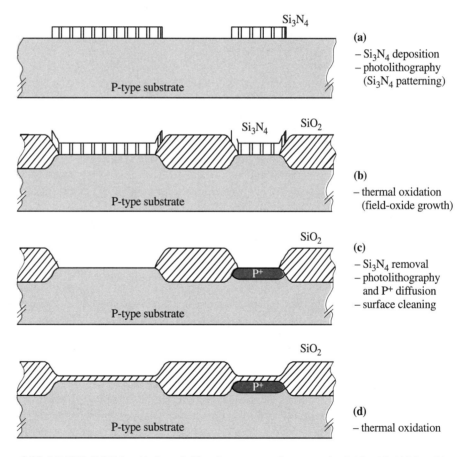

(a)
– Si_3N_4 deposition
– photolithography
  (Si_3N_4 patterning)

(b)
– thermal oxidation
  (field-oxide growth)

(c)
– Si_3N_4 removal
– photolithography
  and P+ diffusion
– surface cleaning

(d)
– thermal oxidation

Figure 2.15 LOCOS (LOCal oxidation of silicon) process used to create the field oxide (thick oxide surrounding the active areas of the thin gate oxide).

Once the field oxide is grown, the silicon nitride is removed. To create the MOS capacitor structure from Fig 2.12, the P^+ area in the silicon is created (using appropriate photolithography and diffusion process steps), and the surface is thoroughly cleaned (Fig. 2.15c) in preparation for thermal oxidation, which will create the thin gate oxide (Fig. 2.15d). The final step would be contact window opening, metal deposition, and patterning by appropriate photolithography processing to obtain the structure from Fig. 2.12.

The process of thermal oxidation cannot be used to create a gate oxide of sufficient quality with GaAs substrates. This is because the quality of the native oxide of GaAs is not good enough to be used as a gate dielectric. Deposition of silicon dioxide onto the GaAs substrate creates a high density of interface traps. These facts practically prevent implementation of the MOS capacitor with GaAs substrates.

■ **Example 2.5 Oxide Growth Kinetics**

The dependence of the thermal oxide thickness (t_{ox}) on the oxidation time and temperature is frequently modeled by the following equation:

$$\frac{t_{ox}}{A/2} = \sqrt{1 + \frac{t + \tau}{A^2/4B}} - 1$$

where A, B, and τ are temperature-dependent coefficients. The values of the coefficients A, B, and τ are given in Tables 2.4 and 2.5, respectively (*Source:* L.E. Katz, "Oxidation," in *VLSI Technology*, S.M. Sze, ed., McGraw-Hill, New York, 1983, pp. 131-167).

How long would it take to grow 0.5 μm of SiO_2 at 920°C in a wet atmosphere? If dry oxidation is applied, what would be the oxide thickness? Comment on the difference. Repeat the above calculations for 1000°C and comment on the results.

Solution: The oxidation time t can be expressed from the model given in the text of the example as

$$t = \left[\left(\frac{t_{ox}}{A/2} + 1 \right)^2 - 1 \right] \frac{A^2}{4B} - \tau$$

The values of the parameters are found in Table 2.4 in the row corresponding to $T = 920$°C. The calculated time is $t = 2.46$ h. If the dry oxidation process is used at the same temperature, the appropriate parameters are found in Table 2.5 in the row for $T = 920$°C. Putting these parameters and $t = 2.46$ h into the equation given in the text of the example, the oxide thickness is calculated to be $t_{ox} = 0.063$ μm $= 63$ nm. The oxide grows much faster in the wet than in the dry ambient.

For the case of $T = 1000$°C, the time of wet oxidation needed to grow 0.5 μm of oxide is found to be $t = 1.26$ h. The same time and temperature used in the process of dry oxidation would grow $t_{ox} = 78$ nm. When the temperature is increased from 920 to 1000°C, the time required to grow 0.5 μm of "wet" oxide is approximately halved. The "dry" oxide grown for the same time is again much thinner, but somewhat thicker than that grown for twice as long at 920°C.

TABLE 2.4 Rate Constants for Wet Oxidation of Silicon

Oxidation Temperature (°C)	A (μm)	Parabolic Rate constant B (μm²/h)	Linear Rate constant B/A (μm/h)	τ (h)
1200	0.05	0.720	14.40	0
1100	0.11	0.510	4.64	0
1000	0.226	0.287	1.27	0
920	0.50	0.203	0.406	0

TABLE 2.5 Rate Constants for Dry Oxidation of Silicon

Oxidation Temperature (°C)	A (μm)	Parabolic Rate Constant B (μm²/h)	Linear Rate Constant B/A (μm/h)	τ (h)
1200	0.040	0.045	1.12	0.027
1100	0.090	0.027	0.30	0.076
1000	0.165	0.0117	0.071	0.37
920	0.235	0.0049	0.0208	1.40
800	0.370	0.0011	0.0030	9.0
700	—	—	0.00026	81.0

2.5.3 Zero Bias and Flat Bands: Definition of the Fundamental MOS-Related Terms

Figure 2.16a illustrates an MOS capacitor with a P-type substrate and an N^+-type polysilicon gate when all the charges are compensated, that is no net charge appears at the capacitor plates. The holes in the P-type substrate are compensated by the negative acceptor ions and the minority electrons, whereas the electrons in the N^+-type gate are compensated by the positive donor ions and the minority holes. It might appear that this is the condition of no voltage applied at the capacitor, however, an examination of the band diagram would indicate that this is not the case.

To construct the energy-band diagram of the MOS capacitor, the procedure used for the case of the P–N junction (Section 2.2.3) has to be developed further to account for the existence of the oxide appearing between the P- and N^+-type silicon regions:

1. As before, the Fermi level lines are drawn first (dashed–dotted lines in Fig. 2.16):
    1.1 If the system is in thermal equilibrium (zero-bias applied), the Fermi level is constant throughout the system. The Fermi level lines in the substrate and the gate have to be matched, as in Fig. 2.16b.
    1.2 If a voltage is applied between the gate and the substrate, the Fermi level (or more precisely, the quasi-Fermi level) lines should be split to express this fact, like in Fig. 2.16d.
2. As in the case of the P–N junction, the conduction and valence bands are drawn for the P-type and N-type neutral regions (away from the oxide–silicon interfaces). The bands are placed appropriately with respect to the Fermi level, so to express

the band diagrams of the P-type and N-type silicon, respectively. This is illustrated in Fig. 2.16b.

3. In the case of the P–N junction, the conduction and the valence band levels would simply be joined by sloped lines to complete the diagram. The oxide that appears between the P- and N$^+$-type regions in the case of the MOS capacitor has a much larger energy gap ($E_g = E_C - E_V$) than the silicon regions. Figure 2.14b illustrates the existence of conduction-band and valence-band discontinuities at the oxide–silicon interface. These discontinuities have to be expressed in the band diagram. The band diagram around the oxide is constructed in the following way (refer to Fig. 2.16c):

3.1 The bands are bent in the P- and N$^+$-type silicon regions toward each other, but they do not come to the same level. There is a difference between the bands (say the mid-gap level E_i) at the N$^+$-gate–oxide interface and at the oxide–P-type substrate interface. This difference expresses the voltage drop across the oxide.

3.2 Lines expressing the discontinuities of the conduction and the valence bands at the oxide–silicon interfaces are drawn. The conduction-band discontinuity is $q\phi_B = 3.2$ eV and the valence-band discontinuity is about 4.7 eV (this makes an oxide energy gap of about $3.2 + 4.7 + 1.12 \approx 9$ eV).

3.3 The conduction and valence bands in the oxide are drawn with straight lines, to indicate that the electric field in the oxide is constant.[3] It is assumed that there is no built-in charge in the bulk of the oxide, which would enforce change in the electric field in the oxide (the oxide charge is modeled as a sheet charge appearing along the oxide–silicon interface).

Fig. 2.16b and c illustrates that there is a potential difference between the N$^+$-type gate and the P-type substrate at zero bias. This is analogous to the potential difference seen in the case of the P–N junction. Although this potential difference in the case of the P–N junction is referred to as the built-in voltage (V_{bi}), in the case of the MOS capacitor it is expressed as the *work-function difference* ($q\phi_{ms}$). This potential difference is related to the existence of the electric field in the oxide and around the interfaces, which further means that there must be some uncompensated charge at the capacitor plates associated with this field. Figure 2.16c illustrates that this charge is due to uncompensated negative acceptor ions on the P-type substrate side, and to uncompensated positive donor ions on the N$^+$-type gate side. The amount of this charge is obviously related to the strength of the electric field, and further to the value of the work-function difference. The work-function difference appears as an important MOS capacitor parameter.

The work function was originally defined for metals, and its physical meaning is best explained for the case of metals. In metals, there is no energy gap and the electrons are found around the Fermi level.

The **work function** is the energy (work) that is needed to remove an average electron (an electron at the Fermi level) from the metal. Thus, the work function expresses the position of the Fermi level with respect to the energy of a free electron in vacuum (vacuum level).

---

[3]Constant electric field E corresponds to a linear electric potential, thus potential energy ($E = -d\varphi/dx \propto dE_{pot}/dx$).

(a) flat bands

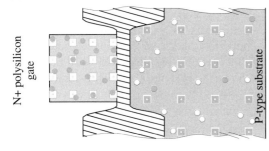

(b) zero bias

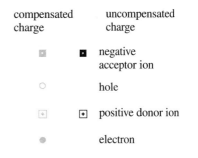

compensated charge | uncompensated charge

■ negative acceptor ion

○ hole

• positive donor ion

● electron

(c) zero bias

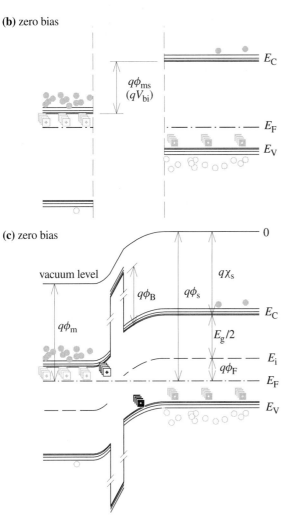

(d) flat bands ($V_G = V_{FB}$)

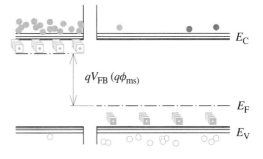

Figure 2.16 Illustration of fundamental MOS-related terms: (a) MOS capacitor cross section, illustrating the existing types of charge, (b) the starting point in construction of the MOS energy-band diagram, (c) the energy-band diagram at zero-bias condition, and (d) the energy-band diagram at flat-band condition.

When dealing with semiconductors, the same definition of the work function has to be used, regardless of the fact that, frequently, there are no electrons at and around the Fermi level. The work functions for different metals are known, and some of them are given in Table 2.6. In the case of semiconductors, the work function depends on the doping type and level, as the position of the Fermi level is changed with the doping. However, the doping does not influence the position of the bands with respect to the vacuum level.

The position of the bottom of the conduction band with respect to the vacuum level is referred to as **electron affinity**, labeled by $q\chi_s$. The electron affinity is equal to the energy needed to remove a free electron with zero kinetic energy (an electron at the bottom of the conduction band) from the semiconductor.

The electron affinities of some important materials are also given in Table 2.6.

Considering the energy-band diagram, as shown in Fig. 2.16c, the work function of a semiconductor ($q\phi_s$) can be related to the electron affinity $q\chi_s$:

$$q\phi_s = q\chi_s + \frac{E_g}{2} + q\phi_F \tag{2.62}$$

where $E_g/2$ is the half-value of the energy gap and $q\phi_F$ is the difference between the mid-gap E_i and the Fermi level E_F.

ϕ_F is referred to as the **Fermi potential**. It is directly related to the difference between the mid-gap and the Fermi level ($q\phi_F = E_i - E_F$). The Fermi potential is an important semiconductor parameter as it expresses the doping type and the level.

The Fermi potential is zero in intrinsic semiconductors as $E_F = E_i$. With the doping, the Fermi level is moved away of the mid-gap, increasing the value of the Fermi potential. It will be shown in Example 2.6 that the Fermi potential is related to the doping level in the following way:

TABLE 2.6 Work Functions and Electron Affinities of the Most Frequently Used Materials at 300 K

Material	Work Function $q\phi_m$ (eV)	Electron Affinity $q\chi_s$ (eV)
Al	4.1	
Pt	5.7	
PtSi	5.4	
W	4.6	
WSi$_2$	4.7	
Si		4.05
GaAs		4.07
Ge		4.0
SiO$_2$		1
N$^+$-Si	4.05	4.05
P$^+$-Si	5.17	4.05

$$\phi_F = \begin{cases} +\frac{kT}{q} \ln \frac{N_A}{n_i} & \text{for P-type} \\ -\frac{kT}{q} \ln \frac{N_D}{n_i} & \text{for N-type} \end{cases} \tag{2.63}$$

where N_A and N_D are the acceptor and donor concentrations, respectively, and n_i is the intrinsic carrier concentration ($n_i = 1.02 \times 10^{10}$ cm^{-3} for silicon at 300 K).

Using Eq. (2.62), the work-function difference can be expressed as

$$q\phi_{ms} = q\phi_m - q\phi_s = q\phi_m - q \left(\chi_s + \frac{E_g}{2q} + \phi_F \right) \tag{2.64}$$

To calculate the work-function difference, the Fermi potential is calculated first for the given doping level (using Eq. 2.63) and, according to Eq. (2.64), combined with the values of $q\chi_s$, E_g, and $q\phi_m$ corresponding to the materials used (Table 2.6). In the case of metal gates, the value of $q\phi_m$ is directly obtained. When silicon gate is used, it may appear necessary to calculate $q\phi_m$ from Eq. (2.62) using an appropriate doping level. Very frequently, however, the gates are very heavily doped (to provide as close an emulation of the metal properties in terms of conductivity as possible), which means the Fermi level is either very close to the bottom of the conduction band (N$^+$-type gate) or very close to the top of the valence band (P$^+$-type gate). Therefore, as given in Table 2.6, the work function of a heavily doped polysilicon gate can be approximated by $q\chi_s$ in the case of N$^+$-type doping and by $q\chi_s + E_g$ in the case of P$^+$-type doping.

It is not very convenient to take the zero-bias condition, considered so far, as a reference point when considering the capacitance–voltage dependence of the MOS capacitor. This is due to the existence of the work-function difference, which creates a built-in field and uncompensated charges. The capacitor plates are not discharged at the zero bias. To achieve the zero-charge condition, the field in the oxide and around the oxide–silicon interfaces should be zero. Zero electric field means no change in the electric potential ($E = -d\varphi/dx \propto dE_{pot}/dx$), in other words *flat bands*. To flatten the bands, it is necessary to split the Fermi levels in the gate and the substrate by an amount that will compensate for the work-function difference, as illustrated in Fig. 2.16d. The Fermi levels are split by applying a voltage between the gate and the substrate.

The voltage needed between the gate and the substrate to flatten the bands is called the *flat-band voltage* (V_{FB}).

From Fig. 2.16d, the flat-band voltage is $V_{FB} = \phi_{ms}$. It is important to note, however, that possible effects of the oxide charge are not shown in Fig. 2.16d. As described in the previous section, there is a charge in the oxide that is typically located close to the oxide–silicon interfaces. The charge that appears close to the oxide–gate interface does not influence the MOS capacitor properties significantly, as it is easily compensated by the charge from the heavily doped silicon gate. The charge sheet appearing close to the oxide–silicon substrate interface, however, influences the mobile carriers in the substrate by its electric field. The electric field of this oxide charge is able to produce significant band bending in the surface area of the silicon substrate. To bring the bands in a flat condition, this field should be compensated by an appropriate gate voltage, as well. In other words, it is necessary to apply a gate voltage to remove any charge attracted to the substrate surface by the oxide charge. If the gate-oxide capacitance is C_{ox}, and the density of the charge is qN_{oc}, the voltage needed is qN_{oc}/C_{ox}. A negative voltage is needed in the case of positive oxide charge (as it would repel the electrons attracted to the surface by the positive oxide charge), and, analogously, a positive voltage is needed in the case of negative oxide charge.

Combining the effects of the work-function difference (ϕ_{ms}) and the oxide charge ($-qN_{oc}/C_{ox}$), the flat-band voltage is expressed as

$$V_{FB} = \phi_{ms} - \frac{qN_{oc}}{C_{ox}} \qquad (2.65)$$

As the capacitor plates are discharged (the bands flat) when $V_G - V_{FB} = 0$, the voltage $V_G - V_{FB}$ will be called the *effective gate bias*. The following two sections consider the influence of different effective gate biases on the capacitance of the MOS structure.

■ **Example 2.6 Fermi Potential (ϕ_F)**

Express the Fermi potential $\phi_F = (E_i - E_F)/q$ in terms of the doping concentrations N_A (N_D) and the intrinsic carrier concentration n_i.

Solution: Consider the case of P-type silicon. Equation (1.67) relates the Fermi level E_F to the doping concentration N_A. From Eq. (1.67),

$$E_V - E_F = kT \ln \frac{N_A}{N_V}$$

To eliminate E_V and involve E_i, it should be observed that $E_i - E_V = E_g/2$, where $E_g/2$ is half the value of the energy gap. The energy gap is related to the intrinsic concentration n_i [Eq. (1.69)], which means $E_i - E_V$ can be expressed, as needed, in terms of n_i. Assuming that $N_C \approx N_V$ in Eq. (1.69), $E_i - E_V$ is obtained as

$$E_i - E_V = \frac{E_g}{2} = kT \ln \frac{N_V}{n_i}$$

Adding the two obtained equations to each other provides the equation for $E_i - E_F$:

$$E_i - E_F = kT \ln \frac{N_A}{n_i}$$

Finally, the Fermi potential is expressed as

$$\phi_F = \frac{E_i - E_V}{q} = \frac{kT}{q} \ln \frac{N_A}{n_i}$$

A similar procedure can be applied to derive the Fermi potential for the case of N-type silicon. In that case the mid-gap line E_i is below the Fermi level, and the Fermi potential appears as negative:

$$\phi_F = \frac{E_i - E_V}{q} = -\frac{kT}{q} \ln \frac{N_D}{n_i}$$

2.5.4 MOS Capacitor in Accumulation, Depletion, and Weak Inversion

Accumulation Mode

Consider first the case of negative effective gate bias, $V_G - V_{FB} < 0$. The negative effective gate bias produces an electric field that attracts holes to the surface of the silicon substrate.

The condition in which the gate voltage attracts the majority carriers to the surface of the silicon substrate is known as the *accumulation mode*.

The density of holes in the surface layer of the silicon substrate is exactly matched by the density of electrons at the gate, induced by the negative gate bias applied. This is illustrated in Fig. 2.17a. The appearance of extra holes in the surface region of the silicon substrate means that the Fermi level in the surface region is closer to the top of the valence band than in the bulk. The energy bands are, therefore, bent upward going from the silicon substrate toward the gate, as shown in Fig. 2.17b. This band bending is due to the difference between the energy bands in the bulk of the silicon and the gate, which is directly related to the effective bias applied ($q V_G - q V_{FB}$). The amount of band bending in the silicon substrate, which is labeled by $q\varphi_s$, is very important in terms of the MOS capacitor properties.

φ_s is the **surface potential** with respect to the potential of the bulk of the silicon substrate, which is taken as the reference level.

The gradient in the energy bands around the oxide–silicon interface and in the oxide expresses the existence of an electric field in the direction from the substrate toward the gate ($E = -d\varphi/dx \propto dE_{pot}/dx$). This is the field that keeps the excess holes in the substrate and the excess electrons in the gate close to the oxide–silicon interface.

Any change in the gate voltage ΔV_G will inevitably produce a change in the band bending. For example, if the voltage is decreased by ΔV_G, the bending is further increased, which means that the Fermi level at the surface of the silicon substrate is moved a bit closer to the top of the valence band. Thus, the density of the holes along the oxide–silicon interface is increased. As the Fermi level is very close to the top of the valence band, only a slight shift in the top of the valence band toward the Fermi level (a slight reduction in the surface potential) is needed to significantly increase the density of the populated energy levels in the valence band by new holes. This is due to the fact that the occupancy probability of the levels in the valence band by holes increases exponentially as the Fermi level moves toward the valence band (Fig. 1.27 in Section 1.5.2). As the change in the surface potential is much smaller than the gate voltage change ΔV_G, almost the entire gate voltage change appears across the oxide. This situation is very much like the situation of a capacitor with metal plates separated by a dielectric equivalent to the gate oxide.

In *accumulation mode*, the surface potential change is negligible compared to the gate voltage changes, which means that a significant part of any gate voltage change appears across the gate oxide. The capacitance per unit area is, therefore, equal to

(a)

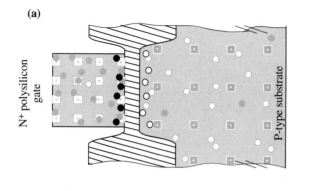

$V_G < V_{FB}$

compensated   uncompensated
charge   charge

▦		negative acceptor ion
○	○	hole
⊡		positive donor ion
●	●	electron

(b)

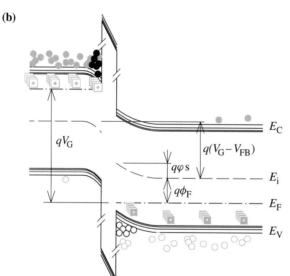

(c)

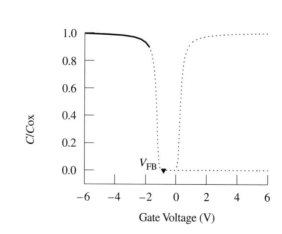

Figure 2.17 MOS capacitor in accumulation: (a) cross section illustrating the type of charge at the capacitor plates, (b) the energy-band diagram, and (c) C–V dependence.

$$C_{\text{ox}} = \frac{\varepsilon_{\text{ox}}}{t_{\text{ox}}} \tag{2.66}$$

Figure 2.17c illustrates that the capacitance of the MOS capacitor in accumulation, being equal to the gate oxide capacitance C_{ox}, is independent on the gate voltage.

Depletion and Weak Inversion

When relatively small positive effective gate bias ($V_G - V_{FB} > 0$) is applied, the electric field produced repels the holes from the surface, creating a depletion layer at the surface of the silicon substrate. The charge that appears at the substrate plate of the capacitor in this case is due to the uncompensated negative acceptor ions, as illustrated in Fig. 2.18a and b. Although these ions are immobile, gate voltage changes must produce related changes in the density of the negative-ion charge. For this to happen, the electric-field lines from the gate have to penetrate through the depletion layer either to further repel the holes (increase in the gate voltage) or to attract the holes back toward the surface (decrease in the gate voltage).

> In *depletion mode*, the surface potential φ_s closely follows the gate voltage changes. This means that almost the entire amount of any gate voltage change appears across the depletion layer at the surface of the semiconductor.

The depletion layer appears as a dielectric additional to the gate oxide. The capacitance associated with the depletion layer is completely analogous to the depletion-layer capacitance at the P–N junction. In this case the depletion-layer capacitance C_d is connected in series to the oxide capacitance C_{ox}, which means that the total MOS capacitance is given as

$$\frac{1}{C} = \frac{1}{C_{\text{ox}}} + \frac{1}{C_d} \tag{2.67}$$

The depletion-layer capacitance depends on the voltage applied—a larger gate voltage produces a wider depletion layer, which means the capacitance is smaller. As a consequence the overall capacitance is reduced when the gate voltage is increased. This mode of MOS capacitor operation is called *depletion mode*.

Figure 2.18c illustrates the reduction of the capacitance in the depletion mode, however, it also illustrates that the capacitor behavior changes as the voltage is further increased. There is a characteristic point that can be explained using the energy-band diagram of Fig. 2.18b. The gate voltage increase above the flat-band voltage level produces band bending in the direction that increases the differences between the top of the valence band and the Fermi level. This is the condition associated with the reduction (and eventual elimination) of the holes from the surface. As a consequence of this band bending, the mid-gap line E_i approaches the Fermi level. The characteristic point mentioned is when the mid-gap line is exactly at the Fermi level at the silicon surface (the situation illustrated in Fig.2.18b). At this point, labeled as V_{MG}, the surface of the silicon substrate is in the intrinsic silicon condition. When the band bending is larger than this (due to a gate voltage larger than V_{MG}),

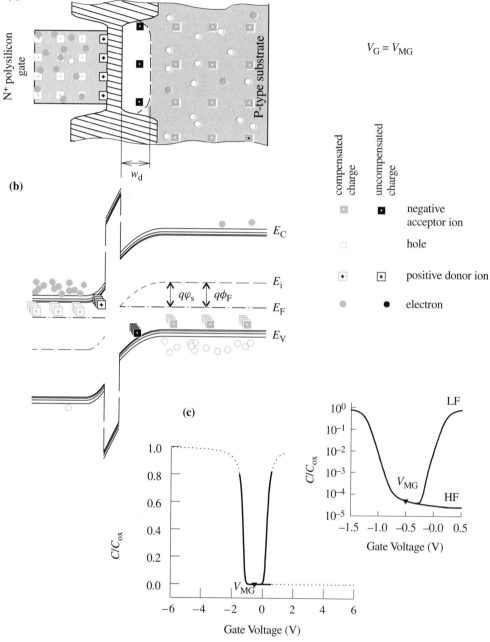

(a)

$V_G = V_{MG}$

N+ polysilicon gate

P-type substrate

w_d

(b)

E_C

E_i

$q\varphi_s$ $q\phi_F$

E_F

E_V

	compensated charge	uncompensated charge	
	▦	■	negative acceptor ion
	○		hole
	⊞	⊡	positive donor ion
	●	●	electron

(c)

C/C_{ox}

1.0
0.8
0.6
0.4
0.2
0.0

V_{MG}

−6 −4 −2 0 2 4 6

Gate Voltage (V)

C/C_{ox}

10^0
10^{-1}
10^{-2}
10^{-3}
10^{-4}
10^{-5}

LF

V_{MG}

HF

−1.5 −1.0 −0.5 −0.0 0.5

Gate Voltage (V)

Figure 2.18 MOS capacitor at the mid-gap point: (a) cross section illustrating the type of charge at the capacitor plates, (b) the energy-band diagram, and (c) C–V dependence.

the mid-gap line E_i crosses the Fermi level at some point, making the Fermi level closer to the bottom of the conduction band than to the top of the valence band. This means that the concentration of the electrons (minority carriers in the P-type substrate) is larger than the concentration of holes at the silicon surface—an *inversion layer* is created at the surface.

The appearance of some mobile charge at the silicon surface means that some of the electric field lines do not need to penetrate through the depletion layer to change the charge as a response to a gate voltage variation. The capacitance increases as the density of the mobile charge (electrons in this case) is increased by the gate voltage. This mode is referred to as *weak inversion*.

It is important to note, however, that the electrons in the inversion layer (minority carriers in the P-type silicon) are due to thermal generation of electron–hole pairs. As the thermal generation is a very slow process, the electrons are unable to respond to fast changes in the gate voltage. The described increase in the capacitance cannot be observed when the frequency of the gate signal is higher than 100 Hz. Figure 2.18c shows that the high-frequency capacitance remains low in the weak inversion mode, as determined by the largest achieved depletion layer-width.

2.5.5 MOS Capacitor in Strong Inversion: Threshold Voltage

The behavior of the MOS capacitor in so-called *strong-inversion mode* is of great importance in view of the MOSFET operation. To understand the difference between the weak and strong inversion, it is necessary to refer to the fact that the tail of the Fermi–Dirac distribution shows an exponential increase in the probability of electrons appearing at the conduction band levels as the difference between the bottom of the conduction band E_C and the Fermi level E_F is reduced.[4] It is also useful to keep in mind that any increase in the gate voltage (ΔV_G) has to be accompanied by a corresponding increase in electron density in the inversion layer (ΔQ_I).

When the Fermi level E_F is not very close to the bottom of the conduction band (weak-inversion mode), the occupancy probability of the electron levels in the conduction band is very small. This means that the electron density increase ΔQ_I, necessary as a response to a gate voltage increase ΔV_G, can be achieved only by a significant band bending to reduce the difference between E_C and E_F. The increased band bending corresponds to a related increase in the surface potential φ_s. A significant part of the increased voltage drop between the gate and the substrate appears between the silicon surface and the bulk of the substrate (the surface potential increase $\Delta \varphi_s$).

As the $E_C - E_F$ difference is reduced, the probability of the electron-level occupancy in the conduction band rises exponentially. This means that a significant increase in ΔQ_I can now be achieved only by a slight reduction of the $E_C - E_F$ difference. The related slight change in the surface potential $\Delta \varphi_s$ is much smaller than the gate voltage change ΔV_G, which means that the increased gate voltage drop (ΔV_G) appears mostly across the gate oxide.

---

[4]The Fermi–Dirac distribution and energy band population are considered in Section 1.5.2, including Fig. 1.27.

In *strong inversion*, the gate voltage changes do not produce significant surface potential changes. It is said that the surface potential φ_s is pinned to a value that is, for convenience, assumed to be equal to $2\phi_F$. Thus, in strong inversion

$$\boxed{\varphi_s \approx 2\phi_F}$$ (2.68)

The gate voltage needed to set the surface potential at $2\phi_F$ is defined as **threshold voltage V_T**.

Figure 2.19a and b illustrates the appearance of electrons at the silicon surface as a response to a gate voltage increase beyond the threshold voltage. The situation in which applied voltage variations produce related variations in the charge located along the oxide interfaces is like the situation in an ordinary metal-plate capacitor with a dielectric equivalent to the gate oxide. The overall MOS capacitance is equal to the gate oxide capacitance, and is therefore voltage independent. This is illustrated in Fig. 2.19c by the solid line labeled as LF. It is important to note that this conclusion is correct as far as the variations in the electron density at the silicon surface (the inversion layer) follow the variations of the gate voltage. At high frequencies of the gate voltage signal, this may not be the case. Although this reservation is not of great importance when the electrons are supplied to the inversion layer by an adjacent N^+-type region (the case of MOSFETs described in Chapter 5), it is essential for MOS capacitors when the source of the electrons is thermal generation. As mentioned, the process of thermal generation is very slow so that at frequencies as low as 100 Hz the electron density variations cannot follow the signal variation. At high frequencies, the electrons in the inversion layer do not influence the capacitance, as they do not respond to gate voltage variations. Figure 2.19c shows that the high-frequency capacitance remains low in the strong-inversion mode, as determined by the largest achieved depletion-layer width.

The onset of the strong-inversion mode is defined by the value of the threshold voltage. Therefore, the threshold voltage becomes an extremely important parameter, and it necessitates a detailed consideration. To begin with, it should be noted that the transition from a significant surface potential change, characteristic of the weak-inversion mode, to an insignificant surface potential change beyond the threshold voltage is by no means ideally abrupt. This observation is important so as not to mystify the meaning and the role of the threshold voltage, that is to clearly show the approximations involved in the concept of the threshold voltage.

The concept of the threshold voltage is used in order to make use of the following approximations:

- The inversion-layer charge is neglected for $V_G \leq V_T$. The only uncompensated charge in the silicon substrate is due to the doping ions in the depletion layer (Q_d).
- The surface potential φ_s and consequently the density of the depletion-layer charge Q_d are constant for $V_G > V_T$. The whole gate voltage increase beyond the threshold voltage ($V_G - V_T$) is spent on creating the inversion-layer charge Q_I:

$$Q_I = (V_G - V_T)C_{ox}$$ (2.69)

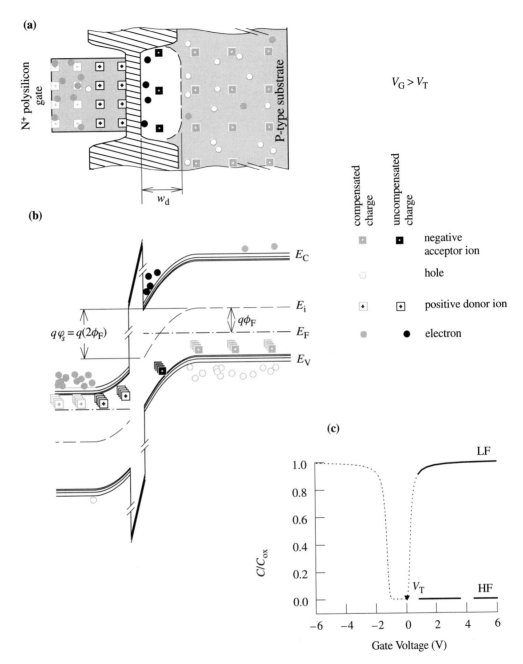

Figure 2.19 MOS capacitor in strong inversion: (a) cross section illustrating the types of charge at the capacitor plates, (b) the energy-band diagram, and (c) low-frequency (LF) and high-frequency (HF) C–V dependence.

The threshold voltage can be expressed in terms of the already introduced parameters of the MOS capacitor: the gate oxide capacitance C_{ox}, the flat-band voltage V_{FB}, the Fermi potential ϕ_F, and the substrate doping level N_A. This can be achieved by deriving an equation that relates the voltage drop across the gate oxide (V_{ox}) to the charge at the silicon-substrate plate of the capacitor ($Q_d + Q_I$):

$$V_{ox}C_{ox} = Q_d + Q_I \tag{2.70}$$

Note that the charge densities Q_d and Q_I are in C/m^2 and the oxide capacitance is in F/m^2 to obtain the voltage V_{ox} in V. The voltage drop across the oxide V_{ox} is given as the difference between the effective gate voltage $V_G - V_{FB}$ on one side of the oxide and the surface potential φ_s on the other side of the oxide. The effective gate voltage has to be used to properly express the fact that $V_{ox} = 0$ (and thus $Q_d = Q_I = 0$) at the flat-band condition (Fig. 2.16d) when $V_G - V_{FB} = 0$ and $\varphi_s = 0$. Therefore,

$$[(V_G - V_{FB}) - \varphi_s]C_{ox} = Q_d + Q_I \tag{2.71}$$

At the onset of strong inversion, the gate voltage is equal to the threshold voltage ($V_G = V_T$), the surface potential is $\varphi_s = 2\phi_F$ [Eq. (2.68)], and the inversion-layer charge is $Q_I = 0$. Applying these conditions to Eq. (2.71), the threshold voltage is obtained as

$$V_T = V_{FB} + 2\phi_F + \frac{Q_d}{C_{ox}} \tag{2.72}$$

This form of the threshold voltage equation involves the density of the depletion-layer charge Q_d, which can be expressed in terms of the surface potential $2\phi_F$ and the doping level N_A. As every acceptor ion in the depletion layer (appearing as uncompensated charge center) carries a unit of negative charge, qN_A expresses the charge in the depletion layer in C/m^3. To express the depletion-layer charge density (in C/m^2), qN_A is multiplied by the depletion layer width w_d:

$$Q_d = qN_Aw_d \tag{2.73}$$

The depletion-layer width w_d can be obtained by solving the Poisson equation, as is done in the case of the abrupt P–N junction. The depletion layer in this case appears only in the P-type substrate, which is analogous to the situation of $N_D \gg N_A$ in the case of the abrupt P–N junction. Therefore, solving the Poisson equation for this case would lead to an equation that is similar to Eq. (2.33). The only difference is that the boundary conditions 0 and $V_{bi} + V_R$ used in the case of the P–N junction would appear as 0 and $2\phi_F$ in this case. Replacing $V_{bi} + V_R$ in Eq. (2.33) by $2\phi_F$, the depletion-layer width at the silicon surface is obtained as

$$w_d = \sqrt{\frac{2\varepsilon_s(2\phi_F)}{qN_A}} \tag{2.74}$$

Combining Eqs. (2.72)–(2.74), the threshold voltage can be expressed in the following form:

$$V_T = V_{FB} + 2\phi_F + \frac{\sqrt{2\varepsilon_s q N_A}}{C_{ox}}\sqrt{2\phi_F} \qquad (2.75)$$

Finally, note that a parameter γ, called the *body factor*, is frequently defined as

$$\gamma = \frac{\sqrt{2\varepsilon_s q N_A}}{C_{ox}} \qquad (2.76)$$

to simplify the threshold voltage equation to the following form:

$$\boxed{V_T = V_{FB} + 2\phi_F + \gamma\sqrt{2\phi_F}} \qquad (2.77)$$

For the case of capacitors made with an N-type silicon substrate, the depletion-layer charge is positive, as it originates from the uncompensated positive donor ions (concentration N_D). Negative voltage is needed at the gate to induce this charge, which means the term $Q_d/C_{ox} = q N_D w_d/C_{ox}$ should appear with a minus sign in the threshold voltage equation. In addition, the Fermi potential [as given by Eq. (2.63)] appears as negative. In the equation for the depletion layer width [Eq. (2.74)] it cannot be used as negative. Careful consideration of the boundary conditions for the electric potential in the process of solving the Poisson equation would indicate that the only thing that matters for w_d is the absolute value of $2\phi_F$. These considerations indicate that the threshold voltage in the case of an N-type silicon substrate has to be modified in the following way:

$$V_T = V_{FB} - 2|\phi_F| - \gamma\sqrt{2|\phi_F|} \qquad \text{N-type substrate} \qquad (2.78)$$

where the body factor γ is given in terms of the donor concentration:

$$\gamma = \frac{\sqrt{2\varepsilon_s q N_D}}{C_{ox}} \qquad (2.79)$$

The unit of the body factor is $V^{1/2}$. This parameter, as well as the whole threshold voltage equation, will be extensively used and discussed further in Chapter 5.

■ **Example 2.7 Calculating the Threshold Voltage**

Technological parameters of an MOS capacitor are given in Table 2.7, together with the values of the relevant physical parameters.

(a) Determine the value of the flat-band voltage.

(b) Calculate the charge density at the onset of strong inversion. What is the type and origin of this charge?

(c) Calculate the value of the body factor.

TABLE 2.7 MOS Technological Parameters

Parameter	Symbol	Value
Substrate doping concentration	N_A	7×10^{16} cm^{-3}
Gate oxide thickness	t_{ox}	30 nm
Oxide charge density	N_{oc}	10^{10} cm^{-2}
Type of gate		N$^+$-polysilicon
Intrinsic carrier concentration	n_i	1.02×10^{10} cm^{-3}
Energy gap	E_g	1.12 eV
Thermal voltage at room temperature	$V_t = kT/q$	0.026
Oxide permittivity	ε_{ox}	3.45×10^{-11} F/m
Silicon permittivity	ε_s	1.04×10^{-10} F/m

(d) Calculate the value of the threshold voltage.

(e) Calculate the charge density in the inversion layer at $V_G = 5$ V.

Solution:

(a) To calculate the flat-band voltage V_{FB} using Eq. (2.65), the work-function difference ϕ_{ms} is needed, and to obtain ϕ_{ms} using Eq. (2.64) the Fermi potential ϕ_F has to be determined first. Using Eq. (2.63),

$$\phi_F = V_T \ln \frac{N_A}{n_i} = 0.41 \text{ V}$$

Using this value of ϕ_F, and reading the values of ϕ_m and χ_s from Table 2.6, the work-function difference can be calculated using Eq. (2.64):

$$\phi_{ms} = \phi_m - \left(\chi_s + \frac{E_g}{2q} + \phi_F \right) = 4.05 - \left(4.05 + \frac{1.12}{2} + 0.41 \right) = -0.97 \text{ V}$$

Finding C_{ox} as $C_{ox} = \varepsilon_{ox}/t_{ox} = 3.45 \times 10^{-11}/30 \times 10^{-9} = 1.15 \times 10^{-3}$ F/m^2, the flat-band voltage is calculated using Eq. (2.65):

$$V_{FB} = \phi_{ms} - \frac{q N_{oc}}{C_{ox}} = -0.96 - \frac{1.6 \times 10^{-19} \times 10^{14}}{1.15 \times 10^{-3}} = -0.974 \text{ V}$$

(b) At the onset of strong inversion, thus $V_G = V_T$, the inversion-layer charge is assumed to be zero. Consequently, the depletion-layer charge Q_d appears as the only uncompensated charge in the silicon substrate. Originating from the uncompensated negative acceptor ions (concentration N_A) in the depletion layer of width w_d, the depletion-layer charge density Q_d (C/m^2) can be expressed as

$$Q_d = q N_A w_d$$

Equation (2.74) gives w_d in terms of N_A and the Fermi potential $2\phi_F$, which after substitution into the above equation leads to

$$Q_d = \sqrt{2\varepsilon_s q N_A (2\phi_F)}$$

Thus,

$$Q_d = \sqrt{2 \times 1.04 \times 10^{-10} \times 1.6 \times 10^{-19} \times 7 \times 10^{22} \times (2 \times 0.41)}$$
$$= 1.38 \times 10^{-3} \text{ C/m}^2$$

(c) The body factor is defined by Eq. (2.76):

$$\gamma = \frac{\sqrt{2\varepsilon_s q N_A}}{C_{ox}} = \frac{\sqrt{2 \times 1.04 \times 10^{-10} \times 1.6 \times 10^{-19} \times 7 \times 10^{22}}}{1.15 \times 10^{-3}}$$
$$= 1.327 \text{ V}^{1/2}$$

(d) The threshold voltage is given by Eq. (2.77):

$$V_T = V_{FB} + 2\phi_F + \gamma\sqrt{2\phi_F} = -0.974 + 2 \times 0.41 + 1.327\sqrt{2 \times 0.41} = 1.05 \text{ V}$$

(e) The assumptions are that the inversion-layer charge density is zero at $V_G = V_T$ and that the whole gate voltage increase beyond the threshold voltage is spent on creating the inversion-layer charge. Thus

$$Q_I = (V_G - V_T)C_{ox} = (5 - 1.05) \times 1.15 \times 10^{-3} = 4.54 \times 10^{-3} \text{ C/m}^2$$

■ Example 2.8 Designing the Threshold Voltage

In CMOS (complementary MOS) integrated circuits, it is necessary to provide equal absolute values of the threshold voltages of MOS structures on P-type and N-type substrates. Determine the value of the donor concentration N_D necessary to provide an N-substrate MOS structure with the absolute value of the threshold voltage equal to the threshold voltage of the P-substrate MOS structure considered in Example 2.7. The values of all other parameters should remain the same.

Solution: Equation (2.78) gives the threshold voltage for the case of N-type silicon substrate:

$$V_T = V_{FB} - 2|\phi_F| - \gamma\sqrt{2|\phi_F|}$$

As the Fermi potential $2\phi_F$, the body factor γ, and the flat-band voltage V_{FB} depend on the donor concentration, it is necessary to express them in terms of N_D. Using Eq. (2.63) for $2\phi_F$, Eqs. (2.65) and (2.64) for V_{FB}, and Eq. (2.79) for γ, the threshold voltage equation is developed as

$$V_T = \phi_m - \chi_s - \frac{E_g}{2q} + V_t \ln\frac{N_D}{n_i} - 2V_t \ln\frac{N_D}{n_i} - \frac{\sqrt{2\varepsilon_s q N_D}}{C_{ox}}\sqrt{2V_t \ln\frac{N_D}{n_i}}$$

where $V_t = kT/q$. Using the values of the known parameters, the above equation is simplified to

$$V_T = -0.56 - 0.026 \ln \frac{N_D}{n_i} - 5.02 \times 10^{-12} \sqrt{N_D} \sqrt{0.052 \ln \frac{N_D}{n_i}}$$

As $\ln N_D/n_i > 0$ (this is because $N_D > n_i$), all the three terms in the above equation are negative, which means that the threshold voltage is negative. To use the absolute value of the threshold voltage, all the minus signs should be changed to pluses. Given that the absolute value of the threshold voltage should be 1.05 V (as obtained in Example 2.7), and developing $\ln(N_D/n_i)$ as $\ln N_D - \ln 1.02 \times 10^{16} = \ln N_D - 36.86$ (where N_D is in m^{-3}), the following equation is obtained:

$$1.05 = 0.56 + 0.026 \ln N_D - 0.026 \times 36.86 + 5.02$$
$$\times 10^{-12} \sqrt{N_D} \sqrt{0.052 \ln N_D - 0.052 \times 36.86}$$

Grouping the terms with the unknown N_D on the left-hand side leads to

$$0.026 \ln N_D + 5.02 \times 10^{-12} \sqrt{N_D} \sqrt{0.052 \ln N_D - 1.917} = 1.448$$

This equation can be solved only iteratively: a guess is made for the value of N_D and the left-hand side (LHS) is calculated and compared to the value on the right-hand side (RHS) (1.448). This comparison provides an indication of whether the value of N_D should be increased or decreased before the next iteration is performed. Once the difference between the LHS and RHS is acceptable, the value of N_D is taken as the solution. Perhaps it makes sense to take the value of the acceptor concentration from Example 2.7 as the initial guess: $N_D = 7 \times 10^{16} cm^{-3} = 7 \times 10^{22} m^{-3}$. The following table illustrates that the LHS value is 2.569, which is higher than the RHS value of 1.448: the concentration N_D should be reduced. The table also illustrates that the next guess of $10^{21} m^{-3}$ is smaller than it should be, whereas the guess of $2 \times 10^{21} m^{-3}$ is slightly larger than the proper concentration. Finally, $N_D = 1.8 \times 10^{21} m^{-3}$ is found to give a quite acceptable value of the LHS (1.441 as compared to the wanted 1.448). Therefore, the solution is taken to be $N_D = 1.8 \times 10^{21} m^{-3} = 1.8 \times 10^{15} cm^{-3}$.

N_D (m^{-3})	LHS
7×10^{22}	2.569
10^{21}	1.380
2×10^{21}	1.454
1.8×10^{21}	1.441

PROBLEMS

Section 2.2 Reverse-Biased P–N Junction

2.1 Which of the following statements, related to a P–N junction, are correct?

- The direction of the built-in electric field in the depletion layer is from the N-type toward the P-type region. T
- At reverse bias, the Fermi level is constant for the entire P–N junction system. F
- The depletion layer acts as a capacitor dielectric. T

- The net (effective) charge density in the depletion layer is zero. F
- A nonzero current of minority carriers flows through the junction at zero bias. T
- The current of minority carriers depends on the voltage applied. F
- At reverse bias, the depletion-layer width is saturated (does not depend on the voltage applied). F
- The net charge outside the depletion layer of a reverse-biased P–N junction is zero. T
- At reverse bias, any electron current is fully compensated by the corresponding hole current. F

2.2 (a) A P–N junction has zero-bias ($V_R = 0$ V) capacitance per unit area $C_d = 0.722$ mF/m^2. What is the depletion-layer width?

(b) Assuming equal and uniform doping on either side of the P–N junction, $N_A = N_D = 5 \times 10^{16}$ cm^{-3}, calculate the number of uncompensated donor ions per unit area.

(c) Realizing that a unit of charge ($q = 1.6 \times 10^{-19}$C) is associated with every uncompensated donor, calculate the density of the capacitor's positive charge. $\boxed{A}$

(d) As $V_R = 0$ V, $Q_C = C_d V_R = 0$ C/m^2 gives a completely different result from the one obtained in part (c). What is the explanation?

- $Q_C = C_d V_R$ cannot be used in this case (the $Q = CV$ relationship is valid only in special cases).

- There is equal density of negative charge, due to the acceptors in the depletion layer, so that the total charge is $Q_C = 0$ C/m^2.

- The built-in voltage V_{bi} is not included, i.e., $Q_C = C_d(V_R + V_{bi})$.

2.3 The $C_d - V_R$ characteristic of a P–N junction capacitor is represented by the following table:

V_R (V)	0.0	1.0	2.0	3.0	4.0	4.5	5.0	5.5
C_d (mF/m^2)	0.722	0.386	0.295	0.248	0.218	0.206	0.197	0.188

Assuming $N_D = N_A = 5 \times 10^{16}$ cm^{-3}, calculate the density of positive charge at $V_R = 5$ V.

2.4 (a) If a sinusoidal signal voltage with zero-to-peak amplitude of 500 mV is superimposed to V_R, calculate the maximum density of positive charge for the P–N junction capacitor of Problem 2.3. Using the result for $V_R = 5$ V, obtained in Problem 2.3, calculate the charge-density increase due to the peak signal voltage.

(b) Using the capacitance definition, the charge-density increase can be approximated as $\Delta Q_C \approx C_d \Delta v_C$, where C_d is assumed to be constant and equal to the zero-signal value $C_d(V_R)$. Calculate ΔQ_C with this method, and compare it to the result from part (a).

2.5 Assume the signal voltage of Problem 2.4 is increased to 5 V (zero-to-peak) so that the instantaneous voltage v_C oscillates between 0 and 10 V.

(a) Using the results of Problems 2.3 and 2.2c, calculate the charge-density decrease due to the negative signal peak.

(b) Calculate the charge-density decrease from $\Delta Q_C \approx C_d \Delta v_C$.

(c) The results from parts (a) and (b) of Problem 2.4 are very close, however, this time the $\Delta Q_C \approx C_d \Delta v_C$ method gives a different result. Why?

- The method in part (a) ignores the negative charge due to the acceptors in the depletion layer.
- The method in part (b) ignores V_{bi}.
- The method in part (b) is meaningless for $\Delta v_C < 0$.
- The $C_d - V_R$ characteristic is nonlinear.

2.6 The N-type and P-type doping levels of a silicon P–N junction are $N_D = N_A = 5 \times 10^{16}$ cm^{-3}. Calculate the built-in voltage V_{bi}

(a) at room temperature,

(b) at 300°C (n_i at 300°C is calculated in Problem 1.32), [A]

(c) at 700° ($n_i = 1.10 \times 10^{18}$ cm^{-3}). Does a negative value for V_{bi} mean anything?

2.7 Calculate V_{bi} for the cases when the P–N junction of Problem 2.6 is implemented in GaAs ($n_i = 2.1 \times 10^6$ cm^{-3}) and Ge ($n_i = 2.5 \times 10^{13}$ cm^{-3}).

2.8 (a) Using the results of Problems 2.6a and 2.7, plot V_{bi} versus E_g for the cases of Ge ($E_g = 0.66$ eV), Si ($E_g = 1.12$ eV), and GaAs ($E_g = 1.42$ eV), respectively.

(b) Derive the $V_{bi}(E_g)$ equation to explain the obtained linear correlation between V_{bi} and E_g. [A]

2.9 A very low doped N$^-$ layer ($N_D = 5 \times 10^{14}$ cm^{-3}) is deposited onto a heavily doped N$^+$ substrate ($N_D = 10^{20}$ cm^{-3}). Calculate the built-in voltage at this N$^-$–N$^+$ junction.

Section 2.3 C–V Dependence of Reverse-Biased P–N Junction: Solving the Poisson Equation

2.10 The N-type and P-type doping levels of an abrupt P–N junction are $N_D = 10^{17}$ cm^{-3}, and $N_A = 10^{15}$ cm^{-3}.

(a) Calculate and compare the zero-bias depletion-layer widths in the N- and P-type regions.

(b) Calculate the maximum electric field in the depletion layer.

(c) Calculate the zero-bias capacitance per unit area.

2.11 The range of operating voltages of an abrupt P–N junction capacitor is 0–10 V. What is the minimum capacitance that can be achieved if the maximum capacitance is 2.5 pF?

2.12 The possibility of varying the capacitance by changing the reverse-bias voltage V_R enables the capacitor of Problem 2.11 to be used as a tuning element in microwave circuits. If the operating point of the capacitor is $V_R = 5$ V, determine the capacitor sensitivity dC/dV_R.

2.13 The donor and acceptor concentration at each side of a P–N junction are given in the following:

	x (μm)		
	−5.0	**0.0**	**5.0**
N_D (cm^{-3})	6×10^{15}	3.5×10^{15}	10^{15}
N_A (cm^{-3})	10^{15}	3.5×10^{15}	6×10^{15}

Determine $\rho_{charge}(x)$ in the depletion layer to demonstrate that this is a linear P–N junction.

(a) Calculate the build-in voltage V_{bi}. $\boxed{A}$

(b) Calculate the zero-bias capacitance per unit area and the capacitance at $V_R = 20$ V.

(c) Calculate N_D of an abrupt P–N junction ($N_A = 10^{20}$ cm^{-3}) so that it has the same zero-bias capacitance as the P–N junction in part (b). Calculate the capacitance at $V_R = 20$ V and compare it to the result in part (b). Comment on the result.

(d) Calculate and compare the maximum zero-bias electric fields at the P–N junctions of parts (b) and (c). $\boxed{A}$

2.14 (a) The charge density at a linear P–N junction can be expressed by $\rho_{charge} = -5 \times 10^8 x$, where x is in m and ρ_{charge} in C/m^3. Design the donor concentration of an abrupt P–N junction ($N_A = 10^{20}$ cm^{-3}) so that the zero-bias maximum field is the same as in the case of the linear junction. Assume $V_{bi} = 0.6$ V for the linear and $V_{bi} = 0.9$ V for the abrupt P–N junction.

(b) Calculate and compare the maximum electric fields at these junctions at $V_R = 20$ V.

Section 2.4 SPICE Parameters and Their Measurement

2.15 Find the appropriate set of parameters for each of the lines in Fig. 2.20.

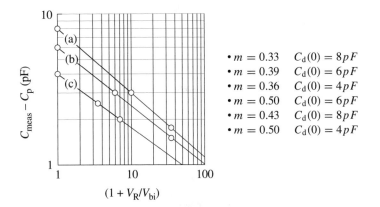

- $m = 0.33$ $C_d(0) = 8pF$
- $m = 0.39$ $C_d(0) = 6pF$
- $m = 0.36$ $C_d(0) = 4pF$
- $m = 0.50$ $C_d(0) = 6pF$
- $m = 0.43$ $C_d(0) = 8pF$
- $m = 0.50$ $C_d(0) = 4pF$

Figure 2.20 Three sets of capacitance–voltage measurements.

2.16 A set of $C_d - V_R$ data is plotted in Fig. 2.21 with three different assumptions for V_{bi}.

(a) Identify which set of data corresponds to each of the V_{bi} values.

(b) Determine the grading coefficient m.

2.17 Using the data from Table 2.8, and assuming parallel parasitic capacitance $C_p = 1$ pF and built-in voltage $V_{bi} = 0.9$ V, obtain the best estimate of the grading coefficient m and the zero-bias capacitance $C_d(0)$. [A]

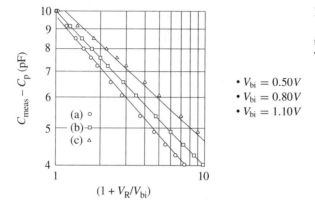

Figure 2.21 A set of measured $C_d - V_R$ points (symbols) is plotted with three different assumptions for V_{bi}. The lines show the best linear fits.

- $V_{bi} = 0.50V$
- $V_{bi} = 0.80V$
- $V_{bi} = 1.10V$

(a) ○
(b) □
(c) △

TABLE 2.8 Capacitance–Voltage Measurements

P–N Junction Capacitance (pF)	Voltage (V)
3.33	−1
2.72	−3
2.44	−5
2.28	−7
2.16	−9

2.18 The capacitance of a P–N junction is $C_d(0) = 3.0$ pF and $C_d(5 \text{ V}) = 1.33$ pF at $V_R = 0$ V and $V_R = 5$ V, respectively. If $V_{bi} = 0.76$ V, determine the grading coefficient m (SPICE parameter).

Section 2.5 Metal–Oxide–Semiconductor (MOS) Capacitor and Thermal Oxide

2.19 Which of the following statements, related to an MOS capacitor, are correct?

- The condition of zero net charge at the MOS capacitor plates is referred to as a flat-band condition.
- There is no field in the oxide at $V_{GS} = V_{FB}$.
- The net charge at the MOS capacitor plates has to be zero at $V_{GS} = 0$.
- The density of inversion-layer charge is expressed as $Q_I = C_{ox}(V_{GS} - V_{FB})$.
- The surface potential φ_s depends exponentially on gate voltage in strong inversion.
- The density of inversion-layer charge at the onset of strong inversion is $Q_I = \gamma \sqrt{2\phi_F}$.

- The capacitance in depletion mode does not depend on the gate voltage applied.
- The inversion-layer capacitance is proportional to $(V_{GS} - V_T)$.

2.20 Which mode of MOS capacitor operation (accumulation, depletion, or strong inversion) is expressed by the energy-band diagram of Fig. 2.22. Knowing that the energy gap of silicon is 1.12 eV, determine the voltage applied between the gate and the silicon substrate.

2.21 The flat-band voltage of an MOS capacitor on N-type substrate is $V_{FB} = -1$ V. If the gate oxide thickness is 5 nm, calculate the value and determine the direction(s) of the electric fields in the gate oxide and at the surface of the semiconductor at zero gate bias ($V_G = 0$).

2.22 Two different MOS capacitors, with different gate oxide thicknesses (3 and 15 nm) have the same density of positive oxide charge ($N_{oc} = 5 \times 10^{10}$ cm^{-2}) close to the oxide–semiconductor interface. Find the flat-band voltage shifts due to this positive oxide charge for these two MOS capacitors. What are the threshold voltage shifts?

2.23 The C–V curves of an MOS capacitor before and after gate-oxide stressing are shown in Fig. 2.23. Find the density of gate oxide charge, N_{oc}, created by this stress.

2.24 What is the work-function difference between heavily doped polysilicon and P-type silicon that is doped with 10^{16} acceptor atoms per cm^3, if the heavily doped polysilicon is

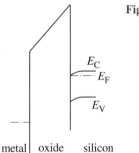

Figure 2.22 MOS energy-band diagram.

E_C
E_F
E_V

metal | oxide   silicon

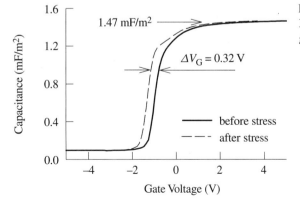

Figure 2.23 C–V curves of an MOS capacitor before and after gate-oxide stressing.

1.47 mF/m²

$\Delta V_G = 0.32$ V

—— before stress
- - - - after stress

(a) P^+ type, $\boxed{A}$

(b) N^+ type.

2.25 (a) The flat-band voltage of an MOS capacitor with an N^+ polysilicon gate is $V_{FB} = -0.25$ V. Assuming zero oxide charge, determine the type and level of the substrate doping.

(b) What would be the flat-band voltage if a P^+ polysilicon gate is used with the same type and level of substrate doping? $\boxed{A}$

2.26 The threshold voltage of an MOS capacitor on a P-type silicon substrate is $V_T = 1.0$ V. What is the type and density of the mobile charge at the silicon surface if 5 V is applied between the gate and the substrate? The gate oxide thickness is 15 nm.

2.27 The threshold voltage of an MOS capacitor on a P-type substrate is $V_T = 0.25$ V and the gate oxide capacitance is $C_{ox} = 6.5$ mF/m^2. How many electrons can be found in a $0.1 \times 0.1~\mu$m^2 of capacitor area if the gate voltage is $V_G = 0.5$ V?

2.28 The threshold and the flat-band voltages of an MOS capacitor are $V_T = -1.0$ V and $V_{FB} = -0.5$ V, respectively. Is this capacitor created on an N-type or P-type semiconductor? What is the density of minority carriers (in C/m^2) at the semiconductor surface when the voltage applied between the metal and semiconductor electrodes is $V_G = -0.75$ V? $\boxed{A}$

2.29 The flat-band voltage and the threshold voltage of an MOS capacitor are $V_{FB} = -3.0$ V and $V_T = -1.0$ V, respectively. The gate oxide capacitance is $C_{ox} = \varepsilon_{ox}/t_{ox} = 3.45 \times 10^{-3}$ F/m^2.

(a) Is this capacitor created on an N-type or P-type semiconductor? Explain your answer.

(b) What is the density of minority carriers (in C/m^2) at the semiconductor surface when the voltage applied between the metal and semiconductor electrodes is $V_G = -2.0$ V?

(c) What is the density of minority carriers (in C/m^2) at the semiconductor surface when no voltage is applied across the capacitor ($V_G = 0$)?

2.30 Calculate the threshold voltage of a P^+ polysilicon gate MOS capacitor on an N-type substrate ($N_D = 5 \times 10^{16}$ cm^{-3}) if the gate oxide thickness is 7 nm. Neglect the gate oxide charge.

2.31 The oxide thickness of an MOS capacitor is 4 nm. The silicon substrate is N-type (doping level $N_D = 7 \times 10^{16}$ cm^{-3}), and the gate is heavily doped N^+-type polysilicon. Assuming zero oxide charge density, determine the mobile charge density at the semiconductor surface, if the voltage applied between the gate and the substrate is -1.5 V.

2.32 Calculate high- and low-frequency strong-inversion capacitances per unit area of an MOS capacitor having 50-nm-thick oxide as the dielectric and P-type substrate doped with $N_A = 10^{15}$ boron atoms per cm^3.

2.33 The accumulation and strong-inversion capacitances of an MOS capacitor, measured by a high-frequency signal, are 9 and 3mF/m², respectively. These capacitances are measured at $V_G = -2$ V and $V_G = 2$ V, respectively.

(a) Is the substrate N or P type?

(b) Determine the gate oxide thickness.

(c) Assuming a uniform substrate doping, calculate the substrate doping concentration.
[A]

2.34 An MOS capacitor with an N⁺ polysilicon gate is biased in strong inversion. What is the surface potential φ_s? Determine the voltage drop across the gate oxide if the gate-to-substrate voltage is 5 V. The doping level of the silicon substrate is $N_A = 5 \times 10^{16}$ cm⁻³.
[A]

2.35 The oxide breakdown electric field is 1 V/nm. Design the oxide thickness of an MOS capacitor so that the breakdown voltage in strong inversion is 5 V. N⁺ polysilicon is to be used for the gate, and the substrate doping is $N_A = 7.5 \times 10^{16}$ cm⁻³. What is the breakdown voltage in accumulation, if the surface potential is neglected in comparison to the breakdown voltage?

2.36 The following relationship between the oxide and semiconductor fields and the density of oxide charge close to the oxide–semiconductor interface can be derived from the integral form of Gauss law:

$$\varepsilon_s E_s - \varepsilon_{ox} E_{ox} = q N_{oc}$$

where both the semiconductor field at the surface (E_s) and the oxide field (E_{ox}) are in the direction toward the substrate. Calculate E_{ox} and E_s for $V_G = V_{FB}$ and (a) $N_{oc} = 0$; (b) $N_{oc} = 5 \times 10^{10}$ cm⁻². [A]

2.37 An MOS capacitor has a P⁺ polysilicon gate, substrate doping of $N_D = 10^{16}$ cm⁻³, and gate oxide thickness of $t_{ox} = 80$ nm. The oxide breakdown electric field is 1 V/nm. Calculate the breakdown voltage in strong inversion if

(a) the oxide charge can be neglected,

(b) as a result of exposure to high electric field, a positive charge with density of $N_{oc} = 5 \times 10^{11}$ cm⁻² is created close to the silicon–oxide interface. [A]

2.38 An MOS capacitor on a P-type substrate with $\phi_{ms} = 1.0$ V is biased by a constant gate voltage $V_G = -7$ V, in order to test the integrity of its 10-nm gate oxide. If this stress creates $N_{oc} = 10^{10}$ cm⁻² of positive charge close to the silicon–oxide interface every hour, how long would it take before the oxide field reaches the critical level of 1 V/nm?

2.39 Five polysilicon-gate N-channel MOSFETs, each with a different gate oxide thickness t_{ox} (45, 47, 50, 53, and 55 nm), are made on P-type silicon substrates having the same doping level $N_A = 5 \times 10^{16}$ cm⁻³. The gate oxide charge Q_{oc} is assumed to be equal for all the MOSFETs as they are processed in a single batch of wafers. Measurements of the threshold voltages V_T are made, yielding the following values: 1.10, 1.16, 1.25, 1.33, and 1.39 V. Discuss the shape and the slope of $V_T(t_{ox})$ dependence, and explain the meaning of the intercept $V_T(t_{ox} = 0)$. Determine the value of the oxide charge Q_{oc} and the metal–semiconductor work-function difference ϕ_{ms}.

REVIEW QUESTIONS

R-2.1 Draw the energy-band diagram of a reverse-biased P–N junction and indicate the minority and the majority carriers. Can minority carriers move through the P–N junction?

R-2.2 Ideally, there should not be any DC current flowing through a capacitor. In the ideal case, can the leakage current of a reverse-biased P–N junction capacitor be zero?

R-2.3 Does the leakage current of a reverse-biased P–N junction depend significantly on the value of reverse-bias voltage?

R-2.4 Why does the capacitance of a reverse-biased P–N junction capacitor depend on the reverse-bias voltage? What reverse-bias voltage provides maximum capacitance?

R-2.5 Abrupt and linear P–N junctions represent two extreme cases. What are the similarities and what are the differences in the derived equations for $C_d - V_R$ dependence? How is the SPICE model related to the these theoretical equations?

R-2.6 Linearization techniques and transformations are used to enable application of the graphic or linear regression technique for SPICE parameter measurement. Why is it not good to apply curve fitting directly to the nonlinear device models?

R-2.7 Is there a net charge at MOS capacitor plates at $V_G = 0$ V? If there is, there must be an electric field in the gate oxide to keep that charge at the capacitor plates. With no gate voltage applied, where can this electric field originate from?

R-2.8 How is the condition of zero charge at MOS capacitor plates referred to? Is there any field in the oxide or the substrate? Is there any potential difference between the surface and the bulk of the silicon substrate?

R-2.9 How is the flat-band voltage expressed in terms of the work-function difference and the oxide charge?

R-2.10 What type of charge appears at MOS capacitor plates when a negative effective gate voltage ($V_G - V_{FB} < 0$) is applied? How is this mode referred to? Assume a P-type semiconductor.

R-2.11 What determines the MOS capacitance in accumulation mode? Does it depend on the voltage applied?

R-2.12 What type of charge appears at MOS capacitor plates when a small positive effective gate voltage ($V_G - V_{FB} > 0$) is applied? How is this mode referred to? Assume a P-type semiconductor.

R-2.13 What determines the MOS capacitance in depletion mode? Does it depend on the voltage applied? Does the surface potential change as the gate voltage is changed?

R-2.14 What type of charge appears at MOS capacitor plates when the gate voltage applied is larger than the threshold voltage ($V_G > V_T$)? How is this mode referred to? Assume a P-type semiconductor.

R-2.15 Write the capacitance-voltage-charge equation ($Q = CV$) for an MOS capacitor in strong inversion to show the relationship of different types of charge in the silicon substrate and the effective gate voltage drop across the gate oxide.

R-2.16 Does surface potential depend significantly on the gate voltage in strong inversion?

R-2.17 **What determines the surface potential in strong inversion? How is it expressed?**

R-2.18 What is the density of the inversion-layer charge (electrons in this case) at the onset of strong inversion?

R-2.19 What is the voltage at the gate that sets the MOS structure at the onset of strong inversion called?

R-2.20 Obtain the threshold voltage equation from the equation written in Question 2.15.

Chapter *3*

DIODES
Forward-Biased P–N Junction and Metal–Semiconductor Contact

Diodes are used for a number of different applications, the most frequently being signal rectification, voltage clamping, reference-voltage circuits, light emitters, photodetectors, and solar cells. Usually, the diodes are based on the P–N junction. Therefore, the integrated-circuit structure of the P–N junction capacitor, shown in Fig. 2.3, represents at the same time the integrated-circuit structure of the P–N junction diode, as illustrated in Fig. 3.1. The variety of applications mentioned above can be achieved by specifically designed P–N junction diodes. The rectifying and reference P–N junction diodes are described in the first three sections of this chapter. Simultaneously, the fundamental diode-related phenomena (including breakdown) are introduced. The light emitters, photodetectors, and solar cells are introduced in Chapter 8, simultaneously with a detailed description of recombination–generation processes. The last section of this chapter is devoted to another type of diode, which is based on metal–semiconductor contact. This type of diode sometimes provides more appropriate characteristics for some applications, and also is an integral part of metal–semiconductor field-effect transistors (MESFETs).

3.1 RECTIFYING DIODES: FUNDAMENTAL EFFECTS AND MODELS

3.1.1 Basic Applications

All the rectifying circuits (the basic rectifying circuit is shown in Fig. 3.2a) necessitate a device that conducts the electric current in one direction only. The current-voltage (I–V) characteristic of this desired device is shown by the dashed line in Fig. 3.2b. The two possible states of this device are "off" and "on." In the "off" state the current is zero for any negative voltage applied (infinitely large resistance, or open circuit); in the "on" state the voltage drop across the device is zero for any positive current flowing through (zero resistance or short circuit).

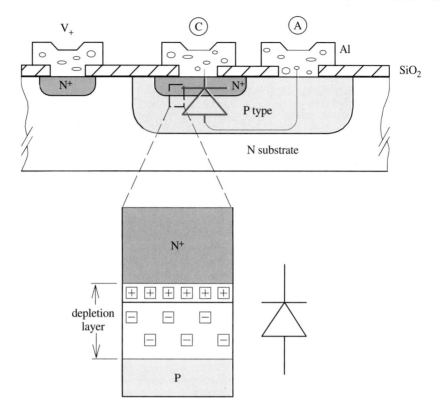

Figure 3.1 Structure of a P–N junction diode in integrated circuits.

The I–V characteristic of the P–N junction diode, shown by the solid line in Fig 3.2b, is close enough to the ideal characteristic to enable design of rectifying circuits usable in a variety of applications. The main difference between the characteristics of the P–N junction diode and the ideal rectifying device is that "on" resistance of the P–N junction drops to a small value at a voltage that is observably larger than zero (around 0.7 V in the case of silicon diodes). Another difference is that the "off" current is not exactly equal to zero, although it is quite small (so much smaller compared to the "on" current that it cannot be observed in the plot shown in Fig. 3.2b). There would be some other "common sense" differences such as limited values of negative voltage or positive current that the diode can withstand.

The diode terminals are referred to as anode and cathode, labeled as *A* and *C*, respectively, in Fig. 3.1. As Fig. 3.1 illustrates, the P-type side of the P–N junction is the anode, while the N-type side is the cathode. When positive voltage is applied between the anode and the cathode, the holes in the P-type and the electrons in the N-type regions are pushed through the P–N junction and further attracted to the negatively biased cathode and positively biased anode, respectively.

The positive voltage applied between the anode and the cathode produces flow of holes and electrons through the P–N junction.

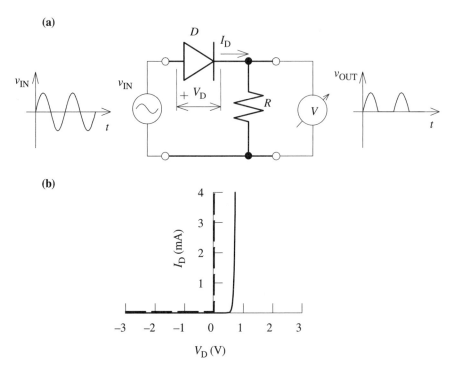

Figure 3.2 (a) Basic rectifying circuit; (b) $I - V$ characteristics of the ideal rectifying device (dashed line) and the P–N junction diode (solid line).

Note that the opposite directions of the hole and electron flows mean that their respective electric currents are in the same direction, from the anode toward the cathode. As the concentrations of holes in the P-type and electrons in the N-type regions (majority carriers) are very high, the current that flows through the P–N junction is significant—the diode is "on." This condition is known as *forward-biased diode*.

On the other hand, when negative voltage is applied between the anode and the cathode, it pulls the holes in the P-type and electrons in the N-type regions (the majority carriers) away from the junction so the majority carriers do not flow through the P–N junction—the diode is "off."

This condition is referred to as the *reverse-biased diode*; it is mentioned in Chapter 1 (isolation using reverse-biased P–N junction), and described in more detail in Chapter 2 (reverse-biased P–N junction capacitor). Although the reverse-bias voltage prevents the majority carriers from flowing through the P–N junction, it consequently enforces the minority carriers to flow through the P–N junction. However, the concentrations of the minority carriers are much smaller than the majority-carrier concentrations, so the reverse-bias current is much smaller, which is the desired rectifying property of the P–N junction used as a diode.

The rectifying property of diodes can be combined with voltage sources to create a variety of clamping circuits. An example is shown in Fig. 3.3. To bring the diode in an "on" state, the anode has to be at a higher potential than the cathode. When this happens, the output voltage cannot exceed the voltage of the voltage source (plus the voltage drop across the "on" diode, ≈ 0.7 V). For any smaller input voltage, the diode is "off," and the input voltage appears directly across the resistor.

An analogously constructed circuit, clamping negative voltages, can be added to the circuit of Fig. 3.3 to create a double clamp circuit. Figure 3.4 illustrates the use of this double clamp as an input protection circuit. The protection circuit does not allow the voltage spikes exceeding the supply voltage values to appear at the sensitive IC input.

3.1.2 Energy-Band and Carrier-Concentration Diagrams

The energy-band model provides deeper insight into the operation of the diode. The energy-band diagrams of zero-biased and reverse-biased P–N junctions are introduced in Section 2.2.3, describing the P–N junction capacitor. For convenience, the energy-band diagrams of Fig. 2.6 are shown again in Fig. 3.5a and b.

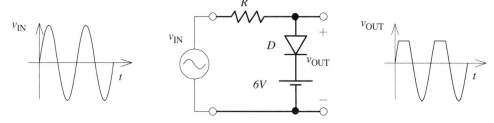

Figure 3.3 A single clamp circuit.

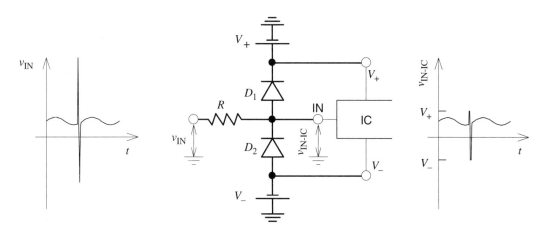

Figure 3.4 A double clamp circuit used as an input protection circuit.

Thermal Equilibrium

Figure 3.5a illustrates that a small number of electrons in the N-type region have high enough energy to pass through the depletion layer of the P–N junction. Similarly, a small number of holes from the P-type side are able to pass through the depletion layer and appear on the N-type side. This current of the majority carriers is, however, exactly compensated for by the current of the minority carriers. There is nothing surprising about this fact as effective current through the P–N junction has to be zero, otherwise the system is not in equilibrium.

Reverse Bias

The effects in the *reverse-biased P–N junction* are illustrated in Fig. 3.5b. Almost the whole reverse-biased voltage (V_R) appears across the depletion layer of the P–N junction, which is now wider than in the case of no voltage applied. The Fermi levels (or more precisely quasi-Fermi levels) in the N-type and P-type semiconductors are now split by $q V_D$ to express the fact that the voltage $V_D = -V_R$ appears between the P-type and N-type regions (this is explained in more detail in Section 2.2.3). As a result, the energy barrier height is increased from $q V_{bi}$ to $q(V_{bi} - V_D) = q(V_{bi} + V_R)$. This means that the majority carriers that would pass through the depletion layer in the zero-bias case are unable to overcome the increased energy barrier. The current of the majority carriers is significantly reduced if not completely blocked. Being the dominant current, the minority-carrier current appears as the effective current flowing through the P–N junction.

The current of the minority carriers flowing through the P–N junction is basically voltage independent. An increase in the reverse-bias voltage V_R would increase the splitting of the Fermi levels and consequently the slope of the energy bands (increased electric field in the depletion layer). It may become a little easier for the electrons to roll down and the holes to bubble up through the depletion layer, however, it is not the slope of the bands in the depletion layer that limits the minority-carrier current—it is always favorable enough for any electron appearing at the top to roll down, and for any hole appearing at the bottom to bubble up to the opposite side. The current of the minority carriers is limited by the number of the electrons or holes appearing at the top or bottom of the bands in the depletion layer.

P–N junctions with higher concentrations of the minority carriers have higher reverse-bias current, which remains basically voltage independent. If the minority-carrier current is labeled by I_S, the current of the reverse-biased diode can be expressed as

$$I_D = -I_S \tag{3.1}$$

Forward Bias

Figure 3.5c illustrates that the energy barrier height is reduced for the *forward-biased P–N junction*. The forward-bias voltage V_D appearing across the depletion layer splits the Fermi levels in the opposite direction, to reduce the barrier height from $q V_{bi}$ to $q(V_{bi} - V_D)$. As an effect, the concentration of electrons in the N-type and holes in the P-type regions with energies higher than the reduced barrier height is significantly increased. This results in a

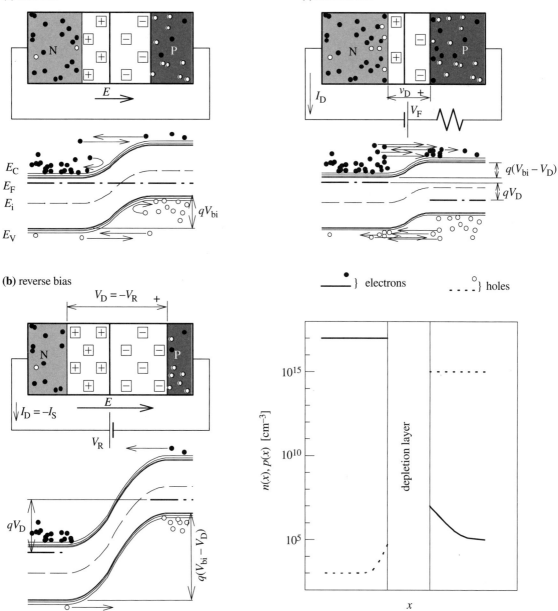

Figure 3.5 Illustration of the effects in a zero-biased (a), reverse-biased (b), and forward-biased (c) diode, including the associated energy-band diagrams and the carrier-concentration diagram in the case of forward bias (c).

significant increase in the number of majority carriers passing through the depletion layer and appearing as minority carriers on the other side. The concentration of the minority carriers (electrons in the P-type and holes in the N-type region) is, consequently, increased along the depletion layer boundary, as shown in Fig. 3.5c.

An excess electron appearing in the P-type region is very likely to meet some of the numerous holes and recombine (remember that the hole is a missing electron in the bond structure of the crystal). Similarly, an excess hole in the N-type region is very likely to recombine with some of the numerous electrons. Some of the minority carriers appearing on the other side get recombined very quickly, while the others enter a lot deeper before they disappear. This process of recombination reduces the minority-carrier concentration from the highest value at the boundary of the depletion layer to the equilibrium level, as shown in the concentration diagram of Fig. 3.5c. The concentration of the majority carriers is not significantly altered, as any recombined carrier is quickly replaced from the power supply.

To see the effect of the carrier recombination, let us follow the path of an electron originating at the negative terminal of the battery. This electron flows into the N-type region of the diode, appearing there as a majority carrier. If it happens that this electron comes to the depletion-layer edge with a large enough energy, it can go through the depletion layer to appear at the P-type side. After making some progress in the P-type region as a minority carrier, it gets recombined with a hole. The recombined hole has to be supplied by a new one from the power supply. The flow of a hole from the positive terminal of the battery to the recombination point is equivalent to the flow of the followed electron through the bonds of the semiconductor atom (valence band) to the positive terminal of the power supply. Therefore, the circuit is closed. The carrier recombination by no means prevents the carriers from completing the electric current circuit; it means only that a part of the circuit is traveled by an electron, while the other part is completed by a hole.

The number of majority carriers able to go over the barrier in the depletion layer increases as the barrier height $q(V_{bi} - V_D)$ is lowered by an increase in the forward-bias voltage V_D. This means that the current of the majority carriers flowing through the depletion layer is increased with an increase in the forward-bias voltage V_D. The diode is said to be forward biased when the voltage V_D is large enough to produce a significant flow of the majority carriers through the P–N junction, that is, when it produces a significant effective diode current (in this case we can neglect the minority-carrier current, which remains constant and small).

To gain an insight into the type of current–voltage dependence of the forward-biased diode, it is necessary to refer to the type of electron or hole energy distribution in the conduction or valence band. The energy distribution of the electrons and holes is according to the Fermi–Dirac distribution, the tails of which are very close to the exponential distribution, as explained in Section 1.5.2 (Fig. 1.27). Therefore, as the energy barrier height in the depletion layer $q(V_{bi} - V_D)$ (Fig. 3.5c) is lowered by the increasing V_D voltage, the number of majority electrons able to go over the barrier increases exponentially. The same occurs with the majority holes.

The current of a forward-biased diode increases exponentially with the voltage. Normalizing the applied voltage by the thermal voltage $V_t = kT/q$, the current–voltage dependence of a forward-biased diode can be expressed as

$$I_D \propto e^{V_D/V_t} \tag{3.2}$$

3.1.3 Basic Current–Voltage (I–V) Equation

Assume the bias arrangement as shown in Fig. 3.6a, and consider the corresponding concentration diagram shown in Fig. 3.6b. The characteristic points labeled in the concentration diagram of Fig. 3.6b are w_p—the depletion layer edge on the P-type side, w_n—the depletion layer edge on the N-type side (appearing with a minus sign as the origin of the x-axis is put at the P–N junction), n_{pe} and p_{ne}—the equilibrium concentrations of the minority carriers, and $n_p(w_p)$ and $p_n(-w_n)$—the minority-carrier concentrations at the edges of the depletion layer.

Following again the path of electrons, starting at the negative battery terminal, we come to the neutral[1] N-type region of the diode. Electrons can move through the neutral N-type region as the electric field from the negative battery terminal pushes them toward the depletion layer (this is the *drift current*). It is very important to clarify the question of how strong this field is. The concentration of electrons in the N-type region (the majority carriers) is huge, $n_n = 10^{17}$ cm^{-3} in the example of Fig. 3.6b. As the drift electron current density is given by $j_n = \sigma_n E = q \mu_n n_n E$ (the differential form of Ohm's law), only small electric field E is needed to produce a moderate $n_n E$ value, thus a moderate current density. Assuming $\mu_n = 625$ cm^2/Vs, we find that the electric field needed to produce 1 μA/μm^2 of drift current density is only $E = 1$ V/mm. If the width of the N-type region is 10 μm, the voltage drop across the neutral region is only 10 mV. It is rightly assumed that almost the whole diode voltage V_D appears across the depletion layer.

Once the electrons approach the depletion-layer edge (the point $x = -w_n$ on the concentration diagram), the part having energies larger than the energy barrier height move through the depletion layer and appear on the P-type side (the point $x = w_p$ on the concentration diagram). In the example of Fig. 3.6b, $n_p(w_p) \approx 10^7$ cm^{-3}, which means that only one in $10^{17}/10^7 = 10^{10}$ electrons from the N-type region have enough energy to overcome the barrier in the depletion layer.

Assuming no loss of electrons in the depletion layer, the electron current at the P-type edge ($x = w_p$) must be equal to the electron current coming from the N-type, say 1 μA/μm^2 as in the above example. If the same electric field and electron mobility are assumed in the P-type region ($E = 1$V/mm and $\mu_n = 625$ cm^2/Vs), the electron drift current at $x = w_p$ is $q \mu_n n_p(w_p)E = 10^{-10}$ μA/μm^2! How can this be? The value is 10^{10} times smaller due to the fact that the concentration $n_p(w_p)$ is 10^{10} smaller than n_n. Should we use different values of the electric field and the electron mobility to avoid this difference?

> The drift current of the minority carriers is negligible compared to the drift current of the majority carries, because the concentration of the minority carriers is much smaller than the concentration of the majority carriers.

---

[1]This region is called neutral because the charge of the mobile carriers (mostly electrons, N type) is exactly compensated by the charge of the originating fixed ions (positively charged donors).

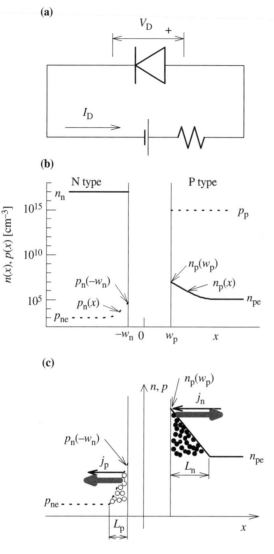

Figure 3.6 (a) Simple biasing circuit showing the definition of the diode current direction I_D and the polarity of the diode voltage V_D; (b) carrier-concentration diagram illustrating the characteristic points and symbols used, and (c) carrier-concentration diagram illustrating the diffusion current of the minority carriers.

What about the fact that "the electron current at the P-type edge ($x = w_p$) must be equal to the electron current coming from the N-type region"? The concentration drops from 10^{17} down to 10^7 cm^{-3}, and the field cannot be increased from 1 V/mm to 10,000 MV/m to maintain the *drift current*. Is there anything else but the electric field that can make a current?

Yes, there is: the concentration gradient, producing the *diffusion current*. There is a gradient in the concentration of electrons in the P-type region—the electron concentration is the highest at the edge of the P-type neutral region and is reduced toward the bulk of the region. The diffusion makes the electrons flow from the higher towards the smaller concentrations, as illustrated in Fig. 3.6c by the thick blue arrow.

The current of the minority carriers is basically diffusion current.

It is not a coincidence that the diffusion current is exactly equal to the drift current coming from the N-type region. If it were smaller, the electrons coming from the N-type region would accumulate at the P-type edge, increasing the concentration gradient and consequently increasing the diffusion current to the level of the drift current. The concentration gradient does not change in time when the two currents are equal—this is referred to as *steady-state condition*.

To derive the I–V equation of the diode, it is sufficient to find the diffusion current of the minority carriers at the edges of the depletion layer. These currents, shown in Fig. 3.6c, are labeled as j_n and j_p, for the cases of electrons and holes, respectively. When the diffusion-current Eq. (1.50) is applied to the minority carriers, it can be written in the following form:

$$j_n = q D_n \frac{\partial n_p(x)}{\partial x}$$

$$j_p = -q D_p \frac{\partial p_n(x)}{\partial x} \tag{3.3}$$

As shown in Fig. 3.6c, the excess minority-carrier concentrations $n_p(x)$ and $p_n(x)$ can be approximated by linear functions, changing from the maximum values at the edges of the depletion layer to the equilibrium values at distances L_n and L_p, respectively. L_n and L_p are considered as parameters, called diffusion length of electrons and holes, respectively. This enables the concentration gradients $\partial n_p(x)/\partial x$ and $\partial p_n(x)/\partial(x)$, and consequently the diffusion current densities to be expressed in the following simplified form:

$$j_n = q D_n \frac{n_{pe} - n_p(w_p)}{L_n}$$

$$j_p = -q D_p \frac{p_n(-w_n) - p_{ne}}{L_p} \tag{3.4}$$

In Eq. (3.4), the concentrations of the minority carriers at the depletion layer edges $n_p(w_p)$ and $p_n(-w_n)$ are the only parameters that significantly depend on the applied voltage. Remember that $n_p(w_p)$ and $p_n(-w_n)$ are related to the electrons and holes in the N-type and P-type regions, respectively, that can overcome the energy barrier in the depletion layer. The energy barrier height is $q(V_{bi} - V_D)$, as shown previously in Fig. 3.5c, which means it depends on the applied voltage V_D. To find out the concentration of electrons having kinetic energies larger than the barrier height $q(V_{bi} - V_D)$, we should count and add all the electrons appearing at the energy levels higher than $q(V_{bi} - V_D)$. We know that the energy distribution of the electrons is given by the Fermi–Dirac distribution, which can be approximated by the exponential function when the Fermi level is inside the energy gap (Section 1.5.2):

$$f = \frac{1}{1 + e^{(E-E_F)/kT}} \approx e^{-(E-E_F)/kT} \tag{3.5}$$

As the function f represents the probability of having an electron with energy E, the total number of electrons having energies larger than $q(V_{bi} - V_D)$ can be found by adding up the probabilities of finding an electron at any single energy level above $q(V_{bi} - V_D)$. As the energy levels in the conduction band are assumed to constitute a continuous energy band, the above-mentioned *adding up* is performed by way of the following *integration*:

$$n_p(w_p) \propto \int_{q(V_{bi}-V_D)}^{\infty} f \, dE = \int_{q(V_{bi}-V_D)}^{\infty} e^{-(E-E_F)/kT} \, dE \tag{3.6}$$

which gives the following result:

$$n_p(w_p) \propto e^{qV_D/kT} = e^{V_D/V_t} \quad \Rightarrow \quad n_p(w_p) = C_A \, e^{V_D/V_t} \tag{3.7}$$

In Eq. (3.7), $V_t = kT/q$ is the thermal voltage ($V_t \approx 25.85$ mV at room temperature), and C_A is a constant which needs to be determined. In order to determine the constant C_A, we can consider the case of zero bias ($V_D = 0$), as we know that the electron concentration is equal to the equilibrium value n_{pe} throughout the P-type region in that case. Therefore, the constant C_A must be chosen so to express the fact that $n_p(w_p) = n_{pe}$ for $V_D = 0$. It is not difficult to see that the constant C_A must be equal to the equilibrium value of the electron concentration in the P-type region n_{pe}, thus

$$n_p(w_p) = n_{pe} \, e^{V_D/V_t} \tag{3.8}$$

In a similar way it can be shown that

$$p_n(-w_n) = p_{ne} \, e^{V_D/V_t} \tag{3.9}$$

Replacing $n_p(w_p)$ and $p_n(-w_n)$ in Eq. (3.4) by Eqs. (3.8) and (3.9), the current densities j_n and j_p are directly related to the voltage applied V_D:

$$j_n = -\frac{qD_n}{L_n} n_{pe} \left(e^{V_D/V_t} - 1\right)$$

$$j_p = -\frac{qD_p}{L_p} p_{ne} \left(e^{V_D/V_t} - 1\right) \tag{3.10}$$

The total current density is obtained as the sum of the electron and the hole current densities:

$$j = j_n + j_p = -\left(\frac{qD_n}{L_n} n_{pe} + \frac{qD_p}{L_p} p_{ne}\right) \left(e^{V_D/V_t} - 1\right) \tag{3.11}$$

Using the fact that the product of the minority- and the majority-carrier concentrations is constant in thermal equilibrium (Eq. 1.21), and that the majority-carrier concentrations are approximately equal to the doping levels N_D and N_A, the equilibrium minority-carrier concentrations are determined from the doping levels as

$$n_{pe} = \frac{n_i^2}{N_A}$$

$$p_{ne} = \frac{n_i^2}{N_D} \tag{3.12}$$

in which case the current density becomes:

$$j = -qn_i^2 \left(\frac{D_n}{L_n N_A} + \frac{D_p}{L_p N_D} \right) (e^{V_D/V_t} - 1) \tag{3.13}$$

Finally to convert the current density j (A/m^2) into the terminal current I_D (in A), the current density is multiplied by the area of the P–N junction A_J. In addition, the sign of the terminal current (positive or negative) needs to be properly used. The convention is that the current is expressed as positive if it flows in the direction that is indicated (as in Fig. 3.6a, for example), and it is expressed as negative if it flows in the opposite direction. On the other hand, the minus sign in Eq. (3.13) for the current density means that the current density vector and the x-axis are in opposite directions, thus the current flows from the P-type toward the N-type side. The current arrow in Fig. 3.6a is consistent with this, but is not consistent with the x-axis direction, therefore the minus sign in Eq. (3.13) should be dropped. With this, the diode current can be expressed as

$$I_D = I_S(e^{V_D/V_t} - 1) \tag{3.14}$$

where I_S is used as a single parameter to replace the following:

$$I_S = A_J qn_i^2 \left(\frac{D_n}{L_n N_A} + \frac{D_p}{L_p N_D} \right) \tag{3.15}$$

In practice, the widths of the effective neutral regions are much smaller than the electron and hole diffusion lengths: $W_{anode} \ll L_n$ and $W_{cathode} \ll L_p$. As a consequence, the minority-carrier concentration will drop to the equilibrium level at the ends of the neutral regions. The concepts shown in the concentration diagrams of Fig. 3.6 and the form of Eq. (3.15) are still valid, so the diffusion lengths can be replaced by the actual widths of the anode and cathode neutral regions:

$$I_S = A_J qn_i^2 \left(\frac{D_n}{W_{anode} N_A} + \frac{D_p}{W_{cathode} N_D} \right) \tag{3.16}$$

The thermal voltage V_t appearing in Eq. (3.14) is approximately equal to 0.02585 V at room temperature. Normally the voltage V_D is either much larger than 0.02585 V (forward bias) or its absolute value is much larger than 0.02585 (reverse bias). This means that $\exp(V_D/V_t)$ is either much larger than 1 (forward bias) or much smaller than 1 (reverse bias). In the case of the reverse bias [$\exp(V_D/V_t) \ll 1$], the diode current is $I_D \approx -I_S$. This expresses the fact that the reverse-bias current does not depend on the voltage applied as discussed in the previous section [Eq. (3.1)]. Consequently, the parameter I_S is called the *saturation current*. In the case of forward bias [$\exp(V_D/V_t) \gg 1$], the diode current is $I_D \approx I_S \exp(V_D/V_t)$. This expresses the fact that the forward-bias current increases exponentially with V_D, again as discussed in the previous section [Eq. (3.2)].

■ **Example 3.1 I_S Current**

It is found that the current of a P–N junction diode at $V_D = 0.7$ V is $I_D = 5.76$ mA. Determine the value of the parameter I_S using the following value for the thermal voltage: $V_t = 25.85$ mV.

Solution: The current of the forward-biased diode is given as

$$I_D = I_S e^{V_D/V_t}$$

therefore,

$$I_S = I_D e^{-V_D/V_t} = 5.76 \times 10^{-3} \times e^{0.7/0.02585} = 10^{-14} \text{ A}$$

■ **Example 3.2 A Diode Circuit**

For the circuit of Fig. 3.2a, find the output voltage v_{OUT} for $v_{IN} = 5$ V. The value of the resistor is $R = 1$ kΩ.

Solution: To begin with, it is obvious that

$$v_{OUT} = v_{IN} - V_D$$

The voltage across the diode can be expressed in terms of the diode current,

$$I_D = I_S e^{V_D/V_t} \Rightarrow V_D = V_t \ln(I_D/I_S)$$

and the diode current I_D is related to the output voltage as $v_{OUT} = I_D R$. Therefore,

$$v_{OUT} = v_{IN} - V_t \ln[v_{OUT}/(I_S R)]$$

The voltage v_{OUT} cannot be explicitly expressed from the above equation. This equation can be solved only by an iterative method: assume v_{OUT} value, calculate the value of the right-hand side (RHS), and compare to v_{OUT}, which is the left-hand side (LHS). The table below gives an example, where 5 V is used as the initial guess, and the obtained RHS value is used as the guess for the next iteration. It can be seen that after three iterations, the LHS and RHS become equal, which is the solution: $v_{OUT} = 4.308$ V.

LHS	RHS
5.000 V	4.304 V
4.304 V	4.308 V
4.308 V	4.308 V

3.1.4 Important Second-Order Effects

The diode theory, presented in the previous sections, introduced the rudimentary diode model [Eq. (3.14)]. This model takes into account only the effects that are of principal importance for the diode operation. However, there is a number of second-order effects that significantly influence the characteristics of real diodes. Consequently, the rudimentary model very frequently fails to accurately fit the experimental data. This section introduces the most important second-order effects, namely carrier recombination and high-level injection effects, parasitic resistance effects and breakdown characteristic.

Recombination Current and High-Level Injection

The following two assumptions were made while developing the rudimentary model in the previous section:

1. No electron-hole recombination occurs in the depletion layer;
2. The electric field in the neutral regions is negligibly small, meaning that there is no voltage drop in the neutral regions.

These assumptions did simplify the modeling, however, most frequently they are not justified in terms of model accuracy.

As for the electron–hole recombination in the depletion layer, it is obvious that some of the electrons and holes will inevitably meet in the depletion layer recombining with each other there. This recombination process enables electrons and holes with kinetic energies smaller than the barrier height $q(V_{bi} - V_D)$ to contribute to the current flow, increasing the total current. Figure 3.5b helps to explain this phenomenon. Obviously, an electron and a hole cannot come to a common point if their kinetic energies (distances from the bottom and the top of the conduction and valence band, respectively) do not add up to $q(V_{bi} - V_D)$, which means such an electron–hole pair cannot possibly get recombined. An important point, however, is that they do not need independently to have energies larger than $q(V_{bi} - V_D)$, as necessary for the diffusion current described in the previous sections. In fact, an electron and a hole with kinetic energies as low as $q(V_{bi} - V_D)/2$ can meet each other in the middle and get recombined. If the recombination current was exclusively due to recombination of the electrons and holes satisfying this minimum energy condition, it would be proportional to $\exp[-(V_{bi} - V_D)/(2V_t)]$, thus proportional to $\exp[V_D/(2V_t)]$.

Whereas the diffusion current is proportional to $\exp[V_D/V_t]$, the maximum recombination current is proportional to $\exp[V_D/(2V_t)]$.

The voltage dependencies of the diffusion and the maximum recombination currents are similar, for they are both exponential-type dependencies. Indeed, they can be expressed in the following common way:

$$I \propto e^{V_D/nV_t} \tag{3.17}$$

where $n = 1$ for the diffusion current and $n = 2$ for the maximum recombination current. More importantly, varying the coefficient n between 1 and 2 would enable any combination of diffusion and recombination currents to be fit.

Furthermore, the introduction of the variable coefficient n eliminates the need for the second assumption used to develop the rudimentary model, namely the assumption of zero voltage drop in the neutral regions. The existence of excess minority electrons and holes (illustrated in Fig. 3.6b and c) means that there must be an electric field associated with this charge. In the case of high injection levels, the voltage drop associated with this field cannot be neglected.

The electric field in the neutral regions leads to voltage drops v_n between the depletion-layer edge and the N-type bulk, and v_p between the P-type bulk and the corresponding edge of the depletion layer. Therefore, the voltage applied V_D is distributed as $v_n + v_{depl} + v_p$, where v_{depl} is the voltage across the depletion layer. It is the voltage drop across the depletion layer that alters the height of the potential barrier in the depletion layer: $q(V_{bi} - v_{depl})$. The assumption used to derive the rudimentary model is that $v_n, v_p \ll v_{depl}$, thus $v_{depl} \approx V_D$.

In the case of high-level injection, the voltage that drops in the neutral regions becomes pronounced, which means the rudimentary model overestimates the current. This is because the barrier height is reduced only by a fraction of qV_D and not by the whole amount of qV_D as assumed in that model. However, the above-introduced coefficient n [Eq. (3.17)] enables this effect to be involved, as the term $qV_D/(nV_t)$ implies that the barrier is reduced by qV_D/n ($n > 1$), which is a fraction of qV_D.

Introduction of the variable coefficient n into the exponential dependence of the diode current on voltage applied, thus

$$I_D = I_S e^{V_D/nV_t} \tag{3.18}$$

provides better fitting to real diode characteristics, which are different from the diffusion model predictions due to the effects of high-level injection and carrier recombination in the depletion layer. The coefficient n is called the *emission coefficient*.

To illustrate the importance of the variable coefficient n, the experimental data of a real diode are fitted with the rudimentary model ($n = 1$) and the model that allows n to be adjusted [Eq. (3.18)] in Fig. 3.7. It is obvious that the rudimentary model (the dashed lines, $n = 1$) does not properly fit the data, while quite satisfactory fitting can be achieved when the emission coefficient is set to $n = 1.23$ (the solid lines).

The emission coefficient n can be introduced into the more general diode equation [Eq. (3.14)], covering both the forward-bias and the reverse-bias regions. The emission coefficient would have the above-described desirable effect in the forward-bias region, and it would not alter the model predictions in the reverse-bias region:

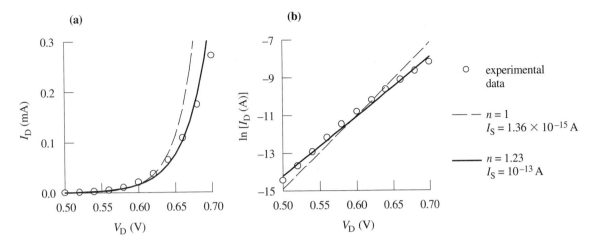

Figure 3.7 Linear–linear (a) and logarithmic–linear (b) plots illustrating best fits obtained by the rudimentary diode model ($n = 1$, dashed lines) and the model with variable emission coefficient n (solid lines). The experimental data (symbols) are for an LM3086 base-to-emitter P–N junction diode.

$$I_D = I_S \left[\exp\left(\frac{V_D}{n V_t} \right) - 1 \right] \tag{3.19}$$

Parasitic Resistance

Figure 3.8 illustrates the existence of parasitic resistances in the structure of the P–N junction diode. The parasitic resistances are especially pronounced at the metal–semiconductor contacts (both the cathode and the anode contacts) and in the neutral regions of the P-type and/or N-type bodies. The effects of all these resistances can be expressed by a single parasitic resistor r_S connected in series with the P–N junction diode itself, as illustrated in Fig. 3.9a.

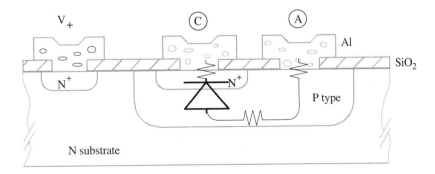

Figure 3.8 Illustration of the parasitic resistance in the structure of the P–N junction diode.

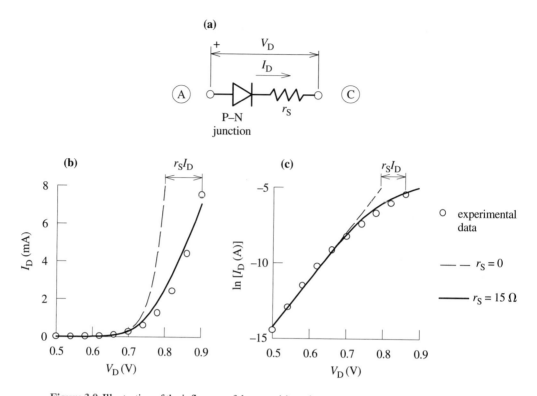

Figure 3.9 Illustration of the influence of the parasitic resistance on P–N junction diode characteristic: (a) electric-circuit model, (b) linear–linear, and (c) logarithmic–linear plots of the diode forward characteristic (the symbols show the experimental data for an LM3086 base-to-emitter P–N junction diode; the difference between the solid lines and the dashed lines shows the effect of the parasitic resistance r_S).

Figure 3.9b and c illustrates that the diode model of Eq. (3.19) (the dashed lines, $r_S = 0$, in Fig. 3.9) fits the experimental data properly only in the region of relatively small diode currents (< 1 mA in this example). Clearly, the voltage drop across the parasitic resistance ($r_S I_D$) is not significant compared to the voltage applied (V_D) when the current I_D is small. However, as the current I_D is increased, the voltage drop $r_S I_D$ becomes pronounced, which means that the voltage drop across the P–N junction itself is smaller than V_D, therefore the current is smaller. Equation (3.19) can be used to model the diode characteristic in this case if $V_D - r_S I_D$ is used instead of V_D. In other words, Eq. (3.19) is applied to the P–N junction diode in the equivalent circuit of Fig. 3.9a, and the resistance r_S is added between the cathode and the anode terminals.

Breakdown

The experimental characteristic of the P–N junction diode given in Fig. 3.10 shows that the reverse-bias current is small ($I_D \approx I_S$) only in the region $-BV < V_D \leq 0$. When the reverse-bias voltage becomes $V_D = -BV$ the current sharply increases. This condition is referred to as *P–N junction breakdown*. The P–N junction breakdown mechanisms, as well as applications of diodes operating in this regime, will be considered in more detail in Section 3.3. Here, the appearance of the breakdown is introduced as a second-order effect that limits the voltage range in which a diode can be used as a rectifier. Although the breakdown voltage of $BV \approx 7$ V (as shown in Fig. 3.10) is typical for the base-to-emitter type diodes, it is by no means representative for all possible P–N junction diodes. The breakdown voltage can be smaller than that, however, more frequently it is much larger than 7 V and can be as large as hundreds or even thousands of volts in specifically designed high-voltage diodes. Obviously, the breakdown voltage is a diode parameter that has to be specified for any particular type of diode.

3.2 SPICE MODELS AND PARAMETERS, STORED-CHARGE CAPACITANCE, AND TEMPERATURE EFFECTS

This section presents the diode models as used in SPICE. Most of the equations used in SPICE are derived in the previous section (static I–V characteristics) or the previous chapter (Section 2.4 describes the SPICE model for the capacitance of reverse-biased P–N junction). This section summarizes the equations and parameters used in SPICE, describes techniques for diode parameter measurement, and introduces the physics and modeling of the stored charge capacitance (this capacitance effectively determines the dynamic properties of the diode). In addition, the temperature effects are briefly described.

3.2.1 Static I–V Characteristic

The SPICE model of the static I–V diode characteristic is summarized in Table 3.1. The diagram in Table 3.1 shows that the diode is modeled as a current source connected in

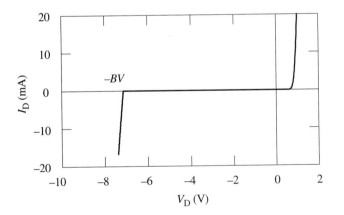

Figure 3.10 Experimental diode characteristic (LM3086 base-to-emitter P–N junction diode) illustrating the breakdown at $V_D = -BV$.

series to a resistor r_S. The resistor r_S represents the parasitic resistances, while the current source is controlled by the voltage drop across the P–N junction (V_{D0}) to describe the I–V characteristic of the P–N junction. The $I_D(V_{D0})$ equation appears in three parts. The first part is for $V_{D0} > -BV$ and it expresses the normal I–V diode characteristic, while the second and the third parts are for $V_D \leq -BV$, expressing the breakdown characteristic of the diode.

The first part of the SPICE $I_D(V_{D0})$ equation is a variation of the previously derived Eq. (3.19). An important change is that the voltage drop across the P–N junction (V_{D0}) is used instead of the terminal diode voltage V_D (obviously, this is to include the effects of the parasitic resistance r_S). The additional term $V_{D0}G_{MIN}$ is not important in terms of the diode characteristic description, it is added to enhance computational efficiency. G_{MIN} is set to a small value (typically 10^{-15} A/V) and normally does not show any observable influence on the diode characteristic. G_{MIN} is a program parameter (not a device parameter), and can be altered by the user.

The three device parameters involved in the first part of the SPICE equation ($V_{D0} > -BV$) are I_S, n, and r_S. Proper values of these parameters need to be set in order to ensure correct simulation. The default values of these parameters (typically, $I_S = 10^{-14}$ A, $n = 1, r_S = 0$) cannot guarantee an acceptable agreement between the model and the real characteristic of any possible type of diode. Figures 3.7 and 3.9 illustrate the importance of properly setting the values of n and r_S, respectively.

Figure 3.11 illustrates the extraction of the I_S, n, and r_S parameters from experimental $I_D - V_D$ data. As the current depends exponentially on the voltage, the $\ln I_D$-V_{D0}

TABLE 3.1 Summary of the SPICE Diode Model: Static I–V Characteristic

		Static Parameters		
Symbol	Usual SPICE Keyword	Parameter Name	Typical Value/ Range	Unit
I_S	IS	Saturation current		A
n	N	Emission coefficient	1–2	
r_S	RS	Parasitic resistance		Ω
BV	BV	Breakdown voltage (positive number)		V
	IBV	Breakdown current (positive number)		A
		Note: $\text{IBV} = \text{IS} \frac{BV}{V_t}$		

Static Diode Model

$$I_D(V_{D0}) = \begin{cases} \text{IS}\,(e^{V_{D0}/N\,V_t} - 1) + V_{D0}G_{MIN} & \text{if } V_{D0} > -BV \\ -\text{IBV} & \text{if } V_{D0} = -BV \\ -\text{IS}\left[e^{-(BV+V_{D0})/V_t} - 1 + \frac{BV}{V_t}\right] & \text{if } V_{D0} < -BV \end{cases}$$

graph is used to linearize the problem. The open symbols show the raw experimental data, that is when the measured V_D voltage is used as the P–N junction voltage V_{D0}. As the voltage drop across r_S (which is $r_S I_D$) is neglected in this case, the voltage V_{D0} is effectively overestimated by $r_S I_D$ (refer to Fig. 3.9). This effect is not pronounced at small currents as $r_S I_D \ll V_{D0}$, however, it becomes observable at high currents. A good initial guess for r_S can be obtained by judging the maximum deviation $r_S I_D$ of the raw experimental data from the straight line extrapolated from the low-current linear portion of the I_D–V_D dependence. The maximum deviation has to be, obviously, divided by the maximum current I_D to obtain r_S. Using the estimated value of r_S, the experimental diode voltage points V_D are transformed into P–N junction voltage points as $V_{D0} = V_D - r_S I_D$. If a straight line is obtained, the value of r_S is taken as the final value. Alternatively, r_S is altered and the process is repeated until a straight line is obtained.

The closed symbols in Fig. 3.11 show the straight line obtained after the voltage effect of the parasitic resistance, $r_S I_D$, is extracted from the raw experimental data. In other words, the closed symbols represent the experimental characteristic of the current source in the SPICE diode model (Table 3.1). As the experimental data are collected in the forward-bias region, where $\exp(V_{D0}/nV_t) \gg 1$, the SPICE $I_D(V_{D0})$ equation is reduced to

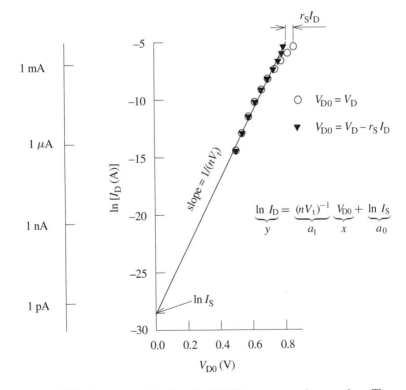

Figure 3.11 Extraction of diode static SPICE parameters, I_S, n, and r_S. The experimental data (symbols) and the fitting (line) are also shown in Fig. 3.9.

$$I_D = I_S e^{V_{D0}/nV_t} \tag{3.20}$$

Therefore, the logarithm of the current $\ln I_D$ linearly depends on V_{D0}:

$$\ln I_D = \frac{1}{nV_t} V_{D0} + \ln I_S \tag{3.21}$$

Figure 3.11 illustrates that the parameters I_S and n are obtained from the coefficients a_0 and a_1 of the linear $\ln I_D - V_{D0}$ dependence as

$$\begin{aligned} I_S &= e^{a_0} \\ n &= \frac{1}{a_1 V_t} \end{aligned} \tag{3.22}$$

The second and the third parts of the SPICE $I_D(V_{D0})$ equation given in Table 3.1 model the breakdown characteristic of the diode. Figure 3.10 shows the diode characteristic in breakdown (V_D around $-BV$). Due to the junction breakdown, the current sharply rises when the reverse-bias voltage $-V_D$ is increased beyond the breakdown voltage $-BV$. This sharp rise in the current is modeled by the exponential dependence shown in the third part of the SPICE $I_D(V_{D0})$ equation of Table 3.1. The second part of this equation defines current IBV as the current at $V_{D0} = -BV$. BV and IBV appear as device parameters. It is obvious from the example of Fig. 3.10 that the breakdown voltage BV can easily be obtained directly from the experimental I_D–V_D characteristic in the breakdown region. In this example, there is no other choice but to set the BV parameter at about 7.2 V. SPICE would accept an independently set IBV parameter, however, this may lead to numerical problems if there is a large discontinuity in the current around the $-BV$ point. The third part of the SPICE $I_D(V_{D0})$ equation is used for any voltage that is smaller than $-BV$. For points very close to $-BV$ (thus $V_{D0} \approx -BV$), the exponential term $e^{-(BV+V_{D0})/V_t} \approx 1$, which means that the current is approximately $-I_S(BV/V_t)$. To avoid discontinuity in the current (and therefore possible numerical problems), the parameter IBV should be set at the value $I_S(BV/V_t)$, as indicated in Table 3.1.

Example 3.3 Measurement of Static SPICE Parameters

A set of measured $I_D - V_D$ values for a P–N junction diode is given in Table 3.2. Obtain SPICE parameters I_S, n, and r_S for this diode.

TABLE 3.2 Current–Voltage Measurements

V_D [V]	0.67	0.70	0.73	0.76	0.80	0.84	0.91	1.00	1.26	1.65
I_D [mA]	0.1	0.2	0.5	1.0	2.0	5.0	10.0	20.0	50.0	100.0

Solution: Let us assume that the parasitic resistance r_S is in the order of 10 Ω. In that case, the voltage drop across the parasitic resistance is $\leq 1 \times 10^{-3} \times 10 = 0.01$ V for currents ≤ 1 mA. This means that the parasitic resistance effect can be neglected (0.01 V is much smaller than ≈ 0.7 V appearing across the P–N junction) for currents ≤ 1 mA. The measured diode current I_D can then be directly related to the measured voltage V_D as $I_D = I_S \exp(V_D/nV_t)$. This exponential equation can be linearized in the following way:

$$\ln I_D = \ln I_S + \frac{1}{nV_t} V_D$$

that is

$$y = a_0 + a_1 x$$

where $y = \ln I_D$, $x = V_D$, $a_0 = \ln I_S$, and $a_1 = 1/nV_t$. The results of this linearization transformation, applied to the first four experimental points ($I_D \leq 1$ mA) from Table 3.2 are given in Table 3.3.

The graphic method, explained in the previous section, can be used to find the coefficients a_0 and a_1 of this linear relationship. Alternatively, these coefficients can be calculated using the numerical linear regression method, described in Section 2.4.2. Applying the system of equations (2.56) to the data of Table 3.3, one obtains:

$$4a_0 + 2.86a_1 = -4.605$$

$$2.86a_0 + 2.0494a_1 = -3.1752$$

The solution of the above system of equations is $a_0 = -19.80$ and $a_1 = 26.08$. The parameters I_S and n can now be calculated as $I_S = \exp(a_0) = 2.52 \times 10^{-9}$ mA $= 2.52 \times 10^{-12}$ A, $n = 1/a_1 V_t = 1.48$.

If the parasitic resistance r_S was zero, the voltage V_D at the highest current $I_D = 100$ mA would be $V_D = 1.48 \times 0.02585 \times \ln(100/2.52 \times 10^{-9}) = 0.93$ V. It can be seen from Table 3.2 that the measured voltage is 1.65 V. The difference $1.65 - 0.93 = 0.72$ V is due to the voltage drop across r_S: $r_S I_D = 0.72$ V. Using this difference, the parasitic resistance is estimated as $r_S = 0.72V/I_D = 0.72/100$ mA $= 7.2$ Ω. If $r_S = 7.2$ Ω is a proper value, the voltage drop across the P–N junction V_{D0}, calculated as $V_D - r_S I_D$ should closely match the values calculated from the diode equation $nV_t \ln(I_D/I_S)$. The results of these calculations are presented in Table 3.4. It can be seen that the theoretical values (the third column) closely match the transformed experimental values (the second column). Therefore, we conclude that $I_S = 2.5 \times 10^{-12}$ A, $n = 1.48$, and

TABLE 3.3 Linearization of Current–Voltage Data

I_D [mA]	$y = \ln I_D$ [$\ln(mA)$]	$x = V_D$ [V]
0.1	−2.303	0.67
0.2	−1.609	0.70
0.5	−0.693	0.73
1.0	0.000	0.76

$r_S = 7.2\ \Omega$ represent a good set of SPICE parameters for the considered diode. If the matching was not good, the value of r_S would be altered to try to improve the matching.

TABLE 3.4 Transformed Current–Voltage Data

I_D (mA)	$V_{D0} = V_D - r_S I_D$ (V)	$V_{D0} = nV_t \ln(I_D/I_S)$ (V)
	$I_S = 2.5 \times 10^{-12}$ A, $n = 1.48$, $r_S = 7.2\ \Omega$	
0.1	0.67	0.67
0.2	0.70	0.70
0.5	0.73	0.73
1.0	0.76	0.76
2.0	0.79	0.78
5.0	0.80	0.82
10.0	0.84	0.84
20.0	0.86	0.87
50.0	0.90	0.91
100.0	0.93	0.93

3.2.2 Dynamic Characteristic: Stored-Charge Capacitance

Sections 2.2, 2.3 and 2.4 describe the capacitance of the reverse-biased P–N junction in detail. This capacitance, referred to as the depletion-layer capacitance, is inherently present in the P–N junction structure. It appears as a parasitic capacitance connected in parallel to the diode, as shown by the large-signal equivalent circuit of the diode presented in Table 3.5. The depletion layer capacitance C_d depends on the voltage applied V_{D0} as described in Sections 2.3 and 2.4. The equation used in SPICE to model the $C_d(V_{D0})$ dependence is also shown in Table 3.5. This equation is directly based on the $C_d(V_R)$ dependence described in detail in Section 2.4 [Eq. (2.52) and Table 2.1]. The difference is that the discussion in Section 2.4 is limited to the reverse-bias region (only reverse-biased P–N junctions are used as capacitors), while here any voltage V_{D0}, positive and negative, has to be included in the equation. In the reverse-bias region, $V_{D0} \approx -V_R$, as the effect of r_S is insignificant due to very small I_D current, and therefore the voltage drop $r_S I_D$ is negligible. This is also correct for positive V_{D0} smaller than about 0.5 V_{bi}. In addition, due to the insignificant I_D current for $V_{D0} < 0.5\ V_{bi}$, the concepts applied to derive the equations for the depletion-layer width and capacitance are still valid. Therefore, the validity of the $C_d(V_R)$ equation is expanded by the transformation $V_R = -V_{D0}$ where V_{D0} is allowed to take values up to 0.5 V_{bi}. SPICE has a different equation to calculate C_d at voltages $V_{D0} > 0.5\ V_{bi}$. However, the depletion-layer capacitance in that region is not very important as the total diode capacitance is dominated by another type of capacitance, stored-charge capacitance, described in the following.

Figure 3.6 illustrates the appearance of excess minority-carrier charge at the sides of the depletion layer in the forward-bias region, when a significant current flows through the diode. The density of this charge directly depends on the value of the current, thus the value of the voltage applied, V_D. Figure 3.12 shows that an increase in the voltage V_D by ΔV_D causes a corresponding increase in the minority-carrier charge (ΔQ_s) stored at the sides of the depletion layer. The charge change ΔQ_s caused by the voltage change ΔV_D,

TABLE 3.5 Summary of the SPICE Diode Model: Dynamic Characteristics

		Dynamic Parameters		
Symbol	**Usual SPICE Keyword**	**Parameter Name**	**Typical Value/ Range**	**Unit**
$C_d(0)$	CJO	Zero-bias junction capacitance		F
V_{bi}	VJ	Built-in (junction) voltage	0.65–1.25	V
m	M	Grading coefficient	$\frac{1}{3}-\frac{1}{2}$	
τ_T	TT	Transit time		s

Large-Signal Diode Model

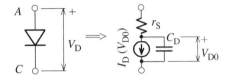

$I_D(V_{D0})$ is given in Table 3.1

$$C_D = C_d + C_s$$

$$C_d = \text{CJO} \left(1 - \frac{V_{D0}}{\text{VJ}}\right)^{-M} \quad \text{(for } V_{D0} < 0.5\,V_{bi})$$

$$C_S = \text{TT}\,\frac{dI_D}{dV_{D0}}$$

that is ΔV_{D0}, expresses the existence of a capacitance $C_s = \Delta Q_s/\Delta V_{D0}$. As the charge Q_s is called stored charge, this capacitance is referred to as *stored-charge capacitance*. The amount of the stored charge Q_s is directly dependent on the value of the current flowing through the diode I_D:

$$Q_s = \tau_T I_D \tag{3.23}$$

The proportionality coefficient τ_T in Eq. (3.23) is called *transit time*. Its physical meaning is described in Section 8.1.3. In SPICE it appears as a diode parameter.

The stored charge capacitance (C_s) is obtained from Eq. (3.23) as

$$C_s = \frac{dQ_s}{dV_{D0}} = \tau_T \frac{dI_D}{dV_{D0}} \tag{3.24}$$

which is the equation used in SPICE as shown in Table 3.5. Both the stored-charge capacitance C_s and the depletion-layer capacitance C_d appear across the P–N junction. Therefore, the total diode capacitance C_D is expressed as a parallel connection of C_s and C_d:

$$C_D = C_d + C_s \tag{3.25}$$

(a)

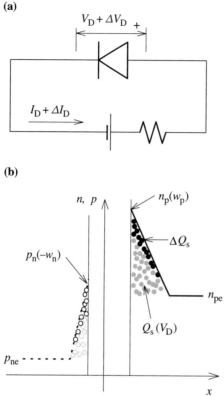

Figure 3.12 Illustration of the stored charge Q_s and the associated capacitance: voltage change ΔV_D (a) leads to change in the stored charge $\Delta Q_s = C_s \Delta V_D$ (b).

(b)

Measurement of the parameters $C_d(0)$, V_{bi}, and m, associated with the depletion-layer capacitance, is discussed in Section 2.4. To obtain the value of the τ_T parameter, high-frequency response or transient measurements are necessary. Figure 3.13 illustrates the effect of the stored-charge capacitance, the importance of setting the value of the associated parameter τ_T, as well as a possible parameter fitting approach.

To explain the effect of the stored-charge capacitance, consider the behavior of the diode when the forward-bias voltage is suddenly switched-off ($t = 50$ ns in Fig. 3.13c). With $\tau_T = 0$ (the dashed lines), the circuit simulation predicts quick exponential decay of the current i_D and the voltage v_D, as determined by the relatively small RC_d time constant. In reality, however, it takes much longer for the diode to turn off (i_D and v_D to drop to zero values). The thermal equilibrium ($i_D = 0$) is not achieved for as long as there is excess minority carrier charge Q_s at the P–N junction. Referring to Fig. 3.12b, the minority-carrier concentrations at the edges of the depletion layer, $n_p(w_p)$ and $p_n(w_n)$, should fall down to the equilibrium levels, n_{pe} and p_{ne}, respectively. This process of C_s discharge determines the time-response and high-frequency behavior of the diode.

The process of C_s discharge should be considered in more detail. When the forward-bias voltage is turned-off ($t = 50$ ns in Fig. 3.13c), the force that pushes the electrons and holes through the depletion layer and drives them further to the power supply is removed. The minority electrons and holes caught at the edges of the depletion layer (the stored charge)

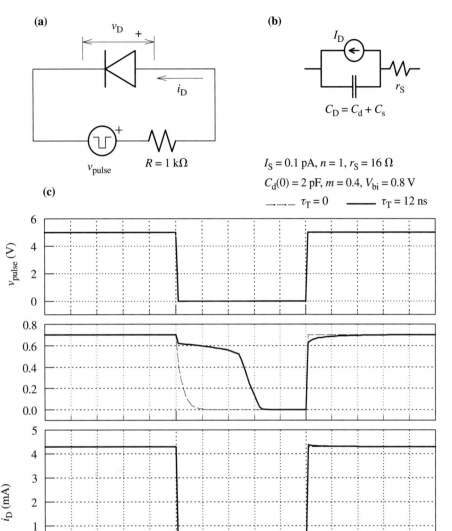

$I_S = 0.1$ pA, $n = 1$, $r_S = 16\ \Omega$

$C_d(0) = 2$ pF, $m = 0.4$, $V_{bi} = 0.8$ V

$\tau_T = 0$ ——— $\tau_T = 12$ ns

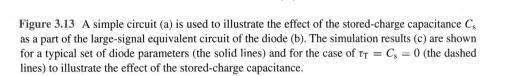

Figure 3.13 A simple circuit (a) is used to illustrate the effect of the stored-charge capacitance C_s as a part of the large-signal equivalent circuit of the diode (b). The simulation results (c) are shown for a typical set of diode parameters (the solid lines) and for the case of $\tau_T = C_s = 0$ (the dashed lines) to illustrate the effect of the stored-charge capacitance.

find it suddenly easier to flow back through the depletion layer returning to their respective original neutral regions. In the spirit of the band diagram of Fig. 3.5c, the excess electrons at the P-type side find it easier to roll down along the bottom of the conduction band back to the N-type region, and the holes at the N-type side find it easier to bubble up along the top of the valence band back to the P-type region. This produces the sudden change in the current i_D from positive to negative, shown in Fig. 3.13c.

The voltage drop v_D is no longer due to the externally applied forward-bias voltage, but to the stored charge at the depletion-layer edges: Q_s/C_s. In this period, the draining current i_D is limited by the external resistor R and the relatively small value of v_D. Consequently, the current $i_D = v_D/R$ appears as nearly constant (Fig. 3.13c). It should be mentioned here that the recombination mechanism continues to take place, which helps (and may even dominate) the process of stored-charge removal. Once the charge Q_s is drained, the current drops exponentially in the process of discharging the depletion-layer capacitance. This period starts at about 73 ns in the example of Fig. 3.13. The time constant associated with this discharging current is approximately $C_d R$.

Figure 3.13 clearly illustrates the difference between two simulated transient responses of the diode, one with the typical value of the parameter τ_T of 12 ns (solid lines), and the other with the stored-charge capacitance neglected by setting $\tau_T = 0$ (dashed lines). The parameter τ_T should be set so that the simulated transient response of a simple circuit, like the one in Fig. 3.13, as closely as possible fits the experimentally measured transient response using the same circuit.

3.2.3 Temperature Effects

The exponential term $\exp(V_{D0}/nV_t) = \exp(qV_{D0}/nkT)$ of the static diode model (Table 3.1) shows the explicit temperature dependence of the diode characteristic. More importantly, there is a very strong implicit dependence through the saturation current I_S. Equation (3.15) shows that I_S is proportional to n_i^2, where the intrinsic-carrier concentration n_i strongly depends on temperature. The intrinsic carrier concentration n_i is described in Section 1.5, where its temperature dependence is explicitly shown by Eq. (1.69). E_g in Eq. (1.69) is the energy gap, which is different for different materials, meaning that diodes based on different materials exhibit different temperature behavior. $E_g = 1.12$ eV for Si, which is the default value taken in SPICE, however, E_g is considered as a parameter and it can be set to a different value to express a different material. It should be noted that E_g depends on the temperature, but this dependence does not significantly influence the diode characteristics.

To include the temperature effects, the saturation current I_S is multiplied in SPICE by a semiempirical factor, as shown in Table 3.6. There is an additional parameter p_t that is called the saturation current temperature exponent, and whose value is typically $p_t = 3$ for silicon diodes.

To illustrate the temperature effects, the diode characteristic (originally shown in Fig. 3.7 for $T = 27°C$) is calculated by the SPICE model for temperatures $T = 100°C$ and $T = -50°C$, and is shown in Fig. 3.14. It can be seen that the temperature basically shifts the I_D–V_D characteristic along the V_D axis by approximately -2 mV/°C. This temperature coefficient is negative, as an increase in the temperature reduces the voltage drop across the diode needed to produce the same current.

TABLE 3.6 Summary of the SPICE Diode Model: Temperature Effects

		Temperature-Related Parameters		
Symbol	Usual SPICE Keyword	Parameter Name	Typical Value/ Range	Unit
E_g	EG	Energy gap	1.12 for Si	eV
p_t	XTI	Saturation current temperature exponent	3	

Temperature-Dependent Diode Model

Static Characteristic[a]

$$I_S(T) = \texttt{IS}\left(\frac{T}{T_{\mathrm{nom}}}\right)^{\texttt{XTI}/\texttt{N}} \exp\left[-\frac{q\,\texttt{EG}}{kT}\left(1 - \frac{T}{T_{\mathrm{nom}}}\right)\right]$$

Dynamic Characteristics[a]

$$V_{\mathrm{bi}}(T) \approx \frac{T}{T_{\mathrm{nom}}}\texttt{VJ} - 2\frac{kT}{q}\ln\left(\frac{T}{T_{\mathrm{nom}}}\right)^{1.5}$$

$$C_d(0, T) = \texttt{CJO}\left\{1 + \texttt{M}\left[400 \times 10^{-6}(T - T_{\mathrm{nom}}) - \frac{V_{\mathrm{bi}}(T) - \texttt{VJ}}{\texttt{VJ}}\right]\right\}$$

[a] IS, VJ, and CJO express the values of these parameters at the nominal temperature T_{nom}.

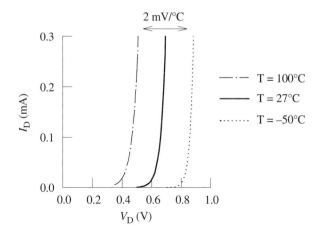

Figure 3.14 Temperature dependence of the diode characteristic

The temperature dependence of the diode capacitance is less significant, and is mainly due to the temperature dependence of the built-in voltage V_{bi} [refer to Eq. (2.4)], although $C_d(0)$ slightly depends on the temperature as well. Table 3.6 also shows the SPICE equations used to calculate the values of V_{bi} and $C_d(0)$ at temperature T, which is different from the nominal temperature T_{nom}.

The operating temperature T is set in SPICE in the same way as the device parameters, with a limitation that a unique operating temperature has to be used for the whole circuit. The nominal temperature is set to $T_{\mathrm{nom}} = 27°C$, although this value can also be changed by the user.

■ **Example 3.4　Temperature Dependence of V_D**

The anode of a P–N junction diode is connected to the positive terminal of a 12-V voltage source through a 1.2-kΩ resistor. This forward-biased diode is used as a temperature sensor. The room temperature ($T = 27°C$) parameters of the diode are $I_S = 10^{-12}$ A, $n = 1.4$, and $r_S = 10$ Ω. What temperature is measured by the diode if the voltage drop across the diode is $V_D = 0.807$ V.

Solution: The voltage drop across the diode changes by -2 mV/°C. We need to find V_D at room temperature to be able to determine the voltage shift, and consequently the temperature shift. The voltage drop across the diode is

$$V_D = nV_t \ln(I_D/I_S) + r_S I_D$$

As the current through the diode is

$$I_D = \frac{V_{supply} - V_D}{R}$$

the voltage drop can be expressed as

$$V_D = nV_t \ln\left(\frac{V_{supply} - V_D}{RI_S}\right) + r_S \frac{V_{supply} - V_D}{R}$$

This equation has to be solved numerically. If we assume $V_D = 0.7$ V, the value of the right-hand side (RHS) of the equation is 0.925 V. Assuming now $V_D = 0.925$ V, the RHS is calculated as 0.923 V. Finally, assuming $V_D = 0.923$ V gives a RHS of 0.923 V as well, which is the voltage drop across the diode at room temperature. Therefore, the voltage shift is $0.807 - 0.923 = -0.116$ V, which gives a temperature shift of -0.116 V/$(-0.002$ V/°C$) = 58°C$. The temperature that is measured by the diode is $27 + 58 = 85°C$.

3.3　REFERENCE DIODES: BREAKDOWN PHENOMENA

This section describes P–N junction diodes used to provide reference voltages. These diodes are referred to as *reference diodes*, or *Zener diodes*. As the reference diodes are operated in the breakdown region, this section provides insight into the P–N junction breakdown phenomena. There are two different types of breakdown: avalanche and tunneling (or Zener). The tunneling is a quantum-mechanical effect, so the section on tunneling will relate to some of the fundamental quantum-mechanical phenomena.

3.3.1　Basic Application

The simplest reference-voltage circuit is shown in Fig. 3.15a. Note that the symbol of the diode is altered to indicate that the diode is used as a reference diode. Figure 3.15b illustrates that the steep (nearly vertical) portion of the I_D–V_D characteristic in the breakdown is used to provide the reference voltage. The dashed load lines in Fig. 3.15b illustrate the effect of input voltage change. An increase in the input voltage shifts the load line increasing the current through the circuit, however, the output voltage remains constant. Similarly, a decrease in the input voltage alters the current only, but not the output voltage, provided the diode remains in the breakdown region.

3.3.2 Avalanche Breakdown

It is already explained that almost the whole reverse-bias voltage applied to a diode drops across the P–N junction depletion layer, giving rise to the electric field E in the depletion layer. The direction of this field is such that it takes the minority carriers through the P–N junction, which make the reverse-bias current of the diode (I_S). Let us consider an electron attracted by the electric field E from the P-type region (Fig. 3.16a). This electron

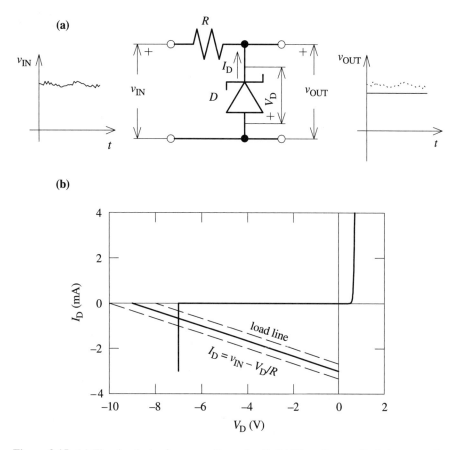

Figure 3.15 (a) The simplest reference-voltage circuit. (b) The reference diode is operated in its breakdown region. Variations of the input voltage (indicated by the dashed load lines) do not change the voltage drop across the diode, it is only the current that is influenced.

is accelerated by the electric field in the depletion layer, which means that the electron gains kinetic energy $E_{kin} = -qEx$ as the distance traveled by the electron (x) increases. As the electron moves through the depletion layer, there is a high probability that it will be scattered by a phonon (vibrating crystal atom) or an impurity atom. In the process of this collision, the electron may deliver its kinetic energy to the crystal or impurity atom. However, if the reverse-bias voltage V_R (and consequently the electric field E) is increased so much that the electron gains a kinetic energy that is large enough to break the covalent bonds when the electron collides with a crystal atom, an electron–hole pair is generated during this collision. This process is illustrated in Fig. 3.16a. Now we have two electrons that can generate two additional electron–hole pairs. The four new electrons can generate an additional four to become eight, then sixteen, and so on in this avalanche process.

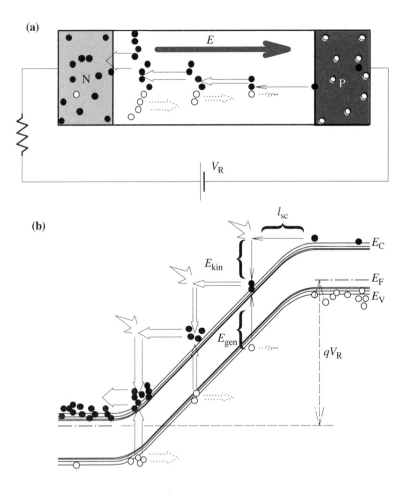

Figure 3.16 Illustration of the avalanche breakdown using a diode cross section (a) and the energy bands (b).

The electron–hole generation process is repeated many times in the depletion layer, as the average distance between two collisions (or also called scattering length) l_{sc} is much smaller than the depletion-layer width. Therefore, once the electric field is strong enough so that the kinetic energy gained between two collisions ($E_{kin} = -qEl_{sc}$) is larger than the threshold needed for electron–hole generation, the avalanche process is triggered and an enormous number of free carriers are generated in the depletion layer. This produces the sudden diode current increase in the breakdown region. Appropriately, this type of breakdown is called *avalanche breakdown*.

The energy-band diagram, shown in Fig. 3.16b, provides a deeper insight into the avalanche mechanism. We have frequently said that electrons roll down along the bottom of the conduction band, while the holes bubble up along the top of the valence band. This rolling and bubbling, however, are not quite as smooth as this simplified model may suggest. At this point we should remember the details of the electron transport along tilted energy bands, explained in Section 1.5.3 (Fig. 1.28). This process is again illustrated in the band diagram of Fig. 3.16b. The horizontal motion of the electrons between two collisions expresses the fact that the total electron energy is preserved. However, the potential energy of the electrons is transformed into kinetic energy when electrons move through the depletion layer as the energy difference from the bottom of the conduction band increases. When the electrons collide and deliver the kinetic energy E_{kin} to the crystal, they fall down on the energy-band diagram. Therefore, the rolling down of electrons and bubbling up of holes are in a staircase fashion.

As mentioned above, the kinetic energy E_{kin} gained between two collisions can be delivered to the crystal as thermal energy. In fact, this is the only possibility when $E_{kin} < E_g$, where E_g is the energy gap. However, as the slope of the bands is increased by increases in the reverse-bias voltage, there will be a voltage V_R at which E_{kin} becomes large enough to move an electron from the valence band into the conduction band. This is the process of electron–hole pair generation. As mentioned above, this process is repeated many times in the depletion layer, where the newly generated electrons also gain enough energy to generate additional electron–hole pairs. This is the avalanche mechanism. Figure 3.16 shows only three steps in the avalanche process, however, in reality there would be many more as the average distance between two collisions (also called the scattering length) l_{sc} is generally much smaller than the depletion layer width. Note that, theoretically, the holes can also generate electron–hole pairs, however, this process is rarely observed as the electron-induced avalanche requires smaller energy and happens at lower voltages.

Understanding the mechanism of the avalanche process gives us the basis to understand the *temperature behavior of the avalanche breakdown*. If the temperature is increased above room temperature, the phonon scattering (Section 1.4.3) is enhanced and the average distance between two collisions (scattering length l_{sc}) is reduced. This means that a larger electric field E is needed to achieve the threshold kinetic energy ($E_{kin} = -qEl_{sc}$).

> An increase in the temperature increases the breakdown voltage. The avalanche breakdown is said to have a positive temperature coefficient.

This temperature dependence of the avalanche breakdown is generally undesirable, as we do not want the reference voltage to be temperature dependent.

We can also understand now that the *avalanche breakdown voltage is concentration dependent*. A diode with higher doping levels in the P-type and N-type regions has a larger built-in electric field (electric field due to the ionized doping atoms in the depletion layer). Therefore less external voltage V_R is needed to achieve the critical breakdown field. Adjusting the doping levels enables diodes with avalanche breakdown voltage as low as ≈ 6 V and as high as thousands of volts to be made.

3.3.3 Tunneling

The tunneling effect is related to the wave properties of electrons, briefly described in Section 1.5. It may be hard to accept the wave–particle duality concept, and it may seem that the particle and wave properties of the electrons should contradict each other. This confusion is due only to a wrong idea that an electron appearing as a particle should precisely be confined in a minute volume of space, probably of a spherical shape. The truth is that the electrons do not have precisely defined dimensions. Figure 3.17 illustrates the electron in a hydrogen atom. All we know about the electron shape is its so-called wave function, which is shown in the one-dimensional graph of Fig. 3.17a and also illustrated by the two-dimensional cross section of Fig. 3.17b. Chapter 12 deals with the wave function in more detail. The meaning of the wave function is not easy to comprehend. However, it is not as hard to accept the fact that the wave function is directly related to the probability of finding the electron at a certain point in space. Figure 3.17 shows that the electron is not localized at distance r_{BOHR} from the atom center, although it is correct that it is the most likely to find it there.

The wave function is different for the electrons in the silicon crystal. Figure 3.18a illustrates the bound electrons (electrons in the valence band) in the P-type region of the diode. The energy gap at the P–N junction separates the valence electrons of the P-type region from the N-type side of the diode. This, however, does not mean that the electron wave function abruptly drops to zero at the energy barrier (the top of the valence band in the depletion layer). This would mean that the electrons can be confined in a strictly defined space, denying them the wave properties they do exhibit. The wave function does have a tail expanding beyond the position of the top of the valence band. Nonetheless, we do not find electrons there, as there are no available energy levels for the electrons in the energy gap.

A different situation is illustrated in Fig. 3.18b. In this case, the width of the energy barrier d is reduced by the increased reverse-bias voltage. This reduced energy barrier width is smaller than the tail of the electron wave function, which means there is some probability that a valence electron from the P-type region is found on the N-type side of the P–N junction. *This is the tunneling* phenomenon.

> The probability of finding an electron (or hole) on the other side of energy barrier is called *tunneling probability*.

The tunneling effect cannot be neglected even though the tunneling probability is typically very small. To explain this, assume a tunneling probability of 10^{-7}. As there are more than 10^{22} electrons per cm^3 in the valence band, the concentration of electrons tunneling through the P–N junction energy barrier is higher than $10^{22} \times 10^{-7} = 10^{15}$ cm^{-3}.

(a)

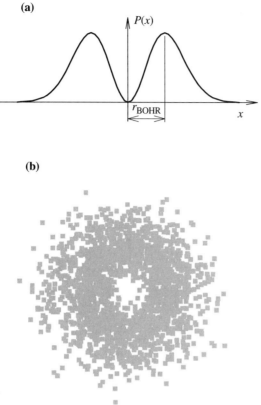

Figure 3.17 One-dimensional (a) and two-dimensional (b) presentation of the wave function of a hydrogen electron.

(b)

This concentration of electrons is high enough to make a significant current when the tunneling electrons are attracted to the positive battery terminal connected to the N-type side of the diode. This current is called *tunneling current*.

If the reverse–bias voltage is further increased, the associated energy-band bending will further reduce the energy barrier width d, increasing the tunneling probability. As the tunneling probability depends exponentially on the energy barrier width, the tunneling current depends exponentially on the reverse-bias voltage. This is illustrated in Fig. 3.18c.

The tunneling breakdown cannot occur unless the depletion layer, and consequently the energy barrier, is very narrow. A narrow depletion layer appears when the doping levels of both the N-type and the P-type region of the diode are very high. In addition, an increase in the reverse–bias voltage V_R does not significantly expand the depletion layer width, w_D. This makes the energy barrier width (d) reduction due to band bending quite significant.

We mentioned before that the doping concentration influences the avalanche breakdown voltage as well. In fact it is the level of doping concentration that determines whether the avalanche breakdown voltage or the tunneling breakdown voltage will be smaller, and therefore determines which type of breakdown will actually occur. If the doping concentration is such that the breakdown occurs at more than about 6 V, the breakdown is generally avalanche-type breakdown. At higher doping levels, the tunneling occurs at voltages lower

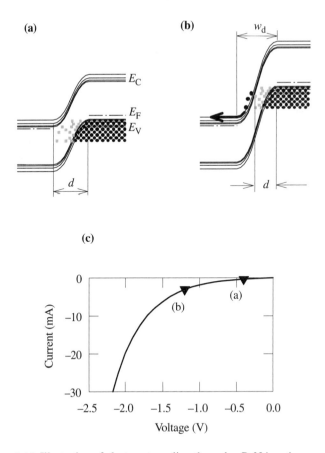

Figure 3.18 Illustration of electron tunneling through a P–N junction: (a) small reverse-biased voltage, no tunneling is observed, (b) the tail of the wave function for valence electrons in the P-type region expands onto the N-type side—this is the phenomenon of electron tunneling through the P–N junction energy barrier, and (c) typical tunneling breakdown I_D–V_D characteristic.

than those needed to trigger the avalanche mechanism, therefore, the tunneling breakdown is actually observed.

Reference silicon diodes with breakdown voltages lower than about 5 V operate in the tunneling mode.

The I_D–V_D characteristic associated with the tunneling breakdown is not as abrupt as the I_D–V_D characteristic of the avalanche breakdown. This can be seen by comparing the I_D–V_D characteristics of Fig. 3.10 and Fig. 3.18c. Consequently, the stability of the reference voltage provided by the diodes above 6 V is a lot better than what can be achieved by the sub-5-V diodes.

Another difference is the temperature coefficient. Whereas the temperature coefficient of the avalanche breakdown is positive, it is negative in the case of tunneling breakdown. An increase in the temperature reduces the tunneling breakdown voltage. This enables the temperature dependence of the reference voltage to be reduced by designing diodes that operate in the mixed, avalanche and tunneling, breakdown mode. The breakdown voltage of these diodes is generally in the 5 to 6 V range.

3.4 SCHOTTKY DIODES: METAL–SEMICONDUCTOR CONTACT

Although the most frequently used, the P–N junction is not the only way of making a semiconductor diode. What is needed to make a diode is a potential barrier that can be reduced by externally applied voltage (to let the carriers flow through) and that can prevent any significant leakage current flow at the reverse bias. In the P–N junctions, this potential barrier (reflected as the built-in voltage) appears due to the difference of the work functions of P-type and N-type semiconductors. Similar potential barrier appears at metal–semiconductor contacts, due to a work-function difference between the metal and the semiconductor. The diodes based on the metal–semiconductor contact are called Schottky diodes.

The Schottky diodes are different from the P–N junctions. The most important difference is that the Schottky diodes operate with a single type of carrier (the majority carriers). The absence of the minority carriers means that the stored-charge effect does not exist, which makes the Schottky diodes suitable for fast switching applications.

This chapter describes the structure and the operation of the Schottky diodes. The energy-band diagrams are used in the explanations as they are by far the best way of comprehending the effects at metal–semiconductor contacts. The fact that metal–semiconductor contacts appear as both rectifying contacts (Schottky diode) and ohmic contacts can be confusing. The last section describes the ohmic contacts to show which technological parameter influences the rectifying/ohmic behavior of the metal–semiconductor contacts.

3.4.1 Schottky Diode Structure and Basic Applications

There are two different contacts of metal to N-type semiconductor in the cross section of Fig. 3.19. The first is the rectifying contact (Schottky diode), which appears between the metal and the low doped N-type semiconductor. The second contact is ohmic contact (nonrectifying), which appears between the metal and the heavily doped N-type semiconductor (labeled as N^+ to express the fact that the doping level is high). The second contact is needed only to provide a contact to the diode cathode, which is the low-doped N-type semiconductor region. The metal at the rectifying contact plays the role of the diode anode.

The N–P junction in Fig. 3.19 provides electric isolation from the substrate and the other devices. This is the already described way of isolation by the reverse-biased P–N junction, frequently used in integrated circuits. The reverse biasing of the N–P junction

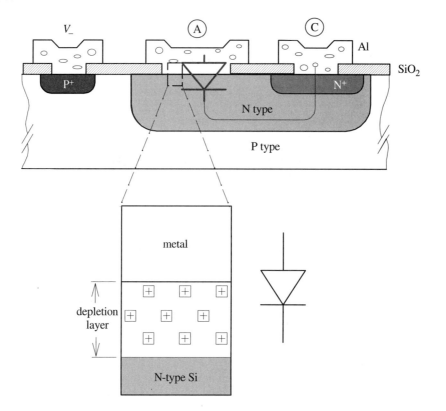

Figure 3.19 Cross section of a Schottky diode.

in Fig. 3.19 is achieved by contacting the P-type substrate to the most negative potential V_-. Schottky diodes are frequently used as discrete components. In that case the structure is different, as the isolation N–P junction is not needed, and the cathode contact can be made as a back contact. To reduce the parasitic resistance in this case, the substrate is heavily doped (N^+), and a surface low–doped N-type layer is deposited to enable the rectifying contact.

In principle, the Schottky diode can be used in the same way as the P–N junction diodes. The Schottky diodes are advantageous in a number of applications. In the rectifying circuits, it is frequently important to minimize the power dissipated in the diode. As Schottky diodes generally have smaller forward-bias voltage V_D compared to the P–N junctions, the power dissipated in the diode $V_D I_D$ is smaller in the case of Schottky diodes. Also, the effect of stored minority-carrier charge does not exist in the Schottky diodes, which makes the turn-off time extremely short. This makes the Schottky diodes very attractive for high-frequency switching and microwave linear applications.

The description of the Schottky diode operation using the band-diagram model, given in the following section, will elucidate the difference between the Schottky and P–N junction diodes.

3.4.2 Energy-Band Diagram of Rectifying Metal–Semiconductor Contacts

It is very useful to consider first the energy band diagrams of an N-type semiconductor and a metal isolated from each other. In Fig. 3.20a, the energy-band diagrams of these two materials are drawn so that the vacuum levels are matched. Remember, the vacuum level is the energy level of a completely free electron (isolated electron in vacuum). The electrons in solids have negative energies with respect to the vacuum level—they need some energy to be able to liberate themselves from the attracting forces in the crystal and become free electrons. The energy-band diagram of the metal shows that a number of electrons are mobilized by the thermal energy, although a lot higher temperature is needed to give the electrons enough energy to leave the metal. This energy, called work function $(q\phi_m)$, is the difference between the vacuum level and the Fermi level E_F. It is useful to remember that the Fermi level represents the energy of an average electron. In other words, the probability of finding an electron with energy corresponding to the Fermi level is 0.5.

As explained in Section 2.5.3, the same definition of the work function has to be used in the case of semiconductors, regardless of the fact that the Fermi level is in the energy gap and there can be no electrons at the Fermi level. In Fig. 3.20a, the Fermi level of the N-type semiconductor is higher than the Fermi level of the metal, which means that the mobile electrons in the semiconductor are at higher energy levels compared to the electrons in the metal.

Thermal Equilibrium

Let us now consider what is going to happen when an electric interaction between the two materials is allowed, as in Fig. 3.20b. In this case, the two materials represent a single system. The electrons from the higher energy levels in the semiconductor move into the metal, creating a depletion layer at the semiconductor surface. The uncompensated positive donor ions in the depletion layer create an electric field, which is associated with the appearance of a potential difference between the surface of the metal and the bulk of the semiconductor. When the thermal equilibrium is established, this potential difference is exactly equal to the initial difference between the Fermi levels, which means the Fermi levels in the semiconductor and the metal are now perfectly aligned to each other. We already know that the Fermi level of a single system in thermal equilibrium is constant.

There is a gap between the metal and the semiconductor in Fig. 3.20b. In reality, it is very likely that there will be a minute gap of the order of atomic distances at the interface. Although this does influence the properties of the metal–semiconductor contact, it is not of essential importance. In terms of easier understanding, it is helpful to assume ideal contact. The ideal metal–semiconductor contact is illustrated in Fig. 3.20c. The energy-band diagram is constructed in the following way:

1. The Fermi level (dash–dot line in Fig. 3.20c) is drawn first.
2. The conduction and valence bands are drawn in the neutral region of the semi-conductor. The bands are placed appropriately with respect to the Fermi level, so

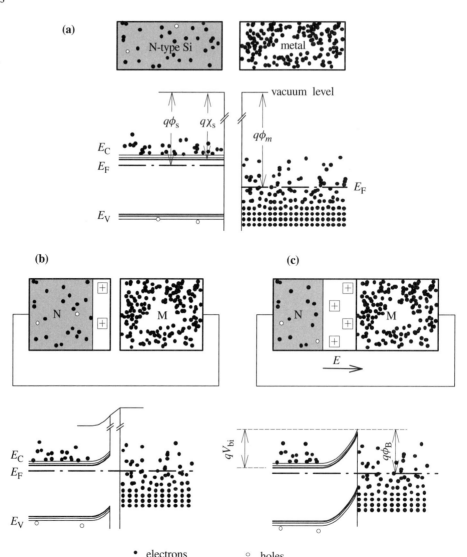

Figure 3.20 Illustration of metal–semiconductor contact in thermal equilibrium: (a) separated N-type silicon and metal, (b) N-type silicon and metal in contact, and (c) full (ideal) contact between N-type silicon and metal.

to express the doping level of the semiconductor (the energy-band diagram in the neutral region should be the same as in Fig. 3.20a).

3. The energy bands are bent in the depletion layer of the semiconductor by an amount that is equal to the original difference between the Fermi levels of the metal and the semiconductor ($q\phi_m - q\phi_s$). This bending represents the built-in voltage inside the semiconductor:

$$q V_{bi} = q\phi_m - q\phi_s = q\phi_{ms} \tag{3.26}$$

4. Inside the metal, no change in the energy-band diagram is practically observed, because the electric field can penetrate only to atomic distances. Consequently, the energy barrier for the electrons in the metal,

$$q\phi_B = q\phi_m - q\chi_s \tag{3.27}$$

is independent of either semiconductor doping or bias applied.[2]

Reverse Bias

Now, we can consider the effects of reverse and forward biases. Figure 3.21a illustrates the case of reverse bias (negative voltage between the metal and the N-type semiconductor). Applying *negative voltage* $V_D = -V_R$ to the metal with respect to the N-type semiconductor means that its *Fermi level is above* the Fermi level of the semiconductor by the amount $q V_R$. Remember that *potential energy* $= -q(electric\ potential)$.

The reverse bias $V_D = -V_R$ increases the energy barrier height for the electrons in the semiconductor to $q(V_{bi} + V_R)$. The increased energy barrier in the depletion layer of the semiconductor effectively prevents the electrons from the semiconductor from moving through the contact.

On the other hand, the barrier for the electrons in the metal ($q\phi_B$) does not change. The minor number of electrons in the metal that are able to overcome the potential barrier and appear in the N-type semiconductor make the *reverse–bias current* of the Schottky diode. Following reasoning analogous to that used in the case of the P–N junction barrier [Eq. (3.6)], we can conclude that the number of electrons able to go over $q\phi_B$ is proportional to $\exp(-q\phi_B/kT)$. Therefore, the reverse-bias current is

$$I_S = I_{S0}e^{-q\phi_B/kT} \tag{3.28}$$

where I_{S0} is a temperature-dependent constant. Given that $q\phi_B$ is bias independent, we conclude that the reverse–bias current of the Schottky diode does not depend on the reverse–bias voltage.[3]

I_S can be considered as a Schottky diode parameter, in which case the reverse-bias current is expressed in the same way as for the P–N junction diode:

$$I_D = -I_S \tag{3.29}$$

---

[2]Both $q\phi_m$ and $q\chi_s$ are doping and bias independent.

[3]A more precise equation for the saturation current and a description of a second-order effect that causes a reverse-bias current increase are given in Section 10.2.

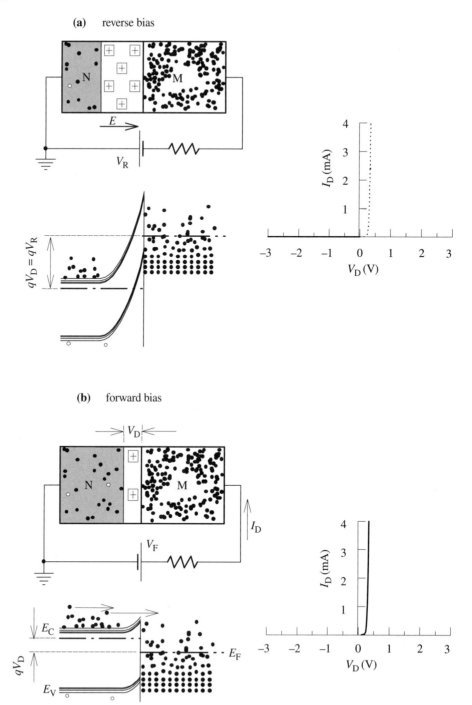

Figure 3.21 Illustration of a reverse-biased (a) and forward-biased (b) Schottky diode.

Forward Bias

Analogously to the case of the P–N junction, forward–bias voltage V_D reduces the energy barrier in the depletion layer to $q(V_{bi} - V_D)$ (Fig. 3.21b). This enables a number of electrons from the semiconductor to overcome the barrier and appear in the metal. The number of electrons that move in the opposite direction, from the metal into the semiconductor, is the same as in the case of reverse bias. However, the current of the electrons from the semiconductor dominates. Moreover, this current increases exponentially with the forward-bias voltage. Again, this is due to the fact that the electrons are exponentially distributed along the energy in the conduction band. As the energy barrier height is reduced by qV_D, the number of electrons able to go over the barrier increases exponentially with V_D.

> Notwithstanding the different physical background of I_S, the forward-bias current of the Schottky diode can be expressed in the same way as the forward-bias current of the P–N junction:

$$I_D = I_S e^{V_D/nV_t} \tag{3.30}$$

Of course, the values of the parameters I_S and n should be adjusted so to fit the characteristic of a particular diode. In that sense, there is no different SPICE model for the Schottky diode. The values of the diode parameters are adjusted appropriately to reflect the characteristics of Schottky diodes.

Although the same mathematical equation can be used to model the I_D–V_D characteristic of the Schottky diodes, there are important practical differences. Typically, the built-in voltage (V_{bi}) is smaller in the case of Schottky diodes.[4] Whereas the current of the P–N junction becomes significant at about 0.7 V, in the case of the Schottky diodes the same current can be achieved by a voltage as low as 0.2 V. This represents an advantage of the Schottky diodes in applications in which the power loss $V_D I_D$ across the diode is critical. This improvement, however, is achieved at the expense of increased reverse-bias (saturation) current I_S. Mathematically, it can be seen that I_S in Eq. (3.30) has to be significantly increased to account for the significant increase in I_D. This can also be concluded from the energy bands. Using a metal with smaller $q\phi_m$ to reduce the built-in voltage (Eq. 3.26) leads to a related reduction of $q\phi_B$ (Eq. 3.27), and hence an increase in I_S (Eq. 3.28). In the extreme case, the asymmetric (rectifying) I_D–V_D characteristic becomes the symmetrical I_D–V_D characteristic of the short circuit.

Another important difference between the P–N junction and the Schottky diode is in the dynamic characteristics. In the case of the P–N junction diode, the electrons from the N-type semiconductor appear as minority carriers in the P-type region after they pass through the junction. This leads to an accumulation of the minority carriers at the sides of the depletion layer—stored charge. When the forward-bias voltage is switched off, the steady-state reverse-bias current cannot be established before the stored charge is

---

[4]Note from Eq. (3.26) that V_{bi} of Schottky diodes can be technologically altered by using metals with different work functions (ϕ_m).

removed.[5] In the case of the Schottky diode (Fig. 3.21b), the electrons from the silicon appear in the *metal* after they pass through the contact. These electrons appear among the huge number of electrons already existing in the metal. They are not minority carriers and do not make any difference.

> The Schottky diode responds much more quickly to voltage changes compared to the P–N junction diode, due to the absence of the stored-charge effects.

3.4.3 Energy-Band Diagram of Ohmic Metal–Semiconductor Contacts

The depletion-layer width in a semiconductor, and therefore the width of the energy barrier at a Schottky contact, depends on the doping level. Mathematically, this can be seen from Eq. (2.34), which gives the depletion layer width in an N-type semiconductor as a part of one–sided ($N_A \gg N_D$) abrupt P–N junction. The same equation would be obtained in the case of metal–N-type semiconductor contact if the Poisson equation was solved. The heavily doped P-type semiconductor in the one-sided P–N junction behaves similarly to the metal in the case of the Schottky diode, as far as the depletion layer is concerned. Equation (2.34) shows that the depletion layer is narrower at higher doping levels N_D. This phenomenon was discussed in Section 2.3.2 [refer to Eq. (2.21) and the associated text].

Figure 3.22 illustrates the contact between metal and heavily doped N-type semiconductor. To express that the doping level is high, the semiconductor is labeled as N^+ type. Due to the high doping level, the energy barrier at the surface of the semiconductor is very narrow. As a consequence, the electrons from the metal can tunnel through the barrier when negative voltage is applied to the metal (Fig. 3.22a). As the applied voltage is increased, the associated splitting of the Fermi levels increases the band bending, further narrowing the energy barrier. This leads to a significant increase in the tunneling current.

> In the case of a very narrow energy barrier at a metal–semiconductor contact, due to high doping level in the semiconductor, the electrons from the metal tunnel through the barrier producing not only a significant reverse–bias current, but also current that strongly depends on the voltage applied.

Figure 3.22b illustrates the case of positive voltage applied to the metal. In this case the electrons from the semiconductor not only go over the reduced barrier (as in the case of the Schottky diode), but also tunnel through the barrier significantly increasing the current.

> Positive voltage at the metal also produces a current flow through the contact that rapidly increases with the voltage increase. The electric characteristic of the contact between a metal and a heavily doped semiconductor is equivalent to the characteristic of a small resistance.

---

[5]The effects of the stored charge on the dynamic characteristics of the P–N junction diode are explained in Section 3.2.2.

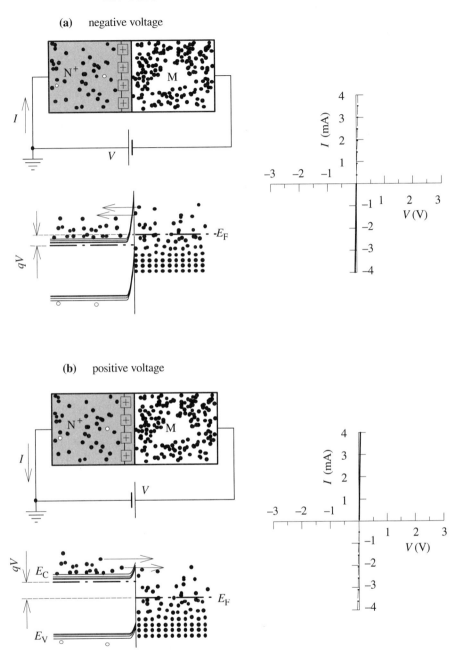

(a) negative voltage

(b) positive voltage

Figure 3.22 Illustration of ohmic metal–semiconductor contact with negative (a) and positive (b) voltage at the metal.

PROBLEMS

Section 3.1 Rectifying Diodes: Fundamental Effects and Models

3.1 Assign each of the band diagrams from Fig. 3.23 to a statement describing the biasing condition.

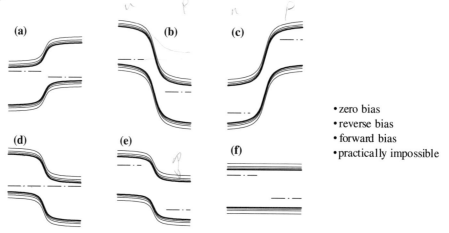

Figure 3.23 Energy-band diagrams.

- zero bias
- reverse bias
- forward bias
- practically impossible

3.2 Assign each of the concentration diagrams from Fig. 3.24 to a statement describing the biasing condition. Electron concentrations are presented with solid lines and hole concentrations are presented with dashed lines.

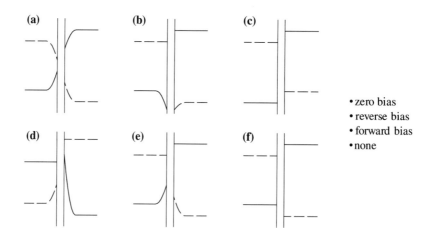

Figure 3.24 Concentration diagrams.

- zero bias
- reverse bias
- forward bias
- none

3.3 Which of the following statements, related to a forward-biased P–N junction, are correct?

- The voltage drop in the neutral regions can be neglected compared to the voltage drop in the depletion layer.

- The electric field in the neutral regions is not large enough to produce a significant drift current of the majority carriers.

- The diffusion current of the minority carriers is significant.
- The number of electrons able to overcome the energy barrier at the P-N junction increases exponentially with the forward-bias voltage.
- The drift current of the minority carriers is significant.
- The forward bias does not influence the barrier height at the P-N junction, as it does not affect the built-in voltage.
- If the forward bias is suddenly switched off, the excess minority carriers continue to flow in the same direction until their concentration drops to the equilibrium level.
- If the forward bias is suddenly switched off, the P-N junction current changes the direction, but continues to flow until the minority carrier concentrations reach the equilibrium levels.

3.4 The donor and acceptor concentration of the diode shown in Fig. 3.1 are $N_D = 10^{20}$ cm^{-3} and $N_A = 10^{16}$cm^{-3}, respectively. Find the minority carrier concentration at the edges of the depletion layer for

(a) forward bias $V_F = V_D = 0.65$ V,

(b) reverse bias $V_R = |V_D| = 0.65$ V. $\boxed{A}$

3.5 The neutral regions of the anode and cathode of the diode from Problem 3.4 are much smaller than the diffusion lengths: $W_{anode} = 4$ μm, $W_{cathode} = 2$ μm. Assuming linear distribution of minority carriers in the neutral regions, find the diffusion current density of the minority electrons and holes at $V_F = 0.65$ V, if the diode is implemented in

(a) silicon ($\mu_n = 1450$ cm^2/Vs, $\mu_p = 500$ cm^2/Vs, $n_i = 1.02 \times 10^{10}$ cm^{-3}),

(b) GaAs ($\mu_n = 8500$ cm^2/Vs, $\mu_p = 400$ cm^2/Vs, $n_i = 2.1 \times 10^6$ cm^{-3}). $\boxed{A}$

3.6 For the diode considered in Problems 3.4 and 3.5, calculate the reverse-bias ($V_R = |V_D| = 0.65$ V) and the forward-bias ($V_F = V_D = 0.65$ V) current densities. Is the forward-bias current equal to the sum of electron and hole diffusion currents calculated in Problem 3.5? Perform the calculations for both Si and GaAs.

3.7 The doping and geometric parameters of a P–N junction diode are $N_D = 10^{20}$ cm^{-3}, $W_{cathode} = 1$ μm $\ll L_p$, $N_A = 5 \times 10^{15}$ cm^{-3}, $W_{anode} = 10$ μm $\ll L_n$, and the junction area $A_J = 500 \times 500$ μm^2. Calculate the saturation current I_S, and then use this result to obtain the forward voltage that corresponds to a current of 100 mA if the semiconductor material is

(a) Si ($\mu_n = 1450$ cm^2/Vs, $\mu_p = 500$ cm^2/Vs, $n_i = 1.02 \times 10^{10}$ cm^{-3});

(b) SiC ($\mu_n = 380$ cm^2/Vs, $\mu_p = 70$ cm^2/Vs, $n_i = 1.6 \times 10^{-6}$ cm^{-3}). $\boxed{A}$

3.8 The saturation current of a P–N junction diode is $I_S = 10^{-11}$ A. The breakdown voltage of this diode can be increased if the doping of the lower-doped N-type region is reduced from $N_D = 5 \times 10^{16}$ cm^{-3} to $N_D = 5 \times 10^{14}$ cm$^{-3} \ll N_A$. How would this reduction of

N_D influence the saturation current? Calculate the new saturation current, assuming that the carrier mobility does not change.

3.9 Find the missing result in the following table:

V_D (V)	0.70	0.72	0.74
I_D (mA)	0.6	?	2.3

3.10 Find the parasitic resistance, if the current of the diode from Problem 3.9 is $I_D = 17$ mA at $V_D = 0.88$ V.

Section 3.2 SPICE Models and Parameters, Stored-Charge Capacitance, and Temperature Effects

3.11 A set of experimental I_D–V_D data, shown by the symbols in Fig. 3.25, is fitted with three different sets of parameters (the lines). Identify which set of parameters corresponds to each of the lines.

3.12 The SPICE parameters of the diode used in the circuit of Fig. 3.6a are: $I_S = 10^{-12}$ A, $n = 1.4$, and $r_S = 10$ Ω. The current flowing through the circuit is found to be $I_D = 3.5$ mA. Knowing that the thermal voltage is $V_t = 26$ mV, determine the voltage between the diode terminals V_D.

3.13 Determine the static SPICE parameters (the saturation current I_S, the emission coefficient n, and the contact resistance r_S) from the following I_D–V_D data:

V_D (V)	0.65	0.70	0.76	0.81	0.89
I_D (μA)	79	264	879	2925	9685

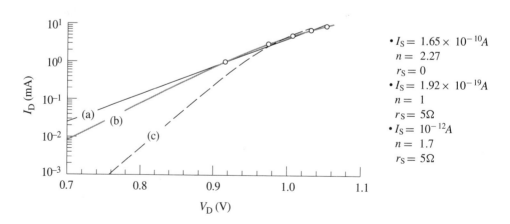

$\bullet\, I_S = 1.65 \times 10^{-10} A$
$n = 2.27$
$r_S = 0$
$\bullet\, I_S = 1.92 \times 10^{-19} A$
$n = 1$
$r_S = 5Ω$
$\bullet\, I_S = 10^{-12} A$
$n = 1.7$
$r_S = 5Ω$

Figure 3.25 A set of measured I_D–V_D points (symbols) is fitted with three different sets of parameters (lines).

3.14 The operating junction temperature of a P–N junction diode is estimated to be $T = 75°C$. To obtain the SPICE parameters for simulations with 75°C as the nominal temperature, two I_D–V_D points are measured at 75°C: $I_{D1} = 0.57$ mA, $V_{D1} = 0.67$ V; $I_{D2} = 1.28$ mA, $V_{D2} = 0.70$ V.

(a) Calculate I_S and n.

(b) What would I_S and n be if these measurements were performed at 0°C? $\boxed{A}$

3.15 A diode circuit is simulated by SPICE with four different sets of parameters. The results for the current flowing through the diode are shown in Fig. 3.26. Identify which set of parameters corresponds to each of the simulation results.

3.16 A current of 1 mA flows through a P–N junction diode. Calculate the associated stored-charge capacitance, if the transit time is $\tau_T = 10$ ns, the emission coefficient of the diode is $n = 1.2$, and the thermal voltage is $V_t = 26$ mV.

3.17 The SPICE parameters of a P–N junction diode are $I_S = 10^{-11}$ A, $n = 1.4$, $r_S = 0$, and $\tau_T = 10$ ns. What is the value of the stored charge if the forward voltage drop across the diode is $V_D = 0.7$ V? $\boxed{A}$

3.18 A P–N junction diode with $\tau_T = 100$ ns conducts a forward-bias current $I_D = 4.3$ mA. At time $t = 0$, the circuit conditions change, so that a reverse-bias voltage $V_E = 20$ V is connected in series with the diode and a resistor $R = 1$ kΩ.

(a) Because of the strong discharging current (≈ 20.7 V/1 kΩ $= 20.7$ mA), the contribution of the recombination to the stored charge removal can be neglected. How long will it take for the stored charge to be removed?

(b) If $V_E = 0$, the diode is discharged with a much smaller current (≈ 0.7 V/1kΩ $= 0.7$ mA), so that the contribution from the recombination cannot be neglected. If it is

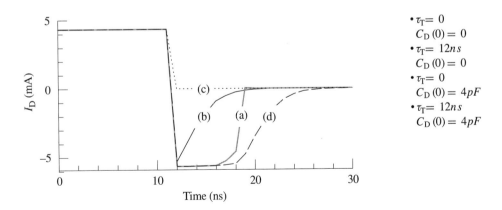

Figure 3.26 SPICE transient analysis of a diode circuit: the diode current is obtained with four different sets of diode parameters.

found that it takes $t_{rs} = 200$ ns for the stored charge to be removed, how much stored charge is removed by recombination?   A

3.19 A forward-biased diode is used as a temperature sensor. To calibrate the sensor, the diode is placed in melting ice and boiling water, and the following forward voltage drops are measured: 0.680 V and 0.480 V, respectively. What is the temperature when the forward voltage drop of the diode is 0.618 V?

3.20 The SPICE equation for the temperature dependence of the saturation current is

$$I_S(T) = I_S \left(\frac{T}{T_{\text{nom}}}\right)^{p_t/n} \exp\left[-\frac{qE_g}{kT}\left(1 - \frac{T}{T_{\text{nom}}}\right)\right]$$

Using the following theoretical dependencies for the intrinsic carrier concentration and the mobility on temperature

$$N_C = A_C T^{3/2}, \qquad N_V = A_V T^{3/2}, \qquad n_i = \sqrt{N_C N_V} e^{-E_g/2kT}$$

$$\mu_{n,p} = C_{n,p} T^{-3/2}$$

and assuming $n = 1$, find the theoretical value of the parameter p_t (saturation-current temperature exponent).   A

Section 3.3 Reference Diodes: Breakdown Phenomena

3.21 Identify the mode of diode operation for each of the I_D–V_D characteristics shown in Fig. 3.27.

3.22 The structure of a silicon reference diode is P$^+$–N–N$^+$. Design the N region so that the avalanche breakdown voltage is $V_{BR} = 12$ V. Assume $V_{bi} = 0.9$ V and that the breakdown (maximum) electric filed for silicon is $E_{max} = 30$ V/μm. The design should specify the doping level (N_D) and the minimum width of the N region (W_N) that ensures that the depletion layer does not become wider than the N region.

3.23 The diode designed in Problem 3.22 is implemented in SiC ($\varepsilon_s = 9.8 \times 8.85 \times 10^{-12}$ F/m, $E_{max} = 300$ V/μm, $V_{bi} = 2.8$ V). What would the breakdown voltage be if (a) W_N is

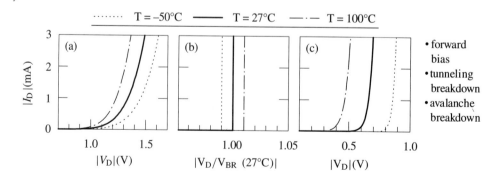

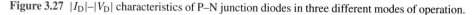

Figure 3.27 $|I_D|$–$|V_D|$ characteristics of P–N junction diodes in three different modes of operation.

made larger than the maximum depletion layer width, (b) the minimum W_N, as calculated in Problem 3.22, is used (in this case, define the breakdown voltage as the reverse-bias voltage that fully depletes the N-type region—punch-through breakdown)? (c) What W_N is needed to achieve the breakdown voltage calculated in part (a) of this problem?   [A]

3.24 The doping concentrations of a silicon P$^+$–N–N$^+$ diode are $N_A = 10^{20}$ cm^{-3}, $N_{D-l} = 10^{15}$ cm^{-3}, and $N_{D-h} = 10^{20}$ cm^{-3}, respectively, while the width of the N-type region is $W_N = 5$ μm. At what voltage does the depletion layer width become equal to W_N (punch-through breakdown)? Is this voltage larger or smaller than the avalanche breakdown voltage ($E_{max} = 30$ V/μm)?

Section 3.4 Schottky Diodes: Metal–Semiconductor Contact

3.25 Each of the graphs in Fig. 3.28 shows a pair of P–N junction and Schottky diode characteristics. Identify the combination that properly groups three characteristics that all belong to one of the two diodes.

3.26 The graphs of Fig. 3.29 show the energy-band diagrams of three different pairs of metal–semiconductor materials. When placed in contact, identify whether each of the systems would create a Schottky or ohmic contact.

3.27 A Schottky diode with N-type silicon ($N_D = 10^{15}$ cm^{-3}) is designed to have built-in voltage $V_{bi} = 0.40$ V and barrier height $\phi_B = 0.65$ V. If the actual donor concentration is 4.66×10^{15} cm^3, what are the built-in voltage and the barrier height?

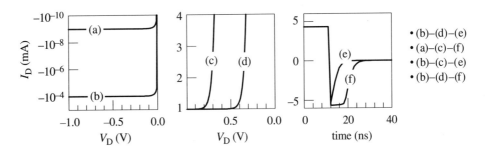

Figure 3.28 Characteristics of a P–N junction and a Schottky diode.

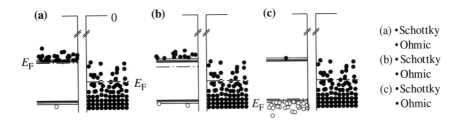

Figure 3.29 Semiconductor–metal energy-band diagrams.

• 0.40 V; 0.65 V   • 0.44 V; 0.65 V   • 0.40 V; 0.69 V
• 0.36 V; 0.65 V   • 0.40 V; 0.61 V   • 0.36 V; 0.69 V

3.28 The energy barrier and the built-in voltage of a Schottky diode are $q\phi_B = 0.65$ eV and $V_{bi} = 0.40$ V, respectively. What are the barrier heights for the electrons in semiconductor and metal, respectively, at

(a) forward bias $V_F = 0.2$ V,

(b) reverse bias $V_R = 5$ V?   [A]

3.29 A Schottky diode is created by depositing tungsten ($q\phi_m = 4.6$ eV) onto N-type silicon. By an appropriate heating, tungsten silicide ($q\phi_m = 4.7$ eV) can be created, becoming effectively the metal electrode of the Schottky diode. Would this heating process increase or decrease the saturation current, and how many times?

3.30 A P–N junction diode and a Schottky diode have the same emission coefficient $n = 1.25$. The forward voltage drop (measured at the same current) of the Schottky diode is 0.5 V smaller. How many times is the reverse-bias saturation current of the Schottky diode larger? The thermal voltage is $V_t = 0.026$ V.

3.31 A Schottky diode, created on N-type silicon ($N_D = 10^{15}$ cm^{-3}), has $V_{bi} = 0.4$ V. What is the depletion-layer capacitance per unit area at

(a) zero bias,   [A]

(b) reverse bias $V_R = 25$ V?

3.32 The saturation current of a Schottky diode can be expressed as (Eq. 10.8)

$$I_S = A_J A^* T^2 e^{-q\phi_B/(kT)}$$

where A_J is the junction area, and A^* is the effective Richardson constant. Assuming $n = 1$, find the theoretical value of the parameter p_t (saturation-current temperature exponent) so that the SPICE equation given in Problem 3.20 can be used for this Schottky diode.

REVIEW QUESTIONS

R-3.1 Consider a forward-biased P–N junction. Is the current of electrons in the N-type region (the majority carriers) due to the drift or diffusion?

R-3.2 Is the electric field in the neutral regions large or small? Why can the voltage drop in the neutral regions be neglected compared to the voltage drop in the depletion layer? Is the electric field in the neutral regions large enough to produce a significant drift current of the majority carriers? Why?

R-3.3 What is the effect of the forward bias on the energy barrier at the P–N junction?

R-3.4 Why is the increase in the number of electrons able to overcome the energy barrier at the P–N junction exponentially dependent on the applied forward-bias voltage?

R-3.5 What happens with the electrons that overcome the energy barrier at the P–N junction and appear in the P-type region? What are these electrons called?

R-3.6 Is the drift current of the minority carriers significant? The diffusion current?

R-3.7 How do the excess minority-carrier concentrations at the edges of the depletion layer depend on the forward-bias voltage? Are these the maximum concentrations of the minority electrons and holes, respectively?

R-3.8 If the forward-bias voltage is suddenly switched off, what happens with the excess minority carriers? In which direction does the current flow immediately after switching the voltage off?

R-3.9 What is the SPICE parameter that accounts for the stored-charge effect?

R-3.10 Can a P–N junction diode be used as a temperature sensor?

R-3.11 How does the avalanche breakdown voltage depend on temperature? Why?

R-3.12 We have seen that energy barriers of sufficient height can block the electron motion. Is the barrier width important as well? Can the energy barrier be so narrow that it lets some electrons through? If yes, what is this effect called?

R-3.13 Make an analogy between the P–N junction and Schottky diode. What is the origin of the built-in voltages in either case? How do typical values compare?

R-3.14 What is the origin and meaning of the barrier potential ϕ_B? Does it depend on the bias applied?

R-3.15 Why do the electrons from the N-type semiconductor not appear as minority carriers after they pass through metal–semiconductor contact?

R-3.16 How does the absence of stored charge influence the Schottky diode characteristics?

Chapter *4*

BASICS OF TRANSISTOR APPLICATIONS

In 1947, a research team at Bell Laboratories invented a solid-state device whose resistance between two output terminals could be controlled by a third input terminal. This three-terminal device was named *transistor*, shortened from the words <u>transfer</u> re<u>sistor</u>. Today, this device is known as a *bipolar junction transistor* (BJT). The BJT was the first minute solid-state device to provide implementations of electrically controlled switch and amplifier of electrical signals. The fact that these two important functions could be performed by a minute device triggered the development of microelectronics.

Many more different types of solid-state transistors appeared after the BJT. A *metal–oxide–semiconductor field-effect transistor* (MOSFET), developed in 1960, would become the most used electronic device ever. It enabled development of digital circuits of *very large-scale integration* (VLSI), the building blocks of computers. Other types of *field-effect transistors* (FETs), like *metal–semiconductor FETs* (MESFETs) and *high-electron mobility transistors* (HEMTs) were subsequently developed for some specific applications, especially high-frequency digital and microwave analog applications.

As a common feature, all types of transistors can perform the following two functions: (1) voltage-controlled switch and (2) voltage-controlled current source. The first function represents the building block of digital circuits and the second function enables construction of different types of linear amplifiers.

Using the concepts of voltage-controlled current source and voltage-controlled switch, this chapter introduces the basic applications of transistors in a generic way. This approach, where the applications are described before the particular devices, clearly defines the desired transistor characteristics for a specific (analog or digital) application. The suitability of a particular type of transistor for a particular application can then be effectively judged against this background.

4.1 ANALOG CIRCUITS

4.1.1 Transistor as a Voltage-Controlled Current Source: The I–V Characteristics

Current source delivers constant current, independent of the voltage drop across the device. The current–voltage characteristic of a current source appears as a horizontal line on the I–V graph. In Fig. 4.1a, the current source is represented by the square symbol with the arrow indicating the direction of the current. The terminals of the current source are labeled by D and S. Consequently, the voltage across the current source is v_{DS} and the current is i_D.

Voltage-controlled current source delivers current that is controlled by a voltage independent of the voltage across the current source itself. In Fig. 4.1a, the controlling voltage

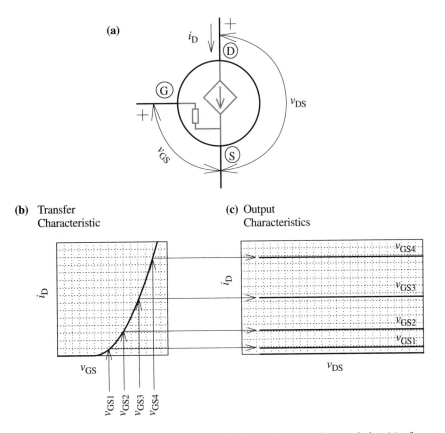

Figure 4.1 Symbol (a), transfer characteristic (b), and output characteristics (c) of a voltage-controlled current source; i_D is the output current, v_{DS} is the output voltage, and v_{GS} is the controlling (input) voltage.

is the one between the terminals labeled as G and S, that is v_{GS}.

The dependence of the *output* current i_D on the *input* voltage v_{GS} is called the *transfer function* or *transfer characteristic*.

In general, the transfer function of a voltage-controlled current source can take any form. In the case of transistors, it is a monotonically increasing function, as any increase in the input (controlling voltage) causes some increase of the output current. Figure 4.1b provides an illustration of a transistor transfer characteristic. It should be mentioned, however, that different types of transistors can exhibit anything between a linear and exponential transfer characteristic.

The output characteristics of a voltage-controlled current source appear as a set of horizontal lines, as in Fig. 4.1c. The set of output characteristics shows that the output current i_D depends on the controlling voltage v_{GS}, but does not depend on the output voltage v_{DS}. Theoretically, a constant value of the current i_D should be observed for any voltage across the current source v_{DS}, and therefore for voltage $v_{DS} = 0$ as well. In the case of transistors (real voltage-controlled current sources), the voltage v_{DS} is the driving force for the current, which means that $i_D = 0$ at $v_{DS} = 0$. This indicates that there is a region of relatively small v_{DS} voltages in which the current does depend on the voltage, and the transistor does not behave as a current source. As this effect is considered in the next section (digital circuits), Fig. 4.7c does not show the transistor characteristics in this region.

4.1.2 Transconductance Amplifier: Input Bias Circuit

The aim of analog circuits is to transfer input signals without (or with as small as possible) distortion. A two-port analog circuit, using *voltage* as input and delivering *current* as output is referred to as a *transconductance amplifier*. Being an analog circuit, the output signal current i_d linearly depends on the input signal voltage v_s. Note that lower case subscripts are used to express the signal voltage and current. The convention employed to distinguish between DC, instantaneous, and signal voltages/currents is given in Table 4,1.

The following linear equation can be used to describe the transconductance amplifier:

$$i_d = g_m v_s \tag{4.1}$$

The input signal voltage v_s is multiplied by a constant (g_m) to appear as output signal current i_d. Relating the output to the input, the constant g_m is in fact the *transfer function*, although it happened to be just a constant. The unit of the transfer function g_m is A/V, because it multiplies volts to convert them into amperes. As this is the unit of conductance, and as g_m relates *output* to *input* quantity, it is called *transconductance*.

In general, transistors have nonlinear characteristics. Frequently, the transistors appear as *normally-off* devices, meaning that the output current is zero at zero input voltage. The output current becomes significant only after an input voltage larger than a "threshold" value is applied. This is the case shown in Fig. 4.1b. The transfer characteristic of a normally-off device exhibits a pronounced nonlinearity. It may seem that the transistors are not suitable for analog applications. To explain this point, assume that a sinusoidal

TABLE 4.1 Symbol Convention for DC, Signal, and Instantaneous Voltages and Currents

Direct Current (DC)	Signal	Instantaneous
	Voltages	
V_{GS}	v_{gs}	v_{GS}
V_S	v_s	v_S
V_{DS}	v_{ds}	v_{DS}
V_{OUT}	v_{out}	v_{OUT}
	Currents	
I_G	i_g	i_G
I_S	i_s	i_S
I_D	i_d	i_D
I_{OUT}	i_{out}	i_{OUT}

DC + Signal = Instantaneous

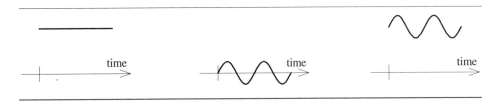

signal voltage v_s is directly applied as the transistor input voltage v_{GS}. Also, define the transistor *operating point* as the point on the transfer characteristic that corresponds to the instantaneous value of the input voltage and the output current: (v_{GS}, i_D). For the case of $v_s = 0$ (no signal applied, or *quiescent condition*), the operating point is at the origin of the graph $(v_{GS} = 0, i_D = 0)$. It is not hard to imagine that the sinusoidal signal voltage would not be converted into sinusoidal output current i_d. This is because the signal voltage would oscillate the operating point through the pronounced nonlinear region of the transfer characteristic.

At higher v_{GS} voltages, when the current increase becomes significant, a limited section of the transfer characteristic appears as nearly linear.

> Transistors can linearly transform *small* input voltage signals into output signal current, provided a narrow linear-like region of the transfer characteristic is used.

To be able to use the suitable region of higher voltages and currents on the transfer characteristic, the operating point in the quiescent condition, labeled as *Q point* in Fig. 4.2b, should be placed in the middle of that linear-like region. This is achieved by providing appropriate constant (DC) input voltage V_{GS} that sets the Q point and superimposing the signal voltage v_s, which oscillates the operating point around the Q point.

> The procedure of setting the quiescent point Q in the middle of nearly linear section of the transfer characteristic is called *input DC biasing*.

A simple input DC-biasing circuit is shown in Fig. 4.2a. It is assumed that the input resistance of the transistor, that is the resistance between the G and S terminals, is much larger than R_1 and R_2 (this can always be achieved by an appropriate choice of R_1 and R_2). In that case, the current flowing into the G terminal is insignificant compared to the current flowing through R_1 and R_2, which is approximately equal to $V_+/(R_1 + R_2)$. The voltage between the G and S terminals of the transistor is then

$$V_{GS} = \frac{R_2}{R_1 + R_2} V_+ \tag{4.2}$$

Obviously, with an appropriate choice of resistors R_1 and R_2, the desired value of the voltage V_{GS} can be provided. Note again that a properly selected V_{GS} would place the Q point in

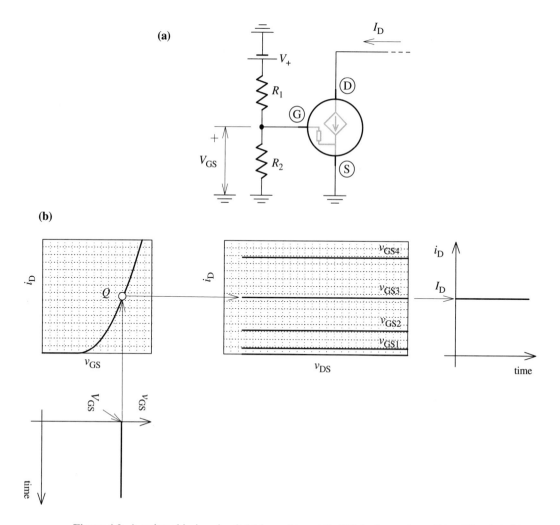

Figure 4.2 A resistor biasing circuit (a) is used to set the DC (quiescent) condition (Q point) (b).

the middle of a nearly linear section, so when a small signal voltage v_s is superimposed, the instantaneous voltage $v_{GS} = V_{GS} + v_s$ remains within this linear-like section of the transfer characteristic.

The resistor divider R_1–R_2 in the circuit of Fig. 4.2a is called the biasing circuit. This kind of biasing circuit is frequently used in amplifier circuits, although more sophisticated biasing circuits exist.

When the resistor divider is used as a biasing circuit, the input signal source cannot be directly connected to the input (G) terminal of the transistor. This is because a voltage source has very small internal resistance, which would virtually short circuit the resistor R_2. In addition, undesirable DC current would be forced into the signal source. To avoid this, a large-value connecting-capacitor C_{C1} is added, as shown in Fig. 4.3a. Due to the high value of the capacitor C_{C1}, the voltage drop across the capacitor changes insignificantly, which means it remains approximately equal to the DC voltage V_{GS} set by the resistor divider. The instantaneous voltage between the G and S terminals of the transistor, being equal to the sum of the voltage across the capacitor and the signal voltage, can be expressed as

$$v_{GS} = V_{GS} + v_s \tag{4.3}$$

This is graphically illustrated in Fig. 4.3b, which shows a sinusoidal signal voltage v_s oscillating the transistor operating point around its quiescent position Q. As the input signal is *small*, the operating point remains within a nearly linear section of the transfer characteristics. This produces sinusoidal signal current i_d at the output. Obviously, the output signal current i_d and the input signal voltage v_s are related as in Eq. (4.1), where the transconductance is determined by the slope of the transfer characteristic at the Q point:

$$g_m = \frac{\Delta i_D}{\Delta v_{GS}}\bigg|_{v_{GS}=V_{GS}} \tag{4.4}$$

The slope of the transistor transfer characteristic directly determines the value of the small-signal transconductance g_m, and therefore the gain of the transconductance amplifier.

4.1.3 Voltage Amplifier: Load Line

The voltage amplifier is a two-port analog circuit that takes *voltage* as the input signal v_s and delivers *voltage* as the output signal v_{out}:

$$v_{out} = A_V v_s \tag{4.5}$$

The transfer function A_V, commonly referred to as the *voltage gain*, should be a constant to provide amplification without distortion.

To begin, let us consider the following important question: **Is it possible to convert the output current of the transconductance amplifier into output voltage simply by connecting a load resistance across the output?** If the resistor connected across the output of the voltage-controlled current source is labeled by R_L, the output voltage should be

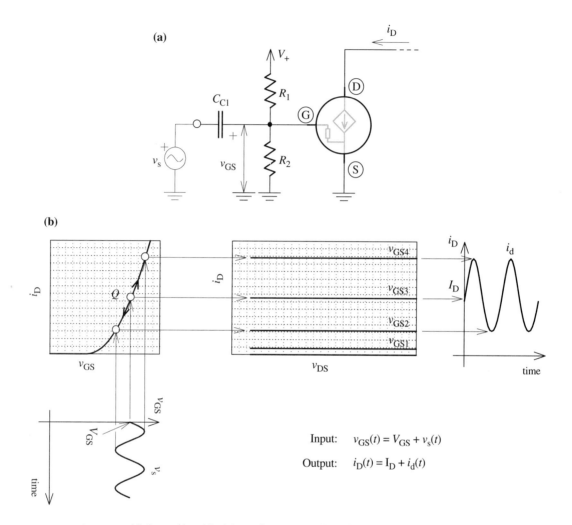

Figure 4.3 (a) A capacitor (C_{C1}) is used to connect the voltage source to the input of the transistor, while isolating it from the DC-biasing voltage; (b) graphic presentation of the transconductance amplification: the input sinusoidal voltage is transformed into output sinusoidal current.

$$v_{\text{out}} = R_{\text{L}} \underbrace{i_{\text{d}}}_{g_{\text{m}} v_{\text{s}}} = \underbrace{g_{\text{m}} R_{\text{L}}}_{A_{\text{V}}} v_{\text{s}} \tag{4.6}$$

Very large values of the output voltage would be achieved by using very large resistance R_{L}, if the current of the voltage-controlled current source i_{d} was really independent on the output voltage [refer to Eq. (4.1)]!

The above reasoning would be correct if the transistors were ideal voltage-controlled current *sources*. It would mean that the transistors could generate high signal voltages out of nothing, and therefore generate power out of nothing. Of course this is impossible. If a loading resistor is directly connected across the output terminals of a transistor, no output signal would be obtained. It was mentioned that the transistors behave as voltage-controlled current sources to a limit. If the voltage that drives the current is not provided ($v_{DS} = 0$), the output current i_D is going to be zero regardless of the value of the input voltage. The transistor is operated at the point of zero v_{DS} voltage, where it does not behave as a voltage-controlled current source.

However, transistors can convert DC voltage and power into signal voltage and power.

The procedure of providing the DC supply voltage in the output circuit of an amplifier is called *output DC biasing*.

Figure 4.4a shows the output DC-biasing circuit suitable for the case of voltage amplifiers. It consists of a resistor R_D and a DC voltage source V_+, connected between the D terminal of the transistor and ground. The output signal voltage is taken from the D terminal of the transistor through a large value capacitor C_{C2}. Similar to the function performed by C_{C1}, this capacitor transfers the signal voltage with minimal attenuation, while it completely blocks the biasing DC voltage that appears across the transistor output.

There are three components that determine the transistor output voltage (v_{DS}) and current (i_D): the transistor itself, the resistor R_D, and the DC voltage source V_+. The output characteristics of the transistor give the $i_D - v_{DS}$ dependence, as governed by the transistor. It is very useful to determine the $i_D - v_{DS}$ dependence as governed by the biasing circuit (R_D and V_+), and plot it on the same graph with the transistor output characteristics. Applying the second Kirchoff's law, it is found that

$$v_{DS} = -R_D i_D + V_+ \qquad (4.7)$$

which can be transformed to express the current i_D:

$$i_D = -\frac{1}{R_D} v_{DS} + \frac{V_+}{R_D} \qquad (4.8)$$

The output DC-biasing circuit establishes a linear relationship between the transistor current i_D and the output voltage v_{DS}, which is referred to as *load line* [Eq. (4.8)].

The load line is plotted in Fig. 4.4b together with the output characteristics of the transistor. The intersection between the load line and the output-characteristic line indicating the transistor current determines the value of the voltage v_{DS}. Consider, for example, the quiescent condition (zero input signal voltage). The transistor current I_D is determined by the input bias voltage V_{GS} as indicated by the Q point on the transfer characteristic. The output transistor characteristics show that setting the current I_D does not set the output voltage V_{DS}, as there is a whole range of possible output voltages. Although the output-biasing circuit also allows for a range of current–voltage values (along the load line), Fig. 4.4b shows that only the unique point Q will satisfy both the transistor and the biasing circuit. Therefore, the output DC voltage V_{DS} appears as an additional component of the quiescent operating point Q.

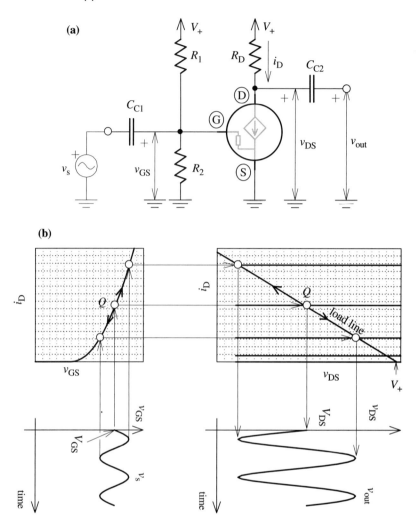

Figure 4.4 (a) Voltage amplifier circuit, (b) graphic presentation of the voltage amplification.

The effect of voltage amplification can now be explained referring to the graphic presentation of Fig. 4.4b. Note that the superposition of the input signal voltage v_s to the input bias voltage V_{GS} oscillates the transistor operating point around its quiescent position Q. For example, the maximum signal voltage causes the maximum current (as seen from the transfer characteristics), which further causes the minimum output voltage v_{out} (as seen from the output characteristics and the load line). Analogously, the minimum signal voltage causes the maximum output voltage v_{out}. Therefore, the signal at the output is inverted, which is due to the fact that the load line has a negative slope. However, this effect is not very important. What is important is the magnitude of the output voltage, which is obviously larger, therefore the voltage amplification is achieved.

To express the magnitude of the output signal voltage v_{out}, apply the second Kirchoff's law to the output circuit:

$$v_{out} = -V_{C_{C2}} - R_D(I_D + i_d) + V_+ \qquad (4.9)$$

where $V_{C_{C2}}$ is the voltage drop across the capacitor, and i_D is replaced by its components $I_D + i_d$. Equation (4.9) can be written in the following form, to separate the DC and the signal terms:

$$\underbrace{v_{out}}_{\text{signal}} = \underbrace{-R_D i_d}_{\text{signal}} \underbrace{-V_{C_{C2}} - R_D I_D + V_+}_{\text{DC}} \qquad (4.10)$$

The DC term in Eq. (4.10) is equal to zero. This is because the capacitor is large and the voltage drop across it does not change significantly, virtually remaining equal to its quiescent value. Equation (4.10) itself shows that $-V_{C_{C2}} - R_D I_D + V_+ = 0$ in the case of $v_{out} = i_d = 0$ (quiescent condition). Therefore, the output signal voltage is related to the output signal current as

$$v_{out} = -R_D i_d \qquad (4.11)$$

Using the dependence of the output signal current i_d on the input signal voltage v_s [Eq. (4.1)], the output voltage can be expressed as

$$v_{out} = -g_m R_D v_s \qquad (4.12)$$

which means that the voltage gain [Eq. (4.5)] of the amplifier is

$$A_V = \frac{v_{out}}{v_s} = -g_m R_D \qquad (4.13)$$

This result is not very much different from the one obtained for the ideal voltage-controlled current source, given by Eq. (4.6). In fact, the only difference is in the minus sign. As noted before, the negative gain is due to the structure of the biasing circuit, which produces a load line with a negative slope. Although the negative gain is acceptable in many applications, it can be converted into a positive gain if a two-stage cascaded amplifier is used.

It is quite expectable that the voltage gain directly depends on the transconductance of the transistor g_m. A larger transconductance means that a larger output current is delivered for the same input voltage, therefore a larger output voltage can be obtain from this current. The above analysis shows that the voltage gain also depends directly on the resistance R_D. Figure 4.4b provides a graphic illustration for this effect. An increase in the resistance R_D would be manifested as a reduction in the load-line slope, which directly converts into larger amplitude of the output signal voltage. Figure 4.4b also illustrates that this would require a larger supply voltage V_+. To see this, note that the load line intersects the v_{DS} axis at $v_{DS} = V_+$ [Eq. (4.8) shows that $i_D = 0$ for $v_{DS} = V_+$]. Therefore, the intersecting point (that is V_+) should be moved toward larger values to accommodate a larger output signal. Otherwise, the output signal would be truncated.

4.2 DIGITAL CIRCUITS

Digital electronic circuits operate with two voltage levels: *low* and *high*. These two voltage levels represent the two digits of the binary system: 0 and 1. We should note here that any number can be expressed by the two digits of the binary system. Digital electronic circuits are used to perform different operations with the binary numbers, as well as to memorize them. It can be said that a set of logic operations represents the basis of digital electronics. The simplest logic function is *inversion*. A digital circuit performing this function is called an *inverter*. The inverter example is used to introduce two basic digital technologies in this section.

4.2.1 Transistor as a Switch

The two voltage levels used in digital circuits are created by operating transistors in two different states, referred to as *OFF* and *ON* states. In other words, the transistor is used as a switch. It is important to note, however, that the transistors in digital circuits appear as *voltage-controlled* switches. The symbol for the voltage-controlled switch, which will be used in this section, is given in Fig. 4.5a.

> The transistor is in an *OFF* state when no current is flowing between the output terminals *D* and *S* (open circuit).

To achieve the *OFF* state, it is enough to provide zero (or very small) controlling voltage v_{GS}. Referring to the transfer characteristic of Fig. 4.1b, it can be seen that the output current i_D is zero for very small v_{GS} voltages. This is again expressed in Fig. 4.5b, in which the small controlling voltage causing the *OFF* state is labeled as v_{GS-LOW}.

In the *ON* state, the switch is closed and the transistor should provide a short circuit between the output *D* and *S* terminals. However, this ideal case can never be achieved with transistors (real voltage-controlled switches). There is always some resistance, no matter how small, associated with a transistor in the *ON* state (closed switch).

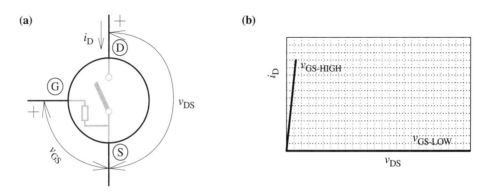

Figure 4.5 Symbol (a) and output characteristics (b) of a voltage-controlled switch.

The transistor is in the *ON* state when a significant current flows between the output *D* and *S* terminals, while the output voltage v_{DS} is small.

This case is also shown in the output characteristic of Fig. 4.5b, in which the controlling voltage causing the *ON* state is labeled as $v_{GS-HIGH}$. The $i_D - v_{DS}$ characteristic is a straight line expressing the resistance of the transistor in the *ON* state.

It is important to emphasize that this resistor-like $i_D - v_{DS}$ characteristic can be observed only at small v_{DS} voltages. The current i_D will not continue to linearly grow for any increase of the voltage v_{DS}. In fact, the current increase will slow down as v_{DS} voltage is increased, and then will completely saturate. When the current saturates, the transistor actually behaves as a current source, as described in the previous section.

In the *ON* mode (high controlling voltage v_{GS} applied), the transistor appears as either

- small-value resistor, at small v_{DS} voltages, or
- current source, at high v_{DS} voltages.

4.2.2 Inverter with Resistive Load

As the transistor mode (switch or current source) is "selected" by the output voltage v_{DS}, it is important to consider what factors influence this voltage. It has been explained in Section 4.1.3 that the output bias circuit sets the output voltage v_{DS}. The bias circuit explained in Section 4.1.3 can be used in digital circuits as well, however, the value of the loading resistor R_D should be adjusted. Referring to Fig. 4.4b, let us set the controlling voltage v_{GS} to the level that corresponds to the maximum v_{GS} voltage in Fig. 4.4b (the top horizontal line on the $i_D - v_{DS}$ characteristics is considered). Assume now a larger value of the loading resistor R_D. The slope of the load line in this case is smaller, intersecting the i_D axis at a smaller V_+/R_D value, while still passing through the point V_+ on the v_{DS} axis. As a consequence, the load line does not intersect the top horizontal current line, which means the circuit solution is not in the region of higher v_{DS} voltages where the transistor behaves as a current source. The transistor is pushed into the switch-mode region (small v_{DS} voltages).

Figure 4.6b shows the case in which the load line intersects the transistor $i_D - v_{DS}$ characteristic (for high controlling voltage v_{GS}) in the region of small v_{DS} voltages (the transistor appears as a closed switch). The output bias circuit is the same as in the case of the amplifier of Fig. 4.4, the only difference being the value of the loading resistor R_D. Note that no input bias circuit is needed in the circuit of Fig. 4.6a, because the input voltage itself moves the operating point from the region of virtually zero i_D current (digital 0) to the region of significant i_D currents (digital 1). When the input voltage is at a low level (digital 0), the transistor is *OFF*, the current i_D is zero, and the output voltage v_{DS} is equal to the supply voltage V_+, which is the digital 1. The quiescent operating point for this case is labeled by Q_{OFF} in Fig. 4.6b. When the input voltage is high (digital 1), the transistor is *ON*, and the output voltage v_{DS} is low (digital 0), as shown by the quiescent operating point Q_{ON} in Fig. 4.6b. Therefore, the circuit of Fig. 4.6a is an *inverter*. The inverting operation is further illustrated in Fig. 4.6c, which also shows that a significant i_D current flows through the inverter when the transistor is *ON*. Obviously, digital circuits based on this technology

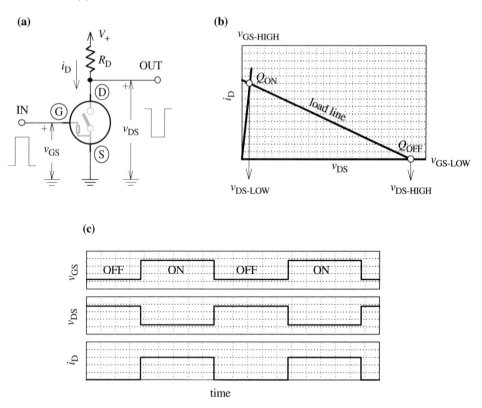

Figure 4.6 Inverter with resistive load: (a) circuit diagram, (b) graphic analysis, and (c) diagrams illustrating the inverting function of the circuit, as well as the fact that a significant current i_D flows through the circuit when the transistor is ON.

dissipate a significant power during their operation. This is a factor that limits the number of digital functions that can be integrated, and limits the reduction of the circuit physical size.

4.2.3 Inverter with Complementary Transistors

To solve the power dissipation problem, the loading resistor R_D in the inverter of Fig. 4.6 is replaced by a *complementary transistor*. The complementary transistor provides "mirror-image" characteristics: a negative input voltage is needed to set the transistor in *ON* mode, the output current flows in the opposite direction, and the output voltage is negative. The voltages and the currents of the complementary transistor are indicated by asterisks in this chapter. Moreover, to avoid confusion with negative voltages, the input voltage of the complementary transistor is converted into positive by making use of the following transformation: $v_{GS}^* = v_G^* - v_S^* = -(v_S^* - v_G^*) = -v_{SG}^*$. Therefore, negative v_{GS}^* appears as positive v_{SG}^* voltage, as illustrated in Fig. 4.7a. The analogous transformation is applied to the output, to obtain positive voltage v_{SD}^*. With these voltage transformations, and change in the current direction, the output characteristics of the complementary transistor appear

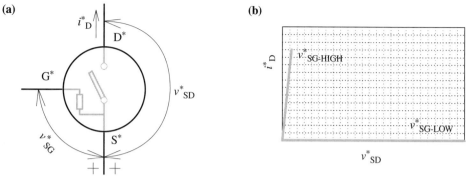

Figure 4.7 The symbol (a) and the output characteristics (b) of a complementary transistor.

identical to the output characteristics of its counterpart. The output characteristics of the complementary transistors are shown in Fig. 4.7b.

To provide nonnegative voltages v_{SG}^* and v_{SD}^*, the complementary transistor is turned round and then placed into the inverter circuit instead of the resistor. This transistor will be referred to as a *loading* transistor. The G^* terminal is connected to the input (G terminal of the *driving* transistor) to complete the inverter circuit. The inverter circuit obtained is shown in Fig. 4.8. Note that the two levels of the input voltage, namely 0 and V_+, correspond to $v_{SG}^* = V_+ - v_{IN} = V_+$ and $v_{SG}^* = V_+ - v_{IN} = 0$, respectively.

To provide graphic analysis of this circuit, the characteristics of the loading circuit (the loading transistor and the voltage source V_+) have to be presented in i_D vs v_{DS} form, and plotted together with the $i_D - v_{DS}$ characteristics of the driving transistor. Applying a procedure analogous to the one described in Section 4.1.3, the voltage v_{DS} can be expressed as

$$v_{DS} = v_{DS}^* + V_+ = V_+ - v_{SD}^* \qquad (4.14)$$

The above transformation turns the characteristics of the loading transistor left to right, as $v_{SD}^* = 0$ appears at the V_+ point on the v_{DS} axis, while $v_{SD}^* = V_+$ appears at the origin of the v_{DS} axis. Therefore, to express the bias circuit in the graphs of Fig. 4.8, the characteristics of the loading transistor (grey lines) are drawn from the V_+ point on the v_{DS} axis to the left. Now, we can analyze the circuit by considering the cases of low and high levels of the input voltage.

In the case of digital 0 at the input (Fig. 4.8a), the driving transistor is *OFF* as its input voltage is $v_{GS} = v_{IN} = 0$. The input voltage of the loading transistor is $v_{SG}^* = V_+ - v_{IN} = V_+$, which sets this transistor in the *ON* mode. Therefore, the output is detached from ground (the driving transistor is *OFF*), and connected to V_+ through the loading transistor. The high level of the output voltage (V_+) means that the input digital 0 is converted to 1 at the output.

The graph of Fig. 4.8a shows the position of the quiescent operating point Q. It is found at the intersection of the load line (the steep $i_D - v_{DS}$ characteristic of the *ON* loading transistor) and the zero-current characteristic of the *OFF* driving transistor. The point Q shows that the output voltage is $v_{DS} = V_+$, while the current flowing through the inverter is $i_D = 0$.

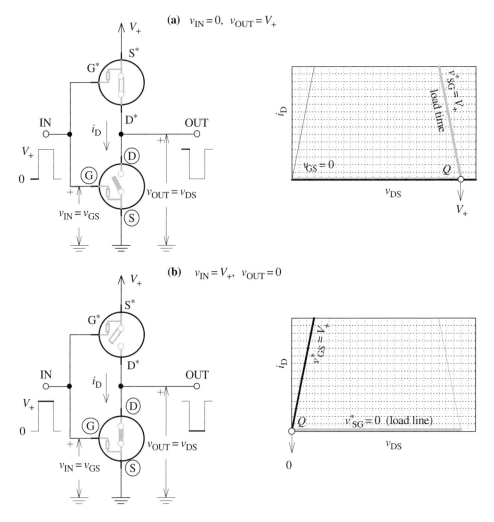

Figure 4.8 Inverter with complementary transistors: (a) $v_{IN} = 0$, (b) $v_{IN} = V_+$.

In the case of digital 1 at the input (Fig. 4.8b), the driving transistor is *ON* as $v_{GS} = v_{IN} = V_+$. The input voltage of the loading transistor is $v_{SG}^* = V_+ - v_{IN} = 0$, which sets this transistor in the *OFF* mode. Therefore, the output is connected to ground through the driving transistor, and detached from V_+ (the loading transistor is *OFF*). In this case, the input digital 1 is converted to digital 0 at the output.

The graph of Fig. 4.8b shows that the load line is now the horizontal zero-current characteristic (the loading transistor is *OFF*), while the steep $i_D - v_D$ characteristic represents the driving transistor. The characteristics intersect at the origin of the graph (Q point), showing that the output voltage is 0 and the current flowing through the inverter is also 0.

The above analysis has shown that the circuit of Fig. 4.8 performs the function of an inverter. The important point to note is that the current flowing through the inverter is zero in either case, digital 0 or digital 1 at the input. This means that digital circuits based on the technology with complementary transistors do not dissipate power to maintain their logic states. It should be pointed out that this does not mean the complementary-transistor circuits do not dissipate power at all. There is always some parasitic capacitance appearing between the output and the ground, which has to be charged and discharged as the output state is changed from 0 to V_+ and vice versa. The charging current flows through the loading transistor, while the discharging current flows through the driving transistor, dissipating some power in the transition period.

REVIEW QUESTIONS

R-4.1 Does the current of a current source depend on the voltage drop across the current source?

R-4.2 Can a current source be controlled by the voltage drop across the current source itself? In other words, must the controlling voltage be independent of the current source?

R-4.3 In principle, a voltage-controlled current source has four terminals, two for the current source and two for the independent controlling voltage. Is it possible to have a voltage-controlled current source with three terminals, where one of the terminals is shared?

R-4.4 What is the function (or characteristic) relating the output current to the input (controlling) voltage called?

R-4.5 Sketch the typical transfer characteristic of a transistor used as a voltage-controlled current source.

R-4.6 Sketch the output characteristics of a transistor used as a voltage-controlled current source. Would you show the current at zero output voltage? Does the transistor appear as a current source in the region of small output voltages?

R-4.7 What is a *normally-off* transistor? Would it deliver any current if a small voltage signal is applied to the input?

R-4.8 Can normally-off transistors be used to linearly transfer small signal voltage into output current? Where should the operating point with no signal applied (quiescent point Q) be placed? How is this achieved?

R-4.9 Can a transistor be operated as a voltage-controlled current source without a supply voltage in the output circuit? What would happen if only the loading element (represented by its internal resistance R_L) was attached to the output of the transistor?

R-4.10 What is the $i_D - v_{DS}$ characteristic of the output bias circuit called? Sketch this characteristic for the case of a bias circuit consisting of a resistor R_D and a voltage source V_+, connected between the D terminal of the transistor and ground. Where does this characteristic intersect the v_{DS} and i_D axes?

R-4.11 Would it be possible to use the transistor as a voltage amplifier if the output current was not completely independent of the output voltage ($i_D - v_{DS}$ characteristics were not perfectly horizontal)? How would a slight increase of the output current influence the voltage gain?

R-4.12 How is the voltage gain influenced by the bias resistor R_D?

R-4.13 What is the other important factor that influences the voltage gain?

R-4.14 Can the same transistor be used as a voltage-controlled current source in an analog circuit and a voltage controlled switch in a digital circuit? If yes, how is the transistor mode (current source or switch) "selected"?

R-4.15 Sketch the output characteristic of the transistor as a closed switch (high-input, low-output voltage). Should it be the characteristic of a short circuit?

R-4.16 Can the output bias circuit used in the voltage amplifier be used to create an inverter? If yes, what is the difference?

R-4.17 Does the inverter with resistive load dissipate power when in a 0–1 input–output state? How about the 1–0 state?

R-4.18 Label the input and output voltage and the direction of the output current so that the characteristics of a complementary transistor look the same as the characteristics of its counterpart.

R-4.19 In the inverter with complementary transistors, is there a stable state when both transistors are in *ON* mode? If no, how is this avoided?

R-4.20 Does the inverter with complementary transistors dissipate any power when in a 0–1 input–output state? How about the 1–0 state? How about the transition periods?

Chapter *5*

MOSFET

MOSFET (metal–oxide–semiconductor field-effect transistor) was fabricated for the first time in 1960, only a year after the beginning of the integrated circuit era in 1959. The MOSFET became the basic building block of very large-scale integrated (VLSI) circuits, therefore becoming the most important microelectronic device. Huge investments have been made in what is known as CMOS technology, a technology used to manufacture circuits consisting of complementary pairs of MOSFETs. Those investments, having been quite favorable, consequently led to the rapid progress in computer and communication integrated circuits that we have seen in the past decades.

However, the application of MOSFETs is not limited to VLSI circuits. MOSFETs play an important role in power-electronic circuits, and are becoming increasingly popular and suitable for microwave applications.

This chapter explains MOSFET principles and characteristics, introduces MOSFET technologies, and describes MOSFET models and parameters used in circuit simulation. The metal–oxide–semiconductor (MOS) capacitor, dealt with in Section 2.5, represents the basis of MOSFET. Therefore, a good grasp of the effects explained in the MOS capacitor section is necessary for an effective understanding of this chapter. Also, the MOSFET involves two P–N junctions, which means the P–N junction concepts introduced in Sections 2.2 and 3.1 need to be understood as well.

5.1 MOSFET PRINCIPLES

5.1.1 MOSFET Structure

As indicated above, the MOSFET is developed from the MOS capacitor. The voltage applied to the gate of the MOS capacitor (refer to Fig. 2.12) controls the state of the silicon surface underneath (this is the P-type Si–SiO_2 interface in Fig. 2.12). Negative gate voltages attract the holes from the P-type silicon to the surface (accumulation), while positive voltages larger than the threshold voltage create a layer of electrons at the surface (inversion).

These two states of the MOS capacitor can be used to make a voltage-controlled switch. To achieve this, the layer of electrons at the surface is contacted at the ends by N^+ regions referred to as *source* and *drain*, as illustrated in Fig. 5.1a. The existence of the electron layer, also referred to as a *channel*, corresponds to the *ON* state of the switch as the electron channel virtually short circuits the source and the drain regions, which are used as the switch terminals. When the gate voltage is below the threshold voltage, the electron layer (the channel) disappears from the surface, and the source and drain N$^+$ regions are isolated by the P-type substrate. This is the *OFF* state of the switch.

The same structure, shown in Fig. 5.1, can be used to create a voltage-controlled current source. This is possible because at higher drain-to-source voltages the current flowing through the channel (*ON* mode) does not increase linearly with the drain-to-source voltage but saturates. The mechanisms of current saturation will be explained in detail in Section 5.1.3. As the saturation current is independent on the voltage drop between the source and drain, the device behaves as a current source. In addition, it is possible to alter the value of this current by the gate voltage, driving the MOSFET between the full *ON* mode (maximum current) and the *OFF* mode (zero current). Therefore, the MOSFET appears as a voltage-controlled current source at higher drain-to-source voltages.

It is obvious from Fig. 5.1 that the MOSFET is essentially a four-terminal device. The four terminals are as follows: the silicon substrate (bulk) (B), the gate (G), the source (S), and the drain (D). The metal contacts to the source, drain, and bulk (S, D, and B, respectively) are shown in Fig. 5.1a. A metal contact to the gate is made behind the source and drain contacts, so it cannot be seen in the cross section shown in Fig. 5.1a.

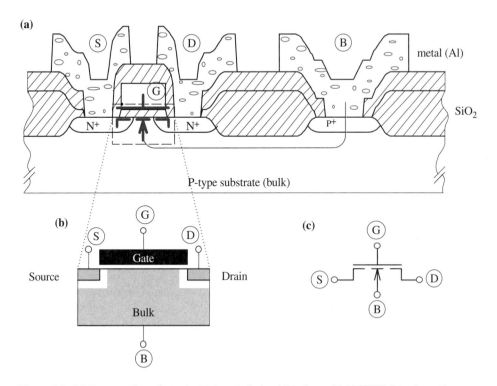

Figure 5.1 (a) Cross section of a typical integrated-circuit N-channel MOSFET, (b) schematic presentation used throughout this chapter, and (c) the circuit symbol.

Very frequently, the bulk and the source are connected together, so that the controlling voltage applied to the gate, and the driving voltage applied to the drain, can be expressed with respect to the common reference potential of the short-circuited source and bulk. In this case the three effective MOSFET terminals, the gate, the drain, and the source, can directly be related to the G, D, and S terminals of the generic transistor of Chapter 4. Sometimes the bulk and the source cannot be short circuited, or a voltage is deliberately applied between the bulk and the source. The effect of the bulk-to-source voltage is called *body effect* and is explained in Section 5.1.2.

Figure 5.1 illustrates one type of MOSFET, called an *N-channel MOSFET*, as the conduction between the source and the drain is provided by the N-type carriers (electrons). It is possible to make a complementary transistor, using N-type substrate and P$^+$ source and drain. In this case, negative gate voltage is needed to create a surface channel of holes; because of this the MOSFET is called a *P-channel MOSFET*.

As explained above, both the N-channel and the P-channel MOSFETs are in an *OFF* state when no voltage is applied to the gate—*normally-OFF* MOSFETs. The N-channel MOSFET is set in *ON* mode by positive gate voltage, while the P-channel MOSFET requires negative gate voltage. MOSFETs can be made *normally ON* by technologically built-in channels. Therefore, there are four different types of MOSFETs, as will be explained in more detail in Section 5.1.4.

5.1.2 MOSFET as a Switch: The Threshold Voltage

This and the subsequent section describe the MOSFET principles in more detail. Again, the energy-band model is used to elucidate the response of electrons and holes to the externally applied voltages. It should be noted that MOSFET effects must be considered in two dimensions. The first dimension is needed to express the gate voltage-related effects, that is to follow the direction from the silicon surface into the silicon bulk (labeled as the x-direction). The second dimension is needed to express the drain-to-source voltage-related effects, that is along the silicon surface (labeled as the y-direction). Because of that, the energy bands have to be shown in both the x- and y-directions. As a convention, the energy bands in the x-direction will be plotted on the left-hand side, while the energy bands in the y-direction will be plotted on the right-hand side in the graphs presented in this chapter. To make their relationship as obvious as possible, most of the graphs will provide two-dimensional energy-band diagrams in the middle.

In this section, the MOSFET is considered as a switch, in order to describe the *ON* and *OFF* states of the MOSFETs. This means the concept of the threshold voltage is considered in considerable detail. The body effect (the effect of the bulk-to-source bias) is also explained in this section. The effects associated with drain-to-source voltage, in particular current saturation, are dealt with in the subsequent section.

Zero Effective Bias (Flat Bands)

As already mentioned, the *ON* and *OFF* states of the MOSFET are controlled by the gate voltage, which directly determines the potential at the surface of the silicon. As in Section 2.5, the potential at the semiconductor surface with respect to its bulk will be denoted by φ_s. It appears as natural to take the zero surface potential as the reference point when different conditions of the silicon surface (as induced by different gate voltages) are considered. The zero surface-potential condition is also known as the *flat-band* condition, as in this case the electric-potential line, hence the potential-energy lines in the energy band diagram, are flat. It was explained in Section 2.5 that the flat bands typically do not occur for zero gate voltage, due to the effects of work-function difference and oxide charge. The gate voltage needed to bring the silicon surface into the flat-band condition was called *flat-band* voltage.

Figure 5.2 illustrates the MOSFET in the flat-band condition. Note that the source and the bulk are short circuited and taken as the reference potential, and that no drain-to-source voltage is applied ($V_{DS} = 0$). The left-hand side and the right-hand side of the first row of diagrams illustrate the MOSFET cross sections in the x- and y-directions, respectively. The central diagram illustrates the MOSFET in three dimensions. The MOSFET is positioned so that the x- and y-directions coincide with the x- and y-directions in the energy-band diagrams presented in the second row of diagrams. Again, the left-hand side and the right-hand side represent the energy-band diagrams along the x- and y-directions, respectively, while the central diagram is a two-dimensional presentation of the conduction band (the valence band is omitted to simplify the diagram). The energy-band diagram along x-direction (the bottom left corner) clearly illustrates the flat-band condition. This diagram is analogous to the diagram given in Fig. 2.16d.

The energy-band diagram along the y-direction (bottom right corner) is shown for the first time. Nonetheless, it can easily be deduced from the energy- band diagram of the N–P junction, given in Figs. 2.6b and 3.5a. Note that the surface structure of the MOSFET (along the y-direction) can be represented by two N–P junctions with the P regions connected to each other. That is, the surface N^+–P–N^+ structure can be split into N^+–P and P–N^+ structures where the P regions are connected to each other. Noting that N^+ denotes only that the doping level in the N^+-type regions is high, the energy-band diagram of the N^+-type source and P-type substrate (along the y-direction) is constructed in the same way as the energy-band diagram of the N–P junction, shown in Figs. 2.6b and 3.5a. The remaining part (P-type substrate and N^+- type drain) is a mirror image of the source-substrate part, and the energy-band diagram is completed by the mirror image of the N–P junction band diagram.

When illustrating the appearance of mobile electrons and holes, it is important to remember that the electron and hole concentrations are expressed by the position of the Fermi level (dash–dotted line) with respect to the energy bands. In the energy-band diagram along the x-direction (bottom left corner), the Fermi level is close to the top of the valence band, which means a number of hole positions in the valence band are actually occupied by holes (open symbols). This is the case in a P-type semiconductor, which is the substrate of the MOSFET. The same situation is seen between the source and drain in the energy-band diagram along the y-direction (bottom right corner). However, in the N^+ regions of the source and drain, the Fermi level is very close to the bottom of the conduction band, which means a large number of electron positions in the conduction band are actually occupied by electrons (closed symbols). The Fermi level along the y-direction is constant, which expresses the fact that no voltage is applied between the drain and the source (thermal equilibrium).

The energy-band diagram along the y-direction (bottom right corner) illustrates that there is an energy barrier that separates the electrons from the source and the drain. If the

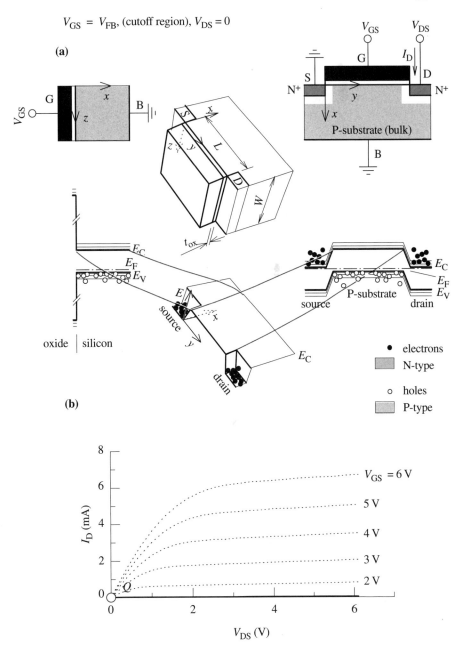

$V_{GS} = V_{FB}$, (cutoff region), $V_{DS} = 0$

(a)

(b)

electrons
N-type
holes
P-type

Figure 5.2 MOSFET in cutoff region: (a) cross-sectional illustrations and energy-band diagrams and (b) the output characteristic (the solid line).

barrier was not there, and the bottom of the conduction band in the drain was lowered,[1] the electrons would start flowing from the source into the drain.

As long as there is an energy barrier between the source and the drain, applying a voltage between the drain and the source (V_{DS}) does not produce any electron current flow. The MOSFET is said to be in cutoff region.

This is expressed by the solid line in Fig. 5.2b, which shows that the drain current I_D is zero for any V_{DS} voltage. The energy-band diagrams in Fig. 5.2a are drawn for the particular case of $V_{DS} = 0$, which is indicated on the output characteristic of Fig. 5.2b by positioning the quiescent point Q at $V_{DS} = 0$.

Strong Inversion

To set the MOSFET in *ON* mode, the energy-potential barrier between the source and the drain has to be reduced. This is achieved by applying positive voltage between the gate and the bulk/source. A part of the applied voltage will drop across the oxide, while the other part will appear as the surface potential φ_s. The appearance of the positive surface potential φ_s is seen on the energy-band diagrams as band bending downward from the reference bulk level (remember, *potential energy* $= -q \times$ *electric potential*). This means that the energy barrier at the surface is being reduced, as can be seen in Fig. 5.3a.

How much should the energy barrier be reduced before the electrons can start flowing between the source and the drain? To answer this important question, let us consider the position of the Fermi level at the silicon surface. To have a significant concentration of electrons at the silicon surface, so that the source and the drain are joined by a channel of electrons, the Fermi level should be closer to the bottom of the conduction band than to the top of the valence band. Normally, the Fermi level is closer to the top of the valence band in a P-type semiconductor.[2] However, as the bands are bent at the silicon surface, the top of the valence band moves away from the Fermi level, while the bottom of the conduction band moves toward the Fermi level (refer to the bottom left energy-band diagram in Fig. 5.3a). Once the Fermi level is closer to the bottom of the conduction band, the occupancy of electron states in the conduction band is more likely than the occupancy of hole states in the valence band, and the concentration of electrons becomes higher than the concentration of holes. At this stage the inversion layer (channel of electrons) is being created. However, the concentration of electrons is not significant before the surface potential reaches the value of $2\phi_F$ (the strong inversion condition). Beyond that point the band bending slows down rapidly and almost saturates, as illustrated in Fig. 5.4. Any further increase in the gate voltage ΔV_{GS} is quickly addressed by an adequate increase in the electron concentration, which forces most of the whole voltage increase ΔV_{GS} to appear across the oxide, leaving the surface potential virtually unchanged.

---

[1]The bottom of the conduction band in the drain can be lowered by applying positive voltage between the drain and the source. Remember that positive electric-potential difference ($V_{DS} > 0$) corresponds to a negative potential-energy difference ($q V_{DS} < 0$).

[2]The relationship between the Fermi level position and the carrier concentration is described in Section 1.5.2.

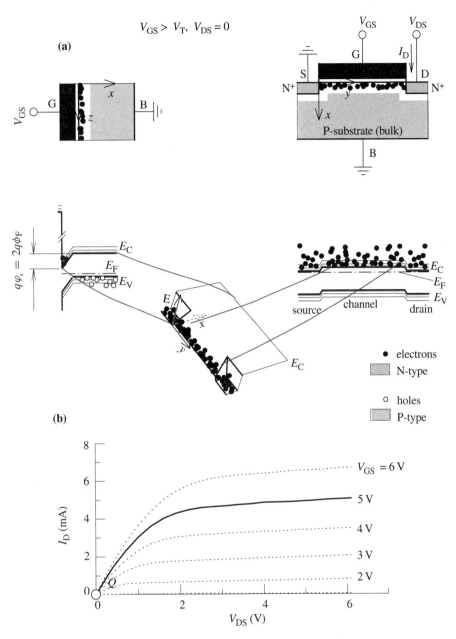

Figure 5.3 MOSFET in strong inversion—the channel of electrons is created: (a) cross-sectional illustrations and energy-band diagrams and (b) the operating point Q is at zero drain current as no driving voltage is applied ($V_{DS} = 0$).

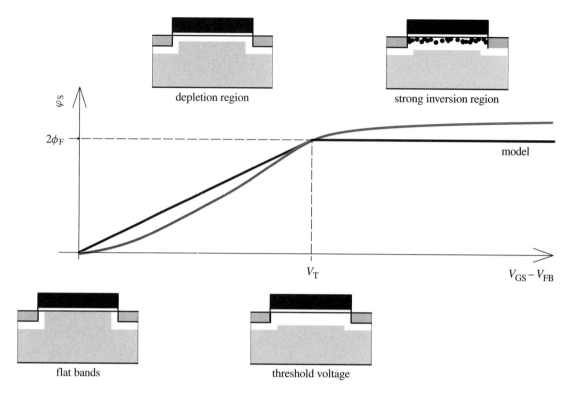

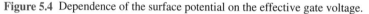

Figure 5.4 Dependence of the surface potential on the effective gate voltage.

The surface potential in strong inversion is pinned at

$$\varphi_s \approx 2\phi_F \tag{5.1}$$

where ϕ_F is the Fermi potential, defined in Section 2.5.3. The gate voltage needed to bring the surface potential to this level is called the *threshold voltage*, V_T.

To analyze the threshold voltage, it is reasonable to assume that the electron concentration in the channel is zero below the threshold voltage, and that it increases linearly with any voltage increase beyond the threshold voltage. However, this does not mean there is no charge at the MOS capacitor plates at the onset of strong inversion ($V_{GS} = V_T$).[3] When positive effective voltage $V_{GS} - V_{FB}$ is applied to the gate, the holes from the silicon surface are repelled, leaving behind uncompensated negative acceptor ions. This charge is called the *depletion-layer charge*. The density of this charge, Q_d, is expressed in C/m^2. If the charge density at the capacitor plates (Q_d in this case) is divided by the capacitance per unit area (the gate-oxide capacitance C_{ox} in this case), the effective voltage drop across the capacitor dielectric is obtained. Therefore,

---

[3]The no-charge condition corresponds to the flat-band condition, thus $V_{GS} = V_{FB}$.

$$\underbrace{\underbrace{V_{GS} - V_{FB}}_{\text{effective gate voltage}} - \varphi_s}_{\text{voltage drop across the gate oxide}} = \frac{Q_d}{C_{ox}} \quad \text{(for } V_{GS} \leq V_T) \tag{5.2}$$

Using the fact that the surface potential is $\varphi_s = 2\phi_F$ at the onset of strong inversion ($V_{GS} = V_T$), the following equation is obtained:

$$V_T - V_{FB} - 2\phi_F = \frac{Q_d}{C_{ox}} \tag{5.3}$$

which leads to the following equation for the threshold voltage:

$$V_T = V_{FB} + 2\phi_F + \frac{Q_d}{C_{ox}} \tag{5.4}$$

It was shown in Section 2.5.5 that the depletion-layer charge density can be calculated as $Q_d = q N_A w_d$, where w_d is the depletion-layer width, which should be obtained from the Poisson equation. The solution of the Poisson equation enables the Q_d/C_{ox} term to be converted into $\gamma \sqrt{2\phi_F}$, where γ is a technological constant called the *body factor*. Therefore, the threshold voltage equation is written in the following form:

$$\boxed{V_T = V_{FB} + 2\phi_F + \gamma \sqrt{2\phi_F}} \tag{5.5}$$

V_{FB}, ϕ_F, and γ are all technological parameters, therefore the threshold voltage is a technological parameter.

> The threshold voltage represents the value of the gate voltage needed to set the silicon surface at the onset of strong inversion.
>
> 1. The density of channel carriers Q_I is negligible for $V_{GS} \leq V_T$, as the strong inversion condition is not reached.
> 2. The density of the channel carriers in the strong-inversion region is determined by the value of the gate voltage in excess of the threshold voltage ($V_{GS} - V_T$) and the gate-oxide capacitance:
>
> $$Q_I = (V_{GS} - V_T)C_{ox} \tag{5.6}$$

Figure 5.3a illustrates the MOSFET in strong inversion (MOSFET in the *ON* mode) with no drain-to-source voltage applied, $V_{DS} = 0$. The bottom right energy-band diagram (*y*-direction) shows the formation of the channel of electrons between the source and the drain. As no drain-to-source voltage is applied, the bottom of the conduction band in the channel is flat in the *y*-direction. As a consequence, there is no flow of electrons, and the drain current I_D is zero. The position of the quiescent point, shown in the output characteristic of Fig. 5.3b, expresses the fact that the drain current is zero for $V_{DS} = 0$, even though the MOSFET is set in the *ON* mode.

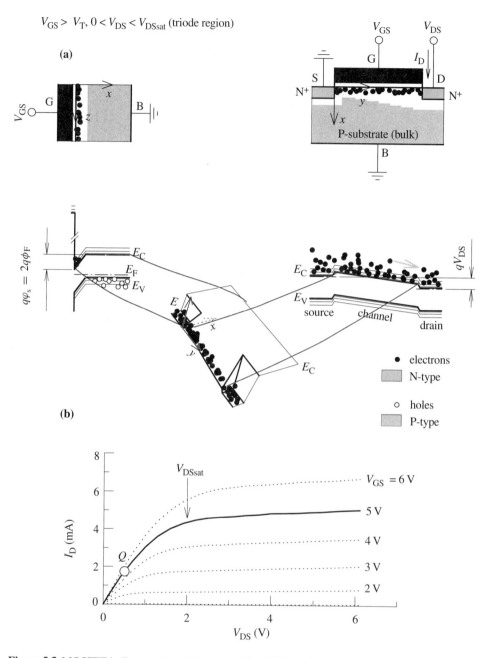

Figure 5.5 MOSFET in linear region: (a) cross-sectional illustrations and energy-band diagrams and (b) the output characteristics.

If the bottom of the conduction band in the drain is lowered with respect to the source, the energy bands in the channel region will be tilted and the electrons in the channel will start rolling down into the drain, as illustrated in Fig. 5.5. The lowering of the bottom of the conduction band in the drain is achieved by externally applied voltage V_{DS}. The bottom right energy-band diagram in Fig. 5.5a shows that the energy difference between the source and the drain is exactly $q V_{DS}$.

The two main factors influencing the value of the current I_D are the slope of the energy bands and the density of electrons in the channel. As the slope is directly proportional to V_{DS}, and the channel electron density is given by Eq. (5.6), the drain current can be expressed as

$$I_D = \beta \ (V_{GS} - V_T) \ V_{DS} \qquad (5.7)$$

where the proportionality factor β involves the oxide capacitance C_{ox}. It will be shown in Section 5.1.3 that the factor β depends on a number of other parameter, in addition to C_{ox}.

Equation (5.7) predicts a linear dependence of the drain current I_D on the drain-to-source voltage V_{DS}. As Fig. 5.5b illustrates, this is correct for small V_{DS} values. This region of V_{DS} voltage is referred to as the *linear region*. The reasons the current saturation occurs at larger V_{DS} voltages are considered in Section 5.1.3.

Body Effect

It has so far been assumed that the source and the bulk of the MOSFET are short circuited ($V_{BS} = 0$). Although the MOSFETs are very frequently used in this way, there are some applications in which the bulk and the source cannot be short circuited, or nonzero voltage is deliberately applied between the bulk and the source. In the case of N-channel MOSFETs, therefore P-type bulk and N-type source, the voltage applied between the bulk and the source should not be positive, as it would bias the bulk-to-source P–N junction in the forward mode, opening a current path between the source and the bulk. Negative bulk-to-source voltages ($V_{BS} < 0$), or equivalently positive source-to-bulk voltages ($V_{SB} > 0$), set the bulk-to-source P–N junction in reverse mode. The reverse bias of the bulk-to-source junction increases the threshold voltage of the MOSFET, which is the effect referred to as the *body effect*.

To understand the body effect, the MOSFET has to be considered in two dimensions: one normal to the semiconductor surface (along the x-axis) and the other along the channel (y-axis). Figure 5.6 provides the energy bands of a MOSFET along those two dimensions for different surface potentials φ_s. In Fig. 5.6a, the surface potential with respect to the bulk is zero (flat bands in the P-type substrate along the x-axis). The effect of the applied V_{SB} voltage is seen in the energy-band diagram along the y-axis. It shows that the applied voltage V_{SB} splits the Fermi level into quasi-Fermi levels of the P-type substrate (E_{FP}) and N-type source (E_{FN}). This is the effect already explained in the section on reverse-biased P–N junction (Fig. 2.6c). As a consequence, the energy barrier between the electrons in the source and the channel is increased by $q V_{SB}$ when compared to the case of $V_{SB} = 0$ (Fig. 5.2).

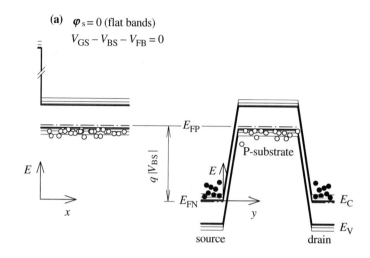

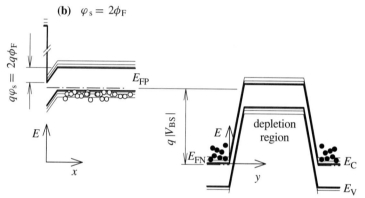

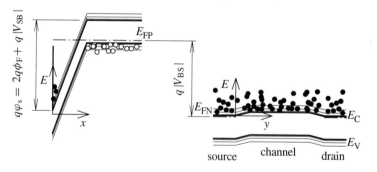

Figure 5.6 Illustration of the body effect: (a) V_{SB} voltage increases the barrier between the electrons in the source and the channel, (b) the surface potential of $2\phi_F$ does not reduce the barrier sufficiently for the electrons to be able to move into the channel, and (c) the surface potential needed to form the channel is $2\phi_F + V_{SB}$.

Figure 5.6b illustrates the case of $\varphi_s = 2\phi_F$. This is the value of the surface potential that corresponds to the strong-inversion mode (the channel of electrons formed) in the case $V_{SB} = 0$. As can be seen from Fig. 5.6b, the surface potential of $\varphi_s = 2\phi_F$ does not reduce the energy barrier between the source and the channel sufficiently in the case of $V_{SB} > 0$. To compensate for the effect of the V_{SB} bias, the surface potential needs to be further increased (therefore, the energy barrier reduced), and that is exactly by V_{SB}. Fig. 5.6c illustrates that the energy barrier between the source and the channel is reduced to the small strong-inversion value when $\varphi_s = 2\phi_F + V_{SB}$.

In the case of $V_{SB} > 0$, the surface potential in strong inversion is $\varphi_s = 2\phi_F + V_{SB}$.

To derive the threshold voltage equation, Eq. (5.2) has to be modified to account for the fact that the bulk is not grounded in this case but voltage $V_{SB} > 0$ is applied. To obtain the voltage drop across the gate oxide, the surface potential φ_s (expressed with respect to the bulk potential) is subtracted from the gate-to-bulk voltage. As the gate-to-bulk voltage is equal to $V_{GS} + V_{SB}$, Eq. (5.2) becomes

$$\underbrace{\underbrace{V_{GS} + V_{SB} - V_{FB}}_{\text{effective gate-to-bulk voltage}} - \varphi_s = \frac{Q_d}{C_{ox}}}_{\text{voltage drop across the gate oxide}} \quad \text{(for } V_{GS} \leq V_T) \quad (5.8)$$

Given that the surface potential is $\varphi_s = 2\phi_F + V_{SB}$ at the onset of strong inversion ($V_{GS} = V_T$), the following equation is obtained:

$$V_T + V_{SB} - V_{FB} - (2\phi_F + V_{SB}) = \frac{Q_d}{C_{ox}} \quad (5.9)$$

which appears to lead to the same threshold voltage equation as Eq. (5.4):

$$V_T = V_{FB} + 2\phi_F + \frac{Q_d}{C_{ox}} \quad (5.10)$$

There is a difference, however, due to the fact that Q_d depends on the surface potential. It has been shown that Q_d/C_{ox} can be converted into $\gamma\sqrt{2\phi_F}$, when $V_{SB} = 0$. In the case of $V_{SB} > 0$, the surface potential is not $2\phi_F$ but $2\phi_F + V_{SB}$, which means Q_d/C_{ox} converts into $\gamma\sqrt{2\phi_F + V_{SB}}$. Therefore, the threshold voltage equation, which involves the effect of V_{SB} voltage, has the following form:

$$\boxed{V_T = V_{FB} + 2\phi_F + \gamma\sqrt{2\phi_F + V_{SB}}} \quad (5.11)$$

The threshold voltage increase ΔV_T, caused by the voltage V_{SB} is

$$\Delta V_T = V_T(V_{SB}) - V_T(V_{SB} = 0) = \gamma(\sqrt{2\phi_F + V_{SB}} - \sqrt{2\phi_F}) \quad (5.12)$$

■ **Example 5.1 Threshold Voltage with $V_{SB} = 0$ and $V_{SB} \neq 0$ (Body Effect)**

An N-channel MOSFET with N^+-type polysilicon gate has oxide thickness of 10 nm and substrate doping $N_A = 5 \times 10^{16}$ cm^{-3}. The oxide charge density is $N_{oc} = 5 \times 10^{10}$ cm^{-2}. Find the threshold voltage if the bulk is biased at 0 and -5 V, respectively. The following constants are known: the thermal voltage $V_T = 0.026$ V, the oxide permittivity $\varepsilon_{ox} = 3.9 \times 8.85 \times 10^{-12}$ F/m, the silicon permittivity $\varepsilon_s = 11.8 \times 8.85 \times 10^{-12}$ F/m, the intrinsic carrier concentration $n_i = 1.02 \times 10^{10}$ cm^{-3}, and the silicon energy gap $E_g = 1.12$ eV.

Solution: To use Eq. (5.11) to calculate the threshold voltage, the Fermi potential ϕ_F, the flat-band voltage V_{FB}, and the body factor γ should be obtained first. According to Eq. (2.63), the Fermi potential is

$$\phi_F = +V_t \ln \frac{N_A}{n_i} = 0.401 \text{ V}$$

Flat-band voltage Eq. (2.65) shows that the work-function difference $q\phi_{ms}$ and the gate-oxide capacitance per unit area C_{ox} are needed as well. The work-function difference is given by Eq. (2.64). In the N^+-type gate, the Fermi level is very close to the bottom of the conduction band, which means the work function of the gate $q\phi_m$ is approximately equal to the electron affinity $q\chi_s$ (Section 2.3), therefore

$$\phi_{ms} = -\frac{E_g}{2q} - \phi_F = -0.961 \text{ V}$$

As the gate-oxide capacitance per unit area is

$$C_{ox} = \varepsilon_{ox}/t_{ox} = 3.45 \times 10^{-3} \text{ F/m}^2$$

the flat-band voltage is obtained as

$$V_{FB} = \phi_{ms} - \frac{q N_{oc}}{C_{ox}} = -0.961 - \frac{1.6 \times 10^{-19} \times 5 \times 10^{14}}{3.45 \times 10^{-3}} = -0.984 \text{ V}$$

The body factor is given by Eq. (2.76):

$$\gamma = \sqrt{2\varepsilon_s q N_A}/C_{ox}$$
$$= \sqrt{2 \times 11.8 \times 8.85 \times 10^{-12} \times 1.6 \times 10^{-19} \times 5 \times 10^{22}}/3.45 \times 10^{-3}$$
$$= 0.375 \text{ V}^{1/2}$$

The threshold voltage at $V_{SB} = 0$ V is calculated as

$$V_T(0) = V_{FB} + 2\phi_F + \gamma\sqrt{2\phi_F + V_{SB}}$$
$$= -0.984 + 2 \times 0.401 + 0.375\sqrt{2 \times 0.401 + 0} = 0.15 \text{ V}$$

To calculate the threshold voltage at $V_{SB} = 5$ V, find the threshold voltage difference ΔV_T using Eq. (5.12):

$$\Delta V_T = 0.375 \left(\sqrt{2 \times 0.401 + 5} - \sqrt{2 \times 0.401}\right) = 0.57 \text{ V}$$

Therefore, $V_T(5 \text{ V}) = V_T(0) + \Delta V_T = 0.72$ V.

5.1.3 MOSFET as a Voltage-Controlled Current Source: Mechanisms of Current Saturation

When the MOSFET is operated as a closed switch ($V_{GS} > V_T$ and $V_{DS} < V_{DSsat}$), the normal electric field from the gate voltage V_{GS} holds the electrons in the inversion layer channel, while the lateral electrical field due to the drain-to-source voltage V_{DS} rolls them into the drain. The channel of electrons expands all the way from the source to the drain, the resistance of which determines the slope of the linear $I_D - V_{DS}$ characteristic.

To use the MOSFET as a current source, I_D should become independent of V_{DS}. This happens at larger drain-to-source voltages ($V_{DS} > V_{DSsat}$), an effect referred to as current saturation. There are two different mechanisms that can cause drain current saturation in MOSFETs. These two mechanisms are considered in the following text.

Channel Pinch-off

As the drain-to-source voltage V_{DS} is increased, the lateral electric field in the channel is increased as well, and may become stronger than the vertical electric field due to the gate voltage. This would first happen at the drain end of the channel. In this situation, the vertical field is unable to keep the electrons at the drain end of the channel as the stronger lateral field sweeps them into the drain. The channel is pinched off at the drain end. The drain-to-source voltage at which this happens is called the *saturation voltage*, V_{DSsat}.

An increase of V_{DS} beyond V_{DSsat} expands the region in which the lateral field is stronger than the vertical field in the channel, effectively moving the pinch-off point closer to the source. This is illustrated in the MOSFET cross section along the y-axis, shown in Fig. 5.7a (upper-right corner). The region created between the pinch-off point and the drain is basically the depletion layer at the reverse-biased drain–substrate junction. Note that we are considering the surface area of the junction, which is influenced by the gate field. Consequently, the surface region of the P–N junction is not in the reverse-bias mode until the drain voltage reaches V_{DSsat}. This is different from the bulk region of the junction, which is in reverse-bias mode for any positive V_{DS} voltage.

The voltage drop across the surface depletion region (the reverse bias) is $V_{DS} - V_{DSsat}$, which is the voltage increase beyond V_{DSsat}. The remaining part of the drain-to-source voltage, which is V_{DSsat}, drops between the pinch-off point and the source. In this region the vertical field is stronger than the lateral field, and the inversion layer (channel of electrons) still exists. It is in fact this part of the source-to-drain region, labeled as channel in Fig. 5.7a, that determines the value of the drain current.

> The voltage drop across the channel of electrons is fixed to V_{DSsat} for $V_{DS} > V_{DSsat}$. As a consequence, the drain current remains fixed to the value corresponding to V_{DSsat}. This effect is called drain current saturation, and the $V_{DS} > V_{DSsat}$ region of the MOSFET operation is referred to as the *saturation region*.

The energy-band diagram of the MOSFET along the y-axis (bottom right diagram in Fig. 5.7a) provides a clearer insight into the effect of current saturation due to channel pinch-off. It shows very steep energy bands in the depletion region, which represents the situation of a very strong lateral field in this region. Electrons do not spend much time on this very

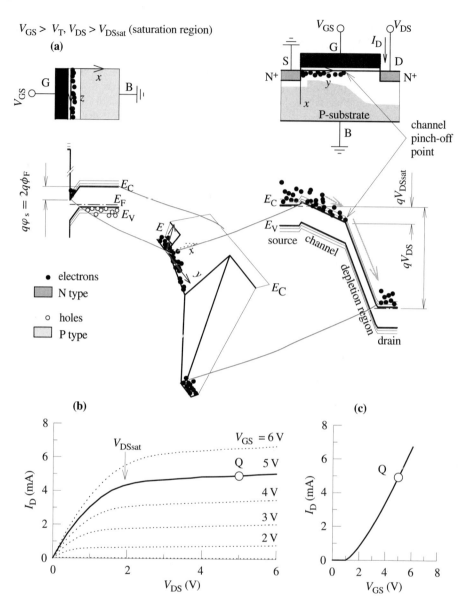

Figure 5.7 MOSFET in saturation region: (a) cross-sectional illustrations and energy-band diagrams, (b) the output characteristics, and (c) the transfer characteristics.

steep part of the bottom of the conduction band; they very quickly roll down into the drain. This part of the source-to-drain region offers little resistance to the electrons. Although an increase of V_{DS} beyond V_{DSsat} continues to lower the position of the conduction band in the drain, Fig. 5.7a illustrates that this does not increase the drain current. The electrons in this shape of energy bands can be compared to a waterfall: the water current depends

on the amount of water before the fall (the channel) and not on the height of the waterfall
($q V_{DS} - q V_{DSsat}$).

1. The depletion region has little influence on the drain current.
2. The value of the drain current is limited by the number of electrons that appear at
   the edge of the depletion region (the pinch-off point).
3. The number of the electrons in the channel, and therefore at the pinch-off point,
   is controlled by the gate voltage and not the drain voltage.
4. As a consequence, the drain current is controlled by the gate voltage, and is
   independent of the drain voltage.
5. The MOSFET acts as a voltage-controlled current source.

$I_D - V_{DS}$ characteristics shown in Fig 5.7b illustrate the effect of drain current saturation.
The solid line shows the complete $I_D - V_{DS}$ dependence in the strong inversion for a fixed
V_{GS} voltage ($V_{GS} = 5$ V in the example of Fig. 5.7b). The characteristic is linear for small
V_{DS} voltages. It deviates from the linear dependence as the increased V_{DS} starts depleting
electrons from the drain end of the channel. The drain-to-source voltage value that pinches
off the channel at the drain end is the *saturation voltage*, V_{DSsat}.

The region of drain-to-source voltages between zero and the saturation voltage ($0 \leq
V_{DS} < V_{DSsat}$) is referred to as the *triode region*. The region of drain-to-source voltages
higher than the saturation voltage ($V_{DS} \geq V_{DSsat}$) is the *saturation region*.

The dotted lines in Fig. 5.7b and the $I_D - V_{GS}$ characteristic of Fig. 5.7c (transfer
characteristic) illustrate the effect of the gate voltage. It is obvious that the current depends
on the gate voltage even in the saturation region. This is because the gate voltage determines
the number of electrons in the channel (in strong inversion), which in turn directly determines
the current. This is a useful property, as it enables the current source (that is MOSFET in
saturation) to be *controlled* by voltage. Figure 5.7b also shows that the saturation voltage
V_{DSsat} is different for different V_{GS} voltages. This is due to the fact that a much smaller
drain-to-source voltage is needed to overcome the effect of a smaller gate voltage and pinch
the channel off at the drain end.

Drift Velocity Saturation

Minimum lateral and vertical dimensions of MOSFETs are continuously being reduced in
order to increase the density and the speed of modern ICs. These MOSFETs are referred to as
short-channel MOSFETs. The operating V_{DS} voltages cannot be proportionally reduced as
the maximum operating voltage has to be kept well above the MOSFET threshold voltage.
As a consequence, short-channel MOSFETs operate with significantly increased lateral
and vertical electric fields in the channel. Although the relative relationship between the
lateral and the vertical electric fields is roughly maintained, these MOSFETs can exhibit
a different type of drain current saturation. It can happen that the current saturates at
drain-to-source voltage smaller than the voltage that would cause channel pinch-off at
the drain end.

To explain this effect, refer to Eq. (1.47) (Section 1.4.1), which shows that the current
is directly proportional to the drift velocity of the carriers. As explained in Section 1.4.1,

the drift velocity follows linear dependence on the lateral electric field in the channel up to a certain level, and saturates if the field is increased beyond that level (Fig. 1.18). The lateral electric field in short-channel MOSFETs can be stronger than the critical velocity-saturating value, while not being stronger than the vertical electric field at the drain end of the channel (the channel is not pinched off). Nonetheless, the drain current will saturate following the carrier velocity saturation in the channel.

Although this is a different mechanism of current saturation, the MOSFET can equally well be used as a voltage-controlled current source.

Related to modeling of this saturation mechanism, it is worth considering the following question: **If the channel is not pinched off, it can be modeled as a resistor between the source and the drain. Should we believe Ohm's law, which says that resistor current linearly depends on the voltage?** As explained in Section 1.4.2, we should not believe Ohm's law, because it is not correct for this case of large lateral electric fields. The velocity saturation and the related current saturation do happen. All we can do with Ohm's law is to alter appropriately the value of the mobility in Eq. (1.48), so that it models this effect properly. The section on MOSFET modeling will describe the mobility models used in SPICE to account for this effect.

5.1.4 Types of MOSFETs

So far, we have considered only one type of MOSFETs, which uses P-type substrate (bulk) and N^+-type source and drain layers, and needs positive voltage at the gate to turn the MOSFET on by creating channel of electrons between the source and the drain. As this type of MOSFET operates with N-type channel (electrons) it is referred to as an *N-channel* MOSFET. It is possible to make a complementary MOSFET using an N-type substrate (bulk) and P^+- type source and drain layers. In this type of MOSFET, the channel connecting the source and the drain in *ON* mode has to be created of holes (P-type carriers), because of which it is called a *P-channel MOSFET*. The terms N-channel and P-channel MOSFETs are frequently replaced by the shorter terms *NMOS* and *PMOS*.

A common characteristic of the above-described N-channel and P-channel MOSFETs is that they are in the *OFF* mode when no gate bias is applied. This is because there is no channel between the source and the drain, and therefore the drain current is zero. Consequently, these MOSFETs are classified as *normally off* MOSFETs. The transfer characteristics of Fig. 5.8 show that the drain current appears for sufficiently large positive gate voltages in the case of an N-channel and negative gate voltages in the case of a P-channel MOSFET. This is because appropriate gate voltages are needed to create the channel of electrons and holes in N-channel and P-channel MOSFETs, respectively. As a consequence, these types of MOSFETs are also referred to as *enhancement MOSFETs*.

MOSFETs can be created with technologically built-in channels. As no gate voltage is needed to set these MOSFETs in the *ON* state, they are called *normally on* MOS-FETs. To turn these types of MOSFETs off, the channels have to be depleted of electrons or holes, so they are also referred to as *depletion-type MOSFETs*. Figure 5.8 illustrates that negative voltage is needed to stop the drain current in the case of the N-channel MOSFET, and similarly positive voltage is needed to set the P-channel MOSFET in the *OFF* state.

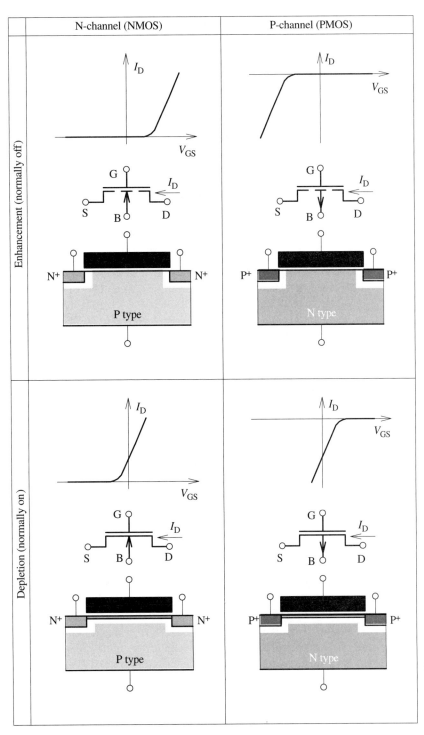

Figure 5.8 Types of MOSFETs.

Defining the threshold voltage as the gate voltage at which the channel is just formed (or depleted), we can say that the threshold voltage of the enhancement type N-channel MOSFETs is positive, while it is negative in the case of the depletion type N-channel MOSFETs. The situation is opposite with the P-channel MOSFETs: negative threshold voltage in the case of the enhancement type, and positive threshold voltage in the case of the depletion type.

The main MOSFET type is the N-channel enhancement type. The P-channel enhancement type MOSFET is used as a complementary transistor in circuits known as *CMOS* (complementary MOS) technology. The N-channel depletion-type MOSFET is used as a kind of complementary transistor in circuits using only N-channel MOSFETs (*NMOS* technology).

5.2 MOSFET TECHNOLOGIES

This section begins with a brief description of an alternative semiconductor doping technique, ion implantation, which enabled the development of modern MOSFET technologies. However, the bulk of the section is devoted to two standard MOSFET technology processes: N-channel MOSFET (NMOS) and complementary MOSFET (CMOS) technologies.

5.2.1 Ion Implantation

Ion implantation is one of two principal ways of introducing doping atoms into a semiconductor substrate. The ion implanter diagram of Fig. 5.9 illustrates the ion-implantation process. The process begins with gas ionization, which creates an ion mixture containing the ions of the doping element. The desired ions are separated according to their atomic mass in a mass separator, the ion beam is then focused, and the ions are accelerated to the desired energy (typically between 10 and 200 keV). The ion beam is scanned across the wafer to achieve uniform doping. When the ions of the doping element hit the wafer, they suffer many collisions with the semiconductor atoms before eventually stopping at some depth beneath the surface. As the target is grounded, to complete the electric circuit, the implanted ions are neutralized by electrons flowing into the substrate.

Although all the beam ions have the same energy, they do not stop at the same distance, since the stopping process involves a series of random events. This is illustrated in Fig. 5.10a. Therefore, the ion-implantation process leads to a bell-shaped profile of the doping atoms, as shown in Fig. 5.10b.

The doping atoms are electrically active only when they replace semiconductor atoms from their positions in the crystal lattice. The implanted atoms generally terminate in interstitial positions. Also, a number of semiconductor atoms will be displaced from their positions, which results in damage of the semiconductor crystal. Because of that, a postimplant anneal must be performed. This anneal should provide sufficient energy to enable the silicon and doping atoms to rearrange themselves back into the crystal structure. Minimum anneal conditions vary between 30 min at approximately 900°C for low-implant levels to 30 min at 1000°C for high-implant levels. As diffusion of the implanted atoms will take place during the anneal, this process is equivalent to the drive-in diffusion process used to redistribute

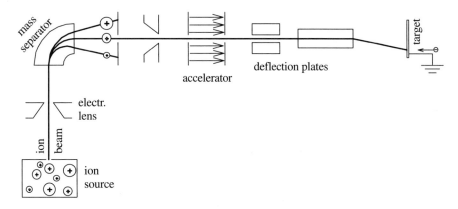

Figure 5.9 Ion-implanter diagram.

the doping atoms (Section 1.3). The change of the ion-implant profile during this annealing is illustrated in Fig. 5.10c.

Two important parameters of the ion-implantation process are the implant energy and the amount of implanted ions (dose). The implant energy determines the depth of the implanted ions. The dose (number of implanted ions per unit area) is determined by the implant time and the current of the ion beam. The current of the ion beam can be measured by measuring the current of electrons flowing from the ground into the substrate to neutralize the implanted ions. Integrating the measured current in time gives the charge implanted into the substrate, which is divided by the unit charge q and the substrate area to obtain the dose. It is obvious that the doping level achieved by ion implantation can very precisely be controlled. As the MOSFET threshold voltage depends on the doping level in the channel region [Eqs. (5.4) and (5.5)], ion implantation enables precise adjustment of the threshold voltage. This has proved one of the most important advantages of ion implantation compared to the diffusion technique. Added flexibility for doping profile engineering appears as an additional advantage of ion implantation. A disadvantage of ion implantation is the complex equipment that is reflected in the cost of the doping process.

There are analytical equations modeling the ion-implantation process. However, their application generally requires numerical techniques to be involved. Because of that, numerical simulation tools are commonly used to design the ion-implantation and subsequent annealing processes, which means establishing the values of the dose, the implant energy, and the annealing time and temperature to obtain the desired doping profile.

5.2.2 NMOS Technology

Section 4.2 introduced the basics of digital circuits. A building block of digital circuits, the inverter with resistive load, is shown in Fig. 4.6. The voltage-controlled switch used in this

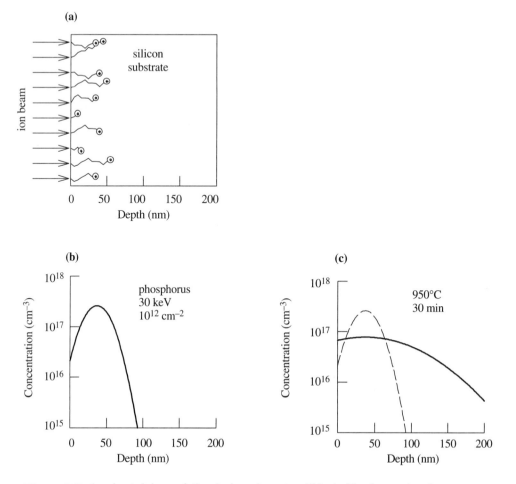

Figure 5.10 Accelerated ions of the doping element collide inside the semiconductor, stopping at scattered depths (a); the doping profile immediately after ion implantation (b) and after annealing (c).

inverter can be implemented by an enhancement-type N-channel MOSFET. As for the load resistor R_D, it could be implemented in MOS technology, however, it would complicate the technology necessitating layers in addition to those used for MOSFETs, and would occupy a relatively large area. It is far more efficient to replace the resistor R_D by the channel resistance of a MOSFET (active load). A depletion-type N-channel MOSFET, connected as in Fig. 5.11, quite efficiently plays tho role of the load element. This shows that complex digital circuits can be built of N-channel MOSFETs only, a technology referred to as NMOS technology.

The cross section and composite layout of the NMOS inverter are shown in Fig. 5.12. They show that the active region of the NMOS inverter is surrounded by relatively thick oxide (called field oxide) and a P^+ diffusion region, which electrically isolate the inverter from the rest of the circuit. Imagine an N^+ region of a neighboring device adjacent to the

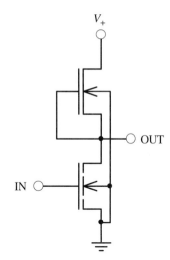

V_+

IN

OUT

Figure 5.11 The circuit of an NMOS inverter.

N^+ drain of the depletion MOSFET, and V_+ metal line running over the space that separates them. If the field oxide and the P^+ region were not there, the two N^+ regions and the metal over them would comprise a turned-on parasitic MOSFET, causing current leakage between the two devices. The thick-field oxide and the increased substrate concentration underneath (P^+ region) increase the threshold voltage of this parasitic MOSFET over the value of the supply voltage V_+, ensuring that it remains off.

Figure 5.12 also shows that the N^+ drain of the enhancement MOSFET and the N^+ source of the depletion MOSFET are merged into a single N^+ region. This is done to minimize the area of the inverter. The circuit would not operate better if separate N^+ regions were made for the drain of enhancement and the source of depletion MOSFETs, respectively, and then connected by a metal line from the top. Quite the opposite, this approach would introduce additional parasitic resistance and capacitance, adversely affecting the circuit performance.

> The principle of IC layer merging is employed whenever possible to minimize the active IC area and maximize its performance.

In the spirit of the layer-merging principle, the gate of the depletion MOSFET is extended to directly contact the N^+ source/drain region. The polysilicon gate area is further extended to serve as the output line of the inverter. In Fig. 5.12, both the input and the output lines are implemented by polysilicon stripes, while the V_+ and ground rails appear as metal (Al) lines. This illustrates that the polysilicon layer can be used as the second interconnection level. When necessary, the N^+ diffusion areas can be used as the third interconnection level. The contact between the metal and the polysilicon levels can be made in a way similar to the contacts between metal and N^+ regions, shown in Fig. 5.12.

The technology sequence used to fabricate the NMOS inverter of Fig. 5.12 is presented in Fig. 5.13a to i. At the beginning of the process sequence, the isolation (field oxide and P^+ region) are created to define the active area. The active area is protected by silicon nitride

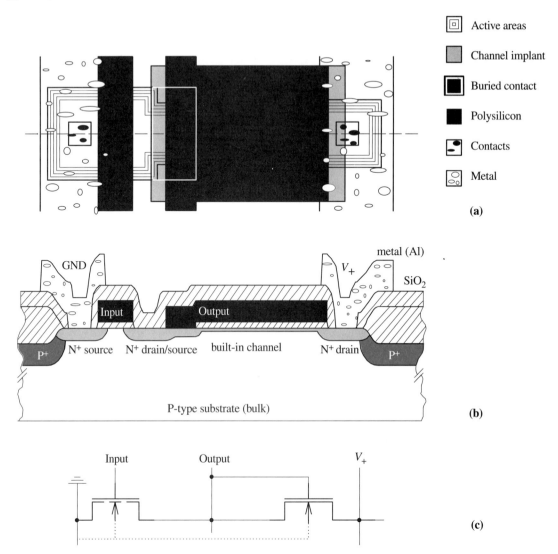

Figure 5.12 The composite layout (a) and cross section (b) of an NMOS inverter are shown along with the circuit diagram (c).

deposited over a thin thermally grown buffer oxide. The silicon nitride layer and the buffer oxide are patterned by a photolithography process, using a mask as shown in Fig. 5.13a. The opaque and transparent areas of the mask correspond to photolithography with positive photoresist. Silicon wafers prepared in such a way are exposed to boron implantation, which creates the P^+ type doping outside the active region. After this, thermal oxidation is applied to grow the thick field oxide. The silicon nitride blocks the oxidizing spices, protecting the active area. This process is known as LOCOS (**loc**al **o**xidation of **si**licon), and is explained in Section 2.5.2 (Fig. 2.15). Figure 5.13b shows that some lateral oxidation occurs, leading to the so-called *bird beak* shape of the field oxide. This has beneficial effects, as it smooths the oxide step, which would otherwise be too sharp and high leading to gaps in the metal layer deposited subsequently. After the field oxide growth, the silicon nitride and the buffer oxide are removed (Fig. 5.13c).

After the definition of the active areas, the second photolithography process is applied to provide selective implantation of phosphorus, creating the built-in channel of the depletion MOSFET (Fig. 5.13d). In this case, the photoresist itself is used to protect the channel area of the enhancement-type MOSFET. In the following steps, the photoresist is removed and the

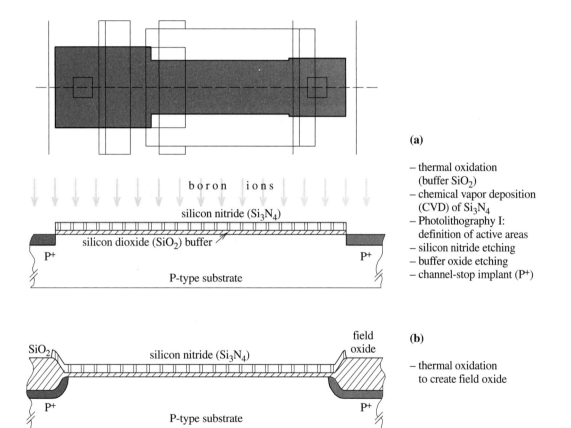

(a)

– thermal oxidation
(buffer SiO$_2$)
– chemical vapor deposition
(CVD) of Si$_3$N$_4$
– Photolithography I:
definition of active areas
– silicon nitride etching
– buffer oxide etching
– channel-stop implant (P$^+$)

(b)

– thermal oxidation
to create field oxide

(c)

– silicon nitride etching
– buffer oxide etching

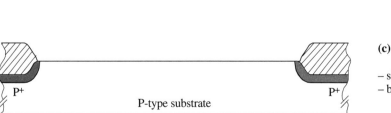

Figure 5.13 (a–i) The NMOS technology process.

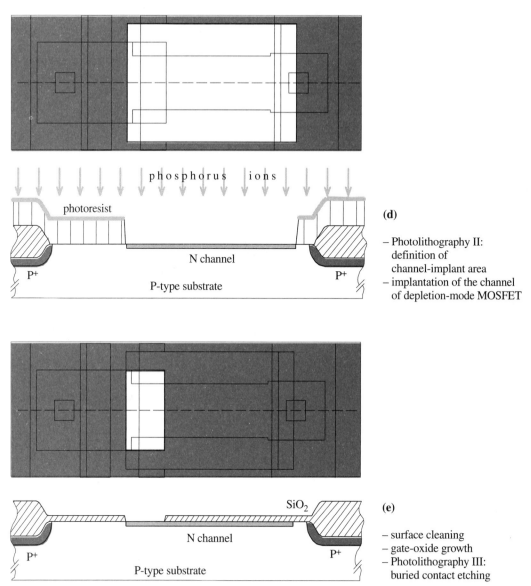

Figure 5.13 Continued

surface thoroughly cleaned to prepare the wafers for gate-oxide growth. Once the gate oxide is grown, the third photolithography is used to etch a hole in the gate oxide (Fig. 5.13e), where the subsequently deposited polysilicon layer should contact the silicon (this is the contact between the gate of depletion MOSFET, its source, and the drain of the enhancement MOSFET). The deposited polysilicon is patterned by the fourth photolithography process and associated polysilicon and gate-oxide etching (Fig. 5.13f). The photolithography mask used for the polysilicon patterning defines the gate lengths of the MOSFETs (the width of

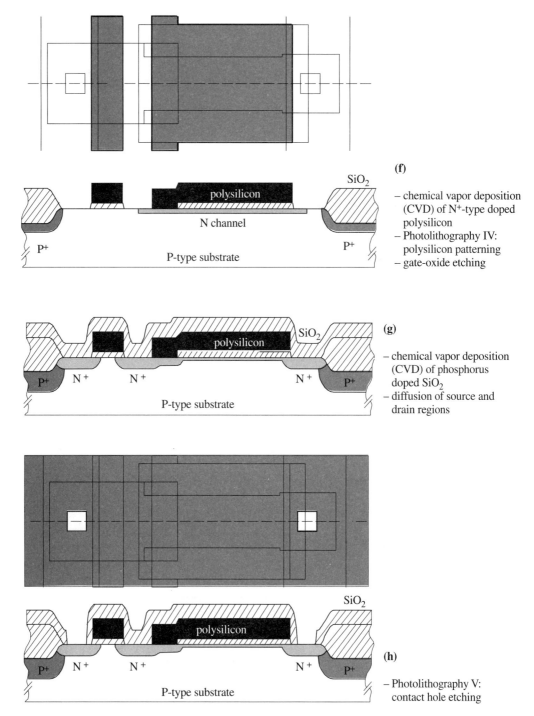

(f)

– chemical vapor deposition (CVD) of N^+-type doped polysilicon
– Photolithography IV: polysilicon patterning
– gate-oxide etching

(g)

– chemical vapor deposition (CVD) of phosphorus doped SiO_2
– diffusion of source and drain regions

(h)

– Photolithography V: contact hole etching

Figure 5.13 Continued

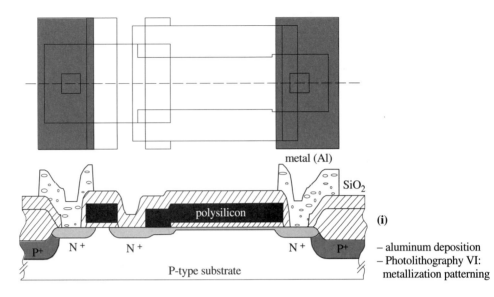

metal (Al)

SiO₂

polysilicon

(i)

P⁺   N⁺   N⁺   N⁺   P⁺

P-type substrate

– aluminum deposition
– Photolithography VI:
 metallization patterning

Figure 5.13 Continued

the polysilicon areas) and the width of N⁺ diffusion regions (the separation between the polysilicon areas).

The alignment of the N⁺ source/drain regions to the MOSFET gate is very important. Although no gap between the gate and the N⁺ source/drain regions is acceptable, large gate-to-N⁺ source/drain overlaps would create large parasitic gate-to-source/drain capacitances, which adversely affects the high-speed performance of the device. If the N⁺ source/drain regions were to be created by a separate mask, the alignment of this mask would create significant problems and/or limitations. The use of polysilicon as a gate material instead of metal (aluminum) enables a self-aligned technique to be employed. Figure 5.13g shows that the N⁺ source/drain regions are obtained by diffusion from phosphorus doped oxide, which is deposited onto the wafer. The polysilicon gates protect the area underneath against phosphorus diffusion, as the phosphorus diffuses into the polysilicon and the substrate areas not covered by the polysilicon (these are the drain/source regions). Some lateral diffusion does occur, which means some gate-to-source/drain overlap is unavoidable. This self-aligned technique cannot be implemented with aluminum instead of the polysilicon, as aluminum deposited over oxide or silicon should not be exposed to temperatures higher than 570°C (this would lead to adverse chemical reactions between the aluminum and the oxide/silicon), while the diffusion process requires temperatures higher than 900°C. The aluminum has smaller resistivity compared to polysilicon, and it may seem that aluminum is a better choice as the gate material. However, the beneficiary effects of the above-described self-aligned technique are so important that they enforce the use of polysilicon as the gate material in modern MOSFET ICs, although the MOSFETs were originally made with aluminum gates.

The phosphorus-doped oxide, originally deposited to provide the N⁺ source/drain diffusion, is also used as the insulating layer between the polysilicon and the subsequently deposited aluminum layer for device interconnection. However, before the aluminum is deposited, contact holes are etched in the insulating oxide layer to provide the necessary contacts to the MOSFET source and drain regions (Fig. 5.13h). After aluminum deposition, the sixth photolithography process is employed to pattern the aluminum layer, as shown in Fig. 5.13i.

5.2.3 Basic CMOS Technology

As described in Section 4.2.3, the use of complementary transistors enables digital circuits with minimized power consumption to be designed. The inverter with complementary transistors of Fig. 4.8 can be implemented with enhancement-type N-channel and P-channel MOSFETs as shown in Fig. 5.14a. This shows that complex digital circuits can be built using complementary pairs of N-channel and P-channel MOSFETs, also referred to as NMOS and PMOS transistors, respectively. This technology is known as *complementary MOS* (CMOS) technology.

As explained in Section 4.2.3, no current flows through the inverter in either steady state, high input/low output, or low input/high output. At high-input voltage, the NMOS is *ON* (gate-to-source voltage higher than its threshold voltage), but the PMOS is *OFF*

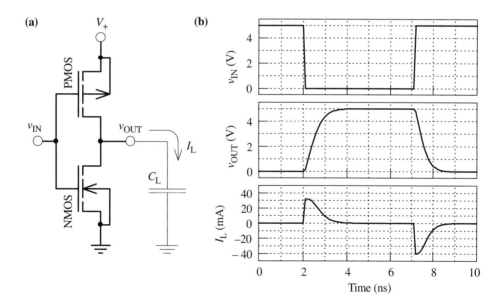

Figure 5.14 The circuit of a CMOS inverter (a) and typical input/output signals (b).

as its gate-to-source voltage ($v_{IN} - V_+$) is small. At low-input voltage, the PMOS is *ON* ($v_{IN} - V_+$ is negative, its absolute value being larger than the PMOS threshold voltage), but the NMOS is now *OFF* as the input voltage is below its threshold voltage. Consequently, no power is needed to maintain any logic state by a CMOS digital circuit. However, CMOS circuits do dissipate power when the logic states are changed. This is because the outputs of the logic cells, like the inverter in Fig. 5.14, are loaded by the parasitic input capacitances of the connected logic cells (represented by C_L in Fig. 5.14). To change the output of the inverter of Fig. 5.14 from a low to a high level, the capacitor C_L has to be charged by current flowing through the PMOS transistor. When the output is changed from a high to a low level, the capacitor C_L has to be discharged through the NMOS transistor. Obviously, in these transition periods, some power is dissipated by the inverter.

The low-power dissipation characteristic of the CMOS circuits has expanded the applications of digital circuits enormously, ranging from battery-supplied entertainment electronics to computer applications as we now know them. The switching speed (maximum operating frequency) was initially a disadvantage of the CMOS circuits, but aggressive MOSFET dimension reduction has led to a dramatic increase in the speed. The dimension reduction has also enabled increased levels of integration, leading to powerful digital ICs. Dimension reduction, or so-called MOSFET down-scaling, is described in more detail in Section 7.1. The CMOS technology has become the dominant electronics technology today.

The composite layout and the cross section of the CMOS inverter implemented in basic N-well technology are shown in Fig. 5.15. The technology is referred to as N-well technology because the N-type bulk needed for the PMOS transistor is implemented as the N-well region diffused into the P-type silicon substrate. The NMOS transistors are created in the P-type substrate itself. As the B terminal of the NMOS transistors (the P-type substrate) is grounded and the B terminal of the PMOS transistors (the N well) is connected to the most positive potential V_+, the N-well–P-substrate junction is reverse biased. Although this provides electrical isolation along the P–N junction, to ensure that no surface leakage occurs, the devices are separated by thick field oxide like the NMOS technology. The N^+-type layer, used to create the source and the drain of the NMOS, is also used to provide contact to the N-well region, which is directly connected to the source of the PMOS. Although no direct connection between the P-type substrate and the source of the NMOS is shown in Fig. 5.15b, the P-type substrate is connected to ground, frequently by a rail enclosing the whole IC area.

There is also P-well CMOS technology, in which PMOS transistors are placed in an N-type silicon substrate, and P-wells are created to place NMOS transistors. The P-well technology was developed before the N-well technology, as it was easier to achieve the desired threshold voltages of NMOS and PMOS transistors. When the use of ion implantation made it possible to adjust the threshold voltages, N-well technology became more popular because the use of the P-type substrate made it compatible with NMOS and bipolar technology (described in Chapter 6).

The technology sequence used to fabricate the CMOS inverter of Fig. 5.14 is presented in Fig. 5.16a to i. At the beginning of the process, N wells are created by ion implantation of phosphorus through an appropriately patterned silicon dioxide masking layer and subsequent annealing (drive-in). Figure 5.16a shows that the masking oxide is removed once

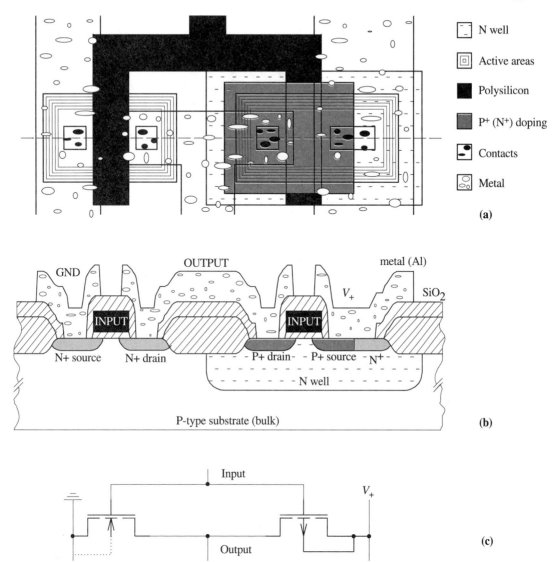

Figure 5.15 The composite layout (a) and cross section (b) of a basic N-well CMOS inverter are shown along with the circuit diagram (c).

the N wells are created. In the next stage, thick field oxide surrounding the active areas is created. As Fig. 5.16b and c illustrate, this is achieved by the same LOCOS process as used in the NMOS technology.

The surface donor concentration N_D in the N well is higher than the acceptor concentration N_A in the P-type substrate. This fact as well as the work-function differences cause a lower NMOS threshold voltage and a higher absolute value of the PMOS threshold voltage

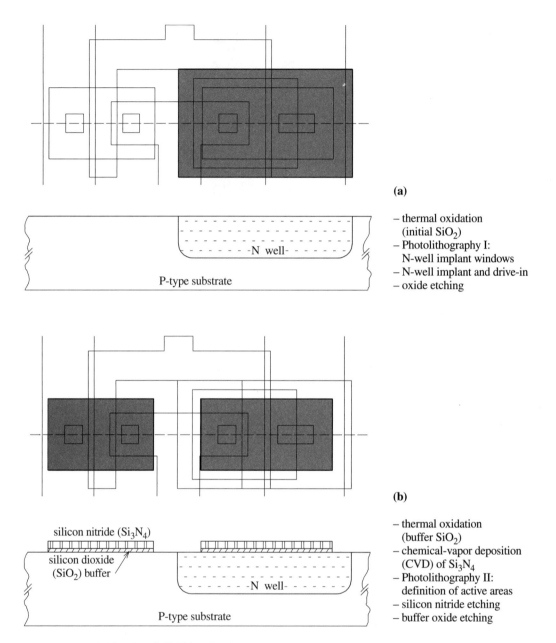

(a)

– thermal oxidation
  (initial SiO_2)
– Photolithography I:
  N-well implant windows
– N-well implant and drive-in
– oxide etching

(b)

– thermal oxidation
  (buffer SiO_2)
– chemical-vapor deposition
  (CVD) of Si_3N_4
– Photolithography II:
  definition of active areas
– silicon nitride etching
– buffer oxide etching

Figure 5.16 (a–i) The N-well CMOS technology process.

(c)

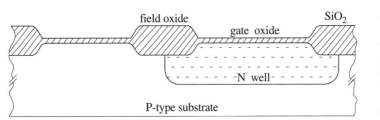

– thermal oxidation
  to create field oxide
– silicon nitride etching
– buffer oxide etching
– surface cleaning
– threshold-voltage
  adjustment implant
– gate-oxide growth

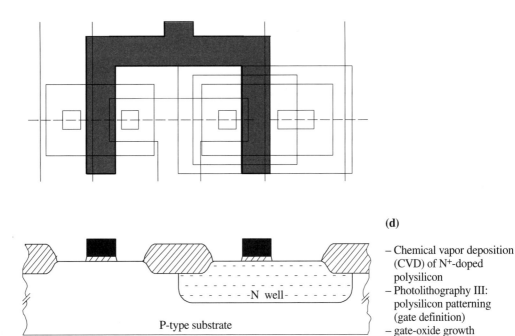

(d)

– Chemical vapor deposition
  (CVD) of N⁺-doped
  polysilicon
– Photolithography III:
  polysilicon patterning
  (gate definition)
– gate-oxide growth

Figure 5.16 Continued

than the desirable values. The performance of CMOS circuits is maximized with a slight positive NMOS threshold voltage and a negative PMOS threshold voltage (enhancement-type MOSFETs) with the absolute value equal to its NMOS counterpart. To adjust the threshold voltages, boron is implanted into the channel areas of both the NMOS and PMOS transistors. This, so-called threshold-voltage adjustment implantation reduces the effective N-type doping level at the surface of the N well, decreasing the absolute value of the PMOS threshold voltage. Also, it increases the P-type doping level at the surface of the P-type substrate, increasing the NMOS threshold voltage. The NMOS and PMOS threshold voltages can be matched perfectly by an appropriately determined dose of the threshold-voltage adjustment implantation.

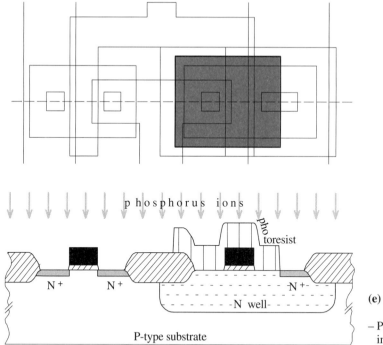

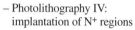

phosphorus ions

photoresist

N⁺   N⁺   N⁺

-N well-

P-type substrate

(e)

– Photolithography IV:
   implantation of N⁺ regions

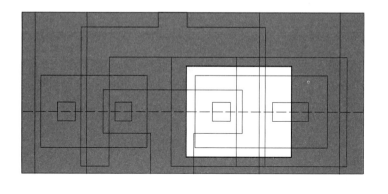

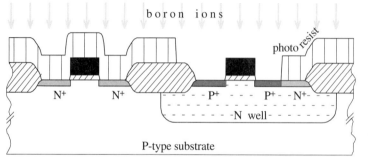

boron ions

photoresist

N⁺   N⁺   P⁺   P⁺   N⁺

- N well-

P-type substrate

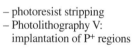

(f)

– photoresist stripping
– Photolithography V:
   implantation of P⁺ regions

Figure 5.16 Continued

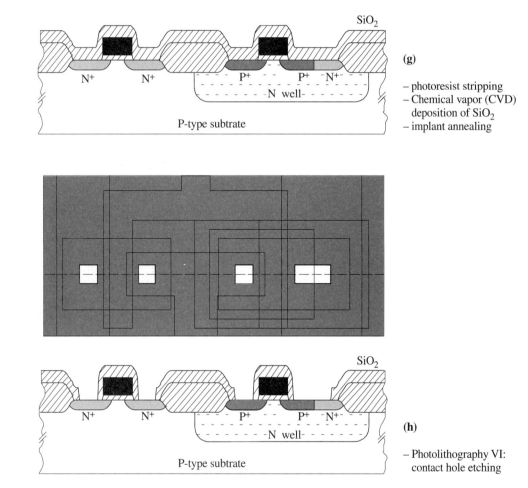

(g)

– photoresist stripping
– Chemical vapor (CVD)
  deposition of SiO_2
– implant annealing

(h)

– Photolithography VI:
  contact hole etching

Figure 5.16 Continued

The gate oxide can be grown either before or after the threshold-voltage adjustment implantation (Fig. 5.16c). Following the gate oxidation, doped polysilicon is deposited and patterned to define the gates of MOSFETs (Fig. 5.16d).

The CMOS technology also makes use of the self-aligned technique to minimize the overlap between the gate and drain/source areas. The difference here, compared to the NMOS technology, is that the doping is achieved by ion implantation of phosphorus (Fig. 5.16e) and boron (Fig. 5.16f) to obtain the N^+ and P^+ regions, respectively. Appropriately patterned photoresist is used to mask the areas that are not to be implanted. As Fig. 5.16g shows, the ion implants are followed by oxide deposition (needed to isolate the polysilicon layer from the subsequent metal layer) and annealing, which is needed to activate the implanted doping ions.

The process finishes in the same way as the NMOS technology, by contact hole etching (Fig. 5.16h), and aluminum deposition and patterning (Fig. 5.16i).

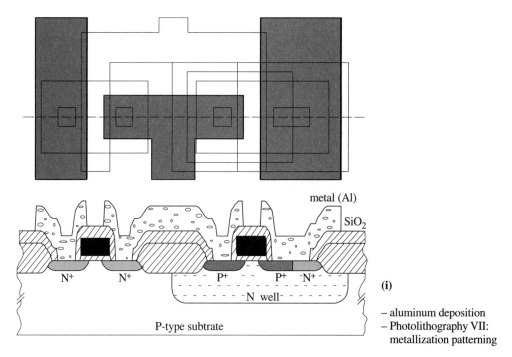

Figure 5.16 Continued

■ **Example 5.2 Threshold-Voltage Adjustment Implant**

Technological parameters of N-well CMOS technology are given in the Table 5.1, together with the values of the relevant physical parameters.

TABLE 5.1 Technological Parameters of N-Well CMOS Technology

Parameter	Symbol	Value
Substrate doping concentration	N_A	10^{15} cm^{-3}
N-well surface concentration	N_D	5×10^{16} cm^{-3}
Gate-oxide thickness	t_{ox}	15 nm
Dose of threshold-voltage adjustment implantation	$\Phi_{implant}$	?
Oxide charge density	N_{oc}	10^{10} cm^{-2}
Type of gate		N$^+$-polysilicon
Intrinsic-carrier concentration	n_i	1.02×10^{10} cm^{-3}
Energy gap	E_g	1.12 eV
Thermal voltage at room temperature	$V_t = kT/q$	0.026
Oxide permittivity	ε_{ox}	3.45×10^{-11} F/m
Silicon permittivity	ε_s	1.04×10^{-10} F/m

(a) Determine the threshold voltages of NMOS and PMOS transistors for $\Phi_{implant} = 0$.

(b) Provided the threshold-voltage adjustment implant is shallow, the threshold voltage shift $|\Delta V_T|$ due to the implant dose $q\Phi_{\text{implant}}$ can be expressed as:

$$|\Delta V_T| = \frac{q\Phi_{\text{implant}}}{C_{\text{ox}}}$$

Determine the dose of threshold-voltage adjustment implant so that the NMOS and PMOS threshold voltages are matched ($V_{\text{T-NMOS}} = |V_{\text{T-PMOS}}|$).

Solution:

(a) The threshold voltage of an NMOS transistor is given by Eq. (2.77) [the same as Eq. (5.5)], while the threshold voltage of a PMOS transistor is given by Eq. (2.78). The threshold voltages are calculated using the procedure given in Example 5.1:

NMOS

$$\phi_F = V_t \ln \frac{N_A}{n_i} = 0.30 \text{ V}$$

$$C_{\text{ox}} = \varepsilon_{\text{ox}}/t_{\text{ox}} = 2.3 \times 10^{-3} \text{ F/m}^2$$

$$\phi_{\text{ms}} = \phi_m - \left(\chi + \frac{E_g}{2q} + \phi_F\right) = -E_g/2q - \phi_F = -0.86 \text{ V}$$

$$V_{\text{FB}} = \phi_{\text{ms}} - \frac{qN_{\text{oc}}}{C_{\text{ox}}} = -0.87 \text{ V}$$

$$\gamma = \sqrt{2\varepsilon_s q N_A}/C_{\text{ox}} = 0.079 \text{ V}^{1/2}$$

$$V_{\text{T-NMOS}} = V_{\text{FB}} + 2\phi_F + \gamma\sqrt{2\phi_F} = -0.21 \text{ V}$$

PMOS

$$\phi_F = -V_t \ln \frac{N_D}{n_i} = -0.40 \text{ V}$$

$$C_{\text{ox}} = \varepsilon_{\text{ox}}/t_{\text{ox}} = 2.3 \times 10^{-3} \text{ F/m}^2$$

$$\phi_{\text{ms}} = \phi_m - \left(\chi + \frac{E_g}{2q} + \phi_F\right) = -E_g/2q - \phi_F = -0.16 \text{ V}$$

$$V_{\text{FB}} = \phi_{\text{ms}} - \frac{qN_{\text{oc}}}{C_{\text{ox}}} = -0.17 \text{ V}$$

$$\gamma = \sqrt{2\varepsilon_s q N_D}/C_{\text{ox}} = 0.561 \text{ V}^{1/2}$$

$$V_{\text{T-PMOS}} = V_{\text{FB}} - 2|\phi_F| - \gamma\sqrt{2|\phi_F|} = -1.47 \text{ V}$$

(b) Without the ion-implant adjustment, the NMOS threshold voltage is negative, making it effectively a depletion-type MOSFET. A boron implant will increase this threshold voltage (due to an increase in the surface P-type concentration), and it will simultaneously reduce the absolute value of the PMOS threshold voltage (due to effective reduction of the surface doping level of the N well). Therefore the threshold voltages after the threshold-voltage adjustment implantation can be expressed as

$$V_{\text{T-NMOS}}(\Phi_{\text{implant}}) = V_{\text{T-NMOS}}(0) + q\Phi_{\text{implant}}/C_{\text{ox}}$$

$$\left|V_{\text{T-PMOS}}(\Phi_{\text{implant}})\right| = \left|V_{\text{T-NMOS}}(0)\right| - q\Phi_{\text{implant}}/C_{\text{ox}}$$

Choosing different values for the implant dose Φ_{implant}, the following results are obtained:

Φ_{implant} (cm^{-2})	ΔV_T (V)	$V_{\text{T-NMOS}}(\Phi_{\text{implant}})$ (V)	$\left|V_{\text{T-PMOS}}(\Phi_{\text{implant}})\right|$ (V)
0	0.00	−0.21	1.47
1.0×10^{11}	0.07	−0.14	1.40
3.0×10^{11}	0.21	0.00	1.26
1.00×10^{12}	0.70	0.49	0.77
1.21×10^{12}	0.84	0.63	0.63

This table illustrates the effect of the adjustment-implant dose on the NMOS and PMOS threshold voltages. The implant dose that matches the threshold voltage can be found directly by observing that each of the threshold voltages should be shifted by $\Delta V_T = [\left|V_{\text{T-PMOS}}(\Phi_{\text{implant}})\right| - V_{\text{T-NMOS}}(\Phi_{\text{implant}})]/2 = (1.47 + 0.21)/2 = 0.84$ V, and finding the dose that corresponds to this threshold voltage shift. As the above table shows, this dose is $\Phi_{\text{implant}} = 1.21 \times 10^{12}$ cm^{-2}.

5.3 MOSFET MODELING

There is a large number of MOSFET models that differ in terms of their accuracy and complexity. Additionally, the models of the principal effects are modified in virtually countless ways to include observed second-order effects. The consideration of MOSFET models, presented in this section, is limited to the equations used in SPICE. Even in SPICE, there are three basic model options, referred to as LEVEL 1, LEVEL 2, and LEVEL 3, plus additional more sophisticated or specific models. The SPICE LEVEL 2 model is a physically based model presented in a number of semiconductor books as the MOSFET model. The SPICE LEVEL 1 model is the simplest MOSFET model, used in many circuit design books. The LEVEL 2 model frequently appears as unnecessarily complex, while the LEVEL 1 model is rarely accurate enough. The SPICE LEVEL 3 model is almost as simple as the LEVEL 1 model (the equations resemble the LEVEL 1 equations) and is almost as accurate as the LEVEL 2 model. Technically, the LEVEL 3 model is the best choice. The use of the LEVEL 3 model may appear unsuitable only because of a lack of understanding of its relationship to the model used in circuit design literature (LEVEL 1) or the model frequently presented in semiconductor physics books (LEVEL 2).

This section directly relates the equations of SPICE LEVEL 1, LEVEL 2, and LEVEL 3 models. The most complex model (LEVEL 2) is derived first, relying on considerations of the principal MOSFET effects. The LEVEL 3 model is derived by simplifying the LEVEL 2 model, which is further simplified to obtain the simplest LEVEL 1 model. The second-order effects included in SPICE are presented for the LEVEL 3 model.

5.3.1 SPICE LEVEL 2 Model

Deriving a MOSFET model means finding an equation to represent the dependence of the drain current I_D on the applied terminal voltages V_{GS}, V_{DS}, and V_{SB}, in general form $I_D = f(V_{GS}, V_{DS}, V_{SB})$. Assuming that all channel electrons terminate at the drain, and neglecting drain–substrate reverse-bias current, the current of the electrons in the channel directly expresses the terminal drain current I_D. The electron flow in the channel is caused by the electric field appearing in the channel due to the drain-to-source bias V_{DS}. This current mechanism, known as *drift current*, is modeled by Ohm's law [Eq. (1.10)]:

$$j = \sigma E \tag{5.13}$$

where j is the current density in A/m^2, E is the electric field, and σ is the conductivity. The conductivity can be expressed in terms of the electron concentration n and electron mobility μ_0 [refer to Eq. (1.16)], which leads to

$$j = q\mu_0 n E \tag{5.14}$$

It appears the current density of the electrons in the channel j, given in units of A/m^2, can simply be multiplied by the channel cross-sectional area to convert it into the drain current I_D, which is expressed in A. As shown in Section 1.1, this would implicitly assume a uniform current density. If the current density changes, it would mean that the average value of the current density is actually taken. If this approach is to be used, all the other differential quantities in Eq. (5.14) should be represented by their average values. Denoting the channel cross section by $x_{ch} W$, and expressing the average value of the electric field by V_{DS}/L_{eff},

$$\underbrace{j\, x_{ch} W}_{I_D\ (A)} = \mu_0\ \underbrace{q n x_{ch}}_{\overline{Q}_I\ (C/m^2)}\ W\ \underbrace{E}_{V_{DS}/L_{eff}} \tag{5.15}$$

the following equation is obtained:

$$I_D = \frac{\mu_0 W}{L_{eff}} \overline{Q}_I V_{DS} \tag{5.16}$$

$\overline{Q}_I$ in the above equations is the average value of the inversion-layer charge density, expressed in C/m^2. For the case of zero, or very small drain-to-source voltage V_{DS}, the inversion-layer charge density is determined only by the gate-to-source voltage V_{GS}. It is equal to zero for $V_{GS} < V_T$, and equal to $(V_{GS} - V_T)C_{ox}$ for $V_{GS} \geq V_T$, as shown by

Eqs. (2.69) and (5.6). Therefore, the inversion-layer charge density is uniform along the channel in the case of small V_{DS}. Replacing $\overline{Q}_I$ in Eq. (5.16) by Q_I from Eq. (5.6), the drain current is obtained as

$$I_D = \frac{\mu_0 W C_{ox}}{L_{eff}}(V_{GS} - V_T)V_{DS} \tag{5.17}$$

which becomes equivalent to the intuitively established Eq. (5.7) when the gain factor β is defined as:

$$\beta = \frac{\mu_0 W C_{ox}}{L_{eff}} \tag{5.18}$$

Equation (5.17) predicts a linear dependence of the drain current on both the gate-to-source and the drain-to-source voltages. The linear $I_D - V_{DS}$ dependence is experimentally observed for small V_{DS} voltages, the region referred to as the *linear region*. An example of experimental $I_D - V_{DS}$ dependence is given in Fig. 5.5b. However, the actual current can deviate from this linear dependence, and even saturate at sufficiently large V_{DS} voltages. As mentioned earlier, the region of V_{DS} voltages between 0 and the drain-to-source voltage at which the drain current saturates (V_{DSsat}) is referred to as the *triode region*.

To model the drain current in the whole triode region (not only its linear part), the influence of V_{DS} voltage on the inversion-layer charge density cannot be neglected. As V_{DS} voltage is increased, the drain end of the inversion layer is being gradually depleted, until it is completely pinched off at the drain end, which is the point of drain current saturation ($V_{DS} = V_{DSsat}$). Therefore, the V_{DS} voltage causes a nonuniform distribution of the inversion-layer charge density along the channel, as illustrated in Fig. 5.17. It also causes a reduction of its average value $\overline{Q}_I$, which is reflected in the current fall below that predicted by the uniform (maximum average) Q_I [Eq. (5.17)].

Q_I reduction toward the drain end of the channel can be modeled through a threshold voltage increase caused by the body effect, which varies in strength along the channel. As described in Section 5.1.2, the increase of strong-inversion surface potential due to the substrate bias is reflected as a threshold voltage increase. In the example of Fig. 5.6b, the surface potential is increased from the usual $2\phi_F$ to $2\phi_F + V_{SB}$ due to the reverse source–bulk bias V_{SB}. In Fig. 5.17, this is the situation at the source end of the channel. At the drain end of the channel, the surface potential in strong inversion is increased further to $2\phi_F + V_{SB} + V_{DS}$, to include the reverse bias of the drain–bulk P–N junction which is $V_{DS} + V_{SB}$. Consequently, the threshold voltage at the drain end of the inversion layer is larger than at the source end (stronger body effect due to the larger body bias, $V_{DS} + V_{SB}$). According to Eq. (5.6), the larger threshold voltage causes a smaller inversion-layer charge density Q_I at the drain end of the channel.

The varying threshold voltage along the channel can uniquely be expressed in terms of the surface potential by generalizing Eq. (5.11) in the following way:

$$V_T = V_{FB} - V_{SB} + \underbrace{2\phi_F + V_{SB}}_{\varphi_s} + \gamma \sqrt{\underbrace{2\phi_F + V_{SB}}_{\varphi_s}} \tag{5.19}$$

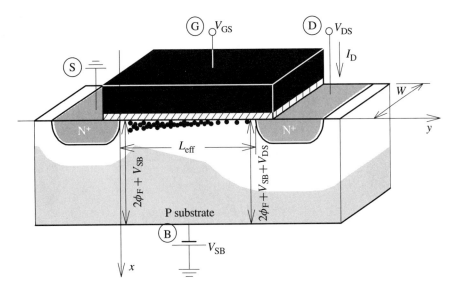

Figure 5.17 N-channel MOSFET diagram, indicating the surface potential at the source and the drain ends of the channel.

As long as φ_s is fixed to $2\phi_F + V_{SB}$, this threshold voltage equation is equivalent to Eq. (5.11). However, if φ_s is allowed to take any value between $2\phi_F + V_{SB}$ (the surface potential at the source end) and $2\phi_F + V_{SB} + V_{DS}$ (the surface potential at the drain end of the channel), the threshold voltage becomes a differential quantity that varies along the channel due to the surface potential variation:

$$v_T(\varphi_s) = V_{FB} - V_{SB} + \varphi_s + \gamma\sqrt{\varphi_s} \tag{5.20}$$

In principle, the average inversion-layer charge density, to be used in Eq. (5.16), should be obtained through the following averaging formula:

$$\overline{Q_I} = \frac{1}{L_{eff}} \int_0^{L_{eff}} Q_I(y)\, dy = \frac{1}{L_{eff}} \int_0^{L_{eff}} [V_{GS} - v_T(\varphi_s)]C_{ox}\, dy \tag{5.21}$$

However, the integration with respect to y (the space coordinate along the channel) is not possible as the surface potential dependence, and for that matter the threshold voltage dependence on y, is not established. Instead, the averaging is performed as follows:

$$\begin{aligned}
\overline{Q_I} &= \frac{1}{(2\phi_F + V_{SB} + V_{DS}) - (2\phi_F + V_{SB})} \int_{2\phi_F + V_{SB}}^{2\phi_F + V_{SB} + V_{DS}} Q_I(\varphi_s)\, d\varphi_s \\
&= \frac{1}{V_{DS}} \int_{2\phi_F + V_{SB}}^{2\phi_F + V_{SB} + V_{DS}} [V_{GS} - v_T(\varphi_s)]C_{ox}\, d\varphi_s
\end{aligned} \tag{5.22}$$

Replacing $v_T(\varphi_s)$ from Eq. (5.20), and solving the above integral, the following result is obtained:

$$\overline{Q}_I = \frac{C_{ox}}{V_{DS}} \left\{ \left(V_{GS} - V_{FB} - 2\phi_F - \frac{V_{DS}}{2} \right) V_{DS} \right.$$
$$\left. -\frac{2}{3}\gamma \left[(2\phi_F + V_{SB} + V_{DS})^{3/2} - (2\phi_F + V_{SB})^{3/2} \right] \right\} \tag{5.23}$$

Inserting the obtained equation for the average inversion-layer charge density $\overline{Q}_I$ into Eq. (5.16), the drain current is obtained as

$$I_D = \beta \left\{ \left(V_{GS} - V_{FB} - 2\phi_F - \frac{V_{DS}}{2} \right) V_{DS} \right.$$
$$\left. -\frac{2}{3}\gamma \left[(2\phi_F + V_{SB} + V_{DS})^{3/2} - (2\phi_F + V_{SB})^{3/2} \right] \right\} \tag{5.24}$$

where the gain factor β, originally defined by Eq. (5.18), is frequently expressed through the so-called transconductance parameter KP:

$$\beta = \frac{\mu_0 W C_{ox}}{L_{eff}} = KP \frac{W}{L_{eff}} \tag{5.25}$$

Equation (5.24) represents the SPICE LEVEL 2 model in the triode region. The plots of this equation, given in Fig. 5.18, show that it can be used only in the triode region. In this region the equation correctly predicts I_D increase with V_{DS}. It cannot be used in the region where it shows a current decrease, which is the saturation region, because its derivation has not included the case of channel pinch-off. As the threshold voltage is larger than the gate-to-source voltage in the pinch-off region, the negative contribution of the $V_{GS} - v_T(\varphi_s)$ term in Eq. (5.22) erroneously reduces the value of the average inversion-layer charge $\overline{Q}_I$, which appears as the current reduction in the saturation region. Nonetheless, the observation that the I_D current, as predicted by Eq. (5.24), reaches maximum at the saturation voltage V_{DSsat} helps us find the value of the saturation voltage. Knowing that the first derivative of I_{DS} is zero at $V_{DS} = V_{DSsat}$ (the maximum of the current I_D), the saturation voltage is obtained as

$$\frac{\partial I_D}{\partial V_{DS}} = 0 \implies V_{DSsat} = V_{GS} - V_{FB} - 2\phi_F$$
$$-\frac{\gamma^2}{2} \left[\sqrt{1 + \frac{4}{\gamma^2}(V_{GS} - V_{FB} + V_{SB})} - 1 \right] \tag{5.26}$$

The SPICE LEVEL 2 model works in the following way: (1) the saturation drain voltage V_{DSsat} is calculated first, using Eq. (5.26); (2) if $V_{DS} < V_{DSsat}$, V_{DS} is used in Eq. (5.24) to calculate the current; (3) if $V_{DS} \geq V_{DSsat}$, V_{DSsat} is used in Eq. (5.24) to calculate the current.

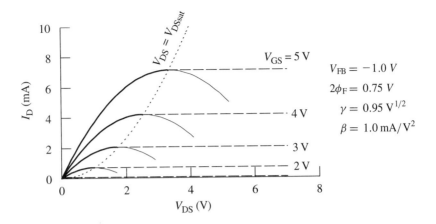

Figure 5.18 MOSFET output characteristics: solid lines: Eq. (5.24); dashed lines, saturation current.

As can be seen from Eq. (5.24), the threshold voltage does not appear as a parameter of the LEVEL 2 model. This is not helpful, as the threshold voltage is very frequently used to characterize MOSFETs. In addition, 3/2 power terms adversely affect computational efficiency of the LEVEL 2 model.

5.3.2 SPICE LEVEL 3 Model: Principal Effects

The LEVEL 3 model can be obtained by simplifying the equation of the LEVEL 2 model. To ensure that 3/2 power terms will disappear, Eq. (5.24) is approximated by the first three terms of the Taylor series:

$$I_D \approx I_D(0) + I_D'(0)V_{DS} + I_D''(0)\frac{V_{DS}^2}{2} \tag{5.27}$$

The first three terms of the Taylor series are taken to achieve a good compromise between accuracy and simplicity. Taking only the first two terms would be even simpler, however, this would lead to linear $I_{DS}(V_{DS})$ dependence, which is obviously not good enough as a MOSFET model. $I_D(0)$, $I_D'(0)$, and $I_D''(0)$ are obtained as follows:

$$I_D(0) = 0$$

$$I_D' = \frac{\partial I_D}{\partial V_{DS}} = \beta(V_{GS} - V_{FB} - 2\phi_F - V_{DS}) - \beta\gamma(2\phi_F + V_{SB} + V_{DS})^{1/2}$$

$$I_D'(0) = \beta(V_{GS} - V_{FB} - 2\phi_F) - \beta\gamma(2\phi_F + V_{SB})^{1/2}$$

$$I_D'' = \frac{\partial^2 I_D}{\partial V_{DS}^2} = \frac{\partial}{\partial V_{DS}}\left(\frac{\partial I_D}{\partial V_{DS}}\right) = -\beta - \tfrac{1}{2}\beta\gamma(2\phi_F + V_{SB} + V_{DS})^{-1/2}$$

$$I_D''(0) = -\beta - \tfrac{1}{2}\beta\gamma(2\phi_F + V_{SB})^{-1/2} \tag{5.28}$$

Putting the obtained $I_D(0)$, $I'_D(0)$, and $I''_D(0)$ into Eq. (5.27), the following drain current equation is obtained:

$$I_D = \beta \left[V_{GS} - \underbrace{(V_{FB} + 2\phi_F + \gamma\sqrt{2\phi_F + V_{SB}})}_{V_T} - \frac{1}{2}\left(1 + \overbrace{\frac{\gamma}{2\sqrt{2\phi_F + V_{SB}}}}^{F_B}\right)V_{DS} \right]V_{DS} \tag{5.29}$$

It can be seen that the threshold voltage, as defined by Eq. (5.11), appears in the drain current equation. Also, a new factor is introduced to additionally simplify this equation. This factor is F_B, defined by

$$F_B = \frac{\gamma}{2\sqrt{2\phi_F + V_{SB}}} \tag{5.30}$$

As the original LEVEL 2 Eq. (5.24) is valid only in the triode region ($0 \leq V_{DS} \leq V_{DSsat}$), the simplified LEVEL 3 Eq. (5.29) is also valid only in the triode region. To find the saturation drain voltage V_{DSsat}, a procedure analogous to the one described in the previous section is used:

$$\frac{\partial I_D}{\partial V_{DS}} = 0 \quad \Rightarrow \quad V_{DSsat} = \frac{V_{GS} - V_T}{1 + F_B} \tag{5.31}$$

The triode-region current I_D reaches maximum for $V_{DS} = V_{DSsat}$. This maximum current is considered as the MOSFET saturation current I_{Dsat}. Putting Eq. (5.31) for V_{DSsat} in place of V_{DS} in the current equation (5.29), the saturation current I_{Dsat} is obtained as

$$I_{Dsat} = I_D(V_{DSsat}) = \frac{\beta}{2(1 + F_B)}(V_{GS} - V_T)^2 \tag{5.32}$$

Although the saturation current is independent of V_{DS}, it does depend on V_{GS}, as shown by Eq. (5.32).

The SPICE LEVEL 3 MOSFET model can be summarized as follows:

$$I_D = \begin{cases} \beta\left[(V_{GS} - V_T)V_{DS} - (1 + F_B)\frac{V_{DS}^2}{2}\right], & \text{if } 0 \leq V_{DS} < V_{DSsat} \\ \dfrac{\beta}{2(1 + F_B)}(V_{GS} - V_T)^2, & \text{if } V_{DS} \geq V_{DSsat} \end{cases} \tag{5.33}$$

where the drain saturation voltage is

$$V_{DSsat} = \frac{V_{GS} - V_T}{1 + F_B} \tag{5.34}$$

while the threshold voltage V_T, the gain factor β, and the factor F_B are given by Eqs. (5.11), (5.25), and (5.30), respectively. Analogous equations apply to the case of P-channel MOS-FETs. The form of these equations is presented in the next section (tables summarizing SPICE models and parameters).

Comparing LEVEL 3 and LEVEL 2 models, it can be concluded that the equations of the LEVEL 3 model are much simpler. This is not achieved at the expense of accuracy, as illustrated in Fig. 5.19. It should be noted that the same set of parameters is used to compare the characteristics in Fig. 5.19. Independent fitting of the LEVEL 2 and LEVEL 3 parameters would produce an even smaller difference between the two models.

5.3.3 SPICE LEVEL 1 Model

The SPICE LEVEL 1 model is obtained when the factor F_B in the LEVEL 3 model is neglected. Replacing F_B in Eqs. (5.33) and (5.34) by zero, the drain current I_D and the drain saturation voltage V_{DSsat} of the SPICE LEVEL 1 model are obtained:

$$I_D = \begin{cases} \beta\left[(V_{GS} - V_T)V_{DS} - \dfrac{V_{DS}^2}{2}\right], & \text{if } 0 \leq V_{DS} < V_{DSsat} \\ \dfrac{\beta}{2}(V_{GS} - V_T)^2, & \text{if } V_{DS} \geq V_{DSsat} \end{cases} \tag{5.35}$$

$$V_{DSsat} = V_{GS} - V_T \tag{5.36}$$

Although the LEVEL 1 model is only slightly simpler than the LEVEL 3 model, the removal of F_B frequently causes significant errors. This is also illustrated in Fig. 5.19, which uses the example of $F_B = 0.95/(2\sqrt{0.75}) = 0.55$. As can be seen, the error of

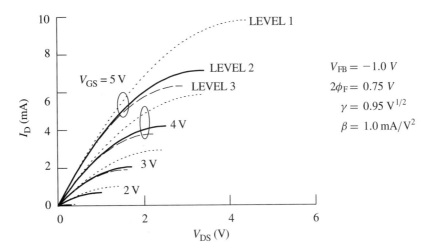

Figure 5.19 Comparison of LEVEL 1, LEVEL 2, and LEVEL 3 models.

LEVEL 1 model is due to the fact that F_B is not negligible compared to 1. Both the drain current and the drain saturation voltage are overestimated. This means even an independent fitting of the parameters would not lead to a good agreement with LEVEL 2 and LEVEL 3 models. While LEVEL 1 model would give better results for smaller F_B factors, the modern MOSFETs tend to have larger rather than smaller F_B.

■ **Example 5.3 Drain Current and Transconductance**

A depletion-type N-channel MOSFET has zero-bias ($V_{SB} = 0$) threshold voltage of $V_T = -2.5$ V. Calculate the drain current of this MOSFET if it is biased with $V_{GS} = 5$ V, $V_{DS} = 10$ V, and $V_{SB} = 0$ V, using SPICE LEVEL 3 and LEVEL 1 models, and compare the results. What is the corresponding transconductance? The MOSFET channel width-to-channel length ratio is 25, $\gamma = 0.85$ V$^{1/2}$, $\phi_F = 0.35$ V, $C_{ox} = 7 \times 10^{-4}$ F/m^2, and $\mu_0 = 1000$ cm^2/Vs.

Solution: The first step is to determine whether the MOSFET operates in the triode or the saturation region. To be able to calculate the saturation voltage V_{DSsat} using Eq. (5.34), the factor F_B needs to be determined:

$$F_B = \gamma/(2\sqrt{2\phi_F}) = 0.85/(2\sqrt{0.70}) = 0.51$$

As the saturation voltage,

$$V_{DSsat} = (V_{GS} - V_T)/(1 + F_B) = [5 - (-2.5)]/(1 + 0.51) = 5.0 \text{ V}$$

is smaller than the applied drain voltage $V_{DS} = 10$ V, the MOSFET is in saturation. Calculating the gain factor

$$\beta = \mu_0 C_{ox} W/L_{eff} = 0.1 \times 7 \times 10^{-4} \times 25 = 1.75 \times 10^{-3} \text{ A/V}^2 = 1.75 \text{mA/V}^2$$

the saturation current, according to the LEVEL 3 model, is obtained as

$$I_D = \frac{\beta}{2(1 + F_B)}(V_{GS} - V_T)^2 = 32.6 \text{ mA}$$

As F_B is neglected in the LEVEL 1 model, the current according to the LEVEL 1 model is

$$I_D = \frac{\beta}{2}(V_{GS} - V_T)^2 = 49.2 \text{ mA}$$

Obviously, the currents is significantly overestimated by the LEVEL 1 model.
The transconductance is defined by Eq. (4.4). Therefore,

$$g_m = \frac{\partial}{\partial v_{GS}}\left[\frac{\beta}{2(1 + F_B)}(v_{GS} - V_T)^2\right]\Bigg|_{v_{GS}=V_{GS}}$$

$$= \frac{\beta}{1 + F_B}(V_{GS} - V_T) = \frac{1.75}{1 + 0.51}(5 + 2.5) = 8.7 \text{ mA/V}$$

■ **Example 5.4 MOSFET in the Linear Region**

A MOSFET operating in its linear region can be used as a voltage-controlled resistor. Determine the sensitivity of the resistance on the gate voltage ($\partial R/\partial V_{GS}$) at $V_{GS} = 5$ V if the depletion-type MOSFET, considered in Example 5.3, is used as a voltage-controlled resistor.

Solution: As the drain current in the linear region is given by

$$I_D = \beta(V_{GS} - V_T)V_{DS}$$

the resistance can be expressed as

$$R = V_{DS}/I_D = \frac{1}{\beta(V_{GS} - V_T)}$$

Therefore, the sensitivity of this voltage-controlled resistor is

$$\frac{\partial R}{\partial V_{GS}} = \frac{\partial}{\partial V_{GS}}\left[\frac{1}{\beta(V_{GS} - V_T)}\right] = -\frac{1}{\beta}\frac{1}{(V_{GS} - V_T)^2}$$

$$= -\frac{1}{1.75 \times 10^{-3}(5 + 2.5)^2} = 10.2 \ \Omega/V$$

5.3.4 SPICE LEVEL 3 Model: Second-Order Effects

This section describes the second-order effects included in the SPICE LEVEL 3 model. The term "second-order" should not be confused with "negligible," as some of these effects are very important in terms of simulation accuracy. The importance of some of the second-order effects would also depend on a particular application. Therefore, the decision to neglect a particular second-order effect can appropriately be made only if the effect is properly understood.

Mobility Reduction with Gate Voltage

This is a second-order effect that can rarely be neglected. The effect is related to the gate voltage and appears even at the smallest drain-to-source voltages. Figure 5.20a shows the transfer characteristic of a MOSFET in the linear region ($V_{DS} = 500$ mV). Due to the small V_{DS} value, the parabolic (V_{DS}^2) term in Eq.(5.33) can be neglected, which leads to the linear-region model, given earlier by Eq. (5.7). The linear $I_D - V_{GS}$ dependence predicted by the model is plotted by the dashed line in Fig. 5.20a. However, experimental data frequently show a deviation from the predicted linear dependence, with the actual drain current falling increasingly below the predicted values as the gate voltage increases. The solid lines in Fig. 5.20 illustrate this effect.

This smaller than expected drain current is due to reduction of channel-carrier mobility. The principal model assumes constant (gate voltage-independent) mobility μ_0 of the carriers

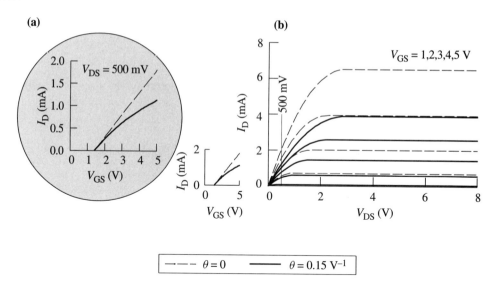

Figure 5.20 Influence of mobility reduction with gate voltage on transfer (a) and output (b) characteristics.

in the channel [refer to Eq. (5.16)]. In reality, the gate voltage-induced vertical field influences the carrier-scattering mechanisms in the channel. The carrier-scattering mechanisms in the very thin inversion layer are multiple and rather complex. However, a number of those scattering mechanisms depend on the inversion-layer thickness, and consequently on the applied gate voltage. As the physically based equations of the mobility dependence on the gate voltage are complex, the following widely accepted semiempirical equation is used in SPICE:

$$\mu_s = \frac{\mu_0}{1 + \theta(V_{GS} - V_T)} \tag{5.37}$$

where the so-called surface-mobility μ_s is now used instead of the low-field mobility μ_0 in the gain factor [Eq. (5.25)]. Accordingly, the transconductance parameter KP is calculated as $KP = \mu_s C_{ox}$, instead of $KP = \mu_0 C_{ox}$. The parameter θ is a SPICE parameter that has to be experimentally determined. It is referred to as the *mobility modulation constant*.

Zero value of the θ parameter effectively eliminates this effect from the SPICE model, as $\mu_s = \mu_0$ in this case. Also, the added $\theta(V_{GS} - V_T)$ term does little in the near-threshold region (small $V_{GS} - V_T$ values). However, at moderate and high gate voltages, the effect becomes pronounced. Figure 5.20b illustrates the importance of this effect for accurate modeling of the output characteristics. Very frequently, it is impossible to achieve an acceptable agreement between the model and the experimental data without the help of the θ parameter.

Velocity Saturation (Mobility Reduction with Drain Voltage)

Channel carrier mobility can also be reduced by a high lateral field in the channel. As explained in Section 1.4 the mobility relates drift velocity to the electric field. At small electric fields $|E|$, the drift velocity v_d increases linearly with the electric field, which leads to the appearance of lateral-field-independent carrier mobility $\mu_s = v_d/|E|$. Long-channel MOSFETs may operate in the low-field region, where the linear $v_d - |E|$ dependence is observed. At higher electric fields, the drift velocity deviates from the linear dependence and even saturates, which is illustrated in Fig. 1.18. As explained in Section 5.1.3, this can be the mechanism responsible for current saturation in short-channel MOSFETs.

The drain-current model is derived from Ohm's law, which assumes a linear dependence of the drift current (thus drift velocity) on the electric field [Eq. (5.14)]. To account for the effect of velocity saturation with electric field increase, the mobility in the drain-current model has to be reduced. As the lateral electric field is proportional to V_{DS}/L_{eff}, the mobility reduction can be expressed in terms of the drain voltage V_{DS} and the effective channel length L_{eff}:

$$\mu_{eff} = \mu_s \Big/ \left(1 + \frac{\mu_s}{v_{max}} \frac{V_{DS}}{L_{eff}}\right) \tag{5.38}$$

Obviously, the surface mobility μ_s is divided by the $1 + \mu_s V_{DS}/(v_{max} L_{eff})$ term to give the so-called *effective mobility* μ_{eff}, which depends on V_{DS}/L_{eff}. It is the effective mobility μ_{eff}, and not μ_s, that is directly used to calculate the transconductance parameter KP and consequently the gain factor β, which appears in the drain-current equation. Therefore, the general form of Eq. (5.25) is

$$\beta = \frac{\mu_{eff} W C_{ox}}{L_{eff}} = KP \frac{W}{L_{eff}} \tag{5.39}$$

The strength of the considered effect is controlled by the v_{max} parameter, which has the physical meaning of *maximum drift velocity*. Although the typical value of v_{max} is 1–2 $\times 10^5$ m/s in silicon, it can freely be adjusted in SPICE, like any other parameter. Setting $v_{max} = \infty$ would completely eliminate this effect from the SPICE MOSFET model, as $\mu_{eff} = \mu_s$ in this case.

The velocity saturation also affects the saturation voltage V_{DSsat}. This effect is not covered by the effective mobility Eq. (5.38). To include this effect, the saturation voltage Eq. (5.34) is modified in SPICE. The modified equation is presented in the next section (Table 5.4).

Finite Output Resistance

The principal model assumes perfectly saturated current (horizontal $I_D - V_{DS}$ characteristics in the saturation region), which means it assumes infinitely large output resistance of the MOSFET used as a voltage-controlled current source ($\Delta V_{DS}/\Delta I_D \to \infty$). Real MOSFETs exhibit finite output resistances. For some applications, it is very important to use the real value of the output resistance during circuit simulation. There are at least two effects that cause an increase of the drain current in the saturation region: (1) *channel length modulation* and (2) *drain-induced barrier lowering (DIBL)*.

The channel length modulation is basically shortening of the actual channel when the saturation happens due to the channel pinch-off at the drain end. The principal model is derived for the triode region, where the channel extends from the source to the drain, so that the average lateral field is $\approx V_{DS}/L_{eff}$. In the saturation, the voltage drop across the channel remains approximately constant (V_{DSsat}), which in a sense means that a constant field V_{DSsat}/L_{eff} is assumed in the channel. This assumption correlates with the fact that the saturation current is constant in the principal model. In reality, the channel becomes shorter as V_{DS} "pushes" the pinch-off point away to a distance L_{pinch} from the drain. This leads to a higher field $V_{DSsat}/(L_{eff} - L_{pinch})$ in the shortened channel, and therefore to a current increase with V_{DS}. This effect is modeled by changing (modulating) the channel length:

$$\beta = \frac{\mu_{eff} W C_{ox}}{L_{eff} - L_{pinch}} = KP \frac{W}{L_{eff} - L_{pinch}} \tag{5.40}$$

where the length of the pinch-off channel L_{pinch} depends on V_{DS} (the SPICE LEVEL 3 equation is given in Table 5.4). This equation is based on the depletion-layer width of an abrupt P–N junction [Eq. (2.33)], with $V_{DS} - V_{DSsat}$ as the voltage drop across the depletion layer, and κ as a fitting parameter (SPICE input parameter). Because the channel length shortening is applied abruptly when V_{DS} becomes larger than V_{DSsat}, the modeled current may not be "smooth" (a first-derivative discontinuity) around the saturation voltage.

DIBL is a more appropriate model for the finite output resistance in the case of short-channel MOSFETs. This effect is due to strong lateral electric field. The principal model assumes that the gate voltage fully controls the surface potential in the channel region. In terms of the energy-band diagram, the assumption is that the channel-carrier density depends on how much the barrier is lowered by the gate voltage: in Fig. 5.2 (bottom right band diagram), the barrier has its full height, and there is no inversion layer; in Fig. 5.3, the gate voltage lowers the barrier and inversion-layer carriers appear in the channel. In reality, the electric field from the drain can also cause some barrier lowering, called *drain-induced barrier lowering*. This "help" from the drain in creating the inversion layer can be modeled by a drain-voltage-induced reduction of the threshold voltage. It was empirically established that the simplest linear relationship is quite satisfactory:

$$V_T = V_{FB} + 2\phi_F + \gamma\sqrt{2\phi_F + V_{SB}} - \sigma_D V_{DS} \tag{5.41}$$

The coefficient expressing the strength of the V_T dependence on V_{DS}, σ_D, is not a SPICE parameter. It is calculated in SPICE by an equation that involves L_{eff}, C_{ox}, and a SPICE parameter η (frequently referred to as the coefficient of the static feedback). This equation is given in Table 5.5. Setting the parameter η to zero eliminates this effect from the model (the dashed lines in Fig. 5.21), while a larger η value expresses a stronger V_T dependence on V_{DS}, therefore a smaller output resistance (solid lines in Fig. 5.21).

Threshold Voltage-Related Short-Channel Effects

The threshold voltage V_T of the MOSFET, as given by the principal model [Eq. (5.11)], is independent on the channel length or width. In reality, this is observed when the channel length and width are much larger than the channel region affected by fringing field at the channel edges (edge effects). However, when the channel length or width is reduced to dimensions that are comparable to the edge-affected region, the threshold voltage experiences dependence on the channel length or width.

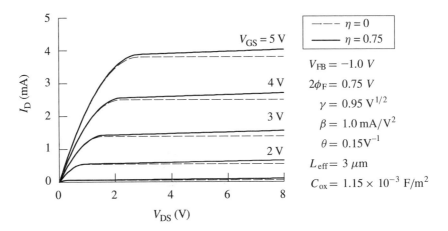

Figure 5.21 Output characteristics with (solid lines) and without (dashed lines) the influence of V_{DS} on V_T.

Figure 5.22 illustrates the edge effects by the electric-field lines (the arrows in the figure) appearing in the MOSFET depletion layer (the clear area). The electric-field lines originate at positive charge centers in the gate or the N-type side of drain/source regions, and terminate at negative charge centers (negative acceptor ions) in the substrate-depletion layer. It can be seen that the electric-field lines at the ends of the channel originate from the source/drain regions and not the gate. Consequently, some charge at the edges of the depletion layer is linked to source and drain charge, and not to the gate charge.

The presence of the source/drain-related edge effect means that less gate charge, and consequently less gate voltage, is needed to create the depletion layer under the channel.

The depletion-layer charge induced by the gate voltage is given by the Q_d/C_{ox} and $\gamma\sqrt{2\phi_F + V_{SB}}$ terms in Eqs. (5.10) and (5.11), respectively, where C_{ox} is the gate-oxide capacitance per unit area and Q_d is the depletion-layer charge per unit of channel area. Q_d

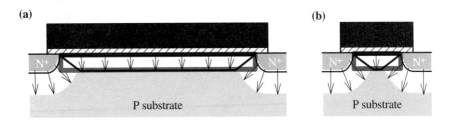

Figure 5.22 Illustration of the threshold voltage-related short-channel effect. A part of the depletion-layer charge under the channel is created by the source and drain electric field (note the arrow origins). This helps the gate voltage to create the depletion layer under the channel, effectively reducing the threshold voltage. The effect is insignificant in long-channel MOSFETs (a), but quite pronounced in short-channel MOSFETs (b).

is related to the total charge Q_d^* (in C) as $Q_d = Q_d^*/(L_{eff}W)$, where $L_{eff}W$ is the channel area. Noting that Q_d^* involves all the charge centers under the channel (the rectangular area in Fig. 5.22), it is easy to see that the Q_d/C_{ox} (and consequently $\gamma\sqrt{2\phi_F + V_{SB}}$) term overestimates the gate voltage needed to create the depletion layer under the channel. The gate voltage should be related to the charge inside the trapezoidal area, and not the whole (rectangular) area under the channel.

In long-channel MOSFETs, like the one in Fig. 5.22a, the edge charge created by the source and drain is much smaller compared to the total charge Q_d^*. This can be seen by comparing the rectangular and the trapezoidal areas in Fig.5.22a. Consequently, the $Q^*/(L_{eff}WC_{ox}) = Q_d/C_{ox} = \gamma\sqrt{2\phi_F + V_{SB}}$ term fairly correctly expresses the voltage needed to create the depletion layer under the channel. As the edge effects are negligible in the long-channel MOSFETs, the threshold voltage does not show dependence on the gate length (Q_d/C_{ox} is gate length independent). In Fig. 5.23, this is the case for MOSFETs with channels longer than approximately 2 μm.

In the case of short-channel MOSFETs (Fig. 5.22b), the charge enclosed in the trapezoidal area is significantly smaller than the charge inside the rectangular area. This is because the edge effect is now pronounced—the source and drain-related fields are creating an observable portion of the depletion layer under the channel.

To model this effect, the depletion-layer charge Q_d^* is multiplied by a *charge-sharing factor* F_s, which can be obtained as the ratio between the trapezoidal and the rectangular areas—this is to convert the total charge in the depletion layer under the channel (Q_d^*) into the charge that is created by the gate (the charge enclosed in the trapezoidal area). As $F_s Q_d^*/(L_{eff}WC_{ox}) = F_s Q_d/C_{ox} = F_s \gamma\sqrt{2\phi_F + V_{SB}}$, the threshold voltage Eq. (5.11) is modified to

$$V_T = V_{FB} + 2\phi_F + F_s\gamma\sqrt{2\phi_F + V_{SB}} \tag{5.42}$$

A number of different equations for the charge-sharing factor F_s have been developed. The equation that is used in SPICE is given in the next section (Table 5.5). SPICE uses the source and drain P–N junction depth x_j, the lateral diffusion x_{j-lat}, and the P–N junction built-in voltage V_{bi} as parameters to calculate the charge-sharing factor F_s.

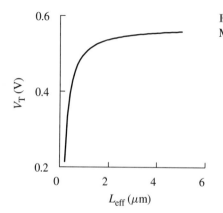

Figure 5.23 Threshold voltage dependence on the MOSFET channel length.

Threshold Voltage Related Narrow-Channel Effects

Threshold voltage can also become dependent on the channel width W, if the channel width is reduced to levels that are comparable to the edge effect regions. Figure 5.24 illustrates the edge effect at the channel ends that determines the channel width. As mentioned earlier, the principal threshold voltage equation is derived under the assumption that the gate voltage creates the depletion-layer charge in the rectangular area under the channel. In reality, the gate voltage depletes a wider region, due to the fringing field effect (Fig. 5.24). This increases the threshold voltage.

To include this effect, the threshold voltage equation is expanded again, its final form being as follows:

$$V_T = V_{FB} + 2\phi_F + \gamma F_s\sqrt{2\phi_F + V_{SB}} - \sigma_D V_{DS} + F_n(2\phi_F + V_{SB}) \qquad (5.43)$$

The new term $F_n(2\phi_F + V_{SB})$ models the threshold voltage increase in narrow-channel MOSFETs. The parameter F_n is calculated from the channel width W, gate-oxide capacitance C_{ox}, and a SPICE parameter δ modulating the strength of this effect. The actual equation is given in Table 5.5. For $\delta = 0$, or wide channels (W large), the parameter F_n approaches zero, eliminating this effect from the threshold voltage equation.

Subthreshold Current

The principal model assumes that the channel-carrier density is zero for $V_{GS} \leq V_T$, appearing abruptly for gate voltages larger than the threshold: $Q_I = (V_{GS} - V_T)C_{ox}$. In reality, the transition from full depletion to strong inversion is gradual (the term "strong" inversion indicates that there would be a "moderate" and even a "weak" inversion). There are mobile carriers in the channel, even for subthreshold gate voltages. Of course, their concentration is very small and is rapidly decaying as the gate voltage is reduced below V_T. Nonetheless, their effect is observable, as they account for the gradual decay of the drain current from the above-threshold levels toward zero.

The vertical field is very low in the subthreshold region, and it does not take a large drain voltage to fully sweep the carriers from the drain end. This creates a concentration

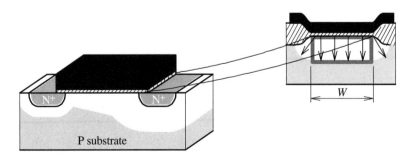

Figure 5.24 Illustration of the narrow channel effect. Fringing electric-field wastes the gate voltage, causing a threshold voltage increase in narrow-channel MOSFETs.

gradient. Because of this, and the low level of the carrier concentration, the diffusion current dominates the drift current in the subthreshold region. Consequently, a diode-like, exponential current–voltage equation is used to model the subthreshold current:

$$I_{D-\text{subth}} = I_{D0} e^{V_{GS}/n_s kT} \tag{5.44}$$

The form of this equation, used in SPICE LEVEL 3, is shown in Table 5.2. The constant I_{D0} is selected to provide a continual transition from the subthreshold to the above-threshold drain current. I_D calculated by the above-threshold model is identical to the subthreshold current at $V_{GS} = V_T + n_s kT$. For smaller gate voltages, the subthreshold model is used, while the above-threshold model is used for larger gate voltages. The coefficient n_s is analogous to the emission coefficient n of diodes. The SPICE LEVEL 3 equation for n_s is given in Table 5.5. There is an input SPICE parameter, NFS, that can be used to control the value of n_s.

5.4 SPICE PARAMETERS AND PARASITIC ELEMENTS

It is necessary to summarize the MOSFET equations used in SPICE and present the hierarchy of MOSFET parameters. This section provides a series of tables designed to enable an intuitive reference to MOSFET parameters. The second part of the section describes the techniques for measurement of the most important parameters.

5.4.1 Summary of SPICE Parameters and Equations

It is not very useful to present the MOSFET SPICE parameters in a single list. Not only would the list be rather long, but the relationships and (in)compatibilities between the parameters could not be clearly expressed. The following hierarchy is used to classify the MOSFET parameters presented in this section:

- Static LEVEL-3 Model (Table 5.2):
  - Principal effects (Table 5.3)
  - Channel-related second-order effects (Table 5.4)
  - Depletion-layer-related second-order effects (Table 5.5, parts I and II).
- Dynamic model (equivalent circuit including parasitic elements) (Table 5.6, parts I and II).

SPICE LEVEL 3 Static Model

The first variables that need to be specified when using a MOSFET are the gate length and width (referred to as geometric variables in Table 5.2). Different MOSFETs can have different gate lengths and widths even in the integrated circuits, where the MOSFETs are made by the same technology process and all the other MOSFET parameters are identical. In SPICE, the gate length and width are considered as device attributes, and are typically stated for every individual MOSFET. However, the gate length and width can also be specified as MOSFET parameters, together with all the other device parameters.

TABLE 5.2 Summary of SPICE LEVEL 3 Static MOSFET Model

Geometric Variables

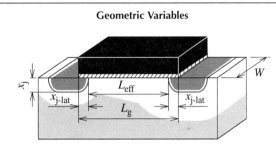

$$L_{\text{eff}} = L_{\text{g}} - 2x_{\text{j-lat}} \quad (x_{\text{j-lat}} \text{ is a parameter; refer to Table 5.5})$$

Symbol	SPICE Keyword	Variable Name	Default Value	Unit
L_{g}	L	Gate length	100×10^{-6}	m
W	W	Channel width	100×10^{-6}	m

Note: L and W can also be specified as parameters

Static Level 3 Model

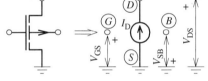

NMOS ($V_{\text{Ts}} = V_{\text{T}} + n_{\text{s}}kT/q$)

PMOS ($V_{\text{Ts}} = V_{\text{T}} - n_{\text{s}}kT/q$)

Sub-V_{T}: $V_{\text{GS}} \leq V_{\text{Ts}}$
Triode: $V_{\text{GS}} > V_{\text{Ts}}$, and $0 < V_{\text{DS}} < V_{\text{DSsat}}$
Satur.: $V_{\text{GS}} > V_{\text{Ts}}$, and $V_{\text{DS}} \geq V_{\text{DSsat}} > 0$

Sub-V_{T}: $V_{\text{GS}} \geq V_{\text{Ts}}$
Triode: $V_{\text{GS}} < V_{\text{Ts}}$, and $0 > V_{\text{DS}} > V_{\text{DSsat}}$
Satur.: $V_{\text{GS}} < V_{\text{Ts}}$, and $V_{\text{DS}} \leq V_{\text{DSsat}} < 0$

$$I_{\text{D}} = \begin{cases} f(V_{\text{GS}}) = \begin{cases} \beta\left[(V_{\text{GS}} - V_{\text{T}})V_{\text{DS}} - (1 + F_{\text{B}})\frac{V_{\text{DS}}^2}{2}\right] & \text{triode region} \\ \frac{\beta}{2(1+F_{\text{B}})}(V_{\text{GS}} - V_{\text{T}})^2, & \text{saturation region} \\ f(V_{\text{GS}} = V_{Ts}) \times e^{-qV_{\text{subth}}/n_{\text{s}}kT}, & \text{sub-}V_{\text{T}} \text{ region} \end{cases} \end{cases}$$

$V_{\text{subth}} = V_{\text{Ts}} - V_{\text{GS}} \geq 0$
$F_{\text{B}} = \frac{\gamma F_{\text{s}}}{2\sqrt{|2\phi_{\text{F}}| + V_{\text{SB}}}} + F_{\text{n}}$ [a]

$V_{\text{subth}} = V_{\text{GS}} - V_{\text{Ts}} \geq 0$
$F_{\text{B}} = \frac{\gamma F_{\text{s}}}{2\sqrt{|2\phi_{\text{F}}| - V_{\text{SB}}}} + F_{\text{n}}$ [a]

	Principal Effects	Channel Related	Depletion Layer Related	All
			Second-Order Effects	
β, V_{DSsat}	Table 5.3	Table 5.4	Table 5.3	Table 5.4
V_{T}, $|2\phi_{\text{F}}|$, γ, F_{s}, F_{n}, n_{s}	Table 5.3	Table 5.3	Table 5.5	Table 5.5

[a]By error, Berkeley SPICE, PSPICE, and HSPICE use factor 4 instead of 2 in front of the square root [*Source:* D. Foty, *MOSFET Modeling with SPICE: Principles and Practice,* Prentice Hall, Upper Saddle River, NY, 1997 (p. 173)].

Generally, there is a difference between *gate length* L_g and *channel length* L_{eff}. The MOSFET diagram given in Table 5.2 illustrates this difference. Although a self-aligned technique is typically used to define the channel as the source and drain regions are created, the lateral diffusion effect leads to the difference between the masking gate and the effective channel. The SPICE input variable is the gate length L_g, while the current–voltage equations use the effective channel length L_{eff}. The relationship between L_g and L_{eff} is shown in Table 5.2. If the lateral-diffusion parameter x_{j-lat} is not specified (set to zero), the gate and channel lengths become equal.

To model the static characteristics, the MOSFET is considered as a voltage-controlled current source. The current is I_D, while the controlling voltages are the voltage drop across the current source V_{DS} and two separate voltages V_{GS} and V_{SB}. Zero current is assumed between the G and S and B and S terminals. Table 5.2 shows the compact form of the LEVEL 3 $I_D(V_{GS}, V_{DS}, V_{SB})$ equation, which appears in three parts, for the three different modes of operation: subthreshold, triode (including the linear mode), and saturation. It also shows the equations used to calculate the factor F_B. The equations for β, V_T, V_{DSsat}, $|2\phi_F|$, γ, F_s, F_n, and n_s, which are obviously needed to calculate $I_D(V_{GS}, V_{DS}, V_{SB})$, are given in Tables 5.3, 5.4, and 5.5. Different equations (given in different tables) are used at different levels of complexity.

Table 5.3 represents the simplest choice, covering the principal effects only. All the parameters shown in Table 5.3 are considered as essential. Although any version of SPICE is expected to have sensible default values of these parameters, no simulation should be trusted unless the values of these parameters are checked and properly specified. Note that the MOSFET gain factor β can be influenced in two ways: one is to specify the transconductance parameter KP and the other is to specify the low-field mobility μ_0 and the gate-oxide thickness t_{ox}. As these two options are mutually exclusive, SPICE will ignore μ_0 and/or t_{ox} when calculating β if KP is specified.

The parameters and equations are common for the enhancement-type (normally-off) and depletion-type (normally-on) MOSFETs. The typical values of the zero-bias threshold voltage, shown in Table 5.3, are for the enhancement-type MOSFETs. If a negative V_{T0} is specified for an N-channel MOSFET, it automatically becomes a depletion-type MOSFET (refer to Fig. 5.8). Analogously, a positive V_{T0} indicates a depletion-type P-channel MOSFET.

As mentioned in Section 5.3.4, some second-order effects significantly influence MOSFET characteristics. In particular, the mobility reduction with the gate voltage (Fig. 5.20) is so important that it can rarely be neglected. Table 5.4 summarizes the three channel-related second-order effects, as they all influence the gain factor β. Additionally, V_{DSsat} is modified for the case of pronounced velocity-saturation effect.

The depletion-layer-related static parameters (Table 5.5) involve the finite-output resistance (η parameter), the gate-oxide charge influence on V_T (N_{oc} parameter), the short-channel (x_j, x_{j-lat}, and V_{bi} parameters), and the narrow-channel (δ parameter) effects. As the gate-oxide charge, and the short-channel and narrow-channel effects modify the zero-bias threshold voltage V_{T0}, the parameters associated with these effects are incompatible with V_{T0} as well as with $|2\phi_F|$ and γ. To properly include these effects, $N_{A,D}$ must be specified, while V_{T0}, $|2\phi_F|$, and γ must not be specified (they are calculated by SPICE). When $N_{A,D}$ is specified, a proper "type of gate" (TPG parameter) should be used to ensure that the zero-bias threshold voltage V_{T0} is properly calculated by SPICE. The static feedback parameter η

TABLE 5.3 Summary of SPICE LEVEL 3 Static Parameters: Principal Effects

Principal Static Parameters

Symbol	SPICE Keyword	Parameter Name	Typical Value NMOS	Typical Value PMOS	Unit
KP (or	KP	Transconductance parameter[a]	1.2×10^{-4}		A/V^2
μ_0 and	Uo	Low-field mobility[b]	700		cm^2/Vs
t_{ox})	Tox	Gate-oxide thickness[b]	20×10^{-9}		m
V_{T0}	Vto	Zero-bias threshold voltage	1	-1	V
$\|2\phi_F\|$	Phi	Surface potential in strong inversion	0.70		V
γ	Gamma	Body-effect parameter	> 0.3		$V^{1/2}$

β, V_T, V_{DSsat}, F_s, F_n and n_s Equations

NMOS	PMOS

$$\beta = \begin{cases} KP\frac{W}{L_{eff}} & \text{if KP is specified} \\ \mu_0\frac{\varepsilon_{ox}}{t_{ox}}\frac{W}{L_{eff}} & \text{if KP is not specified} \end{cases}$$

$$L_{pinch} = 0 \; (\text{KAPPA = 0})$$

$$V_T = \text{Vto} + \text{Gamma}\left(\sqrt{\text{Phi} + V_{SB}} - \sqrt{\text{Phi}}\right) \qquad V_T = \text{Vto} - \text{Gamma}\left(\sqrt{\text{Phi} + V_{BS}} - \sqrt{\text{Phi}}\right)$$

$$V_{DSsat} = \frac{V_{GS} - V_T}{1 + F_B}$$

$$F_s = 1 \; (\text{Xj=0})$$

$$F_n = 0 \; (\text{DELTA=0})$$

$$n_s = 1 + \frac{\gamma F_s(\text{Phi} + |V_{SB}|)^{-1/2}}{2} \; (\text{NFS=0})$$

Constant: $\varepsilon_{ox} = 3.9 \times 8.85 \times 10^{-12}$ F/m

[a,b] Incompatible parameters.

(modeling the effect of finite output resistance) can be used with either group of parameters, provided t_{ox} is specified.

Parameters of Parasitic Elements; Simulating Dynamic Response

There is a number of parasitic elements in the MOSFET structure that can significantly influence the MOSFET characteristics under certain conditions. Perhaps the most important are the parasitic capacitances that directly determine the high-frequency performance of the MOSFET. The large-signal equivalent circuit of the MOSFET, as used in SPICE, is shown in Table 5.6 (PART II). Table 5.6 also lists the parameters associated with all the elements of the equivalent circuit but the current source, which is the only nonparasitic element.

The shown pairs of gate-to-source and gate-to-drain capacitors have different origins, and consequently different models and parameters are associated with these capacitors. The gate-to-source capacitance C_{S2} and the gate-to-drain capacitance C_{D2} are due to overlap between the gate and source/drain regions. SPICE parameters that include these capacitances are C_{GS0} and C_{GD0} (overlap capacitances per unit width). They have to be specified in F/m,

TABLE 5.4 Summary of SPICE LEVEL 3 Static Parameters: Channel-Related Second-Order Effects

		Channel-Related Static Parameters		
Symbol	**SPICE Keyword**	**Parameter Name**	**Typical Value**	**Unit**
KP (or	KP	Transconductance parameter[a]	1.2×10^{-4}	A/V²
μ_0 and	Uo	Low-field mobility[b]	700	cm²/Vs
t_{ox})	Tox	Gate-oxide thickness[b]	20×10^{-9}	m
θ	THETA	Mobility modulation constant	0.1	—
v_{max}	Vmax	Maximum drift velocity	10^5	m/s
κ	KAPPA	Channel length modulation coefficient (needs Nsub)	0.2	—
N_A, N_D	Nsub	Substrate doping concentration	10^{15}	cm⁻³

	β and V_{DSsat} Equations	
NMOS		**PMOS**

$$\Rightarrow \qquad \beta = \mu_{\text{eff}} \frac{\varepsilon_{ox}}{\text{Tox}} \frac{W}{L_{\text{eff}} - L_{\text{pinch}}}$$

$$\mu_{\text{eff}} = \frac{\mu_s}{1 + \mu_s \min(|V_{DS}|, |V_{DSsat}|)/(\text{Vmax} L_{\text{eff}})}$$

$$\mu_s = \frac{\mu_0}{1 + \text{THETA}|V_{GS} - V_T|}$$

$$\mu_0 = \text{KP} \frac{\text{Tox}}{\varepsilon_{ox}}, \text{ if KP is specified; else } \mu_0 = \text{Uo}$$

$$L_{\text{pinch}} = \begin{cases} L_a = \sqrt{\text{KAPPA} \frac{2\varepsilon_s}{q\text{Nsub}} |V_{DS} - V_{DSsat}|} & \text{if Vmax is not specified} \\ \left[\left(\frac{\varepsilon_s}{q\text{Nsub}} \frac{V_{DSsat}}{L_{\text{eff}}} \right)^2 + L_a^2 \right]^{1/2} - \frac{\varepsilon_s}{q\text{Nsub}} \frac{V_{DSsat}}{|L_{\text{eff}}|} & \text{if Vmax is specified} \end{cases}$$

$$\Rightarrow \qquad V_{DSsat} = \begin{cases} \frac{V_{GS} - V_T}{1 + F_B} & \text{if Vmax is not specified[c]} \\ V_{DSsat-corr} & \text{if Vmax is specified} \end{cases}$$

$$V_{DSsat-corr} = V_a + V_b - \sqrt{V_a^2 + V_b^2} \ ^{(d)} \qquad V_{DSsat-corr} = V_a - V_b + \sqrt{V_a^2 + V_b^2} \ ^{d}$$

$$V_a = \frac{V_{GS} - V_T}{1 + F_B}, \qquad V_b = \frac{\text{Vmax} L_{\text{eff}}}{\mu_s} \ ^d$$

Constant: $\varepsilon_{ox} = 3.9 \times 8.85 \times 10^{-12}$ F/m

[a,b] Incompatible parameters.
[c] D. Foty, *MOSFET Modeling with SPICE: Principles and Practice,* Prentice Hall, Upper Saddle River, NJ, 1997 (p. 599).
[d] G. Massobrio and P. Antognetti, *Semiconductor Device Modeling with SPICE,* 2nd ed., McGraw-Hill, New York, 1993 (p. 208).

and SPICE then multiplies the specified values by the channel width W, to convert them into capacitances expressed in F. Assuming overlap of l_{olp}, these parameters can be estimated as $C_{GS0,GD0} = l_{olp} \varepsilon_{ox} / t_{ox}$. Gate-to-bulk overlap capacitance is not shown explicitly in the figure, however it exists, and is connected between points 1 and 2 in the equivalent circuit. The MOSFET cross section along the channel width is shown in Fig. 5.24. The gate-to-bulk overlap capacitance is due to the gate extension outside the effective channel width (W). Assuming overlap of l_{olp}, the parameter of this capacitance can also be estimated as $C_{GB0} = l_{olp} \varepsilon_{ox} / t_{f-ox}$. This parameter, however, means capacitance per unit length, and SPICE multiplies it by the effective channel length L_{eff} to convert the capacitance into F.

TABLE 5.5 Summary of SPICE LEVEL 3 Static Parameters: Depletion-Layer Related Second-Order Effects

PART I

Depletion-Layer-Related Static Parameters[a]

Symbol	SPICE Keyword	Parameter Name	Typical Value	Unit
t_{ox}	Tox	Gate oxide thickness	20×10^{-9}	m
η	ETA	Static feedback	0.7	—
		Note: This parameter can be used with V_{T0}, $2\|\phi_F\|$, and γ; t_{ox} should also be specified.		
N_A, N_D	Nsub	Substrate doping concentration	10^{15}	cm^{-3}
		Note: This parameter has to be specified to include the parameters below.		
N_{oc}	Nss	Oxide-charge density	10^{10}	cm^{-2}
	TPG	Gate material type		—
		Same as drain/source: TPG= 1		
		Opposite of D/S: TPG $= -1$		
		Metal: TPG = 0		
x_j	Xj	P–N junction depth	0.5×10^{-6}	m
x_{j-lat}	Ld	Lateral diffusion	$0.8 \times x_j$	m
V_{bi}	PB	P–N junction built-in voltage	0.8	V
δ	DELTA	Width effect on threshold voltage	1.0	—
	NFS	Subthreshold-current fitting parameter	10^{11}	cm^{-2}

[a] Incompatible parameters: V_{T0}, $\|2\phi_F\|$, and γ.

The gate-oxide capacitance inside the active channel area is included by the capacitors C_{S1} and C_{D1}. These capacitances vary with the applied voltages, and SPICE calculates them accordingly. In the linear region, when the channel expands from the source to the drain, C_{S1} and C_{D2} each makes half of the total gate-oxide capacitance $C_{ox}W L_{eff}$. In saturation, the channel is pinched off at the drain side, and C_{D1} capacitance is smaller. Although no specific parameters are needed to calculate these capacitances, the gate-oxide thickness t_{ox} and the doping level ($\|2\phi_F\|$ or $N_{A,D}$) have to be specified.

In addition to the parasitic capacitors, there are parasitic resistors as well. Although the origins can be different (contact resistances and/or neutral-body resistances), they can be expressed by four parasitic resistors associated with each of the four terminals. The values of these parasitic resistors are direct SPICE parameters.

Finally, the source-to-bulk and drain-to-bulk P–N junctions create parasitic diodes. Although these diodes are normally off, they can create leakage current, and, more importantly, they introduce depletion-layer capacitances. As described in Chapter 2, these capacitances depend on the reverse-bias voltage. The SPICE MOSFET model includes the full diode model (equivalent circuit) for these two P–N junctions. Additionally, each of these junctions is represented by two independent diodes, the bulk and perimeter diode. The zero-bias capacitance (as diode parameter) can be specified per unit area (for the case of the bulk diode D_B) or per unit length (for the perimeter diode D_P). Of course, these parameters necessitate properly specified geometric variables (drain and source diffusion area/perimeter). As an alternative to the complete P–N junction capacitance models, the bulk-to-drain and bulk-to-source capacitances can be specified directly (C_{BD} and C_{BS}

PART II

V_{T}, $2\,|\phi_{\mathrm{F}}|$, γ, F_{s}, F_{n}, and n_{s} **Equations**

NMOS	**PMOS**

$$C_{\mathrm{ox}} = \varepsilon_{\mathrm{ox}}/\mathrm{Tox}$$

⇒ $\quad V_{\mathrm{T}} = V_{\mathrm{T0}} + \gamma\,F_{\mathrm{s}}\Big(\sqrt{|2\phi_{\mathrm{F}}| + V_{\mathrm{SB}}} - \qquad\qquad\qquad V_{\mathrm{T}} = V_{\mathrm{T0}} - \gamma\,F_{\mathrm{s}}\Big(\sqrt{|2\phi_{\mathrm{F}}| + V_{\mathrm{BS}}} -$

$\qquad \sqrt{|2\phi_{\mathrm{F}}|}\Big) - \sigma_{\mathrm{D}}V_{\mathrm{DS}} + F_{\mathrm{n}}(V_{\mathrm{SB}} + 2\phi_{\mathrm{F}}) \qquad\qquad \sqrt{|2\phi_{\mathrm{F}}|}\Big) - \sigma_{\mathrm{D}}V_{\mathrm{DS}} - F_{\mathrm{n}}(V_{\mathrm{BS}} + |2\phi_{\mathrm{F}}|)$

$\qquad V_{\mathrm{T0}} = \phi_{\mathrm{ms}} - \frac{q\mathrm{Nss}}{C_{\mathrm{ox}}} + |2\phi_{\mathrm{F}}| + \gamma\,F_{\mathrm{s}}\sqrt{|2\phi_{\mathrm{F}}|} \qquad\qquad V_{\mathrm{T0}} = \phi_{\mathrm{ms}} - \frac{q\mathrm{Nss}}{C_{\mathrm{ox}}} - |2\phi_{\mathrm{F}}| - \gamma\,F_{\mathrm{s}}\sqrt{|2\phi_{\mathrm{F}}|}$

$$\phi_{\mathrm{ms}} = \begin{cases} -\frac{E_g}{2q} - |\phi_{\mathrm{F}}|, & \text{if TPG} = 1 \\ \frac{E_g}{2q} - |\phi_{\mathrm{F}}|, & \text{if TPG} = -1 \\ \phi_* - |\phi_{\mathrm{F}}|, & \text{if TPG} = 0 \end{cases} \qquad \phi_{\mathrm{ms}} = \begin{cases} \frac{E_g}{2q} + |\phi_{\mathrm{F}}|, & \text{if TPG} = 1 \\ \frac{-E_g}{2q} + |\phi_{\mathrm{F}}|, & \text{if TPG} = -1 \\ \phi_* + |\phi_{\mathrm{F}}|, & \text{if TPG} = 0 \end{cases}$$

$$\sigma_{\mathrm{D}} = 8.15 \times 10^{-22}\,\mathrm{ETA}/(C_{\mathrm{ox}}L_{\mathrm{eff}}^3)^{\mathrm{a}}$$

⇒ $\qquad\qquad\qquad\qquad\qquad \gamma = \frac{1}{C_{\mathrm{ox}}}\sqrt{2\varepsilon_s q\mathrm{Nsub}}$

⇒ $\qquad\qquad\qquad\qquad\qquad |2\phi_{\mathrm{F}}| = 2\frac{kT}{q}\ln\frac{\mathrm{Nsub}}{n_i}$

⇒ $\qquad\qquad\qquad F_{\mathrm{s}} = 1 - \frac{\mathrm{Xj}}{L_{\mathrm{eff}}}\left(\frac{\mathrm{Ld}+w_c}{\mathrm{Xj}}\sqrt{1 - \frac{w_p}{\mathrm{Xj}+w_p}} - \frac{\mathrm{Ld}}{\mathrm{Xj}}\right)^{\mathrm{a}}$

$\quad w_p = \sqrt{\frac{2\varepsilon_s}{q\mathrm{Nsub}}(\mathrm{PB} + V_{\mathrm{SB}})} \qquad\qquad\qquad\qquad w_p = \sqrt{\frac{2\varepsilon_s}{q\mathrm{Nsub}}(\mathrm{PB} + V_{\mathrm{BS}})}$

$$w_c = 0.0631353\mathrm{Xj} + 0.8013929 w_p - 0.0111077 w_p^2/\mathrm{Xj}^{\mathrm{a}}$$

⇒ $\qquad\qquad\qquad\qquad\qquad F_{\mathrm{n}} = \mathrm{DELTA}\,\varepsilon_s\pi/(4C_{\mathrm{ox}}\mathrm{W})^{\mathrm{a}}$

⇒ $\qquad\qquad\qquad\qquad n_{\mathrm{s}} = 1 + \frac{q\mathrm{NFS}}{C_{\mathrm{ox}}} + \frac{\gamma F_{\mathrm{s}}(|2\phi_{\mathrm{F}}|+|V_{\mathrm{SB}}|)^{-1/2} - F_{\mathrm{n}}}{2}{}^{\mathrm{b}}$

Constants:

$\varepsilon_{\mathrm{ox}} = 3.45 \times 10^{-11}$ F/m	$k = 8.62 \times 10^{-5}$ eV/K	$n_i = 1.4 \times 10^{10}$ cm^{-3}
$q = 1.6 \times 10^{-19}$ C	$\phi_* = \phi_{\mathrm{m}} - 4.61V = -.51V$	$\varepsilon_s = 1.044 \times 10^{-10}$ F/m

[a] G. Massobrio and P. Antognetti, *Semiconductor Device Modeling with SPICE*, 2nd ed., McGraw-Hill, New York, 1993 (pp. 205-206).
[b] D. Foty, *MOSFET Modeling with SPICE: Principles and Practice*, Prentice Hall, Upper Saddle River, NJ, 1997 (p. 597).

parameters). In this case, however, the dependence of the capacitance on the reverse-bias voltage is ignored.

5.4.2 Parameter Measurement

This section describes graphic methods for measurement of the most important MOSFET parameters. Although sophisticated software tools can be used to directly fit the nonlinear equations to experimental data, the graphic method is a valuable tool for establishing the initial values of the parameters needed in any nonlinear fitting algorithm. A general consideration of the parameter measurement techniques is given in Section 2.4.

Measurement of V_{T0} and KP

The parameters V_{T0} and KP can be obtained from the linear part of transfer characteristic $I_{\mathrm{D}} - V_{\mathrm{GS}}$. The MOSFET is in the linear mode for small V_{DS} voltages when the quadratic

TABLE 5.6 Summary of SPICE Dynamic MOSFET Model

PART I

Geometric Variables

Symbol	SPICE Keyword	Variable Name	Default Value	Unit
A_D; P_D	AD; PD	Drain diffusion area; . . . perimeter	0; 0	m^2; m
A_S; P_S	AS; PS	Source diffusion area; . . . perimeter	0; 0	m^2; m

Parasitic-Element-Related Parameters

Symbol	SPICE Keyword	Related Parasitic Element	Parameter Name	Typical Value	Unit
R_D	Rd	R_D	Drain resistance	10	Ω
R_S	Rs	R_S	Source resistance	10	Ω
R_G	Rg	R_G	Gate resistance	10	Ω
R_B	Rb	R_B	Bulk resistance	10	Ω
	Rds	Not shown	Drain–source leakage resistance	∞	Ω
t_{ox}	Tox	C_{S1}; C_{D1}	Gate-oxide thickness	20×10^{-9}	m
$\lvert 2\phi_F \rvert$ (or $N_{A,D}$)	Phi (or Nsub)	C_{S1}; C_{D1}	Surface potential (substrate doping)	0.7 (10^{15})	V (cm^{-3})
C_{GD0}	Cgdo	C_{D2}	Gate–drain overlap capacitance per channel width	4×10^{-11}	F/m
C_{GS0}	Cgso	C_{S2}	Gate–source overlap capacitance per channel width	4×10^{-11}	F/m
C_{GB0}	Cgbo	Not shown	Gate-bulk overlap capacitance per channel length	2×10^{-10}	F/m
I_s (or J_s)	IS (or JS)	D_B	Saturation current (current density)	10^{-14} (10^{-8})	A (A/m^2)
V_{bi}	PB/PBSW	D_B/D_P	Built-in voltage	0.8	V
$C_D(0)$	Cj/Cjsw	D_B/D_P	Zero-bias capacitance per unit area/length	2×10^{-4}/ 10^{-9}	F/m^2/ F/m
m	Mj/Mjsw	D_B/D_P	Grading coefficient	$\frac{1}{3}$ - $\frac{1}{2}$	—
C_{BD}; C_{BS}	Cbd; Cbs	D_B/D_P	Drain/source-to-bulk capacitance (incompatible with V_{bi}, $C_D(0)$, and m)		F

term $(1 + F_B)V_{DS}^2/2$ is negligible compared to $(V_{GS} - V_T)V_{DS}$, and for small $V_{GS} - V_T$ voltages when $\mu_{eff} \approx \mu_0$ as the mobility modulation factor $\theta(V_{GS} - V_T) \ll 1$. To simplify the following considerations, let us define a low-field gain factor β_0:

$$\beta_0 = \mu_0 C_{ox} \frac{W}{L_{eff}} \tag{5.45}$$

which is a constant, as opposed to the generally voltage-dependent gain factor β used in Tables 5.2 and 5.4:

$$\beta = \mu_{eff} C_{ox} \frac{W}{L_{eff}} \tag{5.46}$$

PART II

Large-Signal Equivalent Circuit

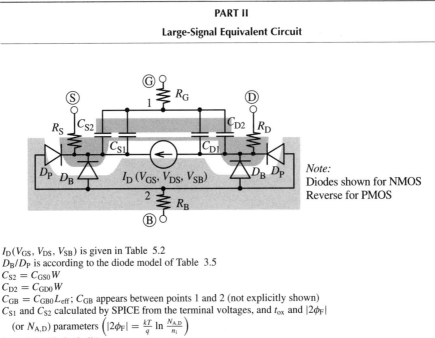

$I_D(V_{GS}, V_{DS}, V_{SB})$ is given in Table 5.2
D_B/D_P is according to the diode model of Table 3.5
$C_{S2} = C_{GSO}W$
$C_{D2} = C_{GDO}W$
$C_{GB} = C_{GBO}L_{eff}$; C_{GB} appears between points 1 and 2 (not explicitly shown)
C_{S1} and C_{S2} calculated by SPICE from the terminal voltages, and t_{ox} and $|2\phi_F|$
  (or $N_{A,D}$) parameters $\left(|2\phi_F| = \frac{kT}{q}\ln\frac{N_{A,D}}{n_i}\right)$
$I_s = J_s A_D$ (drain–bulk)
$I_s = J_s A_s$ (source–bulk)

Therefore, the MOSFET model in the linear region can be written as

$$I_D \approx \beta_0(V_{GS} - V_T)V_{DS} \tag{5.47}$$

Note that the range of V_{GS} voltages in which the $I_D - V_{GS}$ characteristic is approximately linear can be expressed as $V_{DS}(1 + F_B) \ll V_{GS} - V_T \ll 1/\theta$.

Figure 5.25 provides an example of a MOSFET transfer characteristic measured at $V_{DS} = 50$ mV. The linear part appears approximately between $V_{GS} = 0.8$ V and $V_{GS} = 2.0$ V. This part of the transfer characteristic is described by Eq. (5.47) and can be used to obtain $V_{T0} = V_T(V_{SB} = 0)$ and KP. The zero-bias threshold voltage V_{T0} is obtained at the intersection between the V_{GS} axis and the straight line extrapolating the linear part of the transfer characteristic. It can also be obtained analytically, applying Eq. (5.47) to two different measurement points (I_{D1}, V_{GS1}) and (I_{D2}, V_{GS2}), to obtain a system of two linear equations. Eliminating β_0 from these two equations, the zero-bias threshold voltage is obtained as

$$V_{T0} = \frac{V_{GS1} - (I_{DS1}/I_{DS2})V_{GS2}}{1 - I_{DS1}/I_{DS2}} \tag{5.48}$$

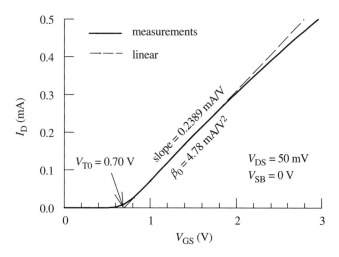

Figure 5.25 MOSFET transfer characteristics in the linear region.

The slope of the linear part of the transfer characteristic is, according to Eq. (5.47), $\beta_0 V_{DS}$. Therefore, β_0 factor is obtained when the slope is divided by the voltage V_{DS}. The transconductance parameter KP can then be calculated as $KP = \beta_0 L_{eff}/W$.

Measurement of γ and 2φF

Measurement of the parameters γ and $2\phi_F$ is based on the dependence of the threshold voltage V_T on the source-to-substrate voltage V_{SB} (the body effect). To collect experimental V_T vs V_{SB} data, the threshold voltage is measured by the procedure described above with the MOSFET biased at different V_{SB} voltages. Figure 5.26a gives an example of transfer characteristics measured with different V_{SB} voltages, which are used to obtain the corresponding V_T voltages.

The equation modeling $V_T(V_{SB})$ dependence is shown in Table 5.3. It can be seen that for a properly chosen $2\phi_F$, V_T vs $(\sqrt{2\phi_F + V_{SB}} - \sqrt{2\phi_F})$ is a linear dependence, the slope of which is γ. Therefore, making an initial guess for $2\phi_F$, the V_T vs $(\sqrt{2\phi_F + V_{SB}} - \sqrt{2\phi_F})$ plot can be used to verify the validity of the assumed $2\phi_F$ value. If the plotted line is not straight, a second guess for $2\phi_F$ is made and the plot is redone. Note that a concave curve indicates that $2\phi_F$ should be increased, while a convex curve indicates that $2\phi_F$ should be decreased. This process is continued until an appropriate straight line is obtained, as illustrated in Fig. 5.26b. The slope of this line is γ.

When the second-order effects from Table 5.5 are to be employed, the substrate doping $N_{A,D}$ and the gate-oxide thickness t_{ox} have to be specified instead of V_{T0}, $2\phi_F$, and γ. If the gate-oxide thickness is not known, it can be obtained from gate-oxide capacitance (C_{ox}) measurements. The gate-oxide capacitance can be measured using a large-area MOSFET biased in accumulation. With the gate-oxide capacitance value, and the already obtained body factor γ, the doping level is calculated using the equation given in Table 5.5.

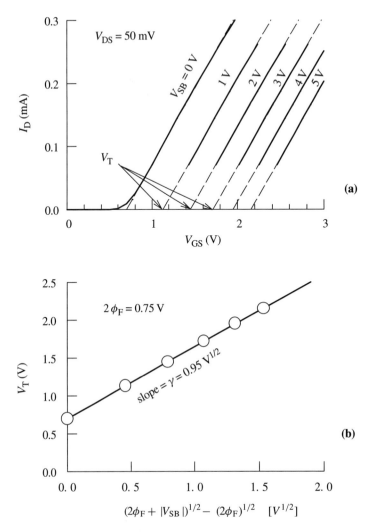

Figure 5.26 Obtaining $V_T(V_{SB})$ data (a) and graphic extraction of the body-effect parameter γ and the surface-inversion potential $2\phi_F$ (b).

Measurement of θ

Figure 5.27a illustrates that the linear model (5.47) overestimates the actual current at higher V_{GS} voltages. The deviation of the measured I_D current from the linear dependence is due to the mobility reduction effect, and is taken into account by the mobility modulation coefficient θ in the SPICE LEVEL 3 model. Therefore, to properly describe the drain current I_D at larger V_{GS} voltages, but still small V_{DS} voltages [$V_{DS}(1 + F_B) \ll V_{GS} - V_T$], an equation more general than Eq. (5.47) is used:

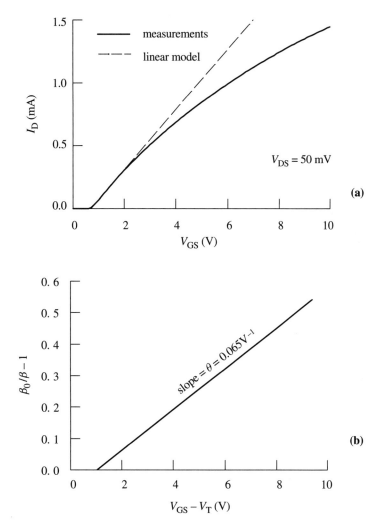

Figure 5.27 MOSFET transfer characteristics at higher V_{GS} voltages (a) and extraction of θ parameter (b).

$$I_D \approx \beta(V_{GS} - V_T)V_{DS} \tag{5.49}$$

The gain factor β in Eq. (5.49) is related to β_0 as

$$\beta = \frac{\beta_0}{1 + \theta(V_{GS} - V_T)} \tag{5.50}$$

Using Eq. (5.49), a set of β values can be calculated from the experimental I_D points measured at different V_{GS} voltages. Of course, the V_{DS} value is known, and the threshold voltage V_T needs to be obtained first. Having the set of β values for different V_{GS} values, and having measured β_0 and V_T as described above, the θ parameter can be obtained from Eq. (5.50). To enable application of the graphic method, Eq. (5.50) is transformed to:

$$\frac{\beta_0}{\beta} - 1 = \theta(V_{GS} - V_T) \tag{5.51}$$

It is obvious that plotting $\beta_0/\beta - 1$ vs $V_{GS} - V_T$ should produce a straight line with the slope equal to the parameter θ, as illustrated in Fig. 5.27b.

Measurement of Effective Length and Parasitic Resistances

Lateral diffusion leads to a difference between the *gate length* L_g and the *effective channel length* L_{eff}. This is illustrated in Table 5.2. Although L_g is the SPICE input parameter, the gate length is not necessarily equal to the *nominal gate length*, specified at the design level. This is due to imperfections of the manufacturing process (over- or underexposed photoresist, over- or underetched polysilicon, etc.).

The difference between the nominal and the effective channel lengths,

$$\Delta L = L_{nom} - L_{eff} \tag{5.52}$$

can be measured electrically if special test structures consisting of MOSFETs with equal widths and scaled lengths are available. As the MOSFET current in the linear region is given by Eq. (5.47), *on resistance* of the MOSFET can be defined and expressed as follows:

$$R_{on} \equiv \frac{V_{DS}}{I_D} = \frac{1}{\beta_0(V_{GS} - V_T)} \tag{5.53}$$

Replacing β_0 by $KP(W/L_{eff}) = KP\,W/(L_{nom} - \Delta L)$ shows that the on resistance is linearly dependent on L_{nom}:

$$R_{on} = \frac{L_{nom} - \Delta L}{KP\,W(V_{GS} - V_T)} \tag{5.54}$$

Figure 5.28 provides an example of measured on resistances for MOSFETs with four different channel lengths and all other parameters identical. Zero on resistance corresponds to $L_{eff} = 0$, or equivalently to $L_{nom} = \Delta L$ [refer to Eq. (5.54)]. The intersection between the extrapolated linear dependence and L_{nom} axis (this is $R_{on} = 0$) directly shows ΔL. Once ΔL is determined, the effective channel length is obtained as $L_{eff} = L_{nom} - \Delta L$, and the gate length can be specified as $L_g = L_{eff} + 2x_{j-lat}$, where x_{j-lat} is also specified as an input parameter.

The above-described method is based on the assumption that the channel resistance (on resistance) is V_{DS}/I_D, which implicitly assumes that the parasitic series resistance is zero. When the parasitic series resistance is not negligible, V_{DS}/I_D does not show the on resistance itself but the on resistance plus the parasitic series resistance: $V_{DS}/I_D = R_{on} + R_{par}$. Obviously, using V_{DS}/I_D data as R_{on} in the case when R_{par} is not negligible leads to errors. The following extended version of Eq. (5.54) should be used in this case:

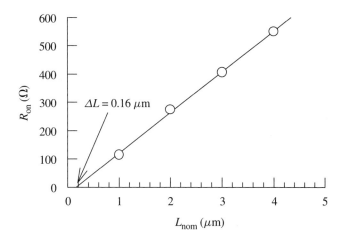

Figure 5.28 Measurement of the difference between the nominal and effective channel lengths.

$$V_{DS}/I_D = \underbrace{\frac{L_{nom} - \Delta L}{KPW(V_{GS} - V_T)}}_{R_{on}} + R_{par} \qquad (5.55)$$

The on resistance [the first term of Eq. (5.55)] depends on the gate voltage, while the parasitic resistance does not. This helps to distinguish between the contributions of the on resistance and the parasitic resistance to V_{DS}/I_D values. It is necessary to measure the set of V_{DS}/I_D vs L_{nom} data for one or more additional gate-to-source voltages V_{GS}. Figure 5.29 shows

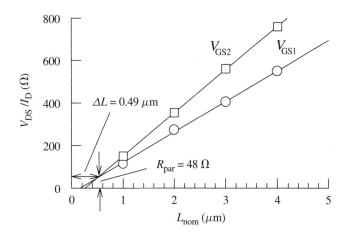

Figure 5.29 Simultaneous measurement of the difference between the nominal and effective channel lengths and the parasitic resistance.

an additional set of data points (labeled as V_{GS2}) added to the plot previously shown in Fig. 5.28. Equation (5.55) shows that different R_{on} resistances at different V_{GS} voltages will cause different V_{DS}/I_D values. However, there is one point where the influence of V_{GS} does not show, and the V_{DS}/I_D value is the same for any V_{GS} voltage. This point is $L_{nom} = \Delta L$; it turns R_{on} into zero for any V_{GS} voltage. Therefore, the straight lines of linear V_{DS}/I_D vs L_{nom} dependencies, measured at different V_{GS} voltages, intersect at a single point, which is $L_{nom} = \Delta L$ and $V_{DS}/I_D = R_{par}$.

Having determined ΔL, the gate length L_g can accurately be specified, as described above. In addition, the determined value of the parasitic series resistance R_{par} shows the combined effect of the source and drain parasitic resistances R_S and R_D. Assuming symmetrical MOSFET, these two parameters can be specified as $R_S = R_D = R_{par}/2$.

■ **Example 5.5 Effective Channel Length**

Two adjacent MOSFETs have the following design dimensions: $L_{nom1} = 1\ \mu m$, $L_{nom2} = 2\ \mu m$, $W_1 = W_2$. The drain currents, measured at $V_{GS} - V_T = 2.5$ V and $V_{DS} = 50$ mV are $I_{D1} = 495\ \mu A$ and $I_{D2} = 180\ \mu A$. Neglecting the parasitic series resistance, determine the effective channel lengths.

Solution: The measurement conditions ($V_{GS} - V_T = 2.5$ V and $V_{DS} = 50$ mV) indicate that the currents I_{D1} and I_{D2} are measured in the linear region. Corresponding on resistances are $R_{on1} = V_{DS}/I_{D1} = 101.0\ \Omega$ and $R_{on2} = 277.8\ \Omega$. As the on resistance is given by Eq. (5.54),

$$R_{on} = \underbrace{\frac{1}{K P W (V_{GS} - V_T)}}_{a}(L_{nom} - \Delta L)$$

the following system of two equations and two unknowns (a and ΔL) can be written:

$$R_{on1} = a(L_{nom1} - \Delta L)$$
$$R_{on2} = a(L_{nom2} - \Delta L)$$

To find ΔL, these two equations are divided:

$$R_{on1}/R_{on2} = (L_{nom1} - \Delta L)/(L_{nom2} - \Delta L)$$

and ΔL expressed as

$$\Delta L = \left(L_{nom2}\frac{R_{on1}}{R_{on2}} - L_{nom1}\right) \Big/ \left(\frac{R_{on1}}{R_{on2}} - 1\right) = 0.43\ \mu m$$

Therefore, the effective channel lengths are $L_{eff1} = L_{nom1} - \Delta L = 1 - 0.4 = 0.6\ \mu m$ and $L_{eff2} = 2 - 0.4 = 1.6\ \mu m$.

■ **Example 5.6 Static Feedback on the Threshold Voltage (η)**

To extract the static feedback on the threshold voltage η (SPICE parameter), the dependence of the threshold voltage on the drain-to-source voltage has been measured, and the following data obtained: $V_{DS} = 2$ V, $V_T = 0.68$ V; $V_{DS} = 3$ V, $V_T = 0.66$ V; $V_{DS} = 4$ V, $V_T = 0.63$ V; $V_{DS} = 5$ V, $V_T = 0.61$ V. Determine the static feedback on the threshold voltage η for this MOSFET. The effective channel length is 2μ m, and the oxide capacitance is 1.726×10^{-3} F/m^2.

Solution: According to Table 5.5 (PART II), the threshold voltage dependence on the drain-to-source voltage is given by

$$V_T = V_{T0} - \sigma_D V_{DS}$$

where V_{T0} is the zero-bias threshold voltage and σ_D is a coefficient that can be determined as the slope of the linear V_T–V_{DS} dependence. Figure 5.30 shows that σ_D is found to be 0.024.

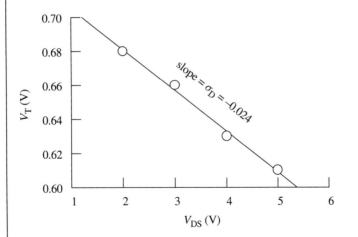

Figure 5.30 Threshold voltage dependence on drain-to-source voltage.

The relationship between σ_D and η is also given in Table 5.5 (PART II). Using that equation, η is calculated as

$$\eta = \sigma_D C_{ox} L_{eff}^3 / 8.15 \times 10^{-22} = 0.024 \times 1.726 \times 10^{-3} (2 \times 10^{-6})^3 / 8.15 \times 10^{-22}$$

$$= 0.41$$

▨ **Example 5.7 Effective Channel Width**

The MOSFET channel width can also differ from the nominal value, which can be significant in narrow-channel MOSFETs. To obtain this difference, the channel conductance $G_{on} = I_D/V_{DS}$ is measured in the linear region for MOSFETs having different channel widths and all the other parameters identical. Determine $\Delta W = W_{nom} - W$ using the following results: $W_{nom} = 4 \ \mu m$, $G_{on} = 0.40 \ \Omega^{-1}$; $W_{nom} = 6 \ \mu m$, $G_{on} = 0.58 \ \Omega^{-1}$; $W_{nom} = 8 \ \mu m$, $G_{on} = 0.76 \ \Omega^{-1}$.

Solution: The channel conductance G_{on} in the linear region is

$$G_{on} = KP\frac{W}{L_{eff}}(V_{GS} - V_T) = \frac{KP}{L_{eff}}(V_{GS} - V_T)(W_{nom} - \Delta W)$$

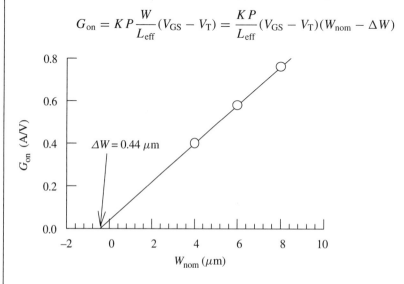

Figure 5.31 Dependence of channel conductance on channel width.

The $G_{on}-W_{nom}$ plot of Fig. 5.31 shows the linear dependence predicted by the above equation. As the zero channel conductance appears at $W_{nom} = \Delta W$, ΔW is found at the intersection between the linear $G_{on}-W_{nom}$ dependence and the W_{nom} axis. From Fig. 5.31, $\Delta W = -0.44 \ \mu m$.

PROBLEMS

Section 5.1 MOSFET Principles

5.1 Figure 5.32 shows five energy-band diagrams, drawn from the oxide–silicon interface into the silicon substrate, and the transfer characteristic of a MOSFET with four labeled points. Identify the four correct band diagrams and relate them to the four points on the transfer characteristic.

5.2 Figure 5.33 shows four energy-band diagrams, drawn from the source to the drain, along the silicon surface. Identify how the energy-band diagrams relate to each of the four points, labeled on the output characteristics of the MOSFET.

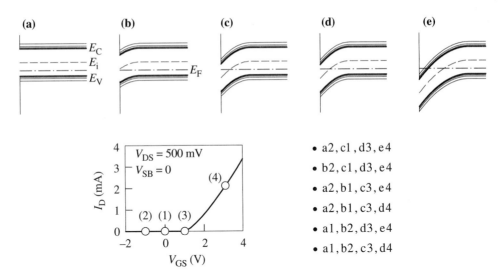

Figure 5.32 Energy-band diagrams and MOSFET transfer characteristics.

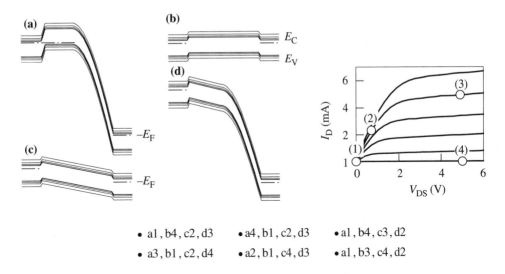

Figure 5.33 Energy-band diagrams and MOSFET output characteristics.

5.3 Which of the following statements, related to MOSFETs, are **not** correct?

- N-type substrate is used to make normally-on P-channel MOSFETs. T
- The net charge at the semiconductor surface is zero at $V_{GS} = V_T$. F
- If a MOSFET is in the linear region, it is also in the triode region. T
- Existence of a significant drain current at $V_{GS} = 0$ V indicates a faulty MOSFET. F

- For a MOSFET in saturation, the channel carriers reach the saturation drift velocity at the pinch-off point. F
- The threshold voltage of an enhancement-type P-channel MOSFET is negative. T
- Positive gate voltage is needed to turn a normally-on P-channel MOSFET off. T
- A MOSFET cannot be in both the triode and saturation region at the same time. T
- The above-threshold current in the MOSFET channel is essentially due to diffusion. F
- Both electrons and holes play significant roles in the flow of drain-to-source current. F

5.4 For an N-channel MOSFET with uniform substrate doping of $N_A = 5 \times 10^{16}$ cm^{-3} and gate oxide thickness of $t_{ox} = 5$ nm, determine the surface potential φ_s and the depletion-layer charge Q_d at

(a) $V_{GS} = V_{FB} = -0.75$ V,

(b) $V_{GS} = -0.5$ V and $V_{GS} = 0$ V (assume zero oxide-charge and interface-trap densities, so that $\varepsilon_{ox}E_{ox} = \varepsilon_s E_s = qN_A w_d$). [A]

(c) $V_{GS} = V_T = 0.2$ V.

(d) $V_{GS} = 0.75$ V. [A]

Plot $\varphi_s(V_{GS})$ and $Q_d(V_{GS})$. ($V_{SB} = 0$.)

5.5 The transconductance of an N-channel MOSFET operating in the linear region ($V_{DS} = 50$ mV, $V_{SB} = 0$) is $g_m = dI_D/dV_{GS} = 2.5$ mA/V. If the threshold voltage is $V_T = 0.3$ V, what is the current at $V_{GS} = 1$ V?

5.6 The substrate doping and the body factor of an N-channel MOSFET are $N_A = 10^{16}$ cm^{-3} and $\gamma = 0.12$ V$^{1/2}$, respectively. If the threshold voltage measured with $V_{SB} = 3$ V is $V_T = 0.5$ V, what is the zero-bias threshold voltage?

5.7 For the MOSFET of Problem 5.6, how many times is the channel resistance increased when V_{SB} is increased from 0 to 3.3 V? The gate and drain voltages are $V_{GS} = 3.3$ V and $V_{DS} = 50$ mV. [A]

5.8 Knowing the following technological parameters, $t_{ox} = 3.5$ nm, $N_D = 5 \times 10^{15}$ cm^{-3}, and $V_{FB} = 0.2$ V, determine the inversion-layer charge density at $V_{GS} = -0.75$ V, $V_{GS} = 0$ V, and $V_{GS} = 0.75$ V for

(a) $V_{BS} = 0$ V.

(b) $V_{BS} = 0.75$ V. [A]

5.9 Body factors of N-channel and P-channel MOSFETs are determined from body-effect measurements as 0.11 and 0.47 V$^{1/2}$, respectively. Determine the substrate doping levels in those MOSFETs. The gate-oxide capacitance is $C_{ox} = 1.726 \times 10^{-3}$ F/m^2.

Section 5.2 MOSFET Technologies

5.10 Ion-implant profiles can be approximated by the Gaussian distribution function

$$N(x) = \frac{\Phi}{\sqrt{2\pi}\,\Delta R_p} \exp\left[-\frac{1}{2}\left(\frac{x - R_p}{\Delta R_p}\right)^2\right]$$

where Φ is the dose, R_p is the range, and ΔR_p is the straggle of the implanted ions. The range and the straggle depend on the energy of the implantation, as shown in Table 5.7. Sketch the implant profiles of boron for three different energies (20, 50, and 100 keV) and dose of 5×10^{11} cm^{-2}. Using the medium energy, change the dose to 10^{11} and 10^{12} cm^{-2} and sketch the implant profiles. Comment on the influence of energy and dose on the implant profiles.

TABLE 5.7 Ion Implant Parameters

			Energy (keV)			
			20	50	100	200
B	R_p	(μm)	0.0662	0.1608	0.2994	0.5297
	ΔR_p	(μm)	0.0283	0.0504	0.0710	0.0921
P	R_p	(μm)	0.0253	0.0607	0.1238	0.2539
	ΔR_p	(μm)	0.0119	0.0256	0.0456	0.0775
As	R_p	(μm)	0.0159	0.0322	0.0582	0.1114
	ΔR_p	(μm)	0.0059	0.0118	0.0207	0.0374

5.11 The maximum operating voltage of an NMOS integrated circuit is 10 V, and the substrate doping level in the field region is $N_A = 5 \times 10^{17}$ cm^{-3}. Determine the minimum field-oxide thickness needed to prevent current leakage between neighboring MOSFETs. Neglect the oxide charge, and consider aluminum gate (the worst case scenario).

Hint: The field oxide can be considered as the gate oxide of a parasitic MOSFET, which should be kept "off" (the maximum operating voltage should be below the threshold voltage) to prevent a possible leakage.

5.12 Figure 5.34 shows the ideal response (no delays and parasitic capacitances to be charged/ discharged) of an NMOS inverter. The power-supply voltage is 5 V.

(a) Find the average power dissipated by this inverter.

(b) Find the average dissipated power if the inverter was implemented in CMOS technology. [A]

5.13 The average power dissipated by a CMOS IC is $P_{diss} = 5$ μW per MOSFET. Progress in technology enables design of CMOS ICs with reduced input capacitance and increased switching frequency, although this may require a reduction in the power-supply voltage. Find P_{diss} for a new design that

(a) halves the input MOSFET capacitances and quadruples the switching frequency, maintaining the power-supply voltage;

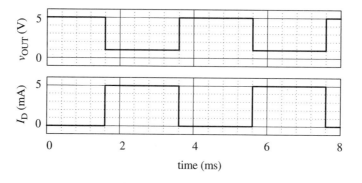

Figure 5.34 Ideal switching response of an NMOS inverter.

(b) reduces the input capacitance five times, increases the switching frequency 25 times and reduces the power supply voltage from 5 V to 1.5 V. [A]

5.14 The designed dose of a threshold-voltage adjustment implant is $\Phi_{implant} = 10^{12}$ cm^{-2}. What is the mismatch between the threshold voltages ($V_{T-NMOS} - |V_{T-PMOS}|$) if the actual dose is 1% higher? The doping element of the adjustment implant is boron, and the gate-oxide thickness is 4 nm. Assume that all the implanted atoms are in the depletion layer.

Section 5.3 MOSFET Modeling

5.15 The solid line in Fig. 5.35, labeled by "N," is for an N$^+$ poly-N-channel MOSFET with the following parameters: $L = 2$ μm, $W = 2$ μm, $t_{ox} = 20$ nm, $N_A = 5 \times 10^{16}$ cm^{-3}, $x_j = 0.5$ μm, $x_{j-lat} = 0.4$ μm, $\mu_0 = 750$ cm^2/Vs, $\delta = 1$, $\theta = 0$, and $\eta = 0$. The other four characteristics are obtained by changing one of the listed parameters. Identify which altered parameter relates to each of the transfer characteristics labeled by 1, 2, 3, and 4.

5.16 One set of output characteristics from Fig. 5.36 is for the nominal MOSFET parameters, as listed in the text of Problem 5.15, while the other three are for either changed values of θ or η or the specified v_{max} parameter. Relate each of the output characteristics to the appropriate set of parameters.

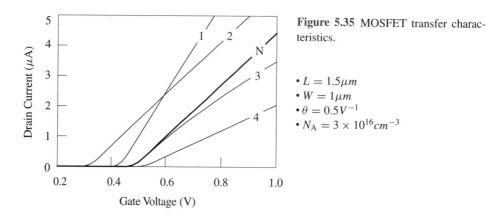

Figure 5.35 MOSFET transfer characteristics.

- $L = 1.5\mu m$
- $W = 1\mu m$
- $\theta = 0.5V^{-1}$
- $N_A = 3 \times 10^{16}cm^{-3}$

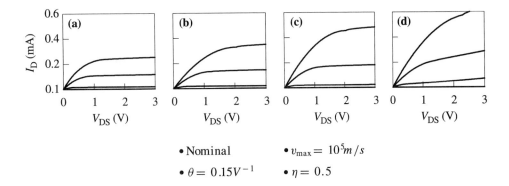

- Nominal
- $\theta = 0.15 V^{-1}$
- $v_{max} = 10^5 m/s$
- $\eta = 0.5$

Figure 5.36 MOSFET output characteristics.

5.17 An N-channel MOSFET with $V_T = 0.25$ V is biased by $V_{GS} = 2.5$ V and $V_{DS} = 500$ mV. The gate-oxide capacitance is $C_{ox} = 2.5$ mF/m^2 and the effective channel length is $L = 1$ μm. Calculate (a) the average lateral field, (b) the average channel conductivity, assuming channel thickness $x_{ch} = 5$ nm, and (c) the current density. Assume $\mu_0 = 750$ cm^2/Vs for the channel-carrier mobility.

5.18 If the channel length of the MOSFET from Problem 5.17 is reduced to $L = 0.2$ μm, maintaining the value of the threshold voltage and the bias conditions, calculate the average lateral electric field. Assuming drift velocity of $v_D = 0.08$ μm/ps at this field (Fig. 1.18), and using the average carrier concentration in the channel ($x_{ch} = 5$ nm), calculate the current density. What is the channel-carrier mobility in this case? $\boxed{A}$

5.19 A P-channel MOSFET has $V_{T0} = 0.2$ V, $\gamma = 0.2$ V$^{1/2}$ (neglect F_B, as $F_B \ll 1$), and $\beta = 5$ mA/V^2. If $V_{GS} = -1$ V, calculate V_{DSsat}, I_{Dsat}, and I_D at $V_{DS} = V_{DSsat}/5$ for

(a) $V_{SB} = 0$ V.

(b) $V_{SB} = -4.1$ V, $|\phi_F| = 0.45$ V. $\boxed{A}$

5.20 (a) Repeat the calculations of Example 5.3 using $\theta = 0.1$ V^{-1} to include the effect of mobility reduction with gate voltage. Use the SPICE LEVEL 3 model.

(b) Find the change of the drain current if the bulk of the MOSFET is biased at $V_{SB} = 5$ V. $\boxed{A}$

5.21 The technological parameters of an N$^+$ poly-gate P-channel MOSFET are $L/W = 10$, $t_{ox} = 4.5$ nm, and $N_D = 10^{16}$ cm^{-3}. Find the drain current and the transconductance at $V_{GS} = -1$ V, $V_{DS} = -50$ mV, and $V_{SB} = 0$ V. Assume $\mu_0 = 350$ cm^2/Vs and $N_{oc} = 0$.

5.22 Considered is an N-channel MOSFET with the gain factor $\beta = 600$ μA/V^2, the drain-bias factor $F_B = 0.7$, and the zero-bias threshold voltage $V_{T0} = 1.1$ V. If the coefficient of the influence of the drain bias on the threshold voltage is $\sigma_D = 0.01$, determine the dynamic output resistance ($r_o = dV_{DS}/dI_D$) of the MOSFET at $V_{GS} = V_{DS}/2 = 5$ V. Channel-length modulation can be neglected.

5.23 The output dynamic resistance of an N-channel MOSFET with $\sigma_D = 0.01$ and negligible channel-length modulation effect is $r_o = 1$ MΩ at $V_{GS} - V_T = 0.5$ V. What is the output resistance at $V_{GS} - V_T = 5$ V if

(a) $\theta = 0$. $\boxed{A}$

(b) $\theta = 0.05$ V^{-1}.

5.24 The length of the channel pinch-off region can be expressed as the depletion layer of an abrupt P–N junction, modulated by a fitting parameter κ:

$$L_{pinch} = \sqrt{\kappa \frac{2\varepsilon_s}{qN_A}(V_{DS} - V_{DSsat})}$$

What is the relative increase (expressed in percents) of the drain current of a $1 - \mu$m MOSFET, when the voltage changes from $V_{DS} = V_{DSsat}$ to $V_{DS} = 5$ V $+ V_{DSsat}$ if $N_A = 5 \times 10^{16}$ cm^{-3} and $\kappa = 0.2$? Assume constant threshold voltage ($\sigma_D = 0$).

Section 5.4 SPICE Parameters and Parasitic Elements

5.25 Obtain the zero-bias threshold voltage V_{T0} and the transconductance parameter KP of an N-channel MOSFET using the data given in Table 5.8. The channel width-to-channel length ratio of the transistor is 100.

TABLE 5.8 I_D–V_{GS} Data

Drain Current (mA)		Gate Voltage (V)
	$V_S = 0, V_B = 0, V_{DS} = 50$ mV	
0.18		1
0.50		1.5

5.26 To determine the value of the body factor γ of an N-channel MOSFET, the threshold voltage dependence on $\sqrt{2\phi_F + |V_{SB}|} - \sqrt{2\phi_F}$ is analyzed. This dependence becomes linear for $2\phi_F = 0.82$ V. Two points of this dependence are given in Table 5.9. Determine the body factor of this MOSFET.

TABLE 5.9 Body Effect Data

V_T (V)	$\left(\sqrt{2\phi_F + |V_{SB}|} - \sqrt{2\phi_F}\right)$ (V$^{1/2}$)
1.0	0.0
2.0	1.0

5.27 For the N-channel MOSFET considered in Problem 5.25, obtain the best estimate of the drain current at $V_{GS} = 5$ V to complete Table 5.10. $\boxed{A}$

TABLE 5.10 I_D–V_{GS} Data

Drain Current (mA)	Gate Voltage (V)
$V_S = 0$, $V_B = 0$, $V_{DS} = 50$ mV	
0.18	1
0.50	1.5
?	5.0
2.49	8.0

5.28 A set of measurements of β vs $V_{GS} - V_T$ is given in Table 5.11. Determine the mobility-modulation coefficient θ used to express the mobility reduction with the gate voltage in the SPICE LEVEL 3 MOSFET model.

TABLE 5.11 β–$(V_{GS} - V_T)$ Data

$V_{GS} - V_T$ (V)	0.5	1.0	3.0	5.0	7.0	9.0
β (mA/V^2)	455	455	385	333	294	263

5.29 The gate of a MOSFET overlaps the source and drain regions by 100 nm each, and the field oxide by 500 nm. The gate-oxide thickness is $t_{ox} = 10$ nm, while the field-oxide thickness is $T_{ox} = 100$ nm. Determine the following SPICE parameters: C_{GD0} (gate-drain overlap capacitance per channel width), C_{GS0} (gate-source overlap capacitance per channel width), and C_{GB0} (gate-bulk overlap capacitance per channel length). A

5.30 The source-bulk and drain-bulk junction depth is $x_j = 100$ nm and the lateral diffusion is $x_{j-lat} = 0.8x_j$. What are the gate-drain and the gate-source overlap capacitances per channel width? The gate-oxide thickness is 8 nm and the gate itself is used as a mask for source/drain implantation (self-aligned structure). What is the total gate capacitance if $L_{gate} = 0.3$ μm and $W = 3$ μm? (Ignore any gate-bulk overlap capacitance.)

5.31 The maximum operating frequency, also called cutoff frequency, of an FET is defined by

$$f_{max} = \frac{g_m}{2\pi(C_{gs} + C_{gd})}$$

where g_m is the transconductance and C_{gs} and C_{gd} are small-signal gate-source and gate-drain capacitances. Find f_{max} at the onset of saturation and $V_{GS} - V_{T0} = 1$ V for an N-channel MOSFET with $L_{gate} = 250$ nm, $L_{eff} = 200$ nm, $W = 20$ μm, $t_{ox} = 5$ nm, $\mu_{eff} = 350$ cm^2/Vs, and $F_B \ll 1$.

REVIEW QUESTIONS

R-5.1 What type of substrate (N or P) is used to make normally-off N-channel MOSFETs? Normally-on N-channel MOSFETs?

R-5.2 What gate-to-source voltage, positive or negative, is needed to turn a normally-on P-channel MOSFET off?

R-5.3 Can a single MOSFET be used as both a voltage-controlled switch (digital operation) and a voltage-controlled current source (analog operation)?

R-5.4 Typically, is the surface potential φ_s zero at $V_{GS} = 0$ V? What is the condition of $\varphi_s = 0$ called?

R-5.5 Can normally-on and normally-off MOSFETs have the same flat-band voltage? Are the electrical conditions (energy bands) in the silicon of normally-off and normally-on MOSFETs equivalent at $\varphi_s = 0$?

R-5.6 Is there any charge at the semiconductor side of an MOS structure at $V_{GS} = V_T$? Is there any mobile-carrier charge?

R-5.7 Is the threshold voltage of a normally-on P-channel MOSFET positive or negative?

R-5.8 Why does source-to-bulk reverse-biased voltage (V_{SB}) increase the threshold voltage? What is this effect called?

R-5.9 Can a MOSFET simultaneously be in both the linear and the triode region?

R-5.10 Do channel carriers face a negligible or infinitely large resistance between the channel pinch-off point and the drain of MOSFET in saturation?

R-5.11 How can the energy bands of a MOSFET in saturation be compared to a waterfall?

R-5.12 How are the devices/gates isolated from each other in NMOS and CMOS ICs?

R-5.13 What are the advantages of IC layer merging?

R-5.14 Why is polysilicon, and not a metal, used as the gate material in modern MOSFET technologies?

R-5.15 Which technology process is more complex, NMOS or CMOS?

R-5.16 What are the advantages of CMOS ICs compared to NMOS ICs?

R-5.17 Why is threshold-voltage adjustment implant needed?

R-5.18 Which SPICE model (LEVEL 1, 2, or 3) would you use to simulate a circuit with MOSFETs? Why?

R-5.19 The mobility reduction with gate voltage is a second-order effect. Can you, typically, neglect it?

R-5.20 What is the effect of neglecting the mobility reduction with the drain voltage?

R-5.21 What is the effect of neglecting the threshold voltage dependence on V_{DS} voltage?

Chapter *6*

BJT

The bipolar junction transistor (BJT) was the first solid-state active electronic device. Before the BJT, electronic amplifiers were based on vacuum tubes. The concepts on which BJTs were based were experimentally and theoretically established by Bardeen, Brattain, and Schokley at the Bell Telephone Laboratories during 1948. The era of semiconductor-based electronics, which has had an enormous influence on the way we live today, actually began with the invention of the BJT.

A number of alternative semiconductor devices have been developed since the first BJTs, notably MOSFETs and MESFETs. Nonetheless, the BJTs are still used, as there are applications in which the BJTs still offer the best performance. In addition, there are applications in which the BJTs are combined with MOSFETs, even in integrated-circuit technology. It should also be noted that the BJT principles are frequently used in a number of specifically designed semiconductor devices.

6.1 BJT PRINCIPLES

This section introduces the BJT principles. The BJT can be used as both a voltage-controlled current source (an analog device) and a voltage-controlled switch (digital device). Both modes of operation are considered to present a complete description of the device. At the end of the section, the BJT is compared to the MOSFET.

6.1.1 Making a Voltage-Controlled Current Source

The essential characteristic of a current source is that its current does not depend on the voltage drop across the current source. In other words, it delivers a constant current at any voltage.

A reverse-biased P–N junction is a semiconductor implementation of a current source. The current through a reverse-biased P–N junction is due to the minority carriers. The energy-band diagram of Fig. 6.1 shows that the minority electrons easily roll down and the minority holes easily bubble up through the depletion layer, making the reverse-bias current. This current is limited by the number of minority electrons and holes appearing at the edges of the depletion layer, and not by the reverse-bias voltage V_{CB}, which sets the energy difference between the P-type and N-type regions. The flow of minority electrons through the depletion layer is analogous to a waterfall, where the water current does not depend on the height of the fall but on the amount of water coming to the edge of the fall.

Being a device whose current does not depend on the voltage applied, the reverse-biased P–N junction exhibits the main characteristic of a current source, at least theoretically. Figure 6.1c and d illustrate the I–V characteristic and the symbol of the reverse-biased P–N junction used as a current source. Of course, a very important question here is whether this

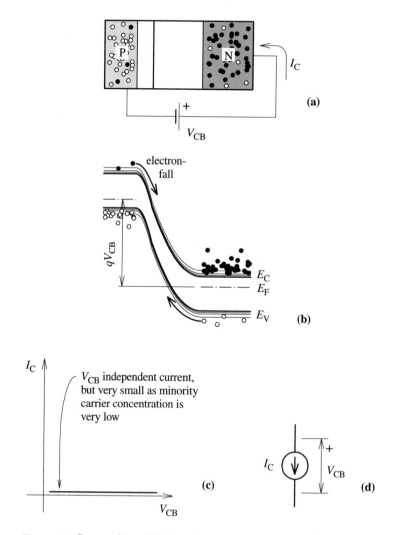

Figure 6.1 Reverse-biased P–N junction as a current source: (a) cross section, (b) energy-band diagram, (c) I–V characteristic, and (d) the current source symbol.

current source is at all useful. It can be argued that its current is too small for any realistic application.

It is true that the P–N junction reverse current is only a leakage current, usually negligible. However, what is important here is the principle of waterfall or "electron-fall" as labeled in Figure 6.1b. The current through the reverse-biased P–N junction can be increased to a significant level by providing more electrons, in the same way as the current of the waterfall increases after a heavy rainfall. In fact, it is necessary to have a way of controlling the number of electrons appearing at the edge of the "fall" so to create a *controlled* current source.

More minority electrons in the P-type region can be created by increased temperature or exposure to light, which would break additional covalent bonds and generate additional electron-hole pairs. This would make a temperature-controlled or light-controlled current source. However, to have an electronic amplifying device, we need a current source that is electrically controlled, say a voltage-controlled current source.

Thinking of a *supply of electrons* that is *controlled by a voltage*, the forward-biased P–N junction appears as a possibility. As the forward bias causes a significant number of electrons to move from the N-type region into the P-type region, the forward-biased P–N junction could be used to supply electrons to the current source. Obviously, this can work only if the two P–N junctions, forward biased (the controlling junction) and reverse biased (the current source), share a common P-type region. This is the case in an NPN BJT structure, illustrated in Fig. 6.2a. The common P-type region is called the *base*. The N type of the forward-biased P–N junction, which emits the electrons, is called the *emitter*, while the N type of the reverse-biased P–N junction is called the *collector*, as it collects the electrons.

Figure 6.2 summarizes the operation of the NPN BJT as a voltage-controlled current source.

> The forward-bias voltage V_{BE} (the input voltage) controls the supply of electrons from the emitter to the depletion layer of the reverse-biased P–N junction ("electron-fall"). The output current depends on the input voltage V_{BE} (shown by the transfer characteristic), but it does not depend on the output voltage V_{CB} (horizontal lines of the output current–voltage characteristics).

Sometimes the voltage between the collector and the emitter V_{CE} is considered as the output voltage. Practically, this does not change much, although V_{CE} involves V_{BE} voltage ($V_{CE} = V_{CB} + V_{BE}$), which is the changing input voltage. This is because the output voltage is at least several volts, which is a much larger value compared to the input voltage (≈ 0.7 V), let alone compared to the changes of the input voltage.

6.1.2 BJT Currents; α and β Current Gains

In the following text, a number of useful observations regarding the components of the principal BJT current as well as the parasitic currents are made, and a number of technological and electrical parameters are defined. The currents and effects discussed are illustrated in Fig. 6.3.

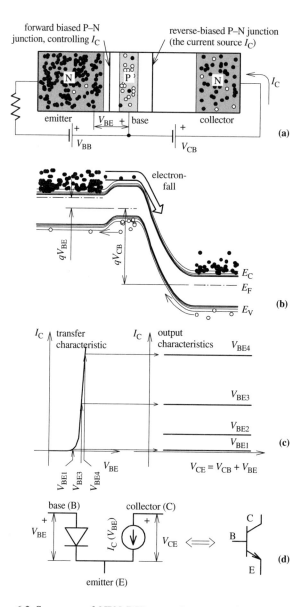

Figure 6.2 Summary of NPN BJT operation as a voltage-controlled current source: (a) cross section showing the three regions, their names, the two junctions, and the biasing arrangement, (b) the energy-band model, (c) the main current–voltage characteristics, and (d) the equivalent circuit (left) and the symbol (right).

1. As the emitter-to-base (E–B) junction is forward biased, electrons are emitted from the N-type emitter into the P-type base (I_{nE} current), but also holes are emitted from the base into the emitter (I_{pE} current). Thus, the total, or terminal emitter current can be expressed as

$$I_E = I_{nE} + I_{pE} \tag{6.1}$$

The useful transistor current is I_{nE}. The ratio between the useful and the total emitter current is called *emitter efficiency*, γ_E:

$$\gamma_E = \frac{I_{nE}}{I_E} \tag{6.2}$$

To maximize the emitter efficiency, I_{nE} (the useful current) should be as large as possible compared to I_{pE}, which should be as small as possible. Although both currents depend exponentially on the forward bias V_{BE}, they also depend on the majority-carrier concentrations, which is electrons in the emitter region and holes in the base region. To maximize I_{nE} and minimize I_{pE}, the doping level of the emitter is as high as possible, while the doping level of the base is as low as possible.

2. Most of the electrons emitted from the emitter pass through the base region to be collected by the reverse-biased P–N junction as collector current I_{nC}. However, some of the electrons are recombined by the holes in the P-type base, contributing to the base and not the collector current. The ratio of electrons successfully transported through the base region is called the *transport factor*:

$$\alpha_T = \frac{I_{nC}}{I_{nE}} \tag{6.3}$$

To maximize the transport factor, the recombination in the base has to be minimized, which is achieved by making the base region as thin as possible. This is well illustrated by a possible argument that two P–N junction diodes with connected anodes (P-type sides) electrically make the structure of the NPN BJT. The problem with such a BJT is that it is useless because of its zero transport factor: all the emitted electrons are recombined in the base, leaving any output (collector) current unrelated to the input current and voltage.

3. The collector current is not only due to the electrons arriving from the emitter. There is a small current due to the minority holes that move from the collector to the P-type base. This current, labeled I_{pC} in Fig. 6.3, is a part of the reverse-bias current of the collector–base junction. The other part is the current of minority electrons that would exist even when no electrons are emitted from the emitter (zero or reverse biased emitter–base junction). The reverse-bias current of the collector–base junction is usually labeled as I_{CB0}. It is a small leakage current, which can most frequently be neglected. It can be noticed only when the BJT is in the *off* mode (both emitter–base and collector–base junctions zero or reverse biased), and is therefore used to characterize the leakage of a BJT in the *off* mode. Neglecting I_{CB0} current, the terminal collector current can be expressed as

$$I_C \approx I_{nC} \tag{6.4}$$

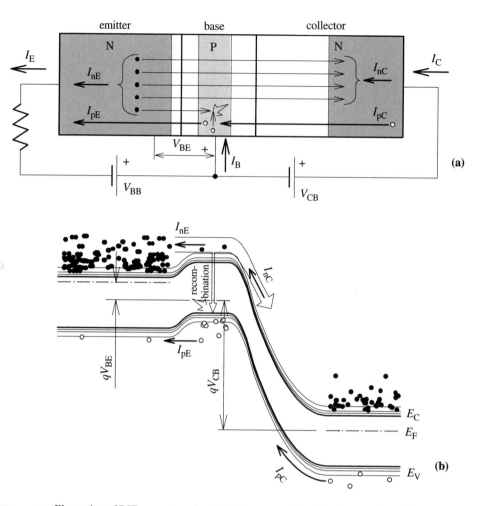

Figure 6.3 Illustration of BJT currents, using the BJT cross section (a) and energy-band diagram (b).

4. As the emitter current I_E depends exponentially on the input bias voltage V_{BE}, so do the currents I_{nE}, I_{nC}, and eventually the output current I_C. The relationship between the output current I_C and the input voltage V_{BE}, which is the transfer characteristic shown in Fig. 6.2c, determines the gain that can be achieved by the BJT. For a voltage-controlled current source, the gain is defined as a *transconductance*, and is expressed in A/V. Alternatively, the input-voltage-related current can be considered as the input quantity, in which case the gain is defined as a current gain. Although the collector is always the output of a BJT used as an amplifier, the input can be either the emitter (in which case the base terminal is common) or the base (in which case the emitter is the common terminal). These two configurations (common base and common emitter) lead to two possible current gain definitions:

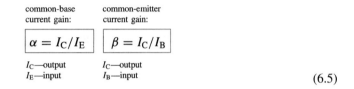

common-base
current gain:

common-emitter
current gain:

$$\alpha = I_C/I_E \qquad \beta = I_C/I_B$$

I_C—output
I_E—input

I_C—output
I_B—input

(6.5)

α and β are very frequently used BJT parameters.

5. As α and β represent current ratios of terminal currents, they can be electrically measured, because of which they are referred to as electrical parameters. The following equation shows that α is directly related to the emitter efficiency γ_E and the transport factor α_T, which are the technological BJT parameters:

$$\alpha = \frac{I_C}{I_E} \approx \frac{I_{nC}}{I_E} = \underbrace{\frac{I_{nC}}{I_{nE}}}_{\alpha_T} \underbrace{\frac{I_{nE}}{I_E}}_{\gamma_E} = \alpha_T \gamma_E \qquad (6.6)$$

The theoretical maximum for γ_E is 1 (no holes injected back into the emitter), and the theoretical maximum for α_T is 1 as well (no electrons recombined in the base). This means that the common-base current gain α cannot be larger than 1. Note that this does not mean the common-base configuration is useless; it cannot provide a real current gain (>1), but it can provide a power gain, for example.

6. α and β factors are related to each other. The following equations show the relationship:

$$\beta = \frac{I_C}{I_B} = \frac{I_C}{I_E - I_C} = \frac{1}{\underbrace{\frac{1}{I_C/I_E}}_{\alpha} - 1} = \frac{1}{\frac{1}{\alpha} - 1} = \frac{\alpha}{1 - \alpha} \qquad (6.7)$$

$$\alpha = \frac{\beta}{\beta + 1} \qquad (6.8)$$

Thus, if α is known, β can be calculated using Eq. (6.7), and vice versa, if β is known, α can be calculated using Eq. (6.8). Typically, $\alpha > 0.99$ (but always < 1), and $\beta > 100$.

■ **Example 6.1 BJT Currents**

The common emitter gain of a BJT operating as a voltage-controlled current source is $\beta = 450$. Calculate the base and the emitter current, if the collector current is 1 mA. What is the common-base current gain α?

Solution: The base current is calculated from the definition of β [Eq. (6.5)] as

$$I_B = \frac{I_C}{\beta} = 2.22 \ \mu A$$

The emitter current is obtained from the first Kirchoff law applied to the BJT:

$$I_E = I_C + I_B = (\beta + 1)I_B = 1.002 \text{ mA}$$

The common-base current gain is

$$\alpha = \frac{\beta}{\beta + 1} = 0.9978$$

6.1.3 BJT as a Switch: The Four Modes of Operation

The voltage-controlled current source is only one possible BJT mode of operation. It is typically referred to as the *active mode*. The BJT can also act as a voltage-controlled switch, operating in *cutoff* (the switch open) and *saturation mode* (the switch closed).

The circuit of Fig. 6.4 is used to explain the appearance of the different modes of operation. This circuit is the BJT implementation of the inverter with resistive load, discussed in Section 4.2.2, Fig. 4.6. The output characteristics of the BJT and the load line are shown in Fig. 6.5 to illustrate the graphic solution of the circuit. As explained in Section 4.1.3, the load line expresses the $I_C(V_{CE})$ relationship determined by the resistor R_C and the voltage source V_+. To derive the load-line equation, we can start from the observation that $V_+ = V_{R_C} + V_{CE}$. To involve the output current I_C, Ohm's law is applied to the resistor ($V_{R_C} = I_C R_C$), which leads to

$$V_{CE} = V_+ - I_C R_C \tag{6.9}$$

Expressing the current I_C, the load line is obtained as $I_C = -V_{CE}/R_C + V_+/R_C$. The slope of the load line is negative ($-1/R_C$), and it intersects the V_{CE} and I_C axes at V_+ and V_+/R_C, respectively. These facts are used to plot the load line on the I_C-vs-V_{CE} graph.

The BJT mode of operation is set by the biasing of the two P–N junctions, base–emitter and base–collector.

The input voltage V_{BE} directly biases the base–mitter junction of the BJT used in the circuit of Fig. 6.4. The base–collector junction, however, is biased through the resistor R_C.

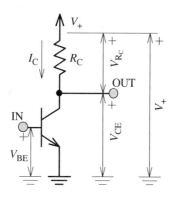

Figure 6.4 A circuit that can set the BJT in cutoff, active, and saturation modes, depending on the input voltage.

Noting that V_{CE} is given by Eq. (6.9), and applying Kirchoff's voltage loop to the BJT ($V_{BC} + V_{CE} - V_{BE} = 0$), the base–collector voltage is expressed as

$$V_{BC} = -V_+ + I_C R_C + V_{BE} \tag{6.10}$$

Cutoff

When the input voltage V_{BE} is zero (or at least smaller than the forward-bias voltage of the base–emitter P–N junction), the emitter emits no electrons, and consequently no electrons can be collected by the reverse-biased collector–base P–N junction. As the output current I_C is zero, the output of the BJT (collector to emitter) appears as an open circuit. V_{BC} voltage is negative ($-V_+ + V_{BE}$), showing that the base–collector junction is reverse biased.

> If none of the two junctions is forward biased, the BJT is in *cutoff mode*. All the terminal currents are zero (neglecting the leakage), and the output is an open circuit (a switch in the *off* mode).

Operating point (1) in Fig. 6.5 illustrates this case.

Normal Active Mode

This is the case described in the previous section: V_{BE} sets the base–emitter junction in forward-bias mode, and the emitter starts emitting electrons that are collected by the positively biased collector. The collector is positively biased with respect to the base when $V_{BC} < 0$. As Eq. (6.10) shows, this is the case when the following condition is satisfied: $I_C R_C + V_{BE} < V_+$.

> If the base–emitter junction is forward biased, and the base–collector junction is reverse biased, the BJT is in *normal active mode*. It performs the function of a voltage-controlled current source.

Operating points (2) and (3) in Fig. 6.5 illustrate this case.

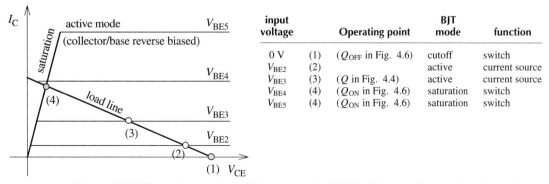

input voltage		Operating point	BJT mode	function
0 V	(1)	(Q_{OFF} in Fig. 4.6)	cutoff	switch
V_{BE2}	(2)		active	current source
V_{BE3}	(3)	(Q in Fig. 4.4)	active	current source
V_{BE4}	(4)	(Q_{ON} in Fig. 4.6)	saturation	switch
V_{BE5}	(4)	(Q_{ON} in Fig. 4.6)	saturation	switch

Figure 6.5 Different input voltages V_{BE} can set the BJT in either cutoff, normal active mode, or saturation mode.

Saturation

It may happen that the above condition is not satisfied, that is a large I_C current may lead to the situation in which $I_C R_C + V_{BE} > V_+$. V_{BC} is positive in this case, which means the collector is at a higher potential than the base, and it does not efficiently collect the electrons injected through the base. The collector current is reduced, and relationships (6.5) do not apply, as

$$I_C < \alpha I_E \qquad I_C < \beta I_B$$

$$I_C\text{—output} \qquad I_C\text{—output}$$

$$I_E\text{—input} \qquad I_B\text{—input} \qquad\qquad (6.11)$$

Note that the V_{BC} voltage cannot be larger than the forward bias voltage of the base–collector junction (≈ 0.7 V). This means the output voltage V_{CE}, which is given by

$$V_{CE} = -V_{BC} + V_{BE} \qquad\qquad (6.12)$$

cannot be smaller than zero. While V_{CE} is positive, it is close to zero, as $V_{BE} \approx 0.7$ V, and $0.7 \geq V_{CB} > 0$. A small V_{CE} voltage means that the I_C current in the circuit of Fig. 6.4 is determined by the voltage source V_+ and the resistor: $I_C = (V_+ - \text{small } V_{CE})/R_C \approx V_+/R_C$.

> The current I_C cannot be increased by the input voltage V_{BE} when the base–emitter is forward biased and the base–collector is not reverse biased. As the output current increase with the input voltage saturates, the BJT is said to be in *saturation mode*. Although the output current is significant, the output voltage V_{CE} is negligible, which is possible because the output resistance is very small (closed switch).

Operating point (4) in Fig. 6.5 illustrates the operation of the BJT in the saturation mode. Note that this operating point appears for the voltage V_{BE4} as well as the larger input voltage V_{BE5}.

A comparison of operating points (4) (saturation) and (1) (cutoff) with operating points Q_{ON} and Q_{OFF} of the inverter presented in Fig. 4.6 directly illustrates how the BJT can be used as a voltage-controlled switch.

The term "saturation" is not consistently used in the case of different types of transistors, namely the BJT and FETs (including the MOSFET). To avoid possible confusion, this fact should be noted and remembered. The BJT in saturation operates as a closed switch (digital device), while the MOSFET in saturation operates as a voltage-controlled current source (analog device). "Saturation" of the MOSFET output current means that it does not increase with the *output* (drain-to-source) voltage. In the case of the BJT, "saturation" means that the output current does not increase with the *input* (base-to-emitter) voltage.

Inverse Active Mode

As each of the two P–N junctions (base–emitter and base–collector) can be either forward or reverse biased, there are four bias possibilities for the BJT. This is illustrated in Fig 6.6.

Three of them are described in the previous text. The fourth bias arrangement is the case in which the base–emitter junction is reverse biased while the base–collector is forward biased. The collector region is the most negatively biased, and it emits electrons that are collected by the emitter as the most positively biased region.

> The mode of reverse-biased base–emitter and forward-biased base–collector junctions is equivalent to the BJT in active mode, with swapped emitter and collector. This mode is called the *inverse active mode*.

If the NPN structure was symmetrical, the inverse mode of operation would be as good as the normal active mode. In real BJTs, the doping level of the collector is the lowest, which means its efficiency (γ_E) is not good when used in the emitter role. Because of this, α and β values in the inverse active mode are small, and no good gain can be achieved in this mode of operation.

■ **Example 6.2 BJT Modes of Operation**

Determine the mode of operation of an NPN BJT with $\beta \approx 450$, if it is known that

(a) $V_{BE} = 0.7$ V, $V_{CE} = 5.2$ V.

(b) $V_{BE} = 0.7$ V, $V_{CE} = 0.2$ V.

(c) $V_{BE} = 0.8$ V, $V_{BC} = 0.3$ V.

(d) $V_{BE} = 0.8$ V, $V_{BC} = -0.7$ V.

(e) $V_{BE} = -0.8$ V, $V_{BC} = 0.7$ V.

(f) $V_{BE} = 0.1$ V, $V_{BC} = -10$ V.

(g) $I_C = 455$ mA, $I_B = 1$ mA.

(h) $I_C = 455$ mA, $I_E = 502$ mA.

Solution:

(a) $V_{BE} = 0.7$ V shows that the base–emitter junction is forward biased. To conclude the biasing of the base–collector junction, the V_{BC} voltage is needed. It is found as

$$V_{BC} = V_{BE} - V_{CE} = -4.5 \text{ V}$$

Negative base-to-collector voltage shows that this P–N junction is reverse biased. With this combination, the BJT is in normal active mode.

(b) In this case, the V_{BC} voltage is positive:

$$V_{BC} = V_{BE} - V_{CE} = 0.5 \text{ V}$$

which in combination with the forward-biased base–emitter junction ($V_{BE} = 0.7$ V) sets the BJT in saturation.

(c) Again, the forward-biased base–emitter junction ($V_{BE} = 0.8$ V) and positive base–collector voltage ($V_{BC} = 0.3$) bias the BJT in saturation mode.

(d) This time, the base–collector junction is reverse biased, which sets the BJT in normal active mode.

(e) This is the reverse situation: the base–emitter is reverse biased ($V_{BE} = -0.8$) while the base–collector is forward biased ($V_{BC} = 0.7$ V). The collector is emitting electrons while the emitter is collecting them. Therefore, the BJT is in inverse active mode.

(f) Theoretically, this could be considered as the normal active mode. However, as $V_{BE} = 0.1$ V is below the forward-bias level of the base–emitter junction, this BJT is practically in cutoff mode.

(g) In the normal active mode, the collector and the base currents are related through the gain factor β:

$$I_C = \beta I_B$$

This BJT satisfies this criterion.

(h) The base current in this case is

$$I_B = I_E - I_C = 47 \text{ mA}$$

and it is obvious that $I_C < \beta I_B$. This means the BJT is in saturation [refer to Eq. (6.11)].

6.1.4 Complementary BJT

Figure 6.7 shows the alternative possibility of making a BJT: the emitter and the collector are a P-type while the base is an N-type semiconductor. This type of transistor is referred to as a *PNP BJT*.

To set the PNP BJT in normal active mode, negative V_{BE} and positive V_{BC} voltages are needed, which is opposite to the case of the NPN BJT. The emitter region is at the highest potential, while the collector is at the lowest potential, which causes holes from the emitter to be emitted and collected by the collector. As the holes make the transistor current, as opposed to electrons in the case of the NPN BJT, the emitter and the collector current

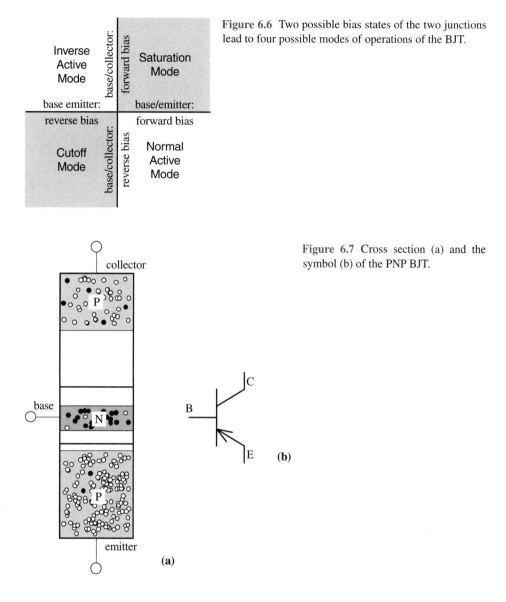

Figure 6.6 Two possible bias states of the two junctions lead to four possible modes of operations of the BJT.

Figure 6.7 Cross section (a) and the symbol (b) of the PNP BJT.

directions are opposite to those in the NPN BJT. The same applies to the base current. Appearing as a mirror image of the NPN, the PNP BJT complements the NPN in some circuit applications.

6.1.5 BJT vs MOSFET

Both the BJT and the earlier introduced MOSFET perform equivalent principal functions: (1) voltage-controlled current source and (2) voltage-controlled switch. Some similarities exist even in the principle of operation. This is perhaps best illustrated by the fact that the energy-band diagram along the channel of a MOSFET in saturation (bottom right diagram of Fig. 5.7a) is very similar to the energy-band diagram of a BJT (Fig. 6.2b). This certainly means that any electrical function implemented in MOSFETs technology can in principle be achieved by BJTs, and vice versa. Extremely important differences exist, however, in the performances and efficiencies achieved by the two possible technologies. At the surface, these may appear as simple quantitative differences, but in practice they appear as qualitative differences. Although it is theoretically possible to build a complex microprocessor in BJT technology, the dramatic developments in information technology would not have occurred with BJTs due to associated yield and power consumption problems. This example illustrates the importance of understanding the differences between the two devices.

The following description of BJT and MOSFET advantages highlights the differences.

BJT Advantages

1. The above-mentioned energy band similarity does not apply to the same areas in both devices. The BJT energy-band diagram of Fig. 6.2b applies to any (x, z) point, assuming the y-axis is in the direction of the electron flow (along the energy-band diagram). As the whole emitter cross-sectional area A_E is effective, a sizable device current can be achieved. In the case of the MOSFET, the energy-band diagram along the channel (bottom right diagram of Fig. 5.7a) applies only in the channel region. As the channel thickness (in the x-direction) is limited by the electric-field penetration into the semiconductor to a couple of nanometers, the channel cross-sectional area $x_{ch} W$ is severely restricted. The BJT structure is advantageous in terms of achieving large device currents, which is important in power applications, both linear and switching.

2. The diode (P–N junction) used as a controlling device in the BJT offers an advantage over the capacitor (MOS structure) used in the MOSFET in terms of the sensitivity of the output current to input voltage. This is the concept of *transconductance*, mathematically expressed as $g_m = dI_{output}/dV_{input}$. A small change in the input V_{BE} voltage, for example from 0.5 to 0.8 V, is sufficient to drive the output current from practically zero to the maximum level. To achieve this with a MOSFET, an input voltage change of at least several volts is necessary. Adding the better current capabilities of the BJT to these observations, the picture of a superior transconductance becomes clear. A higher transconductance of the device is not only related to higher gains of amplifiers, but also to shorter switching times, and superior noise characteristics of both linear and digital circuits.

MOSFET Advantages

1. Here is the other side of the coin: the capacitor (MOS structure) used as the controlling device in MOSFETs, as opposed to the diode (P–N junction) used in BJTs, offers unmatched advantages to the MOSFET:

- It enables the closed-switch mode to be maintained without any power consumption: no input current is needed to support the channel that creates the low-resistance path across the output. This adds to the fact that the other digital state, the open-switch mode, does not require any power consumption either. Using complementary MOSFETs, any logic function can be implemented, and it would not require any power consumption to maintain any logic state. This was explained for the case of an inverter in Sections 4.2.3 and 5.2.3. The BJT dissipates significant power when in saturation (closed-switch function), as this state can be maintained only by significant input and output currents. The problem with the power consumption is not only heat removal (big cooling elements, fans, etc. are needed), but also an extremely low limit in the number of logic cells that can be supported by the current that is possible to supply to a geometrically small IC.

- As the input of the MOSFET is a capacitor, it does not require a biasing resistor–capacitor circuit (refer to Section 4.1.2), which is necessary to limit the current through the input diode of the BJT. Both digital and analog functions can be implemented by circuits consisting of complementary MOSFETs only, with no resistors and capacitors. As large value resistors and capacitors require enormous areas, compared to transistors, this makes the MOSFET technology much more efficient in terms of area usage, again enabling much more complex circuits to be integrated.

2. The MOSFET is a single-carrier transistor (also referred to as a *unipolar transistor*): only electrons matter in N-channel MOSFETs and only holes matter in P-channel MOSFETs. As opposed to this, the holes do matter in the NPN BJT, even though the main transistor current is due to electron flow. The base current of the NPN BJT, which is the input current of the common-emitter transistor, is due to the holes. The fact that both types of carriers are active is reflected in the name of the device: *bipolar* junction transistor. The disadvantage of having both types of carriers in a single circuit (like the base–emitter circuit in BJTs) is the fact that the recombination process, which links the two currents, is relatively slow. This is best illustrated by the appearance of the stored charge capacitance, explained in Section 3.2.2. As the charge stored during the *on* period has to be removed by the recombination process before the diode (and therefore the BJT) turns *off*, the associated delay limits the maximum switching frequency to relatively low values.

Although no general rule can be established, it can be said that BJTs are more suitable for analog applications, especially when high output power is needed, while MOSFETs are much more suitable for digital circuits, especially in terms of achieving ICs able to perform extremely complex functions.

6.2 BIPOLAR IC TECHNOLOGIES

The diffusion layers of bipolar integrated circuits are designed so to optimize the characteristics of the NPN BJT. All other circuit components, including PNP BJTs, are made from the diffusion layers designed for the NPN BJT. In this section, the IC structure of the NPN BJT is described first, and after a brief description of the layer deposition and etching techniques, the implementation of the other circuit components is explained.

6.2.1 IC Structure of NPN BJT

Figure 6.8 illustrates the IC structure of NPN BJT. It is immediately obvious that the active part of the device (N^+PN layers highlighted by the zoom-in rectangle) occupies a small portion of the total cross-sectional area. A large part of the cross-sectional area is taken to satisfy the following two requirements: (1) electrical isolation from the other components of the IC and (2) enabling surface contacts to the three device terminals, base, emitter and collector.

The electrical isolation is provided by the reverse-biased P–N junction, indicated by the dashed line in Fig. 6.8. To create a P–N junction that encloses the device, the P-type substrate (the fourth layer in addition to the three device layers) is needed. Having the P-type substrate at the bottom, the P^+ ring is diffused around the device to cut the bottom N-type layer into so-called N-epitaxial (N-epi)islands. The isolation P^+ ring takes a lot of area, not only because it encircles the device, but also because of the lateral diffusion, which is about 80% of the vertical diffusion, and the vertical diffusion has to be sufficient to penetrate the whole depth of the N-epi layer.

To activate the P–N junction isolation, the P^+ isolation region has to be connected to the lowest potential in the circuit (V_-). This ensures that the isolation P–N junctions are reverse biased.

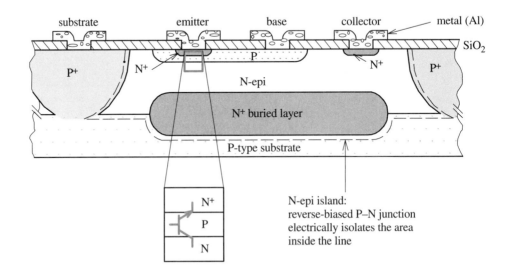

Figure 6.8 Integrated-circuit structure of the NPN BJT.

To provide surface contacts to the device, the P-type layer is extended beyond the N^+ area to contact the base, and the N-epi layer is even wider to provide room for the collector contact. As the aluminum and low-doped N-type silicon make rectifying Schottky contact rather than ohmic contact, the N^+ region is created at the collector contact. Note that this N^+ region is created by the same diffusion process used for the N^+ emitter layer.

The eye-catching N^+ buried layer may not be extending the lateral dimensions of the device, but it certainly makes the technology process more complex. However, if the buried layer was not introduced, the current collected by the active part of the collector (the N-epi region under the N^+ emitter) would have to face the resistance of the relatively long and low-doped N-epi layer before it reaches the collector contact. The N^+ buried layer provides a low-resistive section between the active part of the collector and the collector contact, significantly reducing the collector parasitic resistance. To further reduce this resistance, some bipolar technology processes introduce an additional diffusion process that provides a deep N^+ region connecting the collector contact and the N^+ buried layer. The following are disadvantages of this so-called deep collector diffusion: (1) increased complexity of the technology process and (2) increased lateral dimensions, especially due to the significant lateral diffusion of this deep layer.

6.2.2 Layer Deposition and Etching Techniques

It is obvious that the layers making a semiconductor device can be created in different ways: by doping (diffusion or ion implantation) to create the N-type and P-type silicon layers, by thermal oxidation to create silicon dioxide layers, and by deposition to provide, for example, the interconnecting aluminum layer. In MOSFET technologies, deposition processes are also used to create the insulating silicon dioxide layer, to provide the silicon nitride layer needed in the LOCOS process (Figs. 2.15, 5.13, and 5.16), and for the polysilicon layer creating MOSFET gates. However, it is the bipolar technology that heavily relies on the quality of a deposited layer, which is the so-called N-epi layer.

Although the effects of layer deposition are self-explanatory, the layer-deposition techniques vary from straightforward to very complex and critical ones. It is useful to know the basics of the layer deposition techniques, to be able to understand issues such as cost and quality of different layers used in semiconductor devices. The same holds for the opposite process to layer deposition, which is the layer etching, used in combination with photolithography (Section 1.3.2) to enable layer patterning.

The simplest layer-deposition technique is *evaporation*, frequently used in research laboratories to deposit aluminum. The aluminum has a relatively low melting point, and can easily be melted by passing a high electric current through a tungsten filament. When the aluminum is heated above its melting point, the evaporation of aluminum becomes significant, and the atoms that fall onto the substrate create a thin film. The process is conducted in high vacuum ($<10^{-5}$ torr), using high-purity aluminum pieces, to achieve a clean interface and homogeneity of the deposited aluminum. The needed high vacuum requires the use of two different vacuum pumps: roughing pump and diffusion pump. This is because the high vacuum cannot be achieved by the roughing pump, while the diffusion pump cannot be used before the pressure is reduced to a certain level.

The metal evaporation cannot be used to deposit metal alloys, as the metal components would generally have different melting points. To avoid some reliability problems, a small percentage of silicon has to be added to the aluminum used as metalization in integrated circuits. To deposit this kind of compound material, an alternative technique called *sputtering* is used. In the sputtering process, ions of argon or some other inert gas are created and accelerated to target material to "sputter" away atoms or clusters of atoms, which then may fall onto the substrate surface. Metals can be deposited by *DC sputtering systems* in which the target is biased at a negative potential of several hundred volts, attracting the positive argon ions that hit the target. Dielectric and semiconductor materials, such as silicon dioxide and silicon, can also be deposited by sputtering, however *RF sputtering systems* have to be used in that case.

A better quality of the deposited dielectrics and semiconductors is achieved when the molecules or atoms of these materials are chemically created inside the processing chamber and then fall down to coat the substrate. This process, known as *chemical vapor deposition (CVD)*, relies on an appropriate chemical reaction that is initiated by providing the reacting gases and energy. In standard CVD, the energy is supplied in the form of heat, although the reaction energy may also be supplied by plasma or optical excitation, in which case the processes are referred to as *plasma* and *rapid-thermal processing*, respectively. Examples of chemical reactions used to create silicon, silicon dioxide, and silicon nitride are given below:

$$SiH_4 \rightarrow Si + 2H_2 \qquad (6.13)$$

$$SiH_4 + O_2 \rightarrow SiO_2 + 2H_2 \qquad (6.14)$$

$$3SiH_4 + 4NH_3 \rightarrow Si_3N_4 + 12H_2 \qquad (6.15)$$

The CVD processes are carried out in either atmospheric or low-pressure ambient. In practice, different CVD processes are frequently given different names to highlight the specific application. In particular, the CVD processes used to deposit monocrystalline silicon are referred to as *epitaxial* processes. This is the case for the process used to deposit the low-doped N-epi layer onto the P-type substrate in Fig. 6.8. The deposition of monocrystalline layers requires not only monocrystalline substrates, but also extreme processing conditions, such as very high temperature in conventional epitaxial systems, or ultrahigh-vacuum in the case of *molecular beam epitaxy*.

Etching processes complement the layer-deposition techniques to enable the creation of IC patterns. As explained earlier, the surface selectivity of the etching is achieved by photoresist protection of the areas that are not to be etched. Obviously, the etching process should remove the exposed layer while not affecting the photoresist and the layer underneath. Etching techniques can be divided into two categories, *wet etching* and *dry etching*. The wet etching, consisting of immersion of the wafers into acid baths, is simple and offers inherent vertical selectivity. For example, to etch a layer of SiO_2, HF acid is used, which will dissolve the SiO_2 but will not touch the underlying silicon. The disadvantage of wet etching is the inevitable lateral etch, which means that the photoresist patterns are transferred onto the oxide with a degree of distortion. The dry etching is plasma based, and can be designed to provide anisotropic etching. This not only reduces the distortion, but also enables etching structures with high aspect ratios. The disadvantage of the dry etching is nonexistence of the vertical selectivity, in addition to complex and expensive processing equipment. In the case of SiO_2 etching, the dry etching would not stop after the SiO_2 layer is removed, but would continue into the underlying silicon. Consequently, complex end-point detection systems have to be used to control the dry etching process.

6.2.3 Standard Bipolar Technology Process

In this section, the process sequence used to fabricate the standard NPN structure of Fig. 6.8 is described. Practically, this is the description of the *standard bipolar technology process*, as all other circuit components (resistors, capacitors, diodes, PNP BJTs) are implemented with the layers existing in the standard NPN structure.

The processing sequence begins with thermal oxidation of the P-type silicon substrate (wafers) and subsequent oxide patterning by the first photolithography process to create windows for high-concentration N-type diffusion (N^+). The created N^+ diffusion layer will become the N^+ buried layer after the epitaxial deposition of the low-doped N-type silicon layer (Fig. 6.9b). The epitaxial growth of the low-doped N-type silicon is necessary not only to create the buried layer, but also to enable the creation of the four-layer structure: P-substrate–N-collector–P-base–N^+-emitter. If these four layers were to be created by diffusing one layer into another three times (N-collector into P-substrate, P-base into N-collector, and N^+-emitter into P-base), a hardly achievable and inconveniently low concentration of the initial P-type substrate would be necessary. With the epitaxial process, the concentration of the N-epitaxial (N-epi) layer is independent of the doping level of the underlying P-type substrate, and can be set at an appropriately low level.

After the epitaxial process, P^+ diffusion is employed to define the electrically isolated N-epi islands that carry the individual circuit components. The photolithography mask used for this process, shown in Fig. 6.9c for positive photoresist, illustrates the top view of an N-epi island surrounded by the P^+ isolation diffusion. Vertically, the N-epi islands are still not disconnected at this stage, as the P^+ diffusion still does not reach the underlying P-type substrate as in the final structure shown previously in Fig. 6.8. The reason for doing this diffusion "partway" at this stage is the fact that more diffusion processes follow, which will cause simultaneous diffusion of the doping atoms in the P^+ isolation areas. It is important to limit the overall P^+ diffusion to the level needed to reach the underlying P-type substrate, because any excessive lateral diffusion would mean wasting a significant surface area by the wider P^+ isolation areas, which dominate the surface area of the IC even without any waste.

The definition of the N-epi islands by the P^+ isolation diffusion is followed by the P-type and N^+-type diffusions needed for the P-type base and N^+-type emitter. Of course, the diffusion areas are defined by the associated photolithography processes, labeled as photolithography III and photolithography IV in Fig. 6.9d and e, respectively. Note that photolithography IV opens two windows for the N^+ diffusion: one for the emitter of the NPN BJT and the other for the collector contact.

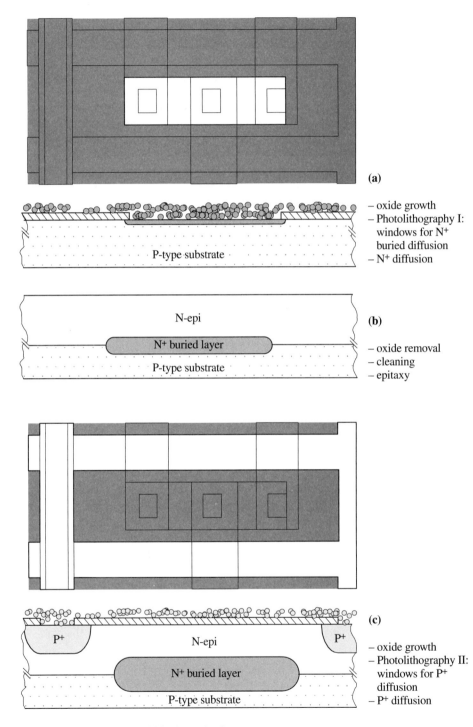

(a)

– oxide growth
– Photolithography I: windows for N⁺ buried diffusion
– N⁺ diffusion

(b)

– oxide removal
– cleaning
– epitaxy

(c)

– oxide growth
– Photolithography II: windows for P⁺ diffusion
– P⁺ diffusion

Figure 6.9 (a–g) The standard bipolar technology process.

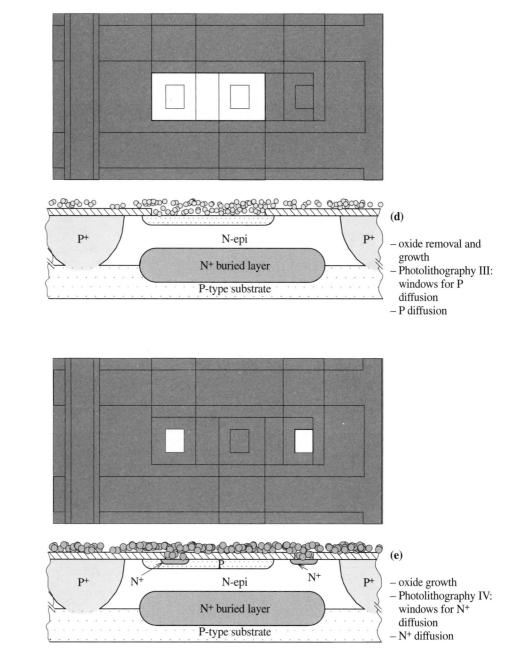

(d)

– oxide removal and growth
– Photolithography III: windows for P diffusion
– P diffusion

(e)

– oxide growth
– Photolithography IV: windows for N+ diffusion
– N+ diffusion

Figure 6.9 Continued

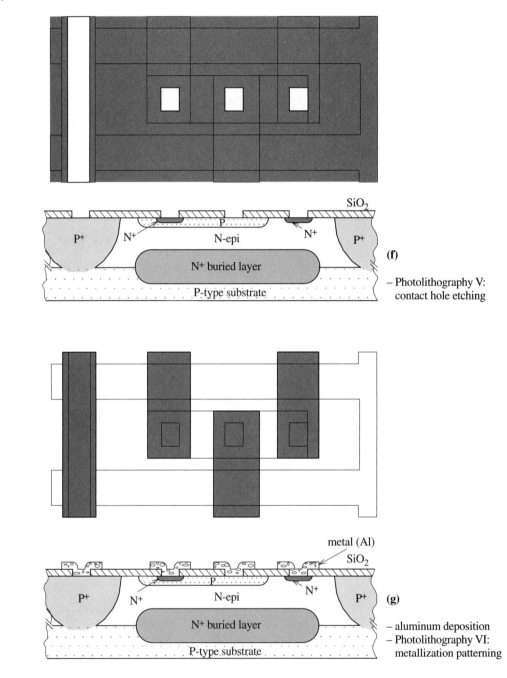

(f)
– Photolithography V:
contact hole etching

(g)
– aluminum deposition
– Photolithography VI:
metallization patterning

Figure 6.9 Continued

At this stage, the NPN BJT is created in the silicon, and what remains is the photolithography V to etch the contact holes through the insulating oxide layer (Fig. 6.9f), then to deposit the aluminum layer and pattern the metalization by the sixth photolithography process (Fig. 6.9g).

Of course, as in the case of MOSFET technologies, a passivation glass is deposited and windows etched over the bonding pads using yet another photolithography process. This part of the technology process is not illustrated in Fig. 6.9.

6.2.4 Implementation of PNP BJTs, Resistors, Capacitors, and Diodes

As already mentioned, the semiconductor layers of the bipolar ICs are designed to maximize the performance of the main component, the NPN BJT. These layers are labeled according to the function they perform in the NPN BJT, as Table 6.1 shows. Theoretically, it is possible to introduce additional layers to optimize or improve the characteristics of the other circuit components, however, this would mean a practically unjustifiable increase in the complexity of the process. Consequently, all other circuit components are designed from the four existing layers, listed in Table 6.1. The text of this section describes the ways these components are implemented in the standard bipolar ICs.

Substrate PNP BJT

It is possible to make a PNP BJT using both existing P-type layers listed in Table 6.1, which are the P-type substrate/P^+ isolation diffusion and the P-type base layers, and the N-epi layer between. This configuration of the PNP BJT, called *substrate PNP*, is illustrated in Fig. 6.10.

Electrically, the P-type substrate/P^+ isolation is a single region, connected to the most negative potential in the circuit (V_-) to ensure the isolation of the N-epi islands.

> The collector of the substrate PNP cannot be connected to an arbitrary point in the circuit— it is automatically connected to V_-.

This severely limits the application of the substrate PNP to the particular case in which the circuit needs a PNP BJT with the collector connected to V_-. As the substrate PNP provides better characteristics than any other PNP in ICs, it should be used as the first option whenever the collector biasing is appropriate.

TABLE 6.1 The Semiconductor Layers of Standard Bipolar ICs

Layer No.	Layer Name	Typical Characteristics
1	P-type substrate/P^+ isolation diffusion	
2	N-epi	10^{15} cm^{-3}, 10 μm
3	P-base diffusion	200 $\Omega/\square$, 2–4 μm
4	N^+-emitter diffusion	5 $\Omega/\square$, 1–2 μm

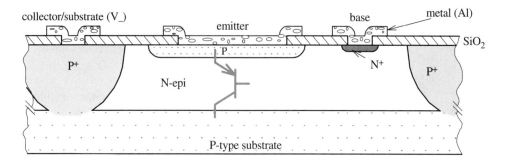

Figure 6.10 Substrate PNP BJT: the collector is connected to the most negative potential in the circuit.

Lateral PNP BJT

Obviously, the P-type substrate/P$^+$ isolation diffusion layer cannot be used, if the application limitation regarding the collector biasing is to be removed. As the only remaining P-type layer is the P-type base diffusion, this layer has to be used for both the emitter and the collector. This is possible by the lateral structure illustrated in Fig. 6.11.

As the collector of this BJT collects only the small part of laterally emitted holes, the transport factor and consequently the current gains α and β are very small. To minimize the waste of emitted holes, the collector P-type region normally surrounds the central emitter P-type region (closed geometry). This is why two P-type collector regions appear on the cross section of Fig. 6.11.

> The characteristics of the lateral PNP BJT are significantly inferior compared to the NPN BJT in ICs, and it should be used only when better circuit performance justifies its use.

Resistors

Unlike the CMOS ICs, which use only complementary pairs of MOSFETs (an example is the inverter of Fig. 4.8), the bipolar ICs generally use resistors and capacitors, such as

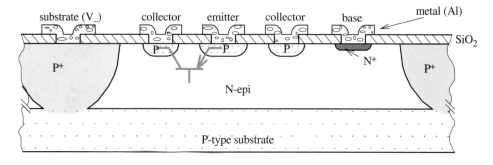

Figure 6.11 Lateral PNP BJT.

the voltage amplifier of Fig. 4.4, and very frequently diodes as well. The general structure of an IC resistor is shown in Fig. 1.2. The resistive body is made of one of the available layers (N-epi, base diffusion, or emitter diffusion), and is isolated from the other resistors and the rest of the IC by a reverse-biased P–N junction. Figure 6.12 illustrates the case of a base-diffusion resistor. The body of the resistor is the P-type base-diffusion layer, which is surrounded by the N-epi layer connected to the most positive potential in the circuit (V_+) to ensure that the resistor current is confined within the P-type region. Connecting the surrounding N-epi layer to V_+ also enables all the base-diffusion resistors to be placed into a single N-epi island, which is a much better solution than using a separate N-epi for every single resistor. The benefits of this so-called layer merging are considered in more detail in the next section. The resistance of the resistor is determined by the surface dimensions L and W, and by the sheet resistance of the layer R_S, as explained in Section 1.1.2.

Resistors can analogously be created using the N^+ diffusion layer and the N-epi, noting that an N-epi island can accommodate only a single N-epi resistor. The N^+ layer offers the smallest sheet resistance, and is therefore the most suitable for small-value resistors.

Large-value resistors are hard to implement in bipolar ICs. The length of a 100-kΩ base-diffusion resistor can exceed the length of the whole IC chip many times. A solution for large-value resistors is to use "snake geometry." Nonetheless, a couple of resistors such as this may take as much as half of the IC active area. This constraint can be relaxed to a certain extent by using "pinch" resistors.

Figure 6.13 illustrates the cross section of a base-diffusion pinch resistor. The idea is not to eliminate the emitter N^+-diffusion layer, which normally comes into the base-diffusion layer, but to use it to reduce the electrically active cross-sectional area of the resistor. If appropriate voltage is applied to the N^+-emitter layer, which means more positive than the terminal voltages V_A and V_B, the N^+-emitter–P-base junction is reverse biased, and the resistor current is forced to flow through the reduced cross-sectional area of the remaining P-type body. In addition, the resistance can be changed to some extent by the voltage applied to the N^+ layer, as a larger reverse-bias voltage will produce a wider depletion layer, reducing the effective cross-section of the resistor.

Pinch resistors can also be created with the N-epi as the resistor body, and the P-type base diffusion layer used to reduce the resistor cross section.

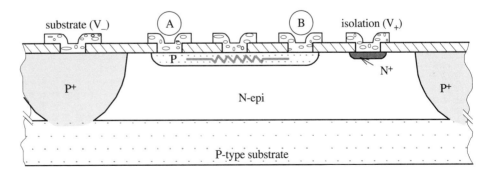

Figure 6.12 The cross section of the base-diffusion resistor.

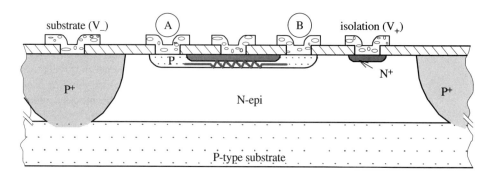

Figure 6.13 Base-diffusion pinch resistor; the emitter region reduces the electrically active cross-sectional area of the resistor.

Capacitors and Diodes

P–N junctions are used for the IC capacitors and diodes. There are three P–N junctions in the four-layer structure of the standard NPN BJT: P-substrate–N-epi, N-epi–P-base, and P-base–N^+-emitter. The application of the P-substrate–N-epi junction is limited by the fact that the P-substrate is connected to the most negative potential (V_-). The P-base–N^+-emitter junction has low breakdown voltage, about 6–7 V, because of the heavy doping in the emitter region. This fact makes it useful as a reference diode. Obviously, there is no choice of reference diodes in the standard bipolar technology, and the reference voltages have to be obtained from reverse-biased base–emitter junctions (6–7 V) and forward-biased P–N junctions (≈ 0.7 V). When a diode with breakdown voltage larger than 6–7 V is needed, the base–collector P–N junction has to be used.

▨ **Example 6.3 PNP BJT in Standard IC Technology**

Draw the cross section of the class B output stage amplifier of Fig. 6.14, implemented in the standard bipolar IC technology.

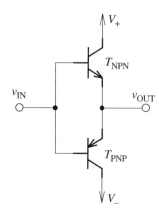

Figure 6.14 The electrical diagram of a class B output stage amplifier.

Solution: T_{NPN} transistor is a standard NPN BJT, the cross section of which is shown in Fig. 6.8. The transistor T_{PNP} is a PNP BJT, and there are two options in the standard bipolar technology: substrate PNP (Fig. 6.10), and lateral PNP (Fig. 6.11). The characteristics of the substrate PNP are superior, however, its collector has to be connected to the lowest potential in the circuit (V_-). This is the case in the circuit of Fig. 6.14, so the choice is the substrate PNP. The cross section of the whole circuit is given in Fig. 6.15.

6.2.5 Layer Merging

Miniaturization of integrated circuits, within the technology limits, generally improves the IC performance:

- Parasitic components, like parasitic resistances and capacitances, are reduced as the length of the conductive lines and the area of the P–N junctions are reduced. This is helpful in terms of different types of limitations, and in particular it improves the circuit response time due to reduced RC time constants.

- Smaller devices mean a smaller active area of the integrated circuit, and therefore improvement in the manufacturing yield. To understand this effect, think of a defect appearing randomly in the silicon crystal—the chance of this defect appearing in the P–N junction region and causing leakage current is higher if the P–N junction area is larger.

- Smaller devices mean that circuits with a larger number of devices can be integrated, while avoiding the zero-yield situation, if not even maintaining the yield at the same level.

Miniaturization has proved a powerful tool for improving the performance and applicability of electronic systems.

Layer merging is a way of reducing the integrated-circuit size by eliminating nonfunctional structures from the IC chip.

The circuit of Fig. 6.16 will help explain the layer-merging principle. Let us concentrate on the transistors T_1 and T_3. These are standard NPN BJTs, and replicating the cross section

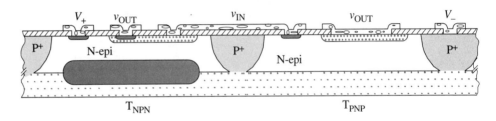

Figure 6.15 The cross section of a class B output stage amplifier.

of Fig. 6.8 twice, the cross section of these two transistors is obtained as in Fig. 6.17a. What happens here is that the central P^+ diffusion region, which takes a lot of space to isolate the two transistors, is electrically short-circuited by the metal line running over the top of the isolation region. What is the point of isolating the two N-epi islands if they have to be short-circuited because the collectors of the two transistors are short-circuited? Straightforward layout design of the IC, easier design automation, and clearer IC layout and cross section could be valid arguments, but they do not stand against the benefits of IC miniaturization, which is obviously helped by merging the two N-epi islands into one.

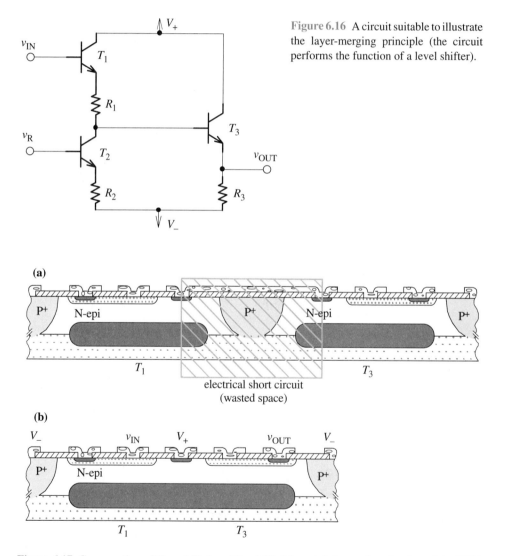

Figure 6.16 A circuit suitable to illustrate the layer-merging principle (the circuit performs the function of a level shifter).

Figure 6.17 Cross section of T_1 and T_3 from Fig. 6.16: (a) straightforward replication of the NPN BJT cross section (no layer merging) leads to an ineffective region marked by the shaded rectangle; (b) after the layer merging, the two transistors are placed into a single N-epi island.

■ **Example 6.4 IC Layer Merging**

Figure 6.18 shows the electrical diagram of a clocked set–reset flip-flop realized in the standard bipolar IC technology, the resistors being implemented as base–diffusion resistors.

(a) Draw the cross section of transistor T_1.

(b) Draw the cross section of transistors T_2 and T_3.

(c) Group the devices into N-epi islands.

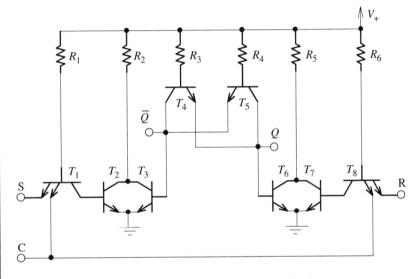

Figure 6.18 The electrical diagram of a clocked set–reset flip-flop

Solution:

(a) We can think of transistor T_1 as two transistors with short-circuited collector and base terminals. Although this is the electrical equivalent of T_1, the layer-merging principle would dictate merging the epi-layers first (to short circuit the collectors), and then merging the P-type base regions (to short circuit the bases). What is left are two N^+-emitter regions appearing in a single P-base and a single N-epi, as illustrated in Fig. 6.19.

(b) The situation with transistors T_2 and T_3 is analogous to transistors T_1 and T_3 in the circuit of Fig. 6.16. The fact that, in this case, the emitters are short circuited is irrelevant from the layer-merging point of view. The bases of T_2 and T_3 are separate, the P-type base regions cannot be merged, and therefore the two emitters have to appear in two different P-type regions, regardless of the fact that they will have to be electrically short-circuited. Therefore, the cross section is analogous to the one given in Fig. 6.17b.

(c) All the resistors can share a single N-epi, which is connected to V_+ to ensure that the P-type region of every single resistor creates a reverse-biased P–N junction with the common N-epi, isolating them electrically from each other. Therefore, the circuit elements can be grouped as follows: N-epi 1 - $R_1, \ldots, R_6$; N-epi 2 - T_1; N-epi 3 - T_2 and T_3; N-epi 4 - T_4; N-epi 5 - T_5; N-epi 6 - T_6 and T_7; and N-epi 7 - T_8.

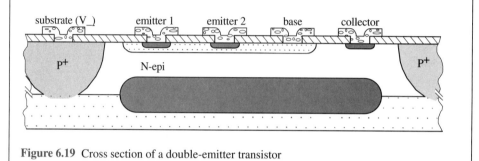

Figure 6.19 Cross section of a double-emitter transistor

◼ Example 6.5 Merging of PNP BJTs

Figure 6.20 shows the electrical diagram of a differential amplifier. Assuming that the resistors are realized as base-diffusion resistors, group the circuit elements into N-epi islands.

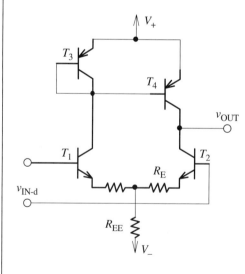

Figure 6.20 The electrical diagram of a differential amplifier.

Solution: This circuit contains PNP BJTs (T_3 and T_4), with separate collectors but a common base. As the N-epi layer plays the role of the base in the case of PNP BJT, the two PNP BJTs (T_3 and T_4) can be placed into a single N-epi that will provide the common base. In general the merging consideration should start with the N-epi layers, but it is not standard to first check if there are BJTs with short-circuited collectors. Furthermore, the bases of T_3 and T_4 are connected to the collector of T_1, which means T_1 can be placed into the same N-epi island.

A further consideration may be given to the fact that the N-epi containing T_1, T_3, and T_4 is at a higher potential than any of the resistor terminals. This would suggest that the resistors can be placed into the N-epi of T_3 and T_4. Although this appears as the optimum solution for the circuit of Fig. 6.20 considered in isolation, this may not be the case when a more complex circuit is considered, where the differential amplifier of Fig. 6.20 is only a small part.

Finally, T_2 needs a separate N-epi island.

6.3 BJT MODELING

The Ebers–Moll model of the BJT provides the principal current–voltage equations. The most important second-order effect (known as Early effect) is added to the basic I–V equations to create what is known as an Ebers–Moll level in SPICE. When the second-order effects related to low- and high-current levels are included, a so-called Gummel–Poon level in SPICE is involved. This section introduces SPICE equations describing the principal and the second-order effects at both levels. It also describes a SPICE large-signal equivalent circuit that includes the parasitic components needed to simulate dynamic BJT response, and makes a direct relationship to the small-signal equivalent circuit most frequently used in circuit-design and analysis books.

6.3.1 Principal Ebers–Moll Model

Although the SPICE equivalent circuit is closely related to the most frequently used small-signal equivalent circuit, it is not as directly related to the physically based version of the Ebers–Moll model, referred to here as the injection version. This section introduces the physically based Ebers–Moll equivalent circuit of the BJT (the injection version), and then transforms it through what is called transport version into the final SPICE version. In this way, the SPICE parameters are related to the physical effects in the BJT.

Injection Version

In the normal active mode, the base–collector junction of the BJT plays the role of a current source controlled by a voltage, or equivalently by the corresponding current. The controlling junction, the base–emitter in the case of normal active mode, can be modeled by a diode. Therefore, the diode and the current source of the upper branch of the circuit given in Fig. 6.21b ($V_{BE} > 0$ and $V_{BC} < 0$) make a proper equivalent circuit of the BJT in the normal active mode. The current of the controlling junction (I_F) is directly related to the current injected into the current-source junction. The current of the current source is labeled $\alpha_F I_F$ to express this fact, where $\alpha_F < 1$ due to the carrier losses related

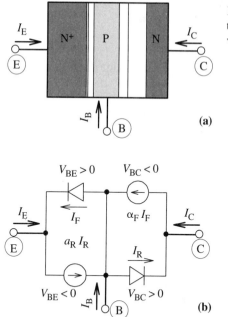

Figure 6.21 SPICE definition of BJT current directions (a) and static BJT equivalent circuit—injection version (b).

to the nonideal emitter efficiency and nonideal transport factor. The current of the controlling junction I_F, of course, depends on the voltage applied to the base–emitter P–N junction V_{BE}:

$$I_F = I_{ES}(e^{V_{BE}/V_t} - 1) \tag{6.16}$$

where I_{ES} is the saturation current of the base–emitter junction.

The BJT model should include all the possible bias arrangements, not only the normal active mode. In the inverse active mode, the roles of the P–N junctions are swapped, and the BJT can be modeled by a circuit that is a mirror image of the circuit modeling the BJT in the normal active mode. The current of the current source is analogously labeled by $\alpha_R I_R$, where the controlling current I_R depends on the base–collector voltage:

$$I_R = I_{CS}(e^{V_{BC}/V_t} - 1) \tag{6.17}$$

I_{CS} being the base–collector saturation current.

Adding the inverse-mode circuit to the normal-mode one, as shown in Fig. 6.21b, does not introduce any adverse effects. If the BJT is in normal active mode, $V_{BC} < 0$ and the current $I_R \approx -I_{CS} \ll \alpha_F I_F \approx I_{ES}\alpha_F \exp(V_{BE}/V_t)$. In fact, the corresponding equation for the collector terminal current, $I_C \approx I_{ES}\alpha_F \exp(V_{BE}/V_t) + I_{CS}$, properly includes the reverse-bias (leakage) current of the base–collector junction.

Moreover, the circuit of Fig. 6.21b automatically includes the two remaining biasing possibilities, the saturation and cutoff modes. In saturation, both V_{BE} and V_{BC} voltages are positive, and both I_F and I_R currents are significant. In the typical case of $V_{BC} < V_{BE}$, the terminal collector current retains the direction as in the case of the normal active mode, however the current intensity is reduced as $I_C = \alpha I_F - I_R$. The voltage drop between the collector and emitter is very small, $V_{CE} = -V_{BC} + V_{BE}$. An increase in V_{BC} causes a further reduction in V_{CE} voltage and I_C current, according to the $I_C - V_{CE}$ characteristic in saturation (Fig. 6.5).

In cutoff, both P–N junctions are reverse biased (V_{BE} and V_{BC} negative), allowing only the flow of the leakage currents I_{ES} and I_{CS}.

Adding the currents from the two branches of the equivalent circuit, the terminal collector and emitter currents can be expressed as

$$I_C = \alpha_F I_{ES}(e^{V_{BE}/V_t} - 1) - I_{CS}(e^{V_{BC}/V_t} - 1) \tag{6.18}$$

$$I_E = -I_{ES}(e^{V_{BE}/V_t} - 1) + \alpha_R I_{CS}(e^{V_{BC}/V_t} - 1) \tag{6.19}$$

while the base terminal current is the balance between the emitter and the collector current:

$$I_B = -I_C - I_E \tag{6.20}$$

This set of equations, known as the Ebers–Moll model, relates all the three terminal currents to the two terminal voltages (V_{BE} and V_{BC}) through the following four parameters: α_F, the common-base current gain of a BJT in the normal active mode; α_R, the common-base current gain of a BJT in the inverse active mode; I_{ES}, the emitter–base saturation current; and I_{CS}, the collector–base saturation current.

Transport Version

The equivalent circuit of the transport version is the same as the injection version of the Ebers–Moll model. The difference is in the way the internal currents are expressed, which are now based on the actual current-source currents, labeled I_{EC} and I_{CC} in Fig. 6.22, rather than the currents injected by the P–N junctions as in Fig. 6.21b. Of course, the relationships between the currents of the controlling P–N junctions and the actual currents of the current sources have to be retained in order to correctly model the BJT. Consequently, the P–N junction currents in Fig. 6.22 cannot be considered as independent, but have to be related to I_{EC} and I_{CC} through the corresponding common-base current gains α_F and α_R.

The transport version relates the I_{EC} and I_{CC} currents to the terminal voltages in the following way:

$$I_{CC} = I_S(e^{V_{BE}/V_t} - 1) \tag{6.21}$$

$$I_{EC} = I_S(e^{V_{BC}/V_t} - 1) \tag{6.22}$$

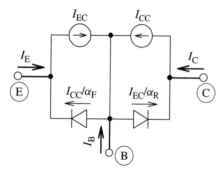

Figure 6.22 Static BJT equivalent circuit—transport version.

Therefore, the terminal currents are given as

$$I_C = I_S(e^{V_{BE}/V_t} - 1) - \frac{I_S}{\alpha_R}(e^{V_{BC}/V_t} - 1) \tag{6.23}$$

$$I_E = -\frac{I_S}{\alpha_F}(e^{V_{BE}/V_t} - 1) + I_S(e^{V_{BC}/V_t} - 1) \tag{6.24}$$

$$I_B = -I_C - I_E \tag{6.25}$$

Comparing the two models, it is obvious that the single saturation current I_S used in these equations is equivalent to neither the base–emitter saturation current I_{ES} nor the base–collector saturation current I_{CS}. The collector current Eqs. (6.23) and (6.18) become equivalent under the following conditions:

$$I_S = \alpha_F I_{ES} \tag{6.26}$$

$$I_S = \alpha_R I_{CS} \tag{6.27}$$

The same conditions also lead to the equivalence of the emitter current Eqs. (6.24) and (6.19). The base–emitter and base–collector saturation currents in real BJTs are different, due to different areas and different doping levels. Obviously, the single I_S current cannot realistically represent both saturation currents at the same time. For realistic simulations, the parameter I_S should be related to the base–emitter junction ($I_S = \alpha_F I_{ES}$) in the case of the normal active mode and to the base–collector junction ($I_S = \alpha_R I_{CS}$) in the case of the inverse active mode.

It appears the choice of a single I_S parameter, instead of two parameters representing the two P–N junctions, complicates parameter measurement and reduces the generality of the model. However, it enables the more general and physically based equivalent circuit of Fig. 6.21b to be related to the equivalent circuit most frequently used in circuit-design and analysis books. This circuit is discussed in the following text.

SPICE Version

A single current source is used to model the BJT in circuit-design and analysis books. The two current sources of Fig. 6.22 can be reduced to one, while effectively maintaining the same relationships between the terminal currents (I_C, I_E, and I_B) and the terminal voltages (V_{BE} and V_{BC}).

The circuit with a single current source is shown in Fig. 6.23. The currents $I_{?1}$ and $I_{?2}$ can be determined so that the terminal I_C and I_E currents are equivalent to the ones in Fig. 6.22:

$$I_C = \underbrace{I_{CC} - \frac{I_{EC}}{\alpha_R}}_{\text{Fig. 6.22}} = \underbrace{I_{CC} - I_{EC} - I_{?1}}_{\text{Fig. 6.23}} \Rightarrow$$

$$I_{?1} = I_{EC}\left(\frac{1}{\alpha_R} - 1\right) = I_{EC}\frac{1 - \alpha_R}{\alpha_R}, \quad I_{?1} = \frac{I_{EC}}{\beta_R} \tag{6.28}$$

$$I_E = \underbrace{-\frac{I_{CC}}{\alpha_F} + I_{EC}}_{\text{Fig. 6.22}} = \underbrace{-I_{CC} + I_{EC} - I_{?2}}_{\text{Fig. 6.23}} \Rightarrow$$

$$I_{?2} = I_{CC}\left(\frac{1}{\alpha_F} - 1\right) = I_{CC}\frac{1 - \alpha_F}{\alpha_F}, \quad I_{?1} = \frac{I_{EC}}{\beta_F} \tag{6.29}$$

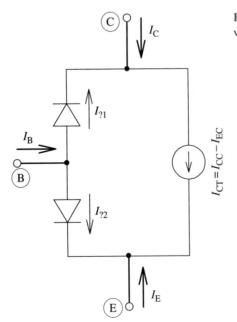

Figure 6.23 Static BJT equivalent circuit—SPICE version.

According to Eq. (6.7), β_F and β_R are common-base current gains of the BJT in the normal active and reverse active modes. With these values of $I_{?1}$ and $I_{?2}$, the terminal currents can be expressed as

$$I_C = I_{CC} - I_{EC} - \frac{I_{EC}}{\beta_R}$$

$$I_E = -I_{CC} + I_{EC} - \frac{I_{CC}}{\beta_F} \tag{6.30}$$

$$I_B = -I_C - I_E$$

Replacing I_{CC} and I_{EC} from Eqs. (6.21) and (6.22), the terminal currents are related to the terminal voltages:

$$
\begin{aligned}
I_C &= I_S(e^{V_{BE}/V_t} - 1) - \left(1 + \frac{1}{\beta_R}\right) I_S(e^{V_{BC}/V_t} - 1) \\
I_E &= -\left(1 + \frac{1}{\beta_F}\right) I_S(e^{V_{BE}/V_t} - 1) + I_S(e^{V_{BC}/V_t} - 1) \\
I_B &= \frac{1}{\beta_F} I_S(e^{V_{BE}/V_t} - 1) + \frac{1}{\beta_R} I_S(e^{V_{BC}/V_t} - 1)
\end{aligned}
\tag{6.31}
$$

These are the final and general equations of the Ebers–Moll model. The three parameters, the saturation current I_S, and the normal active mode and inverse active mode common-base current gains, β_F and β_R, respectively, are all SPICE parameters.

In the case of the normal active mode, $V_{BE}/V_t \gg 1$ and $V_{BC}/V_t \ll -1$, which means $\exp(V_{BE}/V_t) \gg 1$ and $\exp(V_{BC}/V_t) - 1 \approx -1$. This simplifies the general equations to the following form:

$$
\begin{aligned}
I_C &= I_S e^{V_{BE}/V_t} \\
I_E &= -\left(1 + \frac{1}{\beta_F}\right) I_S e^{V_{BE}/V_t} = -\frac{I_C}{\alpha_F} \\
I_B &= \frac{1}{\beta_F} I_S e^{V_{BE}/V_t} = \frac{I_C}{\beta_F}
\end{aligned}
\tag{6.32}
$$

- The output collector current I_C depends exponentially on the input voltage V_{BE} through the I_S parameter and the thermal voltage V_t.

- The emitter and the base currents, I_E and I_B, are related to the collector current through the current gains α_F and β_F, originally defined by Eq. (6.5).

- The minus sign in the I_E equation appears due to the fact that the I_E current direction in the SPICE models is defined into the BJT (Fig. 6.23), which is opposite to the actual direction of the conventional I_E current used in Section 6.1 (Fig. 6.3).

■ **Example 6.6 Ebers–Moll Model for a PNP BJT**

In analogy with the Ebers–Moll model of the NPN BJT, draw the equivalent circuit and write down the general equations of the Ebers–Moll model for the case of PNP BJT. Simplify these equations for the case of a PNP BJT in the normal active mode. In SPICE, the directions of the terminal currents of a PNP BJT are defined to be opposite to their NPN counterparts.

Solution: The PNP BJT is a mirror image of the NPN BJT in the sense that the diode (P–N junction) terminals are swapped over, all the currents are in the opposite directions, and all the voltages are with the opposite polarities. Swapping the diode terminals and reversing the current directions in the circuit of Fig. 6.23, the SPICE equivalent circuit of the PNP BJT is obtained as in Fig. 6.24.

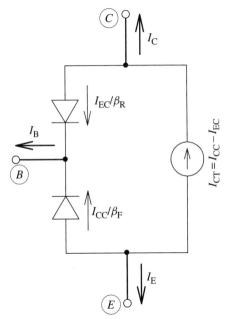

Figure 6.24 SPICE equivalent circuit of PNP BJT.

To avoid using negative voltages, the terminal voltages can be expressed as V_{EB} and V_{CB} rather than $-V_{BE}$ and $-V_{BC}$. With these changes in Eq. (6.31), the Ebers–Moll model of the PNP BJT is obtained as

$$I_C = I_S(e^{V_{EB}/V_t} - 1) - \left(1 + \frac{1}{\beta_R}\right) I_S(e^{V_{CB}/V_t} - 1)$$

$$I_E = -\left(1 + \frac{1}{\beta_F}\right) I_S(e^{V_{EB}/V_t} - 1) + I_S(e^{V_{CB}/V_t} - 1)$$

$$I_B = \frac{1}{\beta_F} I_S(e^{V_{EB}/V_t} - 1) + \frac{1}{\beta_R} I_S(e^{V_{CB}/V_t} - 1)$$

The simplified equations for the case of the normal active mode can be deduced in a similar way:

$$I_C = I_S e^{V_{EB}/V_t}$$

$$I_E = -\left(1 + \frac{1}{\beta_F}\right) I_S e^{V_{EB}/V_t} = -\frac{I_C}{\alpha_F}$$

$$I_B = \frac{1}{\beta_F} I_S e^{V_{EB}/V_t} = \frac{I_C}{\beta_F}$$

■ **Example 6.7 Ebers–Moll Model for Inverse Active Mode**

Simplify Eq. (6.31) for the case of the inverse active mode.

Solution: In this case, $\exp(V_{BC}/V_t) \gg 1$ and $\exp(V_{BE}/V_t) - 1 \approx -1$, which leads to

$$I_C = -\left(1 + \frac{1}{\beta_R}\right) I_S e^{V_{BC}/V_t} = -\frac{I_E}{\alpha_R}$$

$$I_E = I_S e^{V_{BC}/V_t}$$

$$I_B = \frac{1}{\beta_R} I_S e^{V_{BC}/V_t} = \frac{I_E}{\beta_R}$$

■ **Example 6.8 Fundamental BJT Parameters**

The results of measurements performed on an NPN BJT are given in Table 6.2. Calculate the following SPICE parameters: I_S, β_F, and β_R.

TABLE 6.2 Measurement Data

V_{BE} (V)	V_{BC} (V)	I_B (μA)	I_C (mA)
0.80	−5.0	2.6	0.49
−5.0	0.72	353.3	0.90

Solution: Voltages V_{BE} and V_{BC} indicate that the first row of data is for the BJT in the normal active mode, while the second row is related to the inverse active mode. According to Eqs. (6.32),

$$\beta_F = I_C/I_B = \frac{490}{2.6} = 188.5$$

and

$$\ln I_S = \ln I_C - V_{BE}/V_t = \ln 4.9 \times 10^{-5} - \frac{0.80}{0.02585} = -40.87 \Rightarrow I_S$$

$$= 1.78 \times 10^{-18} A$$

In the case of the inverse active mode, the results of Example 6.7 can be used to find β_R:

$$\beta_R = \frac{I_E}{I_B} = \frac{I_C - I_B}{I_B} = \frac{900.0 - 353.3}{353.3} = 1.55$$

Another value for the saturation current can be obtained for the case of the inverse active mode, however, this value is less relevant as the BJT normally operates in the normal active mode.

6.3.2 Second Order Effects: Ebers–Moll Level in SPICE

The output collector current I_C in the normal active mode, as predicted by the principal Ebers–Moll model [Eq. (6.32)], does not depend on the output voltage. This is the case of ideal current source, illustrated by the dotted horizontal lines on the I_C–V_{CE} plot of Fig. 6.25a. The real BJTs, however, do not have perfectly horizontal I_C–V_{CE} characteristics—I_C always increases to some extent with an increase in V_{CE}. The reciprocal value of the slope of the output I_C–V_{CE} characteristic is defined as dynamic output resistance:

$$r_o = \frac{1}{dI_C/dV_{CE}} \tag{6.33}$$

The ideal current source has infinitely large r_o.

In the case of a BJT acting as a controlled current source, r_o is not the same for every input voltage/current. As the input voltage/current is increased, r_o is reduced, which is observed as a more pronounced slope on the corresponding I_C–V_{CE} line. This effect, known as the Early effect, is illustrated in Fig. 6.25a by the solid lines.

A proper inclusion of the real dynamic output resistance is very important when simulating and designing analog circuits. As the Early effect model and parameters are related to the dynamic output resistance, the Early effect appears as the most important second-order effect.

Figure 6.25b illustrates that it is in fact the reverse voltage of the base–collector junction $V_{BC} = -V_{CB}$ that is directly related to the I_C current increase. Practically, however, the difference between V_{CE} and V_{CB}, which is $V_{BE} \approx 0.7$ V, is insignificant at relatively large V_{CE} and V_{CB} voltages needed to observe the Early effect. The increase in the I_C current with an increase in V_{CE}, and therefore V_{CB} voltage, is due to effective shortening of the base width w_{base}, caused by the associated depletion-layer expansion. This is reflected in the alternative name for the Early effect, which is the *base modulation effect*. The narrower base leads to increased saturation current I_S, which causes the I_C increase, as Eq. (6.31) shows.

A physical insight into this effect can be provided by referring to Fig. 3.6c and Eq. (3.16), noting that in the case of NPN BJTs $w_{base} = W_{anode}$ and $w_{emitter} = W_{cathode}$:

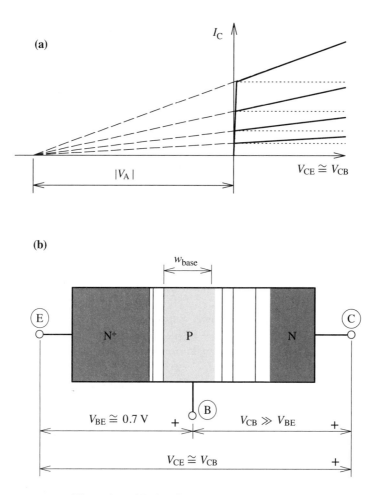

Figure 6.25 Illustration of Early effect: (a) ideal output characteristics (dotted lines) and output characteristics with pronounced Early effect (solid lines) and (b) BJT cross section illustrating base narrowing due to increased base/collector depletion-layer width by increased V_{CB} voltage.

$$I_S = A_E q n_i^2 \left(\frac{D_n}{w_{base} N_A} + \frac{D_p}{w_{emitter} N_D} \right) \tag{6.34}$$

Therefore, the Early effect can be explained by the following sequence of effects, initiated by a V_{BC} (or V_{CE}) increase:

1. depletion layer width at the base–collector junction is increased,
2. w_{base} is reduced,
3. I_S is increased [Eq. (6.34)], and
4. I_C is increased [Eq. (6.31)].

Early suggested a way of modeling the output resistance itself, and its variation with the level of output current, by a single parameter. Figure 6.25a shows that this is possible if it is assumed that the extrapolated I_C–V_{CE} characteristics (the dashed lines) intersect in a single point on the $V_{CE} \approx V_{CB}$ axis, which is known as Early voltage V_A. Obviously, a larger absolute value of the Early voltage means that the output resistance is higher (I_{CE}–V_{CE} lines are closer to the horizontal level), and vice versa. In the ideal case of $r_o \to \infty$, the Early voltage $V_A \to \infty$.

Using the rule of similar triangles, the following relationship can be written with the definition of $|V_A|$ as in Fig. 6.25a:

$$\frac{I_C(|V_{BC}| = 0)}{|V_A|} = \frac{I_C(|V_{BC}|)}{|V_A| + |V_{BC}|} \tag{6.35}$$

which, with regard to the above comments and Eq. (6.32), leads to the following equation for the saturation current:

$$I_S = I_{S0}\frac{|V_A| + |V_{BC}|}{|V_A|} = I_{S0}\left(1 + \frac{|V_{BC}|}{|V_A|}\right) \tag{6.36}$$

I_S becomes V_{BC}-dependent saturation current, used in the Ebers–Moll Eq. (6.31), while the zero-voltage saturation current I_{S0} becomes the SPICE parameter.

An analogous theory applies to the case of the inverse active mode, when the base–emitter junction is reverse biased, the base–collector junction being forward biased. The Early voltage in this case is denoted by V_B.

Another second-order effect included in SPICE at the Ebers–Moll level is due to the parasitic resistances. Very similar to the case of the diode, series resistors r_E, r_B, and r_C are added to the emitter, base, and collector, respectively, to account for the contact resistances and the resistances of the respective regions in the silicon (Fig. 6.26). The resistances r_E, r_B, and r_C are direct SPICE parameters.

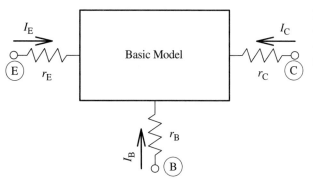

Figure 6.26 The equivalent circuit of the Ebers–Moll model is extended to include the parasitic resistances.

■ **Example 6.9 Early Effect**

It has been determined that the collector current of an NPN BJT increases from 1 to 1.1 mA if the collector-to-emitter voltage is increased from 5 to 10 V. Calculate the Early voltage and the dynamic output resistance of this BJT.

Solution: As two measurement points are available, the following set of two equations can be solved:

$$I_{C1} = I_C(0) \left(1 + \frac{|V_{CB1}|}{|V_A|} \right)$$

$$I_{C2} = I_C(0) \left(1 + \frac{|V_{CB2}|}{|V_A|} \right)$$

The solution can be expressed as

$$V_A = \left(V_{CB2} - V_{CB1} \frac{I_{C2}}{I_{C1}} \right) / \left(\frac{I_{C2}}{I_{C1}} - 1 \right)$$

where $V_{CB1} = V_{CE1} - V_{BE}$, and $V_{CB2} = V_{CE2} - V_{BE}$. Therefore,

$$V_A = \frac{(10 - 0.7) - (5 - 0.7) \times 1.1}{0.1} = 45.7 \text{ V}$$

The reciprocal value of the dynamic output resistance is

$$\frac{1}{r_o} = \frac{dI_C}{dV_{CE}} \approx \frac{dI_C}{dV_{CB}}$$

The first derivative of the I_C–V_{CB} dependence leads to

$$\frac{1}{r_o} = \frac{d}{dV_{CB}} \left[\underbrace{I_{S0} e^{V_{BE}/V_t}}_{I_C(0)} \left(1 + \frac{|V_{CB}|}{|V_A|} \right) \right] = \frac{I_C(0)}{|V_A|}$$

which means the output resistance is

$$r_o = \frac{|V_A|}{I_C(0)}$$

Let us find $I_C(0)$ from the first measurement point:

$$I_C(0) = I_{C1} / \left(1 + \frac{|V_{CB}|}{|V_A|} \right) = 1 / \left(1 + \frac{4.3}{45.7} \right) = 0.914 \text{ mA}$$

The output resistance is now obtained as

$$r_o = \frac{45.7}{0.914} = 50 \text{ k}\Omega$$

6.3.3 Large-Signal and Small-Signal Equivalent Circuits

It is not only the parasitic resistances that are important. The parasitic capacitances are of qualitative importance, as they determine the dynamic characteristics of the BJT, especially at high frequencies. Two P–N junctions are inherently present in any BJT structure, and they bring along the associated capacitances. As described in Chapters 2 and 3, there are two capacitances associated with every P–N junction: the depletion-layer capacitance C_d, and the stored-charge capacitance C_s, which is important only in the case of the forward-biased mode when significant current flows through the P–N junction. The depletion-layer and the stored-charge capacitances of the base–emitter and base–collector junctions appear in parallel with the respective diodes, as shown in Fig. 6.27. Subscripts E and C are used for the base–emitter and the base–collector junctions, respectively. All these capacitances are voltage dependent, as described in Chapters 2 and 3.

There is an additional P–N junction in the standard bipolar IC structure of the BJT, which is the isolating N-epi–P-substrate or collector–substrate junction (Fig. 6.8). This junction is always reverse biased, which means the stored-charge capacitance never becomes important. However, the depletion-layer capacitance is always there, and it can definitely influence the high-frequency characteristics of the BJT. The depletion-layer capacitance of the collector–substrate junction is included in the large-signal equivalent circuit used in SPICE, C_{dS} in Fig. 6.27. It appears connected between the collector and the lowest potential in the circuit (V_-), which is effectively zero level for the signal voltages and currents.

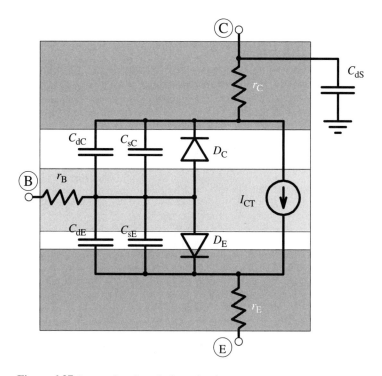

Figure 6.27 Large-signal equivalent circuit.

Let us now consider the specific case of the normal active mode, and transfer the large-signal equivalent circuit of Fig. 6.27 into a small-signal equivalent circuit, frequently used for circuit design and analysis. Table 6.3 summarizes the relationships between the components of the large-signal and the small-signal equivalent circuits.

To begin with, the base–collector junction is reverse biased, so the diode D_C and the stored-charge capacitance C_{sC} can be removed as no diode current flows through this junction. What remains between the base and the collector is the depletion-layer capacitance C_{dC}, which in general depends on the voltage drop across the junction V_{CB} (analogous to the equation given in Table 3.5). However, as we are considering the small-signal situation, we are interested in the voltage V_{CB} that corresponds to the DC operating point Q.[1] Therefore, a single value of the capacitance between the base and the collector can be obtained, and used in the small-signal equivalent circuit. This capacitance, labeled C_μ in the circuit of Fig. 6.28, is known as the Miller capacitance. Its importance lies in the fact that it makes a feedback between the output (collector) and the input (base) when the BJT is used in the common-emitter configuration.

The base–emitter junction is forward biased, and both capacitances are important. The values of the depletion-layer capacitance C_{dE} and the stored-charge capacitance C_{sE} at the DC operating voltage V_{BE} (DC operating point Q) are summed to obtain the capacitance C_π.

The I–V characteristic of the forward-biased diode D_E is not linear, however, in the small range of the small input signal change it can be approximated by a linear segment, which is basically the slope of the I–V characteristic at the operating point. The reciprocal value of this slope is equivalent to the resistance that the small input signal is facing. Therefore, for the small signals, the diode D_E is replaced by its small-signal resistance at the operating point, r_π.

Similarly, the current of the current source I_{CT} is a nonlinear function of V_{BE}, and through the Early effect of V_{CE}. For small signals, however, the I_{CT}–V_{BE} dependence can be approximated by a linear segment $i_c = g_m v_\pi$, where i_c is the small-signal output current, v_π is the small-signal input voltage, and g_m is the slope of the I_{CT}–V_{BE} characteristic at

TABLE 6.3 Relationship between the Components of Large-Signal and Small-Signal Equivalent Circuits

Large-Signal Equivalent Circuit (General Case, Fig. 6.27)	Small-Signal Equivalent Circuit (Normal Active Mode, Fig. 6.28)
D_C	Neglected
C_{sC}	Neglected
C_{dC} at V_{CB}	C_μ
$(C_{dE} + C_{sE})$ at V_{BE}	C_π
D_E at V_{BE}	r_π
I_{CT} at V_{BE}	$g_m v_\pi$ and r_o
C_{dS} at V_{CS}	C_o
r_B	r_B
r_E	r_E
r_C	r_C

[1]Sections 4.1.2 and 4.1.3 describe the relationship between small and instantaneous voltages and currents, as well as the role and the meaning of the DC operating point Q.

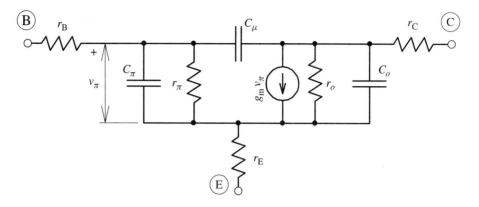

Figure 6.28 Small-signal equivalent circuit.

the operating point. The concept of g_m, which is called *transconductance*, is considered in more detail in Section 4.1.2. The dependence of I_{CT} on V_{CE} shows through the dynamic or small-signal output resistance r_o. The relationship between r_o and $I_{CT} \approx I_C$ is given by Eq. (6.33). Again, the derivative dI_C/dV_{CE} is calculated at the operating point to obtain the value for the small-signal equivalent circuit.

The capacitance C_o is the depletion-layer capacitance C_{dS} at the reverse-bias voltage appearing between the collector and the substrate V_{CS}. Finally, the parasitic resistances r_B, r_E, and r_C are the same in both the large-signal and small-signal equivalent circuits.

■ **Example 6.10** g_m **and** r_π

For the BJT of Example 6.8, calculate the transconductance g_m and the small-signal input resistance r_π at $V_{BE} = 0.80$ V and $V_{BC} = -5$ V.

Solution: This BJT is in normal active mode:

$$g_m = \frac{dI_C}{dV_{BE}} = \frac{d}{dV_{BE}}\left(I_s e^{V_{BE}/V_t}\right) = \frac{1}{V_t}\underbrace{I_s e^{V_{BE}/V_t}}_{I_C}$$

$$g_m = \frac{I_C}{V_t}$$

$$g_m = \frac{0.49 \times 10^{-3}}{0.02586} = 18.95 \text{ mA/V}$$

$$r_\pi = \frac{dV_{BE}}{dI_B} = \beta_F \frac{dV_{BE}}{dI_C} = \beta_F \underbrace{\frac{1}{dI_C/dV_{BE}}}_{1/g_m} = \frac{\beta_F}{g_m}$$

$$r_\pi = \frac{188.5}{18.95} = 9.95 \text{ k}\Omega$$

6.3.4 Second Order Effects: Gummel–Poon Level in SPICE

The common-emitter current gains β_F and β_R, which are SPICE parameters themselves, are constants in the Ebers–Moll model. However, Fig. 6.29 shows that the measured common-emitter current gain β_F of a BJT is different at different current levels. Typically, the common-emitter current gain increases with the collector current, slowly reaching the maximum value at medium currents, and then rather rapidly decreases at high currents. Noting that β_F is plotted versus the logarithm of I_C in Fig. 6.29, we can see that the Ebers–Moll assumption of constant β_F can satisfactorily be used in about a two orders of magnitude wide range of I_C current. However, if the BJT is operated in extreme conditions, very high or very low current levels, the changes of the common-emitter current gain cannot be neglected. There are equations in SPICE, at what is known as the Gummel–Poon level, that account for this type of second-order effects.

According to the Ebers–Moll model, both I_C and I_B are proportional to $\exp(V_{BE}/V_t)$, which results in the expected constant $\beta_F = I_C/I_B$. The diffusion component of the base current does follow the $\exp(V_{BE}/V_t)$ dependence, as shown in Section 3.1.3. However, at low biasing levels, the recombination of the carriers in the bulk and surface depletion layer, as well as other surface leakage mechanisms, leads to an increase of the base current. The increased base current is observed as the β_F reduction at low bias levels. To model this effect, Eq. (6.31) for the base current is modified in the following way

$$I_B = \frac{I_{S0}}{\beta_{FM}}(e^{V_{BE}/V_t} - 1) + C_2 I_{S0}(e^{V_{BE}/(n_{EL}V_t)} - 1)$$

$$+ \frac{I_{S0}}{\beta_{RM}}(e^{V_{BC}/V_t} - 1) + C_4 I_{S0}(e^{V_{BC}/(n_{CL}V_t)} - 1) \tag{6.37}$$

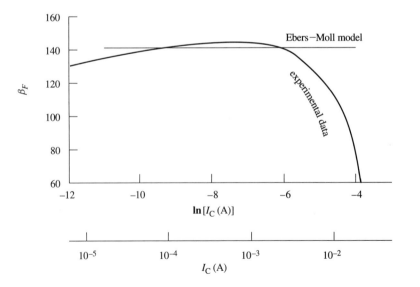

Figure 6.29 The common-emitter current gain at different levels of the collector current. The experimental data are measured on an NPN from a 3086 IC.

Obviously, the $C_2 I_{S0}[\exp(V_{BE}/(n_{EL}/V_t) - 1)]$ and $C_4 I_{S0}[\exp(V_{BC}/(n_{CL}/V_t) - 1)]$ terms are added to include the base increase for the cases of forward-biased base–emitter and forward-biased base–collector junctions, respectively. This introduces four SPICE parameters: C_2, the base–emitter leakage saturation current coefficient; n_{EL}, the base–emitter leakage emission coefficient; C_4, the base–collector leakage saturation current coefficient; and n_{CL}, the base–collector leakage emission coefficient. The current gains β_{FM} and β_{RM} are not additional SPICE parameters, they have the same values as β_F and β_R in the Ebers–Moll model. The subscript M is added to indicate that these are the constant mid-current values of the current gains, and not the variable current gains, as in Fig. 6.29.

To include the effects of high V_{CE} bias (the Early effect), the saturation current I_S was modified. Analogously, I_S can be modified to include the effects of high V_{BE} bias, which causes the collector current to fall below the $\exp(V_{BE}/V_t)$ level, and therefore causes the common-emitter current gain reduction at high bias levels. The fundamental I_S equation shows that the saturation current is inversely proportional to the majority carrier concentrations in the P–N junction regions. In Eq. (6.34), the majority carrier concentrations are assumed to be equal to the doping levels N_A and N_D in the P-type and N-type regions, respectively. In the case of NPN BJT, $1/N_A$ dominates, as the doping level in the base is much lower compared to the emitter ($N_A \ll N_D$, thus $1/N_A \gg 1/N_D$). At high bias levels, however, the contribution of other factors to the majority-carrier (hole) concentration in the base becomes important. These factors are the depletion-layer capacitance, which is related to the Early effect, and the stored-charge capacitance, which causes the drop in I_C.

Gummel and Poon suggested the following equation for the total density of majority carriers in the base:

$$Q_{BT} = \underbrace{Q_{B0}}_{q N_A A_E} + \underbrace{C_{dE} V_{BE} + C_{dC} V_{BC} \frac{A_E}{A_C}}_{\text{depletion-layer charge}}$$

$$+ \underbrace{\frac{Q_{B0}}{Q_{BT}} \tau_F I_S (e^{V_{BE}/V_t} - 1) + \frac{Q_{B0}}{Q_{BT}} \tau_R I_S (e^{V_{BC}/V_t} - 1)}_{\text{stored charge}} \qquad (6.38)$$

where A_E and A_C are the areas of the base–emitter and the base–collector junctions, respectively, and τ_F and τ_R are the normal-mode and inverse-mode transit times. As I_S is inversely proportional to the carrier density in the base, dividing I_S by Q_{BT}/Q_{B0} effectively replaces the doping level N_A by the total charge, which includes the contribution of the depletion-layer and the stored-charge capacitances. Therefore, the collector current I_C in the normal active mode is expressed as

$$I_C \approx \frac{I_{S0}}{q_b} e^{V_{BE}/V_t} \qquad (6.39)$$

where

$$q_b = \frac{Q_{BT}}{Q_{B0}} \qquad (6.40)$$

Defining the following parameters

$$I_{KF} = \frac{Q_{B0}}{\tau_F}, \qquad I_{KR} = \frac{Q_{B0}}{\tau_R} \tag{6.41}$$

$$|V_A| = \frac{Q_{B0}}{C_{dC}} \frac{A_C}{A_E}, \qquad |V_B| = \frac{Q_{B0}}{C_{dE}} \tag{6.42}$$

the factor q_b can be expressed as

$$q_b = \frac{q_1}{2} + \frac{\sqrt{q_1^2 + 4q_2}}{2} \approx \frac{q_1}{2}\left(1 + \sqrt{1 + 4q_2}\right) \tag{6.43}$$

where

$$q_1 = 1 + \frac{V_{BE}}{|V_B|} + \frac{V_{BC}}{|V_A|} \approx 1/\left(1 - \frac{V_{BE}}{|V_B|} - \frac{V_{BC}}{|V_A|}\right) \tag{6.44}$$

and

$$q_2 = \frac{I_{S0}}{I_{KF}}(e^{V_{BE}/V_t} - 1) + \frac{I_{S0}}{I_{KR}}(e^{V_{BC}/V_t} - 1) \tag{6.45}$$

The first equations of q_b and q_1 correspond directly to the original Gummel–Poon model, while the approximate equations correspond to those used in SPICE. Derivation of q_b is given in the following books: (1) R.S. Muller and T.I. Kamins, *Device Electronics for Integrated Circuits*, 2nd ed., Wiley, New York, 1986 (pp. 359–362) and (2) G. Massobrio and P. Antognetti, *Semiconductor Device Modeling with SPICE*, 2nd ed., McGraw-Hill, New York, 1993 (Chapter 2). Both books provide a more detailed description of the Gummel–Poon model, while the second book also lists the equations used in SPICE. I_{KF} and I_{KR} are additional SPICE parameters, which are used to fit the high-level I_C current to the experimental data. The measurement of these parameters is described in the next section. $|V_A|$ and $|V_B|$ are equivalent to the earlier described normal and reverse Early voltages.

Equations (6.40)–(6.45) are general, covering all the possible BJT modes of operation. To generalize the collector current given by Eq. (6.39), the effects of V_{BC} voltage should be added in a way analogous to the case of the Ebers–Moll model. The general I_C equation is given here along with I_B [Eq. (6.37)] and the corresponding I_E equations, to show the complete set of BJT equations at the Gummel–Poon level in SPICE:

$$I_B = \frac{I_{S0}}{\beta_{FM}}(e^{V_{BE}/V_t} - 1) + C_2 I_{S0}(e^{V_{BE}/(n_{EL} V_t)} - 1)$$

$$+ \frac{I_{S0}}{\beta_R}(e^{V_{BC}/V_t} - 1) + C_4 I_{S0}(e^{V_{BC}/(n_{CL} V_t)} - 1)$$

$$I_C = \frac{I_{S0}}{q_b}(e^{V_{BE}/V_t} - e^{V_{BC}/V_t})$$

$$-\frac{I_{S0}}{\beta_{RM}}(e^{V_{BC}/V_t}-1)-C_4 I_{S0}(e^{V_{BC}/(n_{CL}V_t)}-1)$$

$$I_E = -\frac{I_{S0}}{q_b}(e^{V_{BE}/V_t}-e^{V_{BC}/V_t})$$

$$-\frac{I_{S0}}{\beta_{FM}}(e^{V_{BE}/V_t}-1)-C_2 I_{S0}(e^{V_{BE}/(n_{EL}V_t)}-1) \qquad (6.46)$$

The Ebers–Moll equations are included in Eqs. (6.46) of the Gummel–Poon level in SPICE. Default values of the second-order effect parameters reduce Eqs. (6.46) to Eqs. (6.31). $I_{KF} = I_{KR} \rightarrow \infty$ turns q_2 into zero, and $|V_A| = |V_B| \rightarrow \infty$ turns q_1 into unity, which means $q_b = 1$. The default value of C_2 and C_4 is zero, which along with $q_b = 1$ clearly eliminates all the additions in Eqs. (6.46) compared to Eqs. (6.31). As soon as some of the second-order parameters are stated, the Gummel–Poon level in SPICE is automatically activated. SPICE BJT equations are summarized in the next section, which also describes the techniques of parameter measurement.

6.4 SPICE PARAMETERS

The first part of this section provides a classified summary of SPICE parameters and equations, while the second part of the section is devoted to parameter measurement techniques.

6.4.1 Summary of SPICE Parameters and Expressions

As described earlier, there are two levels of model complexity, Ebers–Moll and Gummel–Poon. The two levels differ in the way the common-emitter current gains β_F and β_R are treated. At the Ebers–Moll level, they are considered as constants whose values can be specified by the user. At the Gummel–Poon level, the specified values are considered as the maximum values only (β_{FM} and β_{RM}) while modified I_C and I_B equations account for the variation of the current gain I_C/I_B at different bias conditions. The more complex equations at the Gummel–Poon level include the simpler Ebers–Moll equations. Consequently, and unlike the MOSFET case, the BJT levels do not need to be explicitly specified by the user.

The Ebers–Moll parameters and equations are summarized in Table 6.4. They include the principal effects and the most important second-order effect, which is the Early effect. The Early voltages can be ignored if precise modeling of the output dynamic resistance is not important. For completeness, the equations are given for the case of both NPN and PNP BJT. If the approximation of constant common-emitter current gain is satisfactory and Table 6.4 is used, the following Table 6.5 listing the Gummel–Poon parameters and equation should be ignored. However, Table 6.6 is still relevant, as it summarizes the parasitic elements that may need to be added to the Ebers–Moll parameters for more precise simulation.

In the extreme biasing conditions, when the current gain reduction at very low or high current levels cannot be neglected, Table 6.4 should be replaced by Table 6.5 summarizing the parameters and equations at the more complex Gummel–Poon level.

TABLE 6.4 Summary of SPICE BJT Model: Static Ebers–Moll Level

Ebers–Moll Parameters

Symbol	Usual SPICE Keyword	Parameter Name	Typical Value	Unit
I_{S0}	IS	Saturation current	10^{-16}	A
β_F	BF	Normal common-emitter current gain	150	—
β_R	BR	Inverse common-emitter current gain	5	—
V_A	VA	Normal Early voltage	> 50	V
V_B	VB	Inverse Early voltage		V

Ebers–Moll Model

NPN BJT

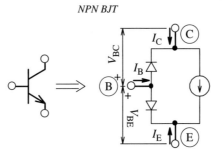

$$I_S = \text{IS}\left(1 - \frac{V_{BC}}{VA} - \frac{V_{BE}}{VB}\right)$$

$$I_C = I_S(e^{V_{BE}/V_t} - 1) - \left(1 + \frac{1}{BR}\right)I_S(e^{V_{BC}/V_t} - 1)$$

$$I_E = -\left(1 + \frac{1}{BF}\right)I_S(e^{V_{BE}/V_t} - 1) + I_S(e^{V_{BC}/V_t} - 1)$$

$$I_B = \frac{1}{BF}I_S(e^{V_{BE}/V_t} - 1) + \frac{1}{BR}I_S(e^{V_{BC}/V_t} - 1)$$

PNP BJT

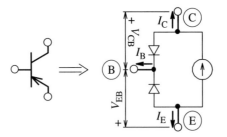

$$I_S = \text{IS}\left(1 - \frac{V_{CB}}{VA} - \frac{V_{EB}}{VB}\right)$$

$$I_C = I_S(e^{V_{EB}/V_t} - 1) - \left(1 + \frac{1}{BR}\right)I_S(e^{V_{CB}/V_t} - 1)$$

$$I_E = -\left(1 + \frac{1}{BF}\right)I_S(e^{V_{EB}/V_t} - 1) + I_S(e^{V_{CB}/V_t} - 1)$$

$$I_B = \frac{1}{BF}I_S(e^{V_{EB}/V_t} - 1) + \frac{1}{BR}I_S(e^{V_{CB}/V_t} - 1)$$

TABLE 6.5 Summary of SPICE BJT Model: Static Gummel–Poon Level

Gummel–Poon Parameters

Symbol	Usual SPICE Keyword	Parameter Name[a]	Typical Value	Unit
I_{S0}	IS	Saturation current	10^{-16}	A
β_{FM}	BF	Maximum normal current gain	150	—
β_{RM}	BR	Maximum inverse current gain	5	—
V_A	VA	Normal Early voltage	> 50	V
V_B	VB	Inverse Early voltage		V
I_{KF}	IKF	Normal knee current	$> 10^{-2}$	A
I_{KR}	IKR	Inverse knee current		A
$C_2 I_{S0}$	ISE	B–E leakage saturation current		A
n_{EL}	NE	B–E leakage emission coefficient	2	—
$C_4 I_{S0}$	ISC	B–C leakage saturation current		A
n_{CL}	NC	B–C leakage emission coefficient	2	—

Gummel–Poon Model

NPN BJT (equivalent circuit as in Table 6.4)

$$\lambda_{BE} = e^{V_{BE}/V_t} - 1 \qquad \lambda_{BC} = e^{V_{BC}/V_t} - 1$$

$$\lambda_{BEL} = e^{V_{BE}/(NE\ V_t)} - 1 \qquad \lambda_{BCL} = e^{V_{BC}/(NC\ V_t)} - 1$$

$$q_1 = \left(1 - \frac{V_{BC}}{VA} - \frac{V_{BE}}{VB}\right)^{-1}, \qquad q_2 = \frac{IS}{IKF}\lambda_{BE} + \frac{IS}{IKR}\lambda_{BC}, \qquad q_b = 0.5q_1\left(1 + \sqrt{1 + 4q_2}\right)$$

$$I_B = \frac{IS}{BF}\lambda_{BE} + ISE\ \lambda_{BEL} + \frac{IS}{BR}\lambda_{BC} + ISC\ \lambda_{BCL}$$

$$I_C = \frac{IS}{q_b}(\lambda_{BE} - \lambda_{BC}) - \frac{IS}{BR}\lambda_{BC} - ISC\ \lambda_{BCL}$$

$$I_E = -\frac{IS}{q_b}(\lambda_{BE} - \lambda_{BC}) - \frac{IS}{BF}\lambda_{BE} - ISE\ \lambda_{BEL}$$

PNP BJT (equivalent circuit as in Table 6.4)

$$\lambda_{EB} = e^{V_{EB}/V_t} - 1 \qquad \lambda_{CB} = e^{V_{CB}/V_t} - 1$$

$$\lambda_{EBL} = e^{V_{EB}/(NE\ V_t)} - 1 \qquad \lambda_{CBL} = e^{V_{CB}/(NC\ V_t)} - 1$$

$$q_1 = \left(1 - \frac{V_{CB}}{VA} - \frac{V_{EB}}{VB}\right)^{-1}, \qquad q_2 = \frac{IS}{IKF}\lambda_{EB} + \frac{IS}{IKR}\lambda_{CB}, \qquad q_b = 0.5q_1\left(1 + \sqrt{1 + 4q_2}\right)$$

$$I_B = \frac{IS}{BF}\lambda_{EB} + ISE\ \lambda_{EBL} + \frac{IS}{BR}\lambda_{CB} + ISC\ \lambda_{CBL}$$

$$I_C = \frac{IS}{q_b}(\lambda_{EB} - \lambda_{CB}) - \frac{IS}{BR}\lambda_{CB} - ISC\ \lambda_{CBL}$$

$$I_E = -\frac{IS}{q_b}(\lambda_{EB} - \lambda_{CB}) - \frac{IS}{BF}\lambda_{EB} - ISE\ \lambda_{EBL}$$

[a]B–E, base–emitter; B–C, base-collector.

TABLE 6.6 Summary of SPICE BJT Model: Parasitic Elements

			Parasitic Element-Related Parameters		
Symbol	**Usual SPICE Keyword**	**Related Parasitic Element**	**Parameter Name[a]**	**Typical Value Range**	**Unit**
r_B	RB	r_B	Base resistance	10	Ω
r_E	RE	r_E	Emitter resistance	2	Ω
r_C	RC	r_C	Collector resistance	15	Ω
$C_{dE}(0)$	CJE	C_{dE}	Zero-bias B–E capacitance		F
V_{biE}	VJE	C_{dE}	B–E built-in voltage	0.8	V
m_E	MJE	C_{dE}	B–E grading coefficient	$\frac{1}{3}-\frac{1}{2}$	—
τ_F	TF	C_{sE}	Normal transit time	10^{-9}	s
$C_{dC}(0)$	CJC	C_{dC}	Zero-bias B–C capacitance		F
V_{biC}	VJC	C_{dC}	B–C built-in voltage	0.75	V
m_C	MJC	C_{dC}	B–C grading coefficient	$\frac{1}{3}-\frac{1}{2}$	—
τ_R	TR	C_{sC}	Inverse transit time	10^{-9}	s
$C_{dS}(0)$	CJS	C_{dS}	Zero-bias C–S capacitance		F
V_{biS}	VJS	C_{dS}	C–S built-in voltage	0.7	V
m_S	MJS	C_{dS}	C–S grading coefficient	$\frac{1}{3}-\frac{1}{2}$	—

Large-Signal Equivalent Circuit

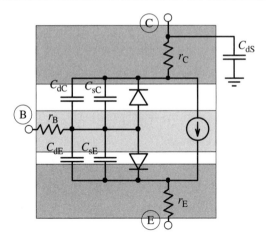

NOTE: The diodes and the current-source direction are shown for NPN BJT. Reverse current direction and diode polarities apply in the case of PNP BJT.

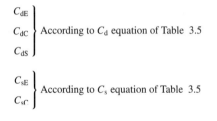

$\left.\begin{array}{c} C_{dE} \\ C_{dC} \\ C_{dS} \end{array}\right\}$ According to C_d equation of Table 3.5

$\left.\begin{array}{c} C_{sE} \\ C_{sC} \end{array}\right\}$ According to C_s equation of Table 3.5

[a]B–E, base–emitter; B–C, base–collector; C–S, collector–substrate.

In either case, Table 6.6 may be relevant as it lists the parameters related to the parasitic elements. This is especially important at high frequencies, in which case the capacitances of Table 6.6 cannot be neglected.

6.4.2 Parameter Measurement

Practical techniques of BJT parameter measurement are shown in this section, using the example of a 3086 NPN biased in the normal active mode. Analogous techniques are employed for measurement of the parameters related to the inverse active mode. The techniques described are useful for obtaining estimated parameter values, which can then be used as the initial values for nonlinear parameter fitting.

Measurement of the Saturation Current and the Current Gain

To find the saturation current I_{S0} and the current gain β_F, a set of measured I_C and I_B values over a range of V_{BE} voltages is needed. The voltage V_{CE} is kept constant, and set at neither too low a value (to avoid the saturation region) nor too high a value (to avoid the influence of the Early effect).

In the normal active mode, I_C and I_B dependencies on V_{BE} are given by Eqs. (6.32). As these are exponential relationships, they can be linearized in the following way:

$$\ln I_C = \ln I_{S0} + \frac{1}{V_t} V_{BE} \tag{6.47}$$

$$\ln I_B = \ln I_{S0} - \ln \beta_F + \frac{1}{V_t} V_{BE} \tag{6.48}$$

Fig. 6.30 shows that the linear $\ln I_C$–V_{BE} and $\ln I_B$–V_{BE} dependencies, with the slope of $1/V_t$, are observed over a wide range of output currents. As Eq. (6.47) shows, the logarithm of the saturation current is obtained as $\ln I_C$ at $V_{BE} = 0$.

It is obvious from Eqs. (6.47) and (6.48) that $\ln I_C$–$\ln I_B = \ln \beta_F$. Therefore, the logarithm of the current gain is obtained as the difference between the $\ln I_C$–V_{BE} and $\ln I_B$–V_{BE} lines.

Measurement of the Early Voltage

It may seem from Fig. 6.25 that the V_A measurement is as simple as the extrapolation of several I_C–V_{CE} lines. However, a number of important points are not immediately obvious. To begin with, Eq. (6.35) cannot directly be used to calculate V_A as the $I_C(V_{BC} = 0)$ point is not in the active region. Using another reference point and assuming that $V_{BC} \approx V_{CE}$, for a convenience, Eq. (6.35) is modified as

$$\frac{I_{C-ref}}{V_{CE-ref} + V_A} = \frac{I_C}{V_{CE} + V_A} \tag{6.49}$$

This equation can further be transformed into the following form:

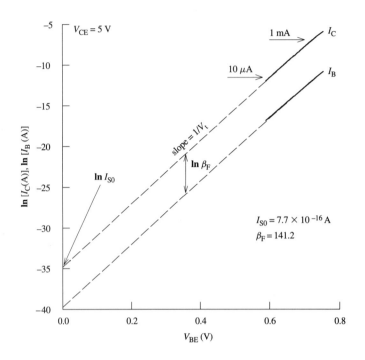

Figure 6.30 Measurement of I_{S0} and β_F.

$$V_{CE} = \underbrace{\frac{V_{CE-ref} + V_A}{I_{C-ref}}}_{\text{slope}} I_C - V_A \qquad (6.50)$$

Only two points, (V_{CE-ref}, I_{C-ref}) and (V_{CE}, I_C) are needed to construct the line defined by Eq. (6.50). It is important, however, that the two points span the entire operating range of V_{CE} voltages.

The range of input voltages V_{BE}, or alternatively input currents I_B, should also cover the entire operating range. The importance of this point is best illustrated by Table 6.7, which shows significant differences between Early voltages obtained by extrapolation of I_C–V_{CE} lines corresponding to different I_B currents. Obviously, if the input current I_B is restricted to a too small value, the Early voltage will be overestimated.

Of course, the results from Table 6.7 open the possibility of establishing a unique V_A, which will properly represent the complete set of I_C–V_{CE} characteristics. As the Early effect is most pronounced at the highest I_B current, the corresponding Early voltage can be used as the first estimate. Using $V_A = 86.0$ V as the initial value, Eq. (6.50) was fitted to the experimental data shown in Fig. 6.31. The best fit was achieved with $V_A = 87.7$ V, which is a value very close to the initial one. Figure 6.31 illustrates that this unique value can properly represent the complete set of output characteristics.

TABLE 6.7 Early Voltages Obtained by Extrapolation of Different I_C–V_{CE} Lines

I_B (μ A)	V_A (V)
20	200.3
40	146.4
60	115.7
80	98.9
100	86.0

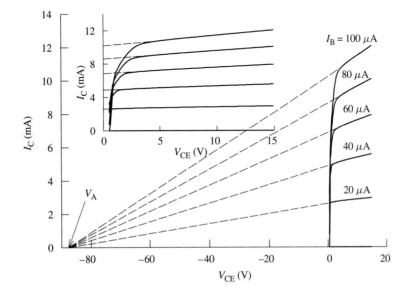

Figure 6.31 Early voltage V_A of a 3086 NPN BJT.

Measurement of the High-Level Knee Current and the Leakage Parameters

The first problem that appears at high current levels is the difference between the applied V_{BE} voltage and the voltage that actually appears across the P–N junction. This difference is due to the voltage drop across the parasitic resistances, as described in Section 3.1.4. In the case of the BJT, the voltage drop across r_B is negligible because the base current is very small, however, the voltage drop across r_E can become pronounced. Although this voltage drop is $r_E I_E$, it can be approximated by $r_E I_C$, as $I_C \approx I_E$. Following the procedure described in Section 3.2.1, the value of r_E is determined to calculate the effective base–emitter voltage $V_{BE} - r_E I_C$ that linearizes the high-current part of the ln I_B–V_{BE} dependence (Fig. 6.32).

Plotting ln I_C versus the effective base–emitter voltage $V_{BE} - r_E I_C$, as in Fig. 6.33, shows that the actual ln I_C still departs from the linear dependence at high current levels. The reduced slope of the ln I_C–V_{BE} dependence causes the reduction in the current gain observed at high I_C currents in Fig. 6.29.

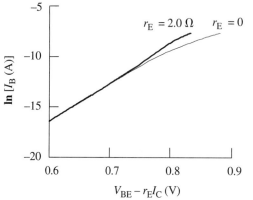

Figure 6.32 The effect of r_E on $\ln I_B - V_{BE}$ dependence.

$r_E = 2.0\ \Omega$ $r_E = 0$

The related parameter in Gummel–Poon Eqs. (6.39) and (6.43)–(6.45) is I_{KF}. To estimate the value of this parameter, $\ln I_C - V_{BE}$ measurements are extended into the high current region. Again, the measurements are performed with properly selected V_{CE} value so that the Early effect is avoided, simplifying Eq. (6.44) to $q_1 \approx 1$. At low and medium current levels, $I_{S0}\exp(V_{BE}/V_t) \ll I_{KF}$, which means Eq. (6.43) reduces to $q_b \approx 1$ because $4q_2 \ll 1$. With $q_b \approx 1$, Eq. (6.39) becomes equivalent to the corresponding Ebers–Moll equation, which predicts the $1/V_t$ slope of the $\ln I_C - V_{BE}$ line. However, at high current levels, $4q_2 \gg 1$ and $2\sqrt{q_2} \gg 1$. This means Eqs. (6.43) and (6.39) can be simplified as

$$q_b \approx \sqrt{q_2} = \sqrt{\frac{I_{S0}}{I_{KF}}}e^{V_{BE}/(2V_t)} \tag{6.51}$$

and

$$I_C \approx \sqrt{I_{S0}I_{KF}}e^{V_{BE}/(2V_t)} \Rightarrow \ln I_C \approx \ln\sqrt{I_{S0}I_{KF}} + \underbrace{\frac{1}{2V_t}}_{\text{slope}} V_{BE} \tag{6.52}$$

Figure 6.33 shows that I_{KF} can be determined in one of the following two ways: (1) as the logarithm of the current at which the high-current line with the slope of $1/2V_t$ intersects the low-current line with the slope of $1/V_t$, and (2) from $\ln\sqrt{I_{S0}I_{KF}}$, which is determined as the value of the high-current line at $V_{BE} = 0$, using the earlier determined value of I_{S0}.

At very low current levels, a significant reduction of the current gain β_F can occur due to the increase of base leakage current, which is modeled by Eq. (6.37) and the associated parameters C_2 and n_{EL}. The leakage component results in a deviation of the $\ln I_B - V_{BE}$ dependence from the line with the $1/V_t$ slope. This situation is not observed in Fig. 6.32, and therefore the parameter C_2 is assumed to be zero. When the effect does appear, the changed slope of the line and its value at $V_{BE} = 0$ are used to estimate n_{EL} and $C_2 I_{S0}$, respectively. This is equivalent to the procedure of n and I_S measurement, described in Section 3.2.1.

The validity of the measured parameters is checked in Fig. 6.34, where the model predictions (the dashed lines) are compared to the experimental data (the solid lines). As mentioned earlier, the estimated parameter values can be used as initial values for nonlinear curve fitting. Figure 6.34 also shows that the curve fitting slightly changed some parameter values to ensure the best fit between the model (open symbols) and the experimental data (the solid lines).

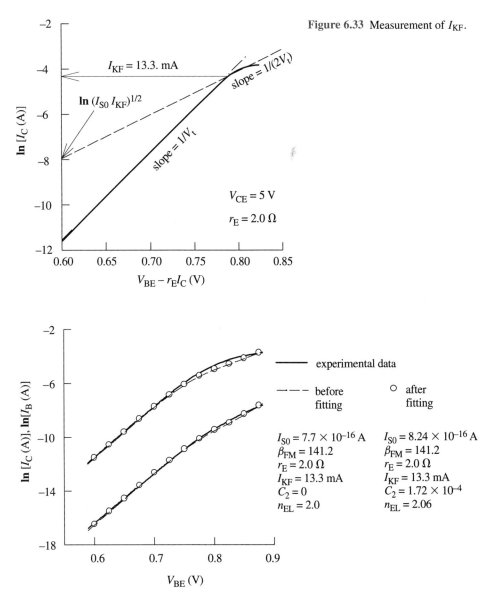

Figure 6.33 Measurement of I_{KF}.

Figure 6.34 Comparison between the experimental data and the SPICE model with parameter values as estimated by the described techniques (before fitting), and with parameter values after nonlinear curve fitting.

6.5 PARASITIC ELEMENTS NOT INCLUDED IN DEVICE MODELS

Although SPICE equivalent circuits and models include the inherently present parasitic elements, such as capacitors and resistors, a number of additional parasitic elements exist in the standard structures of bipolar IC devices. Depending on the specific application conditions, it may be essentially important to externally include these parasitic elements during the simulation. This section describes the most important parasitic elements.

6.5.1 Substrate PNP in the Standard NPN Structure

The equivalent circuit of Fig. 6.27 and Table 6.6 includes all the parasitic resistances and capacitances associated with the four-layer IC structure of the BJT. Note that the capacitor C_{dS} does not exist in discrete three-layer BJTs, but it is added in the SPICE equivalent circuit to cover the case of IC implementation in which the fourth layer introduces an additional P-N junction and the associated capacitance. However, there is a parasitic component that is not included in the SPICE equivalent circuit. This is the substrate, or vertical PNP BJT, whose existence in the four-layer standard NPN BJT structure is illustrated in Fig. 6.35.

Collector-base P–N junction of the parasitic PNP is always reverse biased, as this is the junction that electrically isolates the N-epi island. The emitter–base P–N junction of the parasitic PNP is equivalent to the base–collector P–N junction of the NPN BJT. As Table 6.8 shows, when the NPN is in the normal active mode or cutoff, this junction is also reverse biased, which means the parasitic PNP is in cutoff. In these cases the parasitic PNP can be neglected.

When the NPN is in the saturation or inverse active mode, the common P-base–N-epi junction is forward biased, which sets the parasitic PNP in the normal active mode. In these cases, a part of what would be the NPN collector current flows through the parasitic PNP

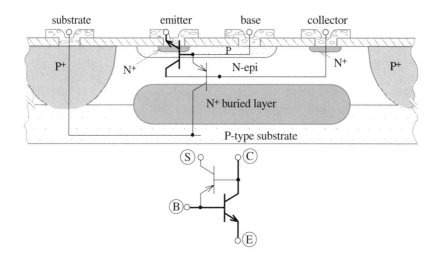

Figure 6.35 Parasitic substrate PNP in the standard NPN structure.

TABLE 6.8 Related Operating Modes of the Standard IC NPN and the Parasitic PNP BJTs[a]

	Standard NPN BJT			Parasitic PNP BJT	
Operating Mode	B–E	B–C	B–E	B–C	Operating Mode
Cutoff	Reverse	Reverse	Reverse	Reverse	Cutoff
Normal active	Forward	Reverse	Reverse	Reverse	Cutoff
Saturation	Forward	Forward	Forward	Reverse	Normal active
Inverse active	Reverse	Forward	Forward	Reverse	Normal active

[a]B–E, base–emitter; B–C, base–collector.

into the substrate contact. The parasitic PNP has rather poor characteristics, the current gain β_F being typically smaller than 5. Nonetheless, $\alpha_F = \beta_F/(1 + \beta_F)$ can still be as high as 0.83, which means that up to 83% of the base terminal current (which is the emitter current of the PNP) is wasted into the substrate.

NPN BJTs are typically biased in the normal active mode in analog circuits. The parasitic PNP is in cutoff in this case and can be neglected. Digital circuits, however, make use of the saturation mode, and some digital circuits use the inverse active mode. The parasitic PNP is in the normal active mode in these cases, and cannot be neglected if precise simulation results are needed. As the equivalent circuit of the NPN BJT does not include the parasitic PNP, it has to be externally added.

6.5.2 Parasitic Elements in Resistor Structures

The resistor is considered as a simple single component in SPICE. The resistors in standard bipolar ICs, however, have associated parasitic resistances and capacitances. Figure 6.36 illustrates the case of a base-diffusion resistor, previously shown in Fig. 6.12.

The P-type–N-epi junction in this case is reverse biased by connecting the N-epi layer to the most positive potential (V_+). As discussed in Section 6.2.4, this is done to isolate the resistors sharing the same N-epi from each other. The reverse-biased P-type–N-epi junction is the base–emitter P–N junction of the parasitic PNP BJT, and as its base–collector junction is also reverse biased, this PNP is always in cutoff mode, which means it can be neglected. This is not the case, however, with the parasitic resistances and capacitances.

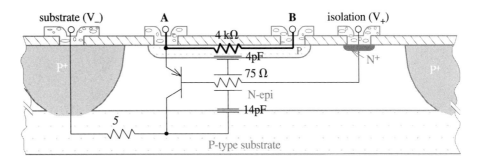

Figure 6.36 Parasitic elements in the structure of a base-diffusion resistor.

The way the two parasitic capacitors are connected to the resistors in the equivalent circuit of Fig. 6.36 shows that the connections are not in single points, neither the left/right nor the middle of the resistors, but rather appear as *distributed* connections. Circuit simulators such as SPICE are not programmed to handle distributed connections. In that case, the distributed connections can be replaced by a series of normal connections, as illustrated in Fig. 6.37. Of course, care should be taken that the values of the cell resistors and capacitors add up to the value of the total resistance or capacitance. If the total resistance is R and there are N cells, each of the resistors in the equivalent circuit should have the value of R/N.

6.5.3 Parasitic Elements in Capacitor Structures

Analogously to the resistor case, the capacitors in standard bipolar ICs have associated parasitic elements, while the capacitor is considered as a single component in SPICE. Figure 6.38 illustrates the case of a base–collector junction capacitor. Although the base–collector junction capacitance is the desired one, its N-epi plate is connected to V_- through the N-epi–P-substrate junction capacitance. The capacitance values shown in Fig. 6.36 indicate that the parasitic capacitance C_p may be larger than the desired one. There is no way that the parasitic capacitance can be neglected when simulating a standard bipolar IC circuit using junction capacitors. In addition, the resistance of the N-epi layer and contact resistances add the parasitic resistor R_p in the equivalent circuit of the base–collector junction capacitor.

Figure 6.37 Equivalent circuit replacing distributed connections of Fig. 6.36.

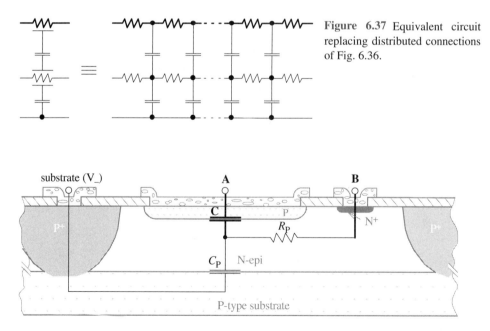

Figure 6.38 Parasitic elements in the structure of a base–collector junction capacitor.

The equivalent circuit of the emitter–base junction capacitor is the same as the one shown in Fig. 6.38. In that case the emitter–base junction creates the capacitance C, the base–collector junction introduces the parasitic capacitance C_p, while the contact resistances and the resistance of the P-type layer are represented by R_p. The capacitance of the N-epi–P-substrate junction is not active in that case, as both the N-epi and P-substrate are connected to DC voltages, V_+ and V_-, respectively.

6.5.4 Parasitic Elements in Diode Structures

The associated parasitic junction capacitance and the parasitic series resistance, appearing in any diode, are included in the device equivalent circuit (Table 3.5). However, the structures of diodes in standard bipolar ICs are more complex, introducing additional parasitics capacitances. Figure 6.39 illustrates the case of the base–collector diode, where the N-epi–P-substrate capacitance connects the anode to the V_- terminal.

The equivalent circuit of the base–emitter diode is similar. The differences are swapped diode terminals and parasitic capacitance C_p of the base–collector junction connecting the cathode to V_+ terminal. Again, the capacitance of the N-epi–P-substrate junction is not active in this case, as both terminals are connected to DC voltages.

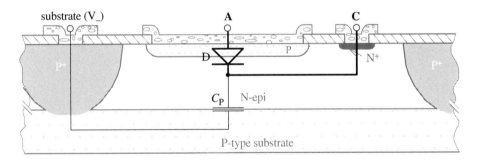

Figure 6.39 Parasitic elements in the structure of a base–collector diode.

PROBLEMS

Section 6.1 BJT Principles

6.1 Find the most suitable description for each of the concentration diagrams shown in Fig. 6.40. Electron concentrations are presented with solid lines and the hole concentrations are presented with dashed lines.

6.2 Figure 6.41 shows four energy-band diagrams, drawn from the emitter to the collector. Identify how the energy-band diagrams relate to each of the four points, labeled on the output characteristics of the BJT.

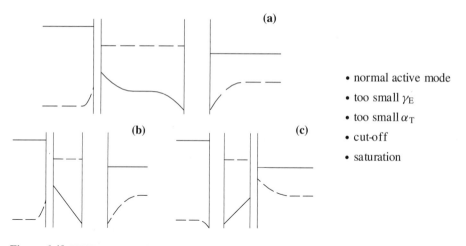

(a)

- normal active mode
- too small γ_E
- too small α_T
- cut-off
- saturation

(b)   **(c)**

Figure 6.40 NPN concentration diagrams

6.3 Assign each of the energy-band diagrams from Fig. 6.42 to the proper description of the BJT type and mode of operation.

6.4 An NPN BJT with base-emitter and base-collector junction areas A_{JE} and A_{JC}, respectively, has the following parameters: the transport factor $\alpha_T = 0.9999$ and the emitter efficiency $\gamma_E = 0.9968$. The ratings of the transistor are the maximum collector-current $I_{max} = 10$ mA, and the maximum collector-base voltage $V_{max} = 25$ V. What would the BJT parameters and ratings be if both junction areas (A_{JE} and A_{JC}) are doubled?

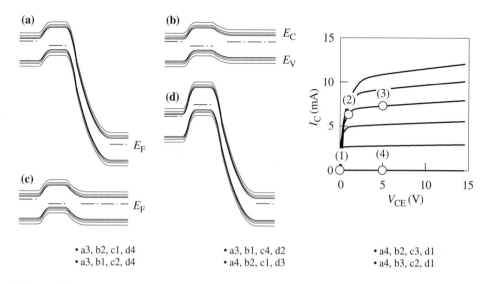

- a3, b2, c1, d4
- a3, b1, c2, d4

- a3, b1, c4, d2
- a4, b2, c1, d3

- a4, b2, c3, d1
- a4, b3, c2, d1

Figure 6.41 Energy-band diagrams and BJT output characteristics.

- $\beta = 604$, $I_{max} = 10$ mA, $V_{max} = 25$ V
- $\beta = 604$, $I_{max} = 20$ mA, $V_{max} = 50$ V
- $\beta = 302$, $I_{max} = 20$ mA, $V_{max} = 12.5$ V

- $\beta = 302$, $I_{max} = 20$ mA, $V_{max} = 25$ V
- $\beta = 151$, $I_{max} = 10$ mA, $V_{max} = 25$ V
- $\beta = 604$, $I_{max} = 20$ mA, $V_{max} = 25$ V

6.5 Which of the following statements, related to BJTs, are **not** correct?

- The relationship $\alpha = \beta/(\beta + 1)$ cannot be used when a BJT is in saturation.

- An increase in V_{BE} drives an NPN BJT from the normal active mode into saturation. As a consequence, the current I_C is increased, but the voltage V_{CE} is decreased.

- The transport factor α_T and the common-base current gain α are different and unrelated parameters.

- The emitter efficiency γ_E depends on the concentration of majority carriers in the base.

- In saturation mode, the collector current I_C does not significantly depend on the input voltage V_{BE}.

- The concentration of the minority carriers in the base is significantly higher compared to the equilibrium level when the BJT is in normal or inverse active mode.

- The concentration of the majority carriers in the base does not affect the transport factor α_T.

- The fact that the collector current I_C does not depend on the output voltage V_{CE}, in normal active mode, is related to the fact that the saturation current of a diode does not depend on the voltage drop across the diode.

6.6 Determine the modes of operation of PNP BJTs on the basis of the following sets of measurements:

(a) $V_{BE} = 0.7$ V, $V_{CE} = -5.2$ V.

(b) $V_{BE} = 0.7$ V, $V_{BC} = -0.7$ V.

(c) $V_{BE} = -0.7$ V, $V_{CE} = -0.2$ V.

(d) $V_{BE} = 0.7$ V, $V_{CE} = -0.2$ V.

(e) $V_{BE} = -5.2$ V, $V_{CE} = 0.7$ V. ▢A

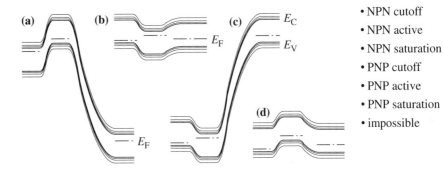

- NPN cutoff
- NPN active
- NPN saturation
- PNP cutoff
- PNP active
- PNP saturation
- impossible

Figure 6.42 Energy-band diagrams of NPN and PNP BJTs.

6.7 The NPN BJT in the circuit of Fig. 6.4 has $\beta = 550$, while $R_C = 1 \text{ k}\Omega$ and $V_+ = 5$ V. Find the minimum base current i_B that ensures that the BJT is in saturation (switch-on mode). Assume $V_{BE} = 0.7$ V.

6.8 A PNP BJT with $\beta = 300$ is used in a circuit similar to the one in Fig. 6.4: the resistor is the same ($R_C = 1 \text{ k}\Omega$), but V_+ is replaced by $V_- = -5$ V. Find the maximum base current $|i_B|$ that ensures that the BJT is in normal active mode. Assume $V_{BE} = -0.7$ V. $\boxed{A}$

Section 6.2 Bipolar IC Technologies

6.9 The NPN BJT of Fig. 6.8 is to be designed for a 10-μm-thick N-epitaxial layer. Simplified design rules can be expressed as follows: the depth of the P$^+$ isolation diffusion should be $x_{j-P^+} = 12 \ \mu$m, minimum size of the diffusion and contact windows is $W_w \times W_w = 5 \times 5 \ \mu$m, minimum spacing between two windows is $S_{w-w} = 5\mu$ m, minimum spacing between a contact edge and the closest P–N junction is $S_{c-j} = 3 \ \mu$m, and minimum spacing between two P–N junctions is $S_{j-j} = 5 \ \mu$m. The lateral diffusion is 80% of the vertical diffusion.

    (a) Calculate the minimum size of the NPN BJT, defined as the distance between the center of the left and the center of the right P$^+$ isolation diffusions in Fig. 6.8.

    (b) Assume that the BJT has a square shape, the side of the square being equal to the minimum distance determined in (a). What percentage of the total BJT area is occupied by the isolation P$^+$ diffusion? $\boxed{A}$

6.10 The average doping levels of a base-diffusion pinch resistor (Fig. 6.13) are $N_{epi} = 10^{14} \text{ cm}^{-3}$, $N_{base} = 10^{16} \text{ cm}^{-3}$ and $N_{emitter} = 10^{20} \text{ cm}^{-3}$. The emitter–base and base–collector junction depths are $x_{jE} = 3 \ \mu$m and $x_{jB} = 5 \ \mu$m, respectively. Assuming abrupt P–N junctions and hole mobility of $\mu_p = 400 \text{ cm}^2/\text{Vs}$, calculate the sheet resistance of this resistor for two different control voltages:

    (a) $V_{EB} = 0$ V.

    (b) $V_{EB} = 5$ V. $\boxed{A}$

6.11 The base–collector junction is to be used as a 0.1-nF capacitor in a bipolar IC. Calculate the needed area of the capacitor if the doping concentrations for the abrupt-junction model are $N_{epi} = 10^{14} \text{ cm}^{-3}$ and $N_{base} = 10^{16} \text{ cm}^{-3}$. The DC bias is $V_R = 5$ V.

6.12 The resistors of the circuits given in Fig. 6.43a are implemented as base-diffusion resistors. (a) Draw the cross section of T_1 and T_2. (b) Draw the cross section of T_4 and T_5. (c) Group the devices into N-epi islands.

6.13 Assuming that the resistors are implemented as base-diffusion resistors, and the diode as a base–collector diode, group the devices of the circuits given in Fig. 6.43b into N-epi islands.

Section 6.3 BJT Modeling

6.14 Generalizing Eq. (6.34) to be applicable to either NPN or PNP BJT, the saturation current of the base–emitter junction can be expressed as

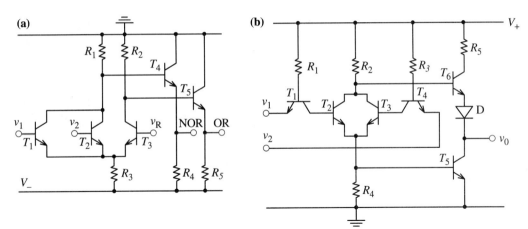

Figure 6.43 Example circuits.

$$I_S = A_E q n_i^2 \left(\frac{D_{\text{base}}}{w_{\text{base}} N_{\text{base}}} + \frac{D_{\text{emitter}}}{w_{\text{emitter}} N_{\text{emitter}}} \right)$$

(a) Derive the equation for the emitter efficiency γ_E. $\boxed{A}$

(b) Calculate γ_E for both an NPN and a PNP BJT, using the following technological parameters: $N_{\text{emitter}} = 10^{19} \text{ cm}^{-3}$, $N_{\text{base}} = 10^{17} \text{ cm}^{-3}$, $w_{\text{emitter}} = w_{\text{base}} = 2 \ \mu\text{m}$, $D_n(10^{19} \text{ cm}^{-3}) = 2.5 \text{ cm}^2/\text{s}$, $D_p(10^{19} \text{ cm}^{-3}) = 1.5 \text{ cm}^2/\text{s}$, $\mu_n(10^{17} \text{ cm}^{-3}) = 800 \text{ cm}^2/\text{Vs}$, $\mu_p(10^{17} \text{ cm}^{-3}) = 320 \text{ cm}^2/\text{Vs}$.

(c) Assuming $\alpha_T = 1$, calculate the common-emitter current gain for both BJTs.

6.15 Calculate the common-emitter current gain (β_R) of an IC NPN BJT operating in inverse active mode. The technological parameters are as follows: $N_{\text{epi}} = 5 \times 10^{14} \text{ cm}^{-2}$, $N_{\text{base}} = 10^{16} \text{ cm}^{-3}$, $w_{\text{epi}} = 10 \ \mu\text{m}$, $w_{\text{base}} = 2 \ \mu\text{m}$, $\mu_n = 1250 \text{ cm}^2/\text{Vs}$, and $\mu_p = 480 \text{ cm}^2/\text{Vs}$. Assume ideal transport factor of the base. $\boxed{A}$

6.16 The parameters of the Ebers–Moll model are $I_{S0} = 10^{-12}$ A, $\beta_F = 500$, and $\beta_R = 1$. Find the values of the emitter, base, and collector currents for

(a) $V_{BE} = 0.65$ V and $V_{CB} = 4.35$ V (normal active mode).

(b) $V_{BE} = 0.65$ V and $V_{CE} = 0.1$ V (saturation mode).

(c) $V_{EB} = 4.65$ V and $V_{BC} = 0.65$ V (inverse active mode). $\boxed{A}$

6.17 For the BJT of Problem 6.16, calculate the transconductance g_m and the small-signal input resistance r_π in normal active mode.

6.18 The measured dynamic output resistance of an NPN BJT is $r_o = 35.7$ kΩ. Calculate the Early voltage if $I_{S0} = 1.016 \times 10^{-16}$ A, and the measurement is taken at $V_{BE} = 0.8$ V. Estimate the resistance at $V_{BE} = 0.7$ V.

6.19 An increase in the reverse base-collector voltage reduces the base width (base modulation effect). At what voltage would the base width be reduced to zero ("punch-through" break-down) if the zero-bias base width is 0.937 μm, and the doping levels are $N_{base} = 10^{16}$ cm^{-3} and $N_{collector} = 5 \times 10^{14}$ cm^{-3}? Assume abrupt junction.

6.20 The breakdown voltage of the BJT considered in Problem 6.19 can be increased by increasing the zero-bias width of the base. What breakdown voltage can be achieved in this way, before the avalanche breakdown of the collector–base junction is reached? Assume $E_{max} = 30$ V/μm. [A]

Sections 6.4 and 6.5 SPICE Parameters and Parasitic Elements Not Included in Device Models

6.21 One set of output characteristics from Fig. 6.44 is for $I_{S0} = 10^{-15}$ A and $\beta_{FM} = 200$ as nominal BJT parameters, while the other three are for either a changed value of β_{FM} or a specified V_A or I_{KF} parameter. Relate each of the output characteristics to the appropriate set of parameters.

6.22 Identify which SPICE parameter should be changed in order to achieve better matching between the model and the experimental results in each of the following cases:

	Model	Experiment
(a)	$r_o = 10$ MΩ	$r_o = 80$ kΩ
(b)	$I_C = 4.9$ mA, $r_\pi = 2.8$ kΩ	$I_C = 4.9$ mA, $r_\pi = 5.2$ kΩ
(c)	$I_C = 1$ mA, $I_B = 10$ μA	$I_C = 6.4$ mA, $I_B = 10$ μA
(d)	$I_{C1} = 1.5$ mA, $I_{B1} = 10$ μA,	$I_{C1} = 1.6$ mA, $I_{B1} = 10$ μA,
	$I_{C2} = 30.0$ mA, $I_{B2} = 10$ μA	$I_{C2} = 19.2$ mA, $I_{B2} = 200$ μA

- increase I_{S0} • increase $C_2 I_{S0}$ • increase β_F
- decrease V_A • decrease I_{KF} • decrease n_{EL}

6.23 Which of the following statements are correct?

- The current gain β_F is the only essential parameter for a BJT in normal active mode.

- The saturation current I_{S0} is the only essential parameter for a BJT in saturation mode.

- The parasitic resistances r_E and r_C are used to set the dynamic output resistance.

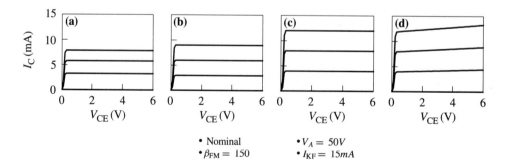

- Nominal   • $V_A = 50V$
- $\beta_{FM} = 150$ • $I_{KF} = 15mA$

Figure 6.44 BJT output characteristics ($I_B = 0$; 20; 40; 60 μA in every diagram).

- The I_{KF} parameter (normal knee current) is important to model the effects at large V_{CE} voltages.

- Neglecting the parasitic elements of the IC base-collector capacitor at low frequencies could not be justified.

- Neglecting the parasitic PNP BJT when the standard IC NPN BJT is in saturation could not be justified.

6.24 Using the data from Table 6.9, estimate the following parameters: I_{S0}, β_F, and I_{KF}. (Use $V_t = 26$ mV.)

6.25 Using the data from Table 6.10, estimate the Early voltage V_A.

6.26 $I_B - V_{BE}$ measurements taken at low bias level are given in Table 6.11. Determine the related parameters $C_2 I_{S0}$ and n_{EL} ($V_t = 26$ mV).

TABLE 6.9 Input and Transfer Characteristics Data

V_{BE} (V)	I_B (A)	I_C (A)	V_{BE} (V)	I_B (A)	I_C (A)
0.60	1.2×10^{-7}	1.2×10^{-5}	0.78	1.3×10^{-4}	7.3×10^{-3}
0.62	2.6×10^{-7}	2.6×10^{-5}	0.80	2.8×10^{-4}	1.2×10^{-2}
0.64	5.5×10^{-7}	5.6×10^{-5}	0.82	6.0×10^{-4}	2.0×10^{-2}
0.66	1.2×10^{-6}	1.2×10^{-4}	0.84	1.3×10^{-3}	3.1×10^{-2}
0.68	2.7×10^{-6}	2.6×10^{-4}	0.86	2.8×10^{-3}	4.8×10^{-2}
0.70	5.8×10^{-6}	5.5×10^{-4}	0.88	6.1×10^{-3}	7.3×10^{-2}
0.72	1.2×10^{-5}	1.1×10^{-3}	0.90	1.3×10^{-2}	1.1×10^{-1}
0.74	2.7×10^{-5}	2.2×10^{-3}	0.92	2.9×10^{-2}	1.6×10^{-1}
0.76	5.9×10^{-5}	4.1×10^{-3}	0.94	6.2×10^{-2}	2.4×10^{-1}

TABLE 6.10 Output Characteristics Data

V_{CE}	I_C (mA)				
(V)	$I_B = 20\ \mu A$	$I_B = 40\ \mu A$	$I_B = 60\ \mu A$	$I_B = 80\ \mu A$	$I_B = 100\ \mu A$
3.0	2.73	5.05	7.07	8.82	10.2
6.0	2.79	5.21	7.34	9.24	11.0
9.0	2.85	5.34	7.54	9.53	11.4
12.0	2.90	5.46	7.74	9.81	11.7
15.0	2.95	5.58	7.93	10.1	12.1

TABLE 6.11 Transfer Characteristic Data

V_{BE} (V)	I_B (nA)	V_{BE} (V)	I_B (nA)
0.60	5.0	0.63	9.50
0.61	6.3	0.64	12.0
0.62	7.6	0.65	15.0

R-6.1 Increased reverse-bias voltage increases the slope of energy bands, that is the electric field, in the depletion layer of a P–N junction. Why doesn't this significantly increase the reverse-bias current?

R-6.2 The equilibrium concentration of electrons in the base of an NPN BJT is n_e. Is the concentration of electrons significantly higher when this BJT is biased in the normal active mode? Is this concentration related to the collector current?

R-6.3 Does the concentration of the majority carriers in the base affect the emitter efficiency γ_E? If yes, how?

R-6.4 Does it affect the transport factor α_T? If yes, how?

R-6.5 Is the transport factor α_T different from the common-base current gain α? Are they related? If yes, how?

R-6.6 Is the common-base current gain α related to the common-emitter current gain β?

R-6.7 Does the relationship between α and β apply in saturation and cutoff?

R-6.8 A change in the base-emitter voltage V_{BE} drives an NPN BJT from the normal active mode into saturation. Is V_{BE} increased or decreased? If the BJT biasing circuit is as in Fig. 6.4, is V_{CE} increased or decreased? How about I_C?

R-6.9 The measured voltage between the base and collector of a BJT connected in a circuit is $V_{BC} = -5$ V. Provided the BJT is not broken down, can you say if this is an NPN or a PNP BJT?

R-6.10 How do the characteristics of a lateral PNP compare to the characteristics of the standard NPN BJT?

R-6.11 Why is the PNP BJT in the standard bipolar IC technology made not with different and complementary layers (mirror image)? In other words, why are the same layers used at the expense of the PNP performance?

R-6.12 What parameters are essential for a proper SPICE simulation of a circuit with a BJT in the normal active mode? Inverse active mode? Saturation?

R-6.13 Low-frequency operation is implicitly assumed in the previous question. Answer that question for the case of high frequency.

R-6.14 Which SPICE parameter is used to adjust the dynamic output resistance?

R-6.15 Is the I_{KF} parameter used to model effects of large V_{CE} voltage or effects of large I_C current?

R-6.16 When do you need to worry about the parasitic PNP BJT?

R-6.17 Could neglecting the parasitic elements of the base-diffusion resistor at low frequencies be justified?

R-6.18 Could neglecting the parasitic elements of the base-collector capacitor at low frequencies be justified?

PART 2

ADVANCED TOPICS

Chapter 7

ADVANCED AND SPECIFIC IC DEVICES AND TECHNOLOGIES

Integrated-circuit (IC) technology has undergone unmatched progress, truly improving peoples' lives. In a way, this is illustrated by the names given to describe different levels of integration: small-scale integration (SSI), medium-scale integration (MSI), large-scale integration (LSI), very large-scale integration (VLSI), and ultralarge-scale integration (ULSI). Although each of these phases appears as an individual *era* in economic terms, remarkably the device and technology principles have not changed. The principles of modern MOSFETs are essentially the same, although critically important features have been added to enable the aggressive dimension down-scaling. The first section of this chapter describes the deep submicron MOSFET, concluding with a consideration of the current issues and trends. The second section is devoted to memory devices, which are specific in terms of device principles, but otherwise have been a standard part and application driver for IC technology development since the very beginning. Finally, the third and fourth sections introduce silicon-on-insulator (SOI) and BiCMOS as advanced technologies likely to foster further progress in IC technology.

7.1 DEEP SUBMICRON MOSFET

7.1.1 Down-Scaling Benefits and Rules

The switching speed (maximum operating frequency) was initially a disadvantage of CMOS circuits. The v_{OUT} vs t diagram of Fig. 5.14b illustrates the times needed to achieve the high/low output level due to the charging/discharging of the load capacitance C_L. These times are determined by the value of the load capacitance, which is the input parasitic capacitance of the subsequent CMOS cells, and the value of the charging/discharging current supported by the MOSFETs. Both these parameters change favorably when the MOSFET channel length L is reduced. A reduction of the channel length by a factor of S would reduce the input capacitance of the CMOS cell S times, due to reduced gate area. It would also increase the charging/discharging current S times as the MOSFET current increases proportionally with the channel length reduction. This would mean an improvement in speed by a factor of S^2. Additionally, this would mean a smaller cell area, thus the possibility of integrating more logic cells to create more powerful ICs. These are extremely motivating, and economically extremely rewarding benefits.

However, a successful reduction of the channel length requires some other device parameters to be appropriately adjusted to avoid possible adverse effects. One of the things that may happen when the channel is shortened is that the drain field may start taking electrons directly from the source (punch-through effect). To prevent this from happening, the substrate doping is increased, which shortens the penetration of the drain field in the substrate. As Eq. (2.33) shows, to reduce a depletion-layer width S times, the doping concentration has to be increased S^2 times. However, the increased substrate doping increases the body factor γ and consequently the threshold voltage, as can be seen from Eqs. (2.79) and (5.11), respectively. A device that turns *ON* at a voltage higher than 25% of the supply-voltage level is generally not acceptable. To keep the threshold voltage down, it is necessary to reduce the thickness of the gate oxide S times. As this increases the input capacitance, to compensate the channel width is also reduced S times. Table 7.1 summarizes these steps as a set of down-scaling guidelines, more frequently referred to as down-scaling rules, and their effects.

However, it was not always possible to avoid all the important adverse effects of scaling down by applying these general rules. The channel length reduction and the associated substrate doping increase result in an increase of the lateral electric field in the channel. Equation (2.20) and Fig. 2.8c illustrate the electric-field dependence on the doping level in the case of an abrupt P–N junction. This increase in the electric field means that the breakdown voltage (Section 3.3) is reduced, reducing the maximum operating voltage. Reductions of the maximum operating voltage of CMOS ICs have already occurred—initially, CMOS ICs could operate at voltages higher than 20 V, then the standard was 5 V, while now we have limitations of 3.3 and 1.5 V, and this will further change in the future. Also, there is a reliability problem associated with high electric fields in the channel, which is discussed in Chapter 11.

7.1.2 Deep-Submicron MOSFET

To reduce the problem of high electric field in the MOSFET channel, the abrupt N^+-drain and source–P-substrate junctions are modified. Lightly doped regions are introduced to linearize the junctions, that is to reduce the maximum doping level at the junctions. An N-channel MOSFET with lightly doped drain and source extensions is shown

TABLE 7.1 General Down-scaling Rules and Their Effects

Channel length	L	$\longrightarrow$	L/S
Channel width	W	$\longrightarrow$	W/S
Gate-oxide thickness	t_{ox}	$\longrightarrow$	t_{ox}/S
Substrate doping	$N_{A,D}$	$\longrightarrow$	$N_{A,D} \times S^2$
Drain current	$I_D \propto W/(Lt_{ox})$	$\longrightarrow$	$I_D \times S$
Input capacitance	$C_{in} \propto WL/t_{ox}$	$\longrightarrow$	C_{in}/S
Maximum switching frequency	$f \propto I_D/C_{in}$	$\longrightarrow$	$f \times S^2$
Cell area	$A \propto WL$	$\longrightarrow$	A/S^2

in Fig. 7.1. As the doping concentration is reduced at the drain-extension–substrate junction, the maximum lateral field appearing at the drain end of the channel is reduced as well. MOSFETs with this type of doping structure are referred to as *lightly doped drain* (LDD) MOSFETs.

The drain and source extensions of the LDD MOSFET are doped by implanting arsenic or antimony ions, which have lower diffusion constants than phosphorus. The lower diffusion constants make it possible to achieve shallow drain/source extensions, which minimizes the punch-through effect. The drain/source extensions are implanted with the polysilicon gate as a mask, to achieve the self-alignment between the gate and the drain/source areas. The spacer layer shown in Fig. 7.1 is created afterward to be used as a mask when the N^+ regions are created by phosphorus implantation. The spacer layer is created by deposition of either oxide or silicon nitride, which fills the polysilicon/silicon-substrate corner, and subsequent etching to remove the deposited layer from the top of the polysilicon and outside the corner region of the silicon substrate.

The channel-length reduction leads to an undesirable increase in the resistance of the gate line. To minimize this effect, the MOSFET structure is further modified by creating a silicide layer at the top of the gate (Fig 7.1). The silicide can be created by depositing titanium, and subsequent annealing, which leads to a reaction between the deposited titanium and the underlying silicon to create a layer of $TiSi_2$. The resistivity of the silicide is much smaller than the resistivity of the polysilicon, which reduces the gate resistance. The silicide is also created in the drain and the source regions to improve the contact between the metalization and the silicon, that is to reduce the contact resistance.

7.1.3 Current Issues and Trends

Down-scaling has dominated microelectronics research, resulting in the rapid progress of CMOS-based electronics that we have seen in the past decades. Many sophisticated processes had to be developed, but the economic background in the form of ensured market demand for ever cheaper and ever more exciting electronic "toys" was there to drive these developments to success. The channel length of the first MOSFETs was tens of microns, and is now below $0.2 \, \mu m$ in production. It may seem that the 100-nm benchmark is just a natural step in the same direction, but the following analysis shows that going beyond that level of dimension reduction is creating issues that are forcing research and industry toward more radical solutions.

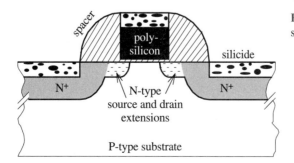

Figure 7.1 Deep-submicron MOSFET structure.

MOSFETs with channel lengths less than 100 nm were experimentally demonstrated as early as 1987 by IBM researchers,[1] followed by a demonstration of a sub-50-nm MOSFET by Toshiba researchers in 1993.[2] Although operational, and potentially applicable in a number of specific areas, these MOSFETs cannot be used as building blocks for CMOS ULSI circuits.

Figure 7.2a shows simulated transfer characteristics of a 100-nm MOSFET with parameters and characteristics similar to the one published by the IBM researchers. As the problems with such a short-channel MOSFET are mostly in the subthreshold region, the left-hand drain-current axis is presented in a logarithmic scale so that the subthreshold current can be seen over several orders of magnitude. According to the basic theory, the subthreshold current is due to diffusion, and should not depend on V_{DS} voltage [Eq. (5.44)]. The MOSFET in the subthreshold region should behave as a BJT in the normal active mode, where the collector (the drain in this case) should collect the carriers emitted into its depletion layer, but should not be generating a current on its own. If that were the case, there would not be a shift between the subthreshold transfer characteristics for $V_{DS} = 0.1$ V and $V_{DS} = 1.0$ V in Fig. 7.2a. The existence of this shift, however, indicates a short-channel effect due to too deep penetration of the drain-to-substrate depletion layer under the gate. The energy-band diagram and the equipotential contours in Fig. 7.3 illustrate that there is too strong an influence from the drain voltage in the region that should be controlled by the gate. There should be an appropriate energy barrier between the source and the channel region for $V_{GS} = 0$ V, however, this barrier is significantly reduced by the drain (the DIBL effect), allowing too many electrons from the source to get over and roll down into the drain (Fig. 7.3a).

The most important problem with a too high "off" current is that the CMOS cells would start consuming significant power, limiting the number of cells that can be integrated. The "off" current in Fig. 7.2a is >1 μA, which means the "leakage current" of a 1-million-transistor IC would be at an unacceptable >1 A level. Quite a few techniques have been under investigation in an attempt to find a commercially viable solution to this problem. Two directions are worth mentioning. One is engineering of the substrate doping profile. For example, a high doping level is used below the surface region, where the gate field is not as strong, with a lower doping at the top to reduce the threshold voltage. Another promising, but more radical direction is the use of a silicon-on-insulator (SOI) structure (the SOI technology is described in Section 7.3).

Although there may be a viable solution for 100-nm MOSFET within the scope of somewhat modified/extended current technology, the following analysis illustrates some ultimate issues and problems as the dimension down-scaling continues. Although quantum-mechanics effects are seen as the ultimate limit (maybe they provide great opportunities!), there are problems well before that, even at the level at which the principles of down-scaling rules still work. Figure 7.2b shows simulated transfer characteristics of the same

---

[1]G.A. Sai-Halasz, M.R. Wordeman, D.P. Kern, E. Ganin, S. Rishton, D.S. Zicherman, H. Schmid, M.R. Polacri, H.Y. Ng, P.J. Restle, T.H.P. Chang, and R.H. Dennard, "Design and experimental technology for 0.1–μm gate-length low-temperature FET's," *IEEE Electron Dev. Lett.*, **EDL-8**, 463–466 1987.

[2]M. Ono, M. Saito, T. Yoshitomi, C. Fiegna, T. Ohguro, and H. Iwai, "Sub-50 nm gate length n-MOSFET's with 10 nm phosphorus source and drain junctions," *IEDM Tech. Dig.*, 119–122, 1993.

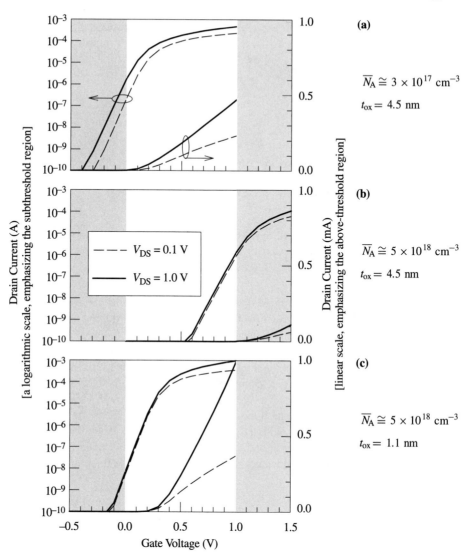

Figure 7.2 Transfer characteristics of 100-nm MOSFETs (the channel width is 1 μm): (a) devices with similar characteristics were experimentally demonstrated as early as 1987, but they exhibit pronounced short-channel effects and unacceptably high "off" current; (b) as known from the down-scaling rules, an increase in the substrate doping eliminates the short-channel effects, but also increases the threshold voltage, in this case above the 1.0 V supply voltage; (c) down-scaling rules require a 1.1 nm gate oxide!

MOSFET when the substrate doping (the dose of the threshold-voltage adjustment implant) is increased 16 times. The short-channel effects are all but eliminated, however, the threshold voltage is increased even beyond 1.0 V, which is set as the maximum supply voltage. Of course, the down-scaling rules require a simultaneous reduction in gate-oxide thickness by

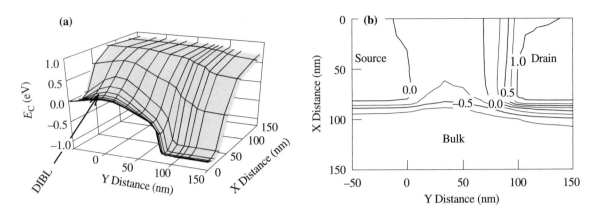

Figure 7.3 Energy-band diagram (a) and equipotential contours (b) for the MOSFET with the transfer characteristics shown in Fig. 7.2 (a): $V_{GS} = 0V$, $V_{DS} = 1.0$ V.

a factor of 4. Comparing the energy-band diagrams and equipotential contours in Figs. 7.3 and 7.4, we can see how the increased substrate doping and reduced gate-oxide thickness eliminate the drain influence in the gate region. Figure 7.2c shows that this does result in acceptable transfer characteristics: low "off" current and low threshold voltage. A problem here is that the oxide thickness needs to be about 1 nm. Even if an almost perfect oxide of 1 nm thickness was technologically achievable, the electrons would tunnel through the 1-nm barrier that such a gate oxide would provide. With a high gate current, the problem with the power dissipation would reappear in just another form.

A possible solution to the problem of gate-oxide thickness is a replacement of the oxide by another material with a higher permittivity. For control over the channel region by the gate voltage, it does not mater whether the oxide thickness is reduced or the permittivity

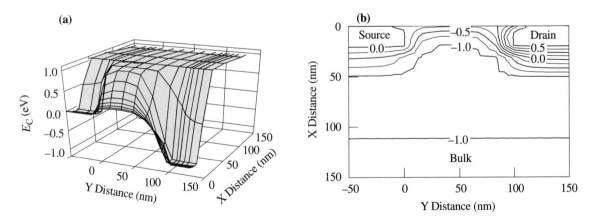

Figure 7.4 Energy-band diagram (a) and equipotential contours (b) for the MOSFET with the transfer characteristics shown in Fig. 7.2c: $V_{GS} = 0$ V, $V_{DS} = 1.0$ V.

is increased. The strength of the field effect is related to the gate-dielectric capacitance per unit area:

$$C = \frac{\varepsilon_d}{t_d} \tag{7.1}$$

There are dielectric materials with permittivities hundreds and even thousands of times higher than the permittivity of the silicon dioxide. Obviously, the same strength of the field effect could be achieved with much ticker dielectric layers. This approach would solve the tunneling limitation, but it would not remove the leakage current problem in the case of many potential gate dielectrics, notably ferroelectric materials. Some high-ε materials, in particular metal oxides such as Ta_2O_5, exhibit acceptable leakage and breakdown field properties. Their big disadvantage has always been a much poorer interface between these films and silicon, compared to the native silicon–silicon dioxide interface. The set of excellent properties of SiO_2 (the quality of interface to Si, the dielectric strength, the large energy-band discontinuities with respect to both the conduction and the valence bands in silicon, hence large barrier heights) and fair blocking of dopant diffusion outweigh the comparative disadvantage of low dielectric permittivity. As a result, the oxide has remained the superior gate material so far. However, a solution will be needed for the tunneling limit, and that is where the metal oxides are expected to play a role.

Another helpful possibility is to replace the polysilicon gates with a suitable metal so that two goals are achieved. Primarily, this is to help reduce the threshold voltage by a selection of a metal with a smaller work function compared to N^+ polysilicon. As the effect of a more negative work-function difference would be a parallel negative shift of the transfer characteristics, this would alleviate the need for aggressive scaling down of the gate dielectric thickness. Second, metal-gate lines would exhibit smaller parasitic gate resistances. However, the gate metal would have to be compatible with the semiconductor processing, in particular the high ion-implant anneal temperatures (the "self-aligning" technology requires the gate to be formed before the drain/source extension implant).

Continuing down-scaling involves a number of other issues as well. The lithography for a mass production of sub-100-nm lines is certainly one of the most important issues. A number of "tricks" have been applied to produce subwavelength lines by the current optical lithography (photolithography). However, photolithography is approaching the ultimate limit, and new solutions will be needed; current research trends involve x-ray lithography and parallelized direct electron-beam writing.

7.2 MEMORY DEVICES

Information storage is a very important function in a variety of electronic systems, and definitely of essential importance in computers. Although the memory function can be implemented with a number of nonsemiconductor devices, the progress made in the area of semiconductor technologies is having a very strong impact on the performance, cost, reliability, and physical size of semiconductor-based memory devices. Consequently, semi-conductor memories are both expanding memory-based applications and replacing non-semiconductor memory devices in a number of existing applications.

The basic characteristics divide the semiconductor memories into two groups: random-access memories (RAM) and read-only memories (ROM). RAMs provide very fast information reading and writing, which can be performed an unlimited number of times. However, RAMs are volatile memories as the information is kept only for as long as the power supply is connected to the RAM devices. Additionally, in a class of RAMs known as dynamic RAMs (as opposed to static RAMs), the information has to be regularly refreshed. ROMs are nonvolatile memories, as no power supply is needed to keep the information. Originally, ROMs could provide only reading of the information that was technologically "written." Now so-called EEPROMs (electrically erasable and programmable ROM) provide the option of multiple information writing (memory programming) in addition to fast information reading. The information writing (memory programming), however, is relatively slow, limiting a possible replacement of RAMs by the modern ROMs.

Modified MOSFET technologies have to be used to manufacture ROMs and dynamic RAMs. This section briefly describes representative memory technologies.

7.2.1 Random-Access Memory (RAM)

As mentioned above, RAMs can be divided into *static* and *dynamic*. The static RAMs are based on bistable electronic circuits, known as flip-flops. As the circuit of a single static RAM cell involves a number of semiconductor components, these memories are not area efficient, and as a consequence the memory size that can be achieved by a single memory IC is very much limited.

The dynamic RAMs are based on the use of a MOS capacitor to store the information, and at least one MOSFET to enable the information reading. A single MOSFET RAM cell is illustrated in Fig. 7.5. This cell memorizes a single **binary digit** (bit), which is one of the two possible states of the memory capacitor C_S: the capacitor empty or the capacitor charged. A number of cells are grouped to create a word, and then the words are organized in an array to define the memory chip area. The MOSFET gates of all the cells making a word are connected to create *read/write select*, or *word* line. Negative voltage at this line will turn all the word's MOSFETs *ON* (note that P-channel MOSFET is used in the cell shown in Fig. 7.5), selecting the particular word for reading or writing. Memory capacitors C_S of all the other words remain isolated, as all the other MOSFETs remain *OFF*. The *read/write* (or *bit*) line will decide which particular cell (bit) of the selected word will be read or written. Reading the information is achieved by sensing the voltage level at the read/write line, while writing the information is achieved by charging or discharging the capacitor C_S.

Although the semiconductor plate of the capacitor C_S is floating when the MOSFET is *OFF*, the charge stored at the capacitor diminishes with time. This is due less to leakage through the capacitor dielectric as to carrier recombination at the semiconductor plate. Because of this effect, the information has to be regularly refreshed. There is a sophisticated circuit as a part of the memory, which reads and refreshes the information at any single bit at regular time intervals. Even though this circuit adds to the complexity of the memory chip, the much smaller area occupied by the dynamic memory cell, compared to its static counterpart, favors the dynamic RAM technology for large memory capacities.

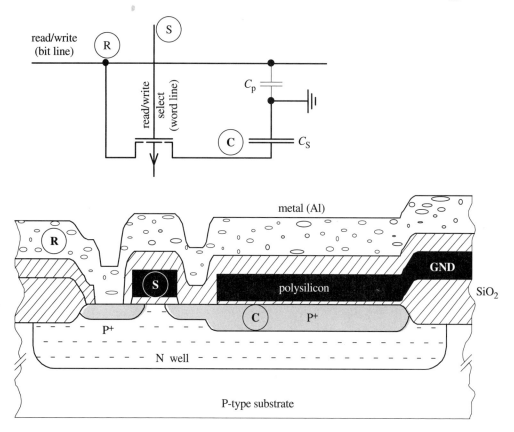

Figure 7.5 A single MOSFET RAM cell.

7.2.2 Read-Only Memory (ROM)

To create nonvolatile memory, which would keep the information even when the power supply is disconnected from the memory element, significantly modified devices have to be used. Memory devices, which provide the option of electrical erasing and programming, are based on trapping electrons in a deep potential well, so deep that they are very unlikely to gain enough thermal energy and escape from the potential well. Figure 7.6a shows the cross section of what is known as a *flash memory* MOSFET. The deep potential well for electron trapping is created by inserting a floating gate between the MOSFET gate (now referred to as the control gate) and the silicon substrate. The discontinuity of the bottom of the conduction band, appearing at the silicon–oxide interface, is used to create the trapping potential well. As this discontinuity is more than 3 eV (Fig. 7.6c), and as the floating gate is completely surrounded by oxide, the electrons appearing in the floating gate would be trapped.

Figure 7.6c shows the energy bands for the case of erased state (no trapped electrons in the floating gate). The electrons that appear in the floating gate in this case are only the

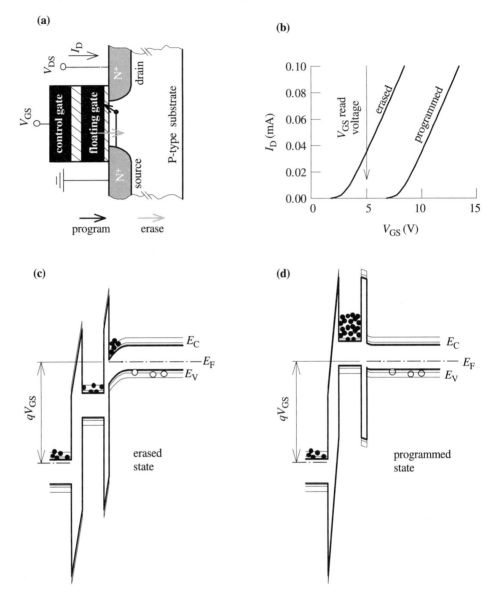

Figure 7.6 Flash-memory MOSFET: the cross section (a), the transfer characteristics (b), and the energy bands in erased (c) and programmed state (d).

normal doping-induced electrons in the N-type polysilicon. The MOSFET is designed so that application of V_{GS} reading voltage turns the MOSFET on, creating channel of electrons between the drain and the source. Figure 7.6b shows that the V_{GS} reading voltage is larger than the threshold voltage of the MOSFET in the erased state.

To program the MOSFET, a large V_{DS} voltage is applied to accelerate the channel electrons to kinetic energies larger than the energy barrier between the silicon and the oxide. So accelerated electrons are referred to as *hot electrons*. A number of these hot electrons will elastically scatter, changing the direction toward the oxide, and with their high kinetic energy will overpass the energy barrier created by the oxide to appear in the floating gate area. With sufficient thickness of the floating gate, most of these electrons will lose their energy inside the floating gate (through nonelastic scattering) to get trapped in the potential well created by the floating gate. Although this process does not seem to be very efficient, with relatively long programming times (in the order of milliseconds), enough electrons can be collected in the potential well of the floating gate to change the MOSFET state. Figure 7.6d shows the energy bands of the MOSFET with electrons trapped in the floating gate. The electric field of these electrons shifts the floating-gate energy bands upward, and as a consequence changes the band bending (and the electric field) at the substrate surface. If enough electrons are trapped in the floating gate, the reading V_{GS} voltage will not be strong enough to compensate for the effect of the trapped electrons, and will not be able to create a channel of electrons at the substrate surface (turn the MOSFET on). Figure 7.6b shows that the threshold voltage of the programmed MOSFET is larger than the V_{GS} reading voltage. The programmed MOSFET is *OFF*, as opposed to the erased MOSFET, which is in *ON* state.

To erase a programmed MOSFET, a large negative voltage is applied to the control gate, which forces the trapped electrons to tunnel back into the silicon substrate through the ultrathin oxide separating the substrate and the floating gate (Fig.7.6a).

The MOSFET can be programmed and erased millions of times before the erasing and programming mechanisms show any observable adverse impact on the MOSFET characteristics. The MOSFET fatigue is a reliability issue, considered in detail in Chapter 11.

7.3 SILICON-ON-INSULATOR (SOI) TECHNOLOGY

A silicon-on-insulator (SOI) structure is very suitable for many applications. A number of improvements can be achieved with SOI CMOS technology. The SOI structure, illustrated in Fig. 7.7a, is typically obtained by one of the following two techniques: (1) **s**eparation of silicon by **im**planted **ox**ygen (SIMOX) and (2) bonded wafers. In the SIMOX process, oxygen is implanted as deeply as possible into the silicon wafer, and annealed at very high temperature to create the buried oxide (SiO_2) layer. The thickness of the buried oxide obtained by SIMOX is typically around 400 nm, while the thickness of the top monocrystalline silicon is around 200 nm. A lot thicker buried oxide and more flexible thicknesses of the top silicon layer can be achieved by the bonded wafer process. In this case, two silicon wafers with pregrown oxides at their surfaces are placed face to face and exposed to further thermal oxidation, which bonds them to each other. After this, one of the wafers is thinned down to the desired top silicon thickness.

To create a CMOS inverter using an SOI substrate, active areas are defined by depositing and patterning silicon nitride (Fig. 7.7b), similar to the basic CMOS technology process. The silicon in the field region (outside the active regions) is etched to approximately half of the original thickness, and then exposed to thermal oxidation (LOCOS). As the field oxide

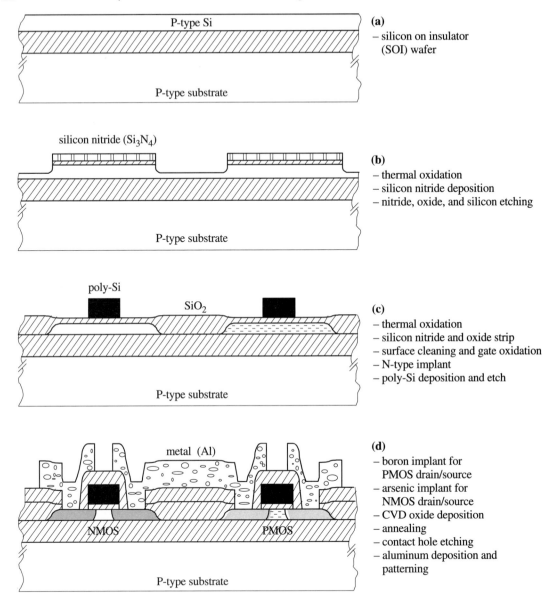

(a)
– silicon on insulator (SOI) wafer

(b)
– thermal oxidation
– silicon nitride deposition
– nitride, oxide, and silicon etching

(c)
– thermal oxidation
– silicon nitride and oxide strip
– surface cleaning and gate oxidation
– N-type implant
– poly-Si deposition and etch

(d)
– boron implant for PMOS drain/source
– arsenic implant for NMOS drain/source
– CVD oxide deposition
– annealing
– contact hole etching
– aluminum deposition and patterning

Figure 7.7 Silicon-on-insulator (SOI) CMOS technology.

grows, it consumes the remaining top silicon in the field region, joining with the buried oxide (Fig. 7.7c). In this way, islands of silicon, isolated from each other by oxide, are created. The remaining processing is very similar to the basic CMOS technology process: polysilicon is deposited and patterned (Fig. 7.7c), P$^+$ and N$^+$ areas implanted to create the sources and the drains of the P-channel and N-channel MOSFETs, and isolation oxide and top metal are deposited and patterned to create the final structure (Fig. 7.7d).

The main advantages of SOI CMOS technology are as follows:

1. The thickness of the oxide in the field region is significantly increased, which significantly reduces the parasitic capacitances created between the interconnecting metal lines and the silicon substrate. Reduced parasitic capacitances result in improved switching speed.
2. N-channel and P-channel MOSFETs are isolated by a dielectric (as opposed to P–N junction), which reduces the leakage current, and more importantly improves the reliability of the IC. Reliability issues are considered in Chapter 11. Here, it can only be mentioned that P^+-source–N-well–P-substrate–N^+-source layers of the N-well CMOS inverter create a parasitic P–N–P–N thyristor structure. This thyristor structure is normally *OFF*, however, if turned *ON* under certain unpredicted conditions, it creates a short circuit between the V_+ and ground rails, permanently damaging the IC. This effect, known as *latch-up*, is one of the primary reliability problems in standard CMOS ICs. The SOI CMOS technology eliminates the N-well–P-substrate junction altogether, eliminating the parasitic thyristor structure, and therefore the latch-up problem.
3. The top silicon layer can be thinned down to tens of nanometers, which means that source and drain regions as shallow as tens of nanometers can reliably be achieved. This reduces the punch-through problem, helping to design more efficient deep-submicron CMOS devices.

7.4 BICMOS TECHNOLOGY

We compared the BJT and the MOSFET in Section 6.1.5 and discovered that no device offers the ultimate advantage. The BJTs proved especially good as output buffers, needed to provide the necessary output power. It is possible to merge the CMOS and standard bipolar IC technologies to have the advantages of both, the CMOS and bipolar circuits, on the same chip. Although this inevitably leads to more complex, and therefore more expensive technology, the benefits gained sometimes justify the cost. The technology that merges bipolar and CMOS IC technologies is referred to as BiCMOS technology.

Figure 7.8 shows the cross section of a CMOS inverter and a standard NPN BJT obtained by a BiCMOS technology process. As can be seen, a P^+ silicon substrate is the starting material, and a P-epitaxial layer is deposited after the buried N^+ diffusion. Using the P-epi is an obvious difference from the standard bipolar technology, which uses N-epi, but it is needed to provide the substrate for the N-channel MOSFETs. The equivalent of the N-epi islands appearing in the standard bipolar technology is achieved by the deep N-type diffusion, which creates the N- well regions needed for the P-channel MOSFETs. Note that P^+ diffusion, which is necessary for the source and drain of the P-channel MOSFETs, is also used to improve the base contact of the NPN BJT.

The buried N^+ diffusion and the deposition of P-epi layer are additional process steps, comparing to the standard N-well CMOS technology. Another addition is the P-type diffusion needed for the base of the NPN BJT. This is a more complex process than the N-well CMOS technology, however, it provides all the devices available in the standard bipolar technology, in addition to the CMOS devices.

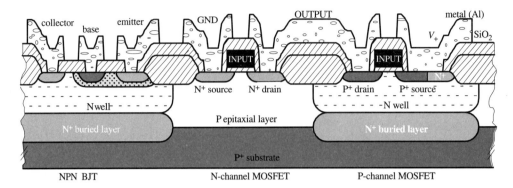

Figure 7.8 The cross section of a CMOS inverter and a standard NPN BJT illustrating a BiCMOS technology.

REVIEW QUESTIONS

R-7.1 Is not it true that MOSFET channel reduction increases the IC speed because it takes less time for the electrons to travel from the source to the drain? If not, why?

R-7.2 Does the MOSFET input capacitance influence the IC speed? How about the MOSFET current?

R-7.3 Punch-through is when the electric field from the drain reaches the source, bypassing the channel. Does this effect depend on the depth of source and drain regions? Does it depend on the substrate doping level?

R-7.4 As the doping level can be increased so much that the field penetration is reduced to several nanometers, why is it not possible (as yet) to develop sub-100-nm CMOS technology?

R-7.5 The threshold voltage is proportional to $\sqrt{N_A}$, where N_A is the substrate doping level. Is this a problem in terms of avoiding the punch-through to develop a sub-100-nm CMOS technology?

R-7.6 The threshold voltage is proportional to t_{ox}, where t_{ox} is the gate-dielectric thickness. Can this fact be used to maintain the threshold voltage value below the operating voltage even in a sub-100-nm CMOS technology? If not, why?

R-7.7 LDD MOSFET became the standard N-channel MOSFET structure. What are the beneficial effects that prevail over the increased technological complexity of this structure?

R-7.8 Reduced channel length means narrower polysilicon gate lines and hence increased gate resistance. Is that a problem? If yes, is there any solution?

R-7.9 Why is it necessary to have both RAMs and ROMs? Could the needed memory function be performed by a single type of memory? Theoretically, is such a type of memory possible?

R-7.10 The principle of information storage is the same in both RAMs and ROMs—electrons are placed in a potential energy well to create a situation corresponding to one binary state, while the energy well is empty in the case of the other binary state. What is the essential difference?

R-7.11 Why is it necessary to refresh dynamic RAMs? How frequently is this done?

R-7.12 What are the advantages of SOI-based CMOS technology?

R-7.13 Is BiCMOS technology always superior to CMOS technology? If not, why?

Chapter *8*

PHOTONIC DEVICES

Electrons can interact with light through the following fundamental mechanisms: (1) *spontaneous light emission*, (2) *light absorption*, and (3) *stimulated light emission*. Each of these three effects is exploited for very useful devices: light-emitting diodes (LED), photodetectors/solar cells, and lasers, respectively. A common name for these devices is *optoelectronic* devices, or alternatively *photonic* devices. These devices cover a range of very important applications: displays, sensors, optical communications, control, etc.

Photons are quanta of light energy: $h\nu$, where h is Planck's constant and ν is the light frequency. To emit a photon, an electron has to give away energy equal to $h\nu$, while after photon adsorbtion an electron gains energy equal to $h\nu$. In semiconductors, electrons lose and gain energy through the processes of *recombination* with the holes and electron–hole *generation*, respectively. Although essential for photonic devices, the recombination–generation processes are of general importance for a sound understanding of semiconductor electronics. The first two sections of this chapter introduce LEDs and photodetectors and also provide an in-depth treatment of the recombination–generation processes. The final section is devoted to lasers.

8.1 LIGHT-EMITTING DIODES (LED): CARRIER RECOMBINATION

The concept of carrier recombination was introduced in Section 1.2.5, and then used in Section 3.1.2 to explain the concentration diagrams of minority carriers at the P–N junction. An electron in the conduction band (a mobile electron) must release some energy to fall down into the valence band and recombine with a hole. In some cases, this energy is released as a photon (spontaneous emission). Therefore, an appropriately designed diode can emit light due to the recombination of the minority carriers. The forward-bias current pushed through the diode determines the rate of recombination, and therefore the light intensity. The basics of *light-emitting diodes* (LED) are introduced in the first part of this section. The subsequent three sections consecutively deepen our insight into the carrier-recombination models and effects. These three sections introduce a recombination model involving the concept of carrier lifetime, extend the continuity equation with the effective recombination term, and introduce the concept of direct and indirect semiconductors, respectively.

8.1.1 LED Application

Light-emitting diodes (LED) are specifically designed and manufactured to produce light when operated in the forward-bias mode. Figure 8.1a shows a simple LED driving circuit, while Fig. 8.1b indicates the operating point on the I_D–V_D characteristic.

The light intensity is directly proportional to the current flowing through the diode:

$$Light\ Intensity \propto I_{DO} = \frac{v_{IN} - V_{DO}}{R} \tag{8.1}$$

If $v_{IN} \gg V_{DO}$, the light intensity is directly proportional to the input voltage v_{IN}. LEDs are very frequently used as binary indicators, that is to visualize the two (*yes* and *no*) logic states. In this case the diodes are operated in two points: zero current (light off) and the optimum (recommended) operating current I_{DO} (light on). The resistor R of the driving circuit is used to adjust the "on" level of the input voltage to the optimum operating current I_{DO}.

The LEDs are not silicon diodes, for reasons that will be described in Section 8.1.4, and their forward-bias voltages are typically significantly larger than 0.7 V. Moreover, LEDs producing different colors are made of different materials and have different forward-bias voltages V_{DO}. To explain this point in more detail, the following section considers the light-emitting mechanism using the energy-band model.

8.1.2 Carrier Recombination

The energy-band diagram of a forward-biased diode was shown in Fig. 3.5c. It is reproduced again in Fig. 8.2b to specifically illustrate the recombination of the minority carriers. As illustrated in this figure, the electrons with high enough energy to overpass the energy barrier at the depletion layer and appear on the P-type side as minority carriers sooner or later recombine with majority holes. Analogously, the holes appearing in the N-type region are recombined with the majority electrons. In the process of electron–hole recombination,

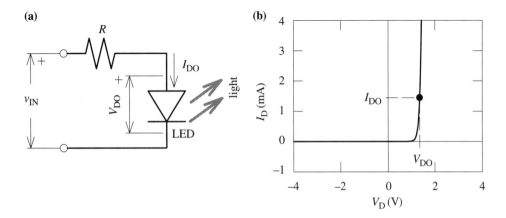

Figure 8.1 LED driving circuit (a) and operating point (b).

the electrons change their energy status, from the high energy levels in the conduction band to the low energy levels in the valence band. The energy difference between the free electron and the recombined electron must be released, and it can be released in the form of a photon. As mentioned earlier, the photon is the quantum of the light energy, given as

$$E_{photon} = h\nu = \frac{hc}{\lambda} \tag{8.2}$$

where h is Planck's constant ($h = 6.626 \times 10^{-34}$ Js), ν is the frequency of light, c is the speed of light ($c = 3 \times 10^8$ m/s), and λ is the light wavelength.

Figure 8.2 illustrates the photons produced by electron–hole recombination. It is obvious that the photon energy $h\nu$ is approximately equal to the energy gap E_g. There is a variety of compound semiconductor materials that provide different energy gap values suitable for *different colors* of the visible light. Figure 8.3 shows the energy-gap values needed to produce different colors of light. The energy-gap values of the visible LEDs vary from 1.6 to 3.2 eV. These are significantly larger values than the energy gap of silicon. As a consequence, energy barriers at these P–N junctions are significantly higher than in the case of silicon. This means that the 0.7 V cannot reduce the barrier height sufficiently for the electrons and holes to be able to overpass it, and in turn produce the forward-bias current.

Visible LEDs are made of semiconductor materials with larger energy gaps than silicon, and exhibit proportionally larger turn-on voltages.

Another important characteristic of the LEDs is *light intensity*, which is directly related to the number of photons emitted per unit time, $P_{opt}A_J/h\nu$.[1] If every recombined minority carrier was emitting a photon, $P_{opt}A_J/h\nu$ would be equal to the number of injected (and

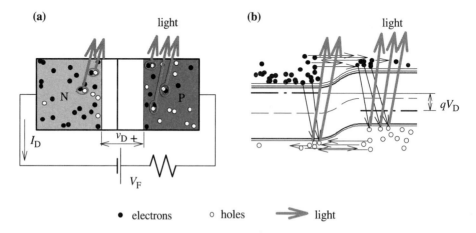

Figure 8.2 Cross section (a) and energy-band diagram (b) of a forward-biased LED, illustrating emission of photons due to electron–hole recombination.

---

[1] P_{opt} is the optical power density in W/m², A_J is the junction area, and $h\nu$ is the energy of a single photon.

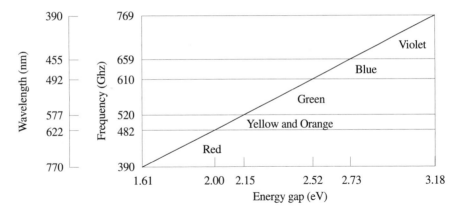

Figure 8.3 Different energy gaps are needed to produce LEDs emitting different colors of light.

consequently recombined) minority carriers, I_D/q. However, only a fraction η_Q of the recombination events results with light emission, so that

$$\frac{P_{opt}}{h\nu} A_J = \frac{\eta_Q}{q} I_D \qquad (8.3)$$

The parameter η_Q expresses the efficiency of an LED, and is referred to as the *radiative recombination efficiency*.

Given a certain radiative recombination efficiency, the light intensity is directly proportional to the concentration of electron–hole pairs recombined per unit time. This is called the *recombination rate*, and is expressed in units of concentration by time—$m^{-3} s^{-1}$. As both electrons and holes are needed for recombination, the recombination rate is directly proportional to the concentration of available electrons and the concentration of available holes. In the P-type region, there are plenty of available holes, and the recombination rate is basically limited by the electron concentration (the minority carriers). In the N-type region, there are plenty of electrons, hence the recombination rate is limited by the concentration of holes (again, the minority carriers).

Attempting to relate the recombination rate to the minority-carrier concentrations, let us consider the following paradox. Assume a P-type semiconductor with some concentration of electrons (minority carriers) and some (does not mater how small) recombination rate. As an electron by electron is recombined, after some time (does not mater how long) all the electrons will disappear through the recombination, purifying the material to an ideal P-type semiconductor! This paradox shows that the recombination cannot be considered in isolation from the opposite process, thermal generation. As emphasized in Section 1.2.5, the recombination and generation rates are equal in thermal equilibrium, so that the concentration of the minority carriers remains constant.

However, forward-biased diodes are not in thermal equilibrium, and excess minority carriers are injected in both the P-type and N-type regions. This is explained in Section 3.1.3 and illustrated by the concentration diagrams of Fig. 3.5c and Fig. 3.6b. In the areas of increased minority-carrier concentrations (inside and at the edges of the depletion layer), the recombination rate is higher than the rate of thermal generation. This means that when the injection of the minority carriers is stopped (the forward-bias voltage is switched-off), the higher recombination than generation rate reduces the minority-carrier concentration. As the minority-carrier concentration approaches the equilibrium level, the recombination rate approaches the value of the thermal generation rate, and when they become equal the concentration value is stabilized, which is the thermal equilibrium.

The above considerations indicate that the difference between the recombination and the thermal generation rates can directly be related to the difference between the actual minority-carrier concentration and its thermal equilibrium value. The difference between the recombination and thermal generation rates can be called the *effective recombination rate*, labeled by r_n and r_p for the cases of minority electrons and holes, respectively. The difference between the actual minority-carrier concentrations and their thermal equilibrium values can be called *excess minority-carrier concentrations*, labeled by $\delta n_p(x)$ and $\delta p_n(x)$:

$$\delta n_p(x) = n_p(x) - n_{pe}$$
$$\delta p_n(x) = p_n(x) - p_{ne} \tag{8.4}$$

The effective recombination rate is directly proportional to the excess minority-carrier concentration:

$$r_n = \frac{n_p(x) - n_{pe}}{\tau_n} = \frac{\delta n_p(x)}{\tau_n}$$

$$r_p = \frac{p_n(x) - p_{ne}}{\tau_p} = \frac{\delta p_n(x)}{\tau_p} \tag{8.5}$$

The proportionality coefficients in Eq. (8.5), τ_n and τ_p, are called *minority electron and hole lifetimes*, respectively. They are expressed in units of time, and incorporate all other influences on the recombination rate except the excess minority-carrier concentration. Note that a higher recombination rate translates into a shorter minority-carrier lifetime.

Equation (8.5) represents the mathematical model for the recombination. The next section shows how it can be used to determine the minority-carrier distributions $n_p(x)$ and $p_n(x)$ and provides further insight into the switching characteristics of the P–N junction diodes. To conclude this section, we refer to the above-mentioned relationship between the light intensity of LEDs and the recombination rate. Equation (8.5) shows that the effective recombination rate is directly proportional to the excess minority-carrier concentration. The level of excess minority-carrier recombination is directly related to the diode current. Therefore, we can confirm the validity of Eq. (8.1), relating the light intensity of LEDs directly to the diode current.

8.1.3 Continuity Equation with the Effective Recombination Term

The continuity equation is one of the fundamental equations of semiconductor physics. It can be used to determine time variations and spatial distributions of the current carriers. It was introduced in Section 1.3.4 with the example of charge-neutral particles used to dope the semiconductor material by diffusion. The physical meaning of the continuity equation was used later to introduce the Poisson equation in Section 2.3.1. The basic principle of the continuity equation was expanded during the derivation of the Poisson equation to include the effects of uncompensated charge centers (source/drain of the electric-field lines). The effects of electric-filed line termination (negative charge center) are analogous to the influence of carrier recombination on the particle current density. In this section, the continuity equation is applied to the electrons and holes, and extended to include the effects of electron–hole recombination.

Figure 8.4a illustrates the continuity equation when there is no carrier recombination. The continuity equation basically states that any change of electron concentration ($\partial n_p/\partial t \neq 0$) in a closed space element is due to the difference in electron current density flowing into and out of the space element ($\partial j_n/\partial x$). If the current density is uniform along the x-axis ($\partial j_n/\partial x = 0$), the electron concentration in the considered space element does not change in time ($\partial n_p/\partial t = 0$). Mathematically, this is expressed in the following way:

$$\frac{\partial n_p}{\partial t} = \frac{1}{q}\frac{\partial j_n}{\partial x} \qquad (8.6)$$

This is the continuity equation applied to the minority electrons. Comparing this form of the continuity equation to the one given by Eq. (1.32) for neutral particles, it can be seen that the only difference is that the electric current density j_n in units of $A/m^2 = C/(m^2 s)$ is divided by the electron charge $-q$ to become analogous to the particle current density J expressed in units of $1/(m^2 s)$.

To include the effects of recombination, consider Fig. 8.4b. In this case the electron concentration in the considered space element changes not only because of the current density gradient ($\partial j_n/\partial x$), but also because of the effective recombination r_n. The recombination contribution to the electron concentration change is exactly equal to the recombination rate. Adding the recombination effect, the continuity equation applied to the minority electrons becomes

(a) **(b)**

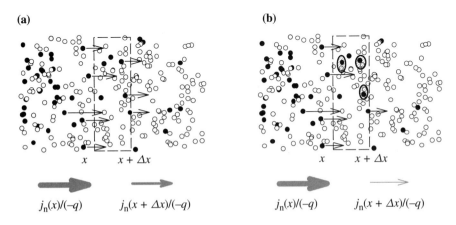

Figure 8.4 Illustration of the continuity equation without effective recombination (a) and with the effective recombination term included (b). Note that the electron and hole distributions are like those found in the P-type region of the forward-biased diode (Fig. 3.5c).

$$\frac{\partial n_p}{\partial t} = \frac{1}{q}\frac{\partial j_n}{\partial x} - r_n \tag{8.7}$$

Replacing the effective recombination rate by Eq. (8.5), the continuity equation can be expressed in the following form:

$$\frac{\partial n_p}{\partial t} = \frac{1}{q}\frac{\partial j_n}{\partial x} - \frac{n_p(x,t) - n_{pe}}{\tau_n} \tag{8.8}$$

Analogously, the continuity equation applied to the minority holes is

$$\frac{\partial p_n}{\partial t} = -\frac{1}{q}\frac{\partial j_p}{\partial x} - \frac{p_n(x,t) - p_{ne}}{\tau_p} \tag{8.9}$$

The three-dimensional forms of electron and hole continuity equations are as follows:

$$\frac{\partial n_p}{\partial t} = \frac{1}{q}\left(\frac{\partial j_n}{\partial x}x_u + \frac{\partial j_n}{\partial y}y_u + \frac{\partial j_n}{\partial z}z_u\right) - r_n = \frac{1}{q}\nabla j_n - r_n$$

$$\frac{\partial p_n}{\partial t} = -\frac{1}{q}\left(\frac{\partial j_p}{\partial x}x_u + \frac{\partial j_p}{\partial y}y_u + \frac{\partial j_p}{\partial z}z_u\right) - r_p = -\frac{1}{q}\nabla j_p - r_p \tag{8.10}$$

Concentration Diagram of the Minority Carriers

The concentration diagrams of the minority carriers in a forward-biased P–N junction are shown in Fig. 3.5c and Fig. 3.6b. Here, the continuity equation is solved to obtain the mathematical equation for $n_p(x)$ (electron concentration in the P-type region).

To begin with, note that we are considering the case of a constant current flowing through the P–N junction (steady state), which means the electron concentration $n_p(x)$ does not change in time: $\partial n_p(x)/\partial t = 0$. Next, note that the current of the minority electrons j_n is the diffusion current given by Eq. (3.3). Putting $\partial n_p(x)/\partial t = 0$, and replacing j_n in the continuity equation (8.8) by Eq. (3.3), leads to the following equation:

$$0 = D_n\frac{\partial^2 n_p(x)}{\partial x^2} - \frac{n_p(x) - n_{pe}}{\tau_n} \tag{8.11}$$

This equation can be transformed into the following form:

$$\frac{\partial^2 [n_p(x) - n_{pe})]}{\partial x^2} - \frac{n_p(x) - n_{pe}}{L_n^2} = 0 \tag{8.12}$$

where

$$\boxed{L_n^2 = D_n\tau_n} \tag{8.13}$$

L_n is called the *diffusion length* of electrons, and is same quantity used previously in Eq. (3.4) and illustrated in Fig. 3.6c. Equation (8.13) links the diffusion length to the diffusion coefficient D_n and the lifetime τ_n of the minority carriers. In the second-derivative term of Eq. (8.12), $n_p(x)$ is expanded into $n_p(x) - n_{pe}$ in order to obtain a second-order differential equation with a single dependent variable. This variable is in fact the excess minority-carrier concentration $\delta n_p(x) = n_p(x) - n_{pe}$. The expansion of $n_p(x)$ into $n_p(x) - n_{pe}$ is possible because $\partial^2 n_{pe}/\partial x^2 = 0$.

To solve Eq. (8.12), its characteristic equation is written

$$s^2 - \frac{1}{L_n^2} = 0 \tag{8.14}$$

and solved, which gives the following two roots: $s_{1,2} = \pm(1/L_n)$. The general solution of Eq. (8.12) is then expressed as

$$n_p(x) - n_{pe} = A_1 e^{s_1 x} + A_2 e^{s_2 x} = A_1 e^{x/L_n} + A_2 e^{-x/L_n} \tag{8.15}$$

where A_1 and A_2 are integration constants to be determined from the boundary conditions. One boundary condition is $n_p(\infty) - n_{pe} = 0$, which turns the constant A_1 to zero (if A_1 were not zero, $n_p(\infty) - n_{pe}$ would be infinitely large, which is physically impossible). Using the other boundary condition, which is $n_p(w_p) - n_{pe}$ according to Fig. 3.6b, the constant A_2 is determined and inserted into the general solution. With this, the solution of equation (8.12) is obtained as

$$\underbrace{n_p(x) - n_{pe}}_{\delta n_p(x)} = \underbrace{[n_p(w_p) - n_{pe}]}_{\delta n_p(w_p)} e^{-(x-w_p)/L_n} \tag{8.16}$$

The obtained solution shows that the minority electron concentration in the P-type region changes exponentially from the boundary value $[n_p(w_p)]$ at the depletion layer edge $(x = w_p)$ to its equilibrium value $[n_p(x \to \infty) = n_{pe}]$. As $n_p(w_p) = n_{pe} \exp(V_D/V_t)$ [Eq. (3.8)], $n_p(w_p) \gg n_{pe}$ in the case of forward bias and $n_p(w_p) \approx 0$ in the case of reverse bias. Plots of $n_p(x)$ appear in Fig. 3.6b for the case of forward bias and in Fig. 8.9 for the case of reverse bias.

To derive the equation for the diode current, $n_p(x)$ was approximated by a linear dependence (Fig. 3.6c) to express the first derivative $\partial n_p(x)/\partial x$ needed in the diffusion equation (3.3) in the simple form used in Eq. (3.4). How good is this approximation? Let us expand the obtained $n_p(x)$ function [Eq. (8.16)] into the Taylor series:

$$\delta n_p(x) = \delta n_p(w_p) - \delta n_p(w_p)\frac{x - w_p}{L_n} + \delta n_p(w_p)(\frac{x - w_p}{L_n})^2 - \cdots \tag{8.17}$$

The linear approximation of Fig. 3.6c means that the quadratic and all higher order terms in the Taylor expansion of the real $n_p(x)$ function are neglected. The first derivative of the first two terms of the Taylor expansion gives

$$\frac{\partial \delta n_p(x)}{\partial x} = \frac{\partial n_p(x)}{\partial x} = -\frac{\delta n_p(w_p)}{L_n} = \frac{n_{pe} - n_p(w_p)}{L_n} \tag{8.18}$$

which is exactly the term used in Eq. (3.4).

Relationship between the Minority-Carrier Lifetime and the Transit Time τ_T

The transit time τ_T is a SPICE parameter used to model the effect of stored-charge capacitance, as explained in Section 3.2.2. In the following, it is shown that the transit time is related to the minority carrier lifetime(s).

Let us express the transit time using its defining Eq. (3.23):

$$\tau_T = \frac{Q_s}{I_D} \tag{8.19}$$

where Q_s is the stored charge due to the excess minority carriers and I_D is the diode current.

Next, let us assume N$^+$–P junction, meaning that the doping level of the N-type is much higher than the doping level of the P-type region. With this assumption, we can neglect the current due to the minority holes as the minority electron current is much larger [$n_{pe} \gg p_{ne}$ in Eqs. (3.10) and (3.11)]. In this case, the diode current density is $j \approx j_n$, which means the current is

$$I_D = A_J j_n \tag{8.20}$$

where A_J is the area of the diode junction. Replacing j_n by Eq. (3.4), the diode current is expressed as

$$I_D = A_J q D_n \frac{n_{pe} - n_p(w_p)}{L_n} \tag{8.21}$$

Let us now express Q_s in terms of $n_p(w_p)$, n_{pe}, and L_n, as well, to replace it together with the above equation for I_D in Eq. (8.19). Figures 3.6c and 3.12b illustrate Q_s. To obtain Q_s, the average excess minority-carrier concentration $[n_p(w_p) - n_{pe}]/2$ is multiplied by the charge that every electron carries (this gives the average charge in C/m^3), and also multiplied by the volume $A_J L_n$, to obtain the charge in C:

$$Q_s = -q A_J L_n \frac{n_p(w_p) - n_{pe}}{2} \tag{8.22}$$

Replacing I_D and Q_s in Eq. (8.19) from Eqs. (8.21) and (8.22), respectively, the transit time is obtained as

$$\boxed{\tau_T = \frac{L_n^2}{2D_n}} \tag{8.23}$$

As approximate values of the diffusion coefficient D_n and the diffusion length L_n are frequently given, this equation is often used to estimate the transit time value, which appears as a SPICE parameter. To relate the transit time to the lifetime, replace L_n in the above equation by Eq. (8.13):

$$\tau_T = \frac{\tau_n}{2} \tag{8.24}$$

Therefore, the transit time is directly related to the minority-carrier lifetime. According to Eq. (8.24), the transit time is approximately equal to the half-value of the minority-electron lifetime in N⁺–P junctions. It can be similarly shown that in the case of P⁺–N junctions, the transit time is approximately equal to $\tau_p/2$. Obviously, the transit time is a combination of the minority hole and electron lifetimes in the diodes where the above approximations cannot be applied.

8.1.4 Direct and Indirect Semiconductors

E–x energy band diagrams, like the one used in Fig. 8.2, illustrate the basic concepts of the carrier recombination process. It may seem surprising, but it is not only the change in electron energy that is important for the recombination process. Another important factor is the value of electron momentum p, or equivalently the wave vector $k = p/(h/2\pi)$, before and after the recombination event. The simplest situation occurs when the electron momentum is zero in both the conduction band (thus before the recombination) and the valence band (after the recombination event). This is possible if the conduction band minimum and the valence band maximum appear at $k = 0$. This assumption was implicitly made in the case of E–k diagrams shown in Fig. 1.29. There are semiconductors whose E–k diagrams satisfy this condition. One example is GaAs, as the simplified E–k diagram of Fig. 8.5a illustrates. Actually, quite a few compound semiconductors exhibit this type of E–k diagram, however with different E_g values.

> Semiconductors with the conduction-band minimum and the valence-band maximum appearing for the same wave vector k are called *direct semiconductors*.

In the process of *direct recombination*, the electron momentum does not change, and the electron energy can be given away as a photon ($E = h\nu$).

There are semiconductors with more complicated types of E–k diagrams. The single most important example is silicon, whose simplified E–k diagram is shown in Fig. 8.5b.

> Semiconductors with the conduction-band minimum and the valence-band maximum appearing for different wave vectors k are called *indirect semiconductors*.

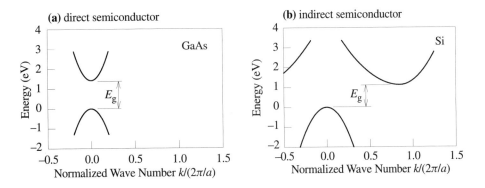

Figure 8.5 Simplified E–k diagrams of GaAs (a) and Si (b); a is the crystal-lattice constant.

A recombination event in indirect semiconductor necessitates a change in both electron energy and electron momentum. Recombination in indirect semiconductors is typically assisted by trap energy levels, appearing because of the existence of different types of crystal defects or the presence of impurity atoms. These defects/traps are called *recombination centers*. The energy levels of the recombination centers enable the electrons and holes to recombine in two steps. For example, (1) an electron exchanges the momentum with another particle (a phonon,[2] for example) to move from the conduction-band minimum to the trap energy level where its momentum is zero, while the energy remains the same, and (2) the electron releases some energy as it drops to the valence-band maximum. As phonons are typically involved in the recombination process, the electron energy is typically released to the phonons (that is, it increases the crystal-lattice temperature) and not as light. As a consequence, the indirect semiconductors cannot be practically used to create LEDs.

8.2 PHOTODETECTORS AND SOLAR CELLS: EXTERNAL CARRIER GENERATION

In addition to emitting light, diodes can absorb light to generate electrons and holes. This process of light conversion into electric current is not only useful for electronic light detection, but also for conversion of solar power into electric power, using specifically designed P–N junctions called solar cells. Following brief descriptions of the applications of diodes as photodetectors and solar cells, the process of light-induced electron–hole generation is considered in more detail. After that, the continuity equation is completed by adding a term to include the external-carrier generation. In the final section, the continuity equation is used to derive a photocurrent equation.

8.2.1 Basic Applications

Photodetector diodes, or photodiodes, are typically used in the reverse-bias region, as illustrated in Fig. 8.6. In the dark, the current–voltage characteristics of photodiodes are the same as the characteristics of rectifying diodes. This means that only the leakage current flows in the reverse-bias region. When exposed to light, the reverse current of the photodiode increases proportionally to the light intensity. This current is referred to as *photocurrent* (I_{photo}). Figure 8.6b illustrates that the photocurrent, similar to the normal reverse-bias current, does not depend on the reverse-bias voltage.

The circuit of Fig. 8.6a converts the light intensity into voltage V_o. In the dark, the current through the circuit is approximately zero, and therefore the voltage drop across R is zero as well. The load line in Fig. 8.6b intersects the diode characteristics at 0 mA and -7 V, which is the assumed reverse-bias voltage; therefore, $V_o = V_R - V_D = 7 - 7 = 0$. As the light intensity and consequently the photocurrent increase, the voltage drop across the resistor, that is the output voltage V_o, increases as well. This is accompanied by corresponding

---

[2]The concept of phonons is introduced in Section 1.4.3.

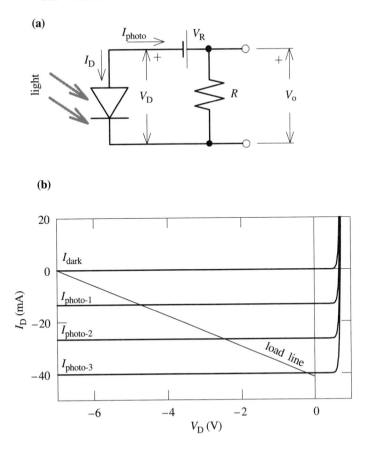

Figure 8.6 (a) A simple circuit biasing a photodetector diode; (b) the current–voltage characteristics of the photodiode and the load line provide graphic analysis of the circuit.

reduction in the reverse-bias of the photodiode. The load line in Fig. 8.6b illustrates that the maximum output voltage is approximately limited to the reverse-bias voltage V_R. If the light intensity is increased beyond this point, the diode is pushed toward the forward-bias region, where the appearance of the normal forward-bias current, which flows in the opposite direction, limits the output voltage increase.

As opposed to the photodetectors, the diodes used as solar cells operate in the forward-bias region. Figure 8.7a shows that the diode as a solar cell is directly connected to a loading element (resistor R).

Two extreme biasing conditions of the solar-cell diode are short circuit ($R = 0$) and open circuit ($R = \infty$). At short circuit ($V_{DO} = 0$), the only current flowing through the diode is the photocurrent. Although this condition is useful to measure the value of the photocurrent, it produces no power as $V_{DO}I_{DO} = 0$.

With some load resistance R, the voltage drop across the diode V_{DO} becomes positive, while the current I_{DO} still remains negative. The negative power value $P_d = V_{DO}I_{DO}$

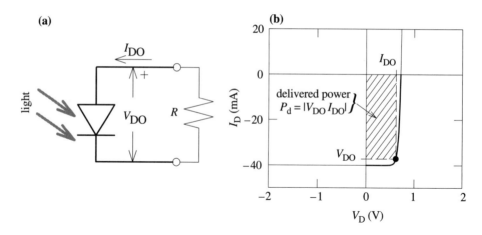

Figure 8.7 The solar-cell diode is directly connected to a loading element: (a) the electric circuit; (b) the operating point (I_{DO}, V_{DO}) is in the quadrant of negative currents and positive voltages (forward bias).

indicates that the diode acts as a power generator. The power generated by the diode is indicated by the shaded area in Fig. 8.7b.

The current I_{DO} in the circuit of Fig. 8.7a remains negative because the photocurrent dominates the normal forward-bias current. As the source of the forward bias is the photocurrent, the total current I_{DO} can never become positive. In the extreme case, the normal forward-bias current can become equal to the photocurrent making the total current I_{DO} equal to zero, which is the open circuit condition. Although the open circuit condition provides the maximum voltage drop V_{DO}, it produces no power as, again, $V_{DO}I_{DO} = 0$.

It is obvious that the value of the load resistance R directly influences the amount of the power delivered. The maximum delivered power corresponds to a single value of the load resistance R. However, any change in the photocurrent (that is the light intensity) will change the value of the load resistance, which maximizes the power.

8.2.2 Carrier Generation

The photodetector and solar-cell diodes operate by the mechanism of light induced electron–hole generation. In this process, which is opposite to light emission, the photon energy $h\nu$ is used to destroy a covalent bond liberating an electron and creating a hole. If the generated electron and hole diffuse to or are generated in the depletion layer of the diode, the existing electric field sweeps them away before they get a chance to recombine, creating the photocurrent.

The photodetector and the solar-cell application circuits shown in Fig. 8.6a and Fig. 8.7a, respectively, are presented in Fig. 8.8a and b with the diode symbol replaced by cross sections illustrating the electron–hole generation by the light. It is shown that the electrons and holes generated in the depletion layer move toward their respective majority-carrier regions, like the minority carriers, due to the electric field (E) direction in the depletion layer. Therefore, the current due to the light-generated electrons and holes adds up to the thermal reverse-bias current I_S. The light also generates electrons and holes in the neutral N-type and P-type regions, which can diffuse to the depletion layer, contributing to the photocurrent.

Clearly, to maximize photodiode sensitivity, the depletion layer should be as wide as possible. As explained in Section 2.3, the depletion-layer width depends on the doping level: $w_{depl} \propto 1/\sqrt{N_{A,D}}$ in the case of an abrupt P–N junction. Therefore, the lowest doping level that can technologically be achieved is the most favorable in terms of maximizing the depletion-layer volume. The most common photodetectors are made with such a layer between the P-type and N-type regions, which are needed as the diode terminals (the anode and the cathode). To distinguish this very lightly doped, almost *intrinsic* region, from the N-type and P-type regions, it is labeled by I, and consequently, the diode is specifically referred to as a *PIN photodiode*. The thickness of the I region is such that it is completely depleted.

The energy-band diagrams of the photodetector and the solar-cell diodes, also given in Fig. 8.8, provide a deeper insight into the carrier generation mechanisms and their exploitation for light detection and solar power conversion. Remembering that electrons roll down and the holes bubble up along the energy bands, we can clarify the first important

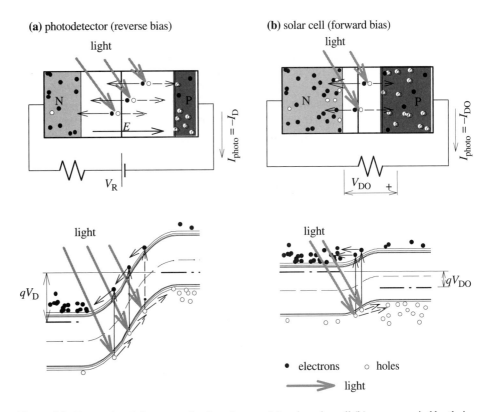

Figure 8.8 Energy-band diagrams of a photodetector (a) and a solar cell (b), accompanied by their respective application circuits, where the diode symbols are replaced by diode cross sections.

point, which is the direction of the photocurrent flow. This is a simple question as far as the photodetector circuit is concerned, and the energy bands are not needed to answer it: the reverse-bias voltage V_R (that is the associate electric field E) drives the electrons generated in the depletion layer toward the neutral N-type region and the holes toward the neutral P-type region. The situation is not as obvious in the case of the solar-cell circuit. **What does drive the light-generated electrons and holes if there is only a resistor attached to the diode?**

The energy-band diagram of Fig. 8.8b shows that there is a slope in the energy bands, and that the electrons generated in the depletion layer roll down toward the N-type region and the holes bubble up toward the P-type region. This energy-band slope (that is electric field) is due to the ionized doping atoms in the depletion layer (the built-in electric field) as explained in Section 2.2.3. The splitting of the Fermi levels in the photodetector and the solar-cell diodes (Fig. 8.8a and b, respectively) is in the opposite directions, indicating the reverse bias of the photodetector and the forward bias of the solar cell. However, the energy-band slopes in the depletion layers are in the same direction, which means that the photocurrents are in the same direction as well. It may appear that current generation efficiency is better in the photodetectors, however, it should be kept in mind that the photodetector and the solar-cell diodes are differently designed. To provide a quick response, the capacitance of the photodetector is minimized by minimizing the diode area and increasing the depletion-layer width by inserting the "I" region. The solar cells are essentially large area P–N junctions to maximize the current generated by the incoming light.

The second point that can easily be explained by the energy-band diagrams is the independence of the photocurrent on the reverse-bias voltage applied (refer to the current–voltage characteristics of Fig. 8.6b). Although an increase in the reverse-biased voltage does increase the steepness of the energy bands in the depletion layer, the photocurrent is not increased because it is not limited by the steepness of the energy bands. The built-in bending of the bands (the built-in electric field) alone is good enough for every generated electron or hole to easily roll down or bubble up through the depletion layer. What limits the photocurrent is the rate of electron–hole pairs generated by the light. This number is increased by the light intensity, which is expressed by the direct dependence of the reverse-bias current on the light intensity (Fig. 8.6b).

The third point that becomes obvious from the energy-band diagrams is that light with photon energies smaller than the energy gap E_g of the semiconductor cannot possibly move an electron from the valence band into the conduction band to generate a free electron–hole pair. From the condition $E_g = h\nu_{min} = hc/\lambda_{max}$, the maximum wavelength of the light that can generate electrons and holes is found as

$$\lambda_{max} = \frac{hc}{E_g} \tag{8.25}$$

The energy gap of the silicon corresponds to $\lambda_{max} = 1.1 \ \mu$m. As almost the complete spectrum of solar radiation is below this maximum wavelength, silicon is an excellent material for solar cells. The fact that silicon is an indirect semiconductor (Fig. 8.5b) does not prevent light absorption: when the photon energy is taken by a valence electron, it jumps up to a defect energy level and subsequently moves into the conduction band changing the momentum. The solar cells are most frequently made of silicon. Like the LEDs, the photodetectors are made of different materials to maximize sensitivity to the light of a nominated color while minimizing sensitivity to other colors.

8.2.3 General Form of the Continuity Equation

Analogously to the effect of the effective recombination, the external generation can change the concentration of electrons (holes) per unit time $\partial n/\partial t$ ($\partial p/\partial t$). To include the effect of carrier recombination, the recombination rate term r_n (r_p) was included in the continuity equation [Section 8.1.3, Eq. (8.7)]. This expresses the reduction in carrier concentration per unit time due to the effective recombination. If there is an external source of carrier generation, like the light, the carrier concentration is increased. This increase per unit time is analogously called the *external generation rate*, and is labeled by g_n and g_p for electrons and holes, respectively.

Very frequently, the external generation rate is not uniform inside the semiconductor. In the example of carrier generation by light, most of the light is absorbed close to the semiconductor surface, causing a rapid decay of light intensity inside the semiconductor. It is very frequently necessary to express the generation rate as a function of the space coordinate: $g_n(x)$, and $g_p(x)$.

Adding the external generation rate $g_n(x)$ to Eq. (8.8), the complete form of the continuity equation applied to the minority electrons is obtained:

$$\frac{\partial n_p}{\partial t} = \frac{1}{q}\frac{\partial j_n}{\partial x} + g_n(x) - \frac{n_p(x) - n_{pe}}{\tau_n} \tag{8.26}$$

The continuity equation can be applied to the majority electrons and holes as well as to the minority carriers. To provide the general form of the continuity equation, the indexes p and n denoting minority-carrier concentrations can be omitted. Therefore, the general forms of the one-dimensional continuity equations for the electrons and holes are as follows:

$$\frac{\partial n}{\partial t} = \frac{1}{q}\frac{\partial j_n}{\partial x} + g_n(x) - \frac{n - n_e}{\tau_n}$$
$$\frac{\partial p}{\partial t} = -\frac{1}{q}\frac{\partial j_p}{\partial x} + g_p(x) - \frac{p - p_e}{\tau_p} \tag{8.27}$$

Note that the electron and hole concentrations n and p are functions of x and t: $n = n(x, t)$ and $p = p(x, t)$. Similarly, the current densities j_n and j_p can in general vary in time t and space x: $j_n = j_n(x, t)$, and $j_p = j_p(x, t)$.

The general three-dimensional continuity equations for the electrons and holes are

$$\frac{\partial n}{\partial t} = \frac{1}{q}\nabla j_n + g_n(x, y, z) - \frac{n - n_e}{\tau_n}$$
$$\frac{\partial p}{\partial t} = -\frac{1}{q}\nabla j_p + g_p(x, y, z) - \frac{p - p_e}{\tau_p} \tag{8.28}$$

The electron and hole concentrations and the electron and hole current densities are now four-dimensional functions: $n = n(x, y, z, t)$, $p = p(x, y, z, t)$, $j_n = j_n(x, y, z, t)$, and $j_p = j_p(x, y, z, t)$.

The general form of the continuity equation is used in numerical programs developed for semiconductor device simulation. Solution of the continuity equation gives the time and spatial dependence of the electron and hole concentrations. However, to solve the continuity equation, the current density functions are needed.

8.2.4 Photocurrent Equation

In this section, a photocurrent equation is derived, assuming uniform carrier generation in the area of interest (around the P–N junction). Let us first deal with the photocurrent due to the carrier generation in the depletion layer. As the strong electric field in the depletion layer immediately separates the electrons and holes, we basically need to convert the generation rate g_n into photocurrent. The generation rate g_n expresses the **number** of electron–hole pairs generated in a unit of the depletion-layer volume per second. The generation rate multiplied by the unit charge, qg_n, expresses the **charge** generated in a unit of the depletion-layer volume per second. As the electrons generated in the entire volume of the depletion layer contribute to the photocurrent, qg_n is multiplied by the depletion-layer volume $A_J w_d$, which gives the electric charge generated per unit time: $qg_n w_d A_J$. The charge generated per unit time (C/s = A) is the photocurrent:

$$I_{photo-d} = q A_J g_n w_d \qquad (8.29)$$

The carriers generated in the neutral regions, far from the P–N junction, recombine as there is neither electric field nor concentration gradient to produce drift or diffusion current. With $\partial j_n/\partial x = 0$ and $\partial n_p/\partial t = 0$, we find from the continuity equation (8.26) that $n_p - n_{pe} = g_n \tau_n$. This means that the steady-state level of the excess electron concentration, due to a uniform generation rate g_n, is equal to $g_n \tau_n$. However, at the edge of the depletion layer, the concentration is $n_p(w_p) = n_{pe} \exp(V_D/V_t) \approx 0$. This means there is a concentration gradient around the P–N junction, as illustrated in Fig. 8.9. In the dark, the concentration gradient is small, leading to a small reverse-bias diffusion current I_s, which is the already well-known saturation current of the diode. The carrier generation lifts the concentration in the neutral region to $g_n \tau_n + n_{pe}$, increasing the concentration gradient and therefore the reverse-bias diffusion current. To determine the component of the photocurrent due to the diffusion of excess electrons, $I_{photo-n}$, we can again assume a linear concentration gradient in the diffusion equation (3.3): $\partial n_p/\partial x \approx (g_n \tau_n + n_{pe})/L_n$. Clearly, this leads to the following results

$$I_n = q A_J D_n \underbrace{\frac{n_{pe}}{L_n}}_{I_{S-n}} + q A_J D_n \underbrace{\frac{g_n \tau_n}{L_n}}_{I_{photo-n}} \qquad (8.30)$$

Given that $L_n^2 = D_n \tau_n$, the photocurrent can be expressed as

$$I_{photo-n} = q A_J g_n L_n \qquad (8.31)$$

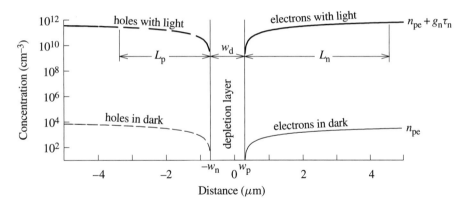

Figure 8.9 Minority-carrier concentration diagrams for a reverse-biased P–N junction.

An analogous equation would be obtained for the diffusion current of holes in the N-type region: $I_{\text{photo}-\text{p}} = q A_J g_p L_p$, where $g_p = g_n$ if the carrier generation is uniform.

Adding the three components of the photocurrent together, the total photocurrent becomes

$$I_{\text{photo}} = q A_J g_n (w_d + L_n + L_p) \tag{8.32}$$

It should be noted that the drift photocurrent $I_{\text{photo}-\text{d}}$ responds almost instantly to changes in light intensity. On the other hand, the response of the diffusion photocurrents $I_{\text{photo}-\text{n}}$ and $I_{\text{photo}-\text{p}}$ is limited by the rate of establishing the concentration profiles. For fast-response photodetectors, it is desirable to have $w_d \gg L_n + L_p$, so that the drift photocurrent dominates. To achieve this, the PIN structure is used. The very low doping of the "intrinsic" region enables a very wide depletion layers, which not only helps to satisfy the condition $w_d \gg L_n + L_p$, but also increases the magnitude of the photocurrent. The width of the "intrinsic" region is designed so that it is fully depleted at very small reverse-bias voltages. As the total depletion layer width is dominated by the "intrinsic" region, this means that the widening of the depletion layer with the reverse-bias voltage is negligible, so the photocurrent is approximately voltage independent (a light-controlled current source).

■ **Example 8.1 Concentration Diagrams for a Reverse-Biased P–N Junction and Diffusion Photocurrents**

Solve the continuity equation to obtain the equations for the concentration gradients plotted in Fig. 8.9, and then use the result to derive the diffusion photocurrents.

Solution: For the case of electrons, we use Eq. (8.26). For the steady-state case, $\partial n_p / \partial t = 0$:

$$0 = \frac{1}{q}\frac{dj_n}{dx} + g_n - \frac{n_p(x) - n_{pe}}{\tau_n}$$

where $j_n = qD_n dn_p/dx$ [Eq. (3.3)]. Given that $L_n^2 = D_n\tau_n$, we have

$$\frac{d^2[n_p(x) - n_{pe}]}{dx^2} - \frac{n_p(x) - n_{pe}}{L_n^2} + \frac{g_n}{D_n} = 0$$

The only difference between this equation and Eq. (8.12) is the added constant g_n/D_n. Because of that, the general solution of this equation can be expressed as the solution of Eq. (8.12), plus an additional constant:

$$n_p(x) - n_{pe} = A_1 e^{x/L_n} + A_2 e^{-x/L_n} + A_3$$

As before, A_1 has to be zero in order to have a finite concentration for $x \to \infty$. A_3 is not an independent constant, it has to be set so to ensure that $n_p(x) - n_{pe} = A_2 \exp(-x/L_n) + A_3$ is a solution of the differential equation. Replacing the solution expressed in this form back into the differential equation, we obtain

$$\underbrace{\frac{A_2}{L_n^2}e^{-x/L_n}}_{d^2[n_p(x)-n_{pe}]/dx^2} - \underbrace{\frac{A_2}{L_n^2}e^{-x/L_n} - \frac{A_3}{L_n^2}}_{[n_p(x)-n_{pe}]/L_n^2} + \frac{g_n}{D_n} = 0$$

$$-\frac{A_3}{L_n^2} + \frac{g_n}{D_n} = 0$$

$$A_3 = g_n\frac{L_n^2}{D_n} = g_n\tau_n$$

Therefore,

$$n_p(x) - n_{pe} = A_2 e^{-x/L_n} + g_n\tau_n$$

where A_2 has to be determined so that the boundary condition $n_p(w_p)$ is satisfied:

$$n_p(x) - n_{pe} = [n_p(w_p) - n_{pe} - g_n\tau_n]e^{-(x-w_n)/L_n} + g_n\tau_n \qquad (x \geq w_p)$$

In the case of a reverse-biased P–N junction, $n_p(w_p) = \exp(V_D/V_t) \approx 0$:

$$n_p(x) = (n_{pe} + g_n\tau_n)\left[1 - e^{-(x-w_n)/L_n}\right] \qquad (x \geq w_p)$$

This is the equation for the concentration profile of minority electrons, plotted in Fig. 8.9. Clearly, if there is no external generation ($g_n = 0$), the equation is simplified to

$$n_p(x) = n_{pe}\left[1 - e^{-(x-w_n)/L_n}\right] \qquad (x \geq w_p)$$

Using this result in the diffusion-current equation

$$I_n = qA_J D_n\frac{dn_p(x)}{dx}\bigg|_{x=w_n}$$

the following current equation is obtained:

$$I_n = q A_J D_n \frac{n_{pe} + g_n \tau_n}{L_n} = \underbrace{q A_J \frac{D_n n_{pe}}{L_n}}_{I_{S-n}} + \underbrace{q A_J g_n L_n}_{I_{photo-n}}$$

which is the same result for the photocurrent as Eq. (8.31).

The corresponding equations for the minority holes are

$$p_n(x) = (p_{ne} + g_p \tau_p)\left[1 - e^{(x+w_p)/L_p}\right] \qquad (x \le -w_n)$$

$$I_p = -q A_J D_p \frac{p_{ne} + g_p \tau_p}{L_p} = \underbrace{q A_J \frac{D_p p_{ne}}{L_p}}_{I_{S-p}} + \underbrace{q A_J g_p L_p}_{I_{photo-p}}$$

■ **Example 8.2 Light Absorption and Drift Photocurrent**

Due to the absorption of light in the semiconductor material, the uniform carrier generation g cannot always be used. The light absorption causes an exponential decay of the light intensity from the surface inside the semiconductor. Consequently, the external generation rate exponentially depends on the distance from the semiconductor surface:

$$g_n(x) = g_n(0)e^{-\alpha x}$$

where α is the optical absorption coefficient. Assuming that 10-μm thick N-type region is at the surface of the PIN diode ($w_N = 10\ \mu$m), calculate the photocurrent generated in the depletion layer. The surface generation rate is $g_0 = 10^{19}\ \mathrm{cm}^{-3}\ \mathrm{s}^{-1}$ and the optical absorption coefficient is $\alpha = 0.01\ \mu\mathrm{m}^{-1}$. The width of the fully depleted "intrinsic layer" is $w_d = 100\ \mu$m, and the junction area is $A_J = 1\ \mathrm{mm}^2$.

Solution: As g_n is not uniform in this case, it cannot simply be multiplied by $q A_J w_d$ to obtain the photocurrent. However, we can find the average generation rate inside the depletion layer:

$$\bar{g}_n = \frac{1}{w_d} \int_{w_N}^{w_N+w_d} g_n(x)\, dx = \frac{1}{w_d} g_n(0) \int_{w_N}^{w_N+w_d} e^{-\alpha x}\, dx$$

Note that the depletion layer edges are at w_N and $w_N + w_d$, which are the integration limits. Solving the above integral leads to

$$\bar{g}_n = \frac{g_n(0)}{w_d \alpha}[e^{-\alpha w_N} - e^{-\alpha(w_N+w_d)}]$$

The photocurrent is then

$$I_{photo} = q A_J w_d \bar{g}_n = q A_J \frac{g_n(0)}{\alpha}[e^{-\alpha w_N} - e^{-\alpha(w_N+w_d)}]$$

$$I_{photo} = 91.5\ \mu\mathrm{A}$$

■ Example 8.3 A Photoconductor

A semiconductor bar with negligible free carrier concentration in the dark is exposed to a light source.

(a) Assuming a uniform external generation rate of $g_n = 5 \times 10^{18}$ cm^{-3} s^{-1}, determine the conductivity of the semiconductor. Use equal electron and hole mobilities $\mu_n = \mu_p = 100$ cm^2/Vs and equal electron and hole lifetimes $\tau_n = \tau_p = 10\ \mu$s.

(b) Determine how long it takes for the conductivity to drop $e = 2.71$ times after the light source is switched off.

Solution:

(a) To solve this problem, the continuity equation for electrons is used and simplified to express the following facts: (1) a steady-state condition is considered ($\partial n/\partial t = 0$), (2) there is no current flowing through the semiconductor ($\partial j_n/\partial x = 0$) (note that even when contacts are made at the end of the semiconductor bar to make a photoconductor, the current that can flow through the structure does not change along x, which means the condition $\partial j_n/\partial x = 0$ is still satisfied), and (3) the carrier concentration in the dark is negligible, $n_e = 0$. With these simplifications, Eq. (8.27) becomes

$$g_n = \frac{n}{\tau_n}$$

This result expresses the fact that the generation rate is equal to the recombination rate. The electron concentration can now be calculated as

$$n = g_n \tau_n = 5 \times 10^{13}\ \text{cm}^{-3}$$

As the same concentration of holes is generated, the conductivity is

$$\sigma = q(\mu_n n + \mu_p p) = 0.16(\Omega\ \text{m})^{-1}$$

(b) When the light is switched off, the external generation rate becomes zero. The effective recombination now works to bring the semiconductor in thermal equilibrium, reducing the carrier concentration. The time derivative of the electron concentration in the continuity equation is not equal to zero:

$$\frac{dn}{dt} = -\frac{n}{\tau_n}$$

The partial derivative $\partial n/\partial t$ is replaced by the ordinary derivative, as in this case the electron concentration depends on time only. To solve the above differential equation, the variables are separated

$$\frac{dn}{n} = -\frac{dt}{\tau_n}$$

and integration of the both sides is performed:

$$\ln n = -t/\tau_n + A$$

The integration constant A can be transformed into $-\ln n(0)$ to enable the electron concentration to be expressed as

$$n = n(0)e^{-t/\tau_n}$$

The new form of the integration constant $n(0)$ is equal to the electron concentration before the light has been switched off. Therefore, $n(0) = 5 \times 10^{13}$ cm^{-3}. The constant $n(0)$ basically represents the initial electron concentration, which decays exponentially. It is obvious that the electron concentration, and therefore the conductivity, is reduced by $e = 2.71$ times after $t = \tau_n = 10\,\mu$s. This in a sense provides a definition for the lifetime τ_n.

8.3 LASERS

The word *laser* is an acronym for "light amplification by stimulated emission of radiation." The distinguishing characteristic of lasers is emission of strong narrow beams of monochromatic light.[3] Lasers are widely used, with many applications that are familiar to almost everybody. An important application of semiconductor lasers is to generate the monochromatic light that carries information through optical-fiber communication systems.

8.3.1 Stimulated Emission, Inversion Population, and Other Fundamental Concepts

The following introduces semiconductor lasers in a way that does not require preliminary study of gas lasers. The semiconductor lasers are most similar to the LEDs introduced in Section 8.1. Like the LEDs, the current of a forward-biased P–N junction causes recombination of excess minority carriers, which results in light emission. The difference is that the laser light is monochromatic, which results from the process of *stimulated emission*, as opposed to the spontaneous emission in the case of LEDs.

To better understand the process of stimulated emission, it is good to know some of the fundamental properties of photons, which distinguish them as particles from electrons. It was mentioned in Section 1.5 that no more than one electron can occupy a single electron state (Pauli exclusion principle). Therefore, if there is an electron in a particular state, the probability that another electron will get into that state is 0—the electrons "shy away" from each other. As opposed to this behavior, the photons "flock" into a single state. When there

---

[3]Monochromatic light is single-wavelength light or, practically, a very narrow band of wavelengths.

are *n* identical photons, the probability that one more photon will enter the same state is *enhanced* by factor $(n + 1)$. *The probability that an atom will emit a photon with particular energy* $h\nu$ *is increased by the factor* $(n + 1)$ *if there are already n photons with this energy.*[4] According to this property, electrons are classified as Fermi particles (described by the Fermi–Dirac distribution), while photons are classified as Bose particles as they obey a different Bose-Einstein distribution.

Let us put mirrors at two ends of the P–N junction that emits light due to a forward-bias current. The emitted light will reflect from the mirrors, so that the intensity of the light in the direction normal to the mirrors becomes dominant. More importantly, the presence of this light with frequency ν will increase the probability of minority excess electrons falling from the conduction band down to the valence band emitting light with the same frequency ν. As the presence of more $h\nu$ photons causes a further increase in the emission of light with frequency ν, a chain reaction is triggered leading to what is known as *stimulated emission*. Clearly, the stimulated emission can amplify a small-intensity incoming light beam to produce a large-intensity light beam, as illustrated in Fig. 8.10. This is called *optical amplification.*

To make use of the generated light, one of the mirrors is made slightly transparent so that a highly directional, monochromatic beam of light gets out of the device. One more problem needs to be solved and the laser is operational. The concentration of excess electrons in the conduction band needs to be maintained at a high level, otherwise as the photons get out,

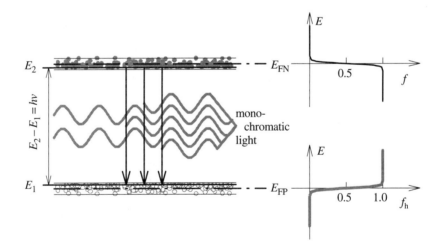

Figure 8.10 Stimulated emission: incoming photons trigger electron–hole recombination to generate more photons with the same energy $h\nu$. Population inversion: electron and hole distributions (f and f_h) relate to separate quasi-Fermi levels (E_{FN}, and E_{FP}), which appear inside the bands to indicate that the concentrations of both electrons and holes are very high.

[4]*Source:* R. Feynman, R. Leighton, and M. Sands, *The Feynman Lectures on Physics—Quantum Mechanics*, Addison-Wesley, Reading, MA, 1966 (p. 47).

the excess electrons will be spent and the light generation will die away. This is achieved by maintaining the forward-bias current above the needed level (*threshold current*), so that the concentration of the injected minority electrons into the P-type region is sufficiently higher than the equilibrium level. This condition is referred to as *population inversion* .

In the nonequilibrium case, the concentrations of electrons and holes are expressed by two separate quasi-Fermi levels (E_{FN} and E_{FP}). The example of Fig. 8.10 illustrates the case of very high concentrations of both electrons and holes, so high that both E_{FN} and E_{FP} appear inside the energy bands. It can be shown that this is necessary for stimulated emission to exceed absorption. There are three necessary factors for a stimulated recombination of an electron-hole pair: (1) the existence of photons with energy $h\nu$, (2) the existence of an electron at energy level E_2 in the conduction band, and (3) the existence of a hole at level E_1 in the valence band, where $E_2 - E_1 = h\nu$. Therefore, the rate of stimulated recombination can be expressed as

$$r_{(st)} \propto f(E_2, E_{FN}) f_h(E_1, E_{FP}) I(h\nu) = f_2(1 - f_1) I(h\nu) \qquad (8.33)$$

where $I(h\nu)$ is the density of $h\nu$ photons, while $f(E_2, E_{FN}) = f_2$ and $f_h(E_1, E_{FP}) = 1 - f_1$ are the probabilities of having an electron and a hole at E_2 and E_1, respectively. The presence of photons, however, can cause electron–hole generation due to absorption by electrons at E_1, which subsequently move to E_2. The generation rate g is proportional to the probability of having an electron at E_1, $f_1 = 1 - f_h(E_1, E_{FP})$, and a hole at E_2, $1 - f_2 = 1 - f(E_2, E_{FN})$:

$$g \propto f_1(1 - f_2) I(h\nu) \qquad (8.34)$$

The condition for stimulated emission to exceed absorption is

$$r_{(st)} > g \qquad (8.35)$$

which according to Eqs. (8.33) and (8.34), and assuming equal proportionality coefficients, can be expressed as

$$f_2(1 - f_1) > f_1(1 - f_2) \qquad (8.36)$$

This condition can be transformed as follows:

$$f_2 - f_2 f_1 > f_1 - f_1 f_2 \ \rightarrow \ f_2 > f_1 \qquad (8.37)$$

$$\frac{1}{1 + \exp[(E_2 - E_{FN})/kT]} > \frac{1}{1 + \exp[(E_1 - E_{FP})/kT]} \qquad (8.38)$$

$$\frac{E_1 - E_{FP}}{kT} > \frac{E_2 - E_{FN}}{kT} \qquad (8.39)$$

and finally,

$$E_{FN} - E_{FP} > h\nu = E_2 - E_1 \tag{8.40}$$

Therefore, the stimulated emission can exceed the absorption only if strong nonequilibrium concentrations of electrons and holes are maintained, such that the difference between the respective quasi-Fermi levels is larger than the energy of the emitted photons.

8.3.2 A Typical Heterojunction Laser

As mentioned in Section 8.1.4, the recombination in indirect semiconductors (like silicon) does not lead to significant light emission. Therefore, the semiconductor lasers are made of direct semiconductors, typically III–V and II–VI compound semiconductors. In addition to this, it is practically impossible to reach the condition given by Eq. (8.40) by an ordinary P–N junction. Also, it is necessary to confine the emitted light beam inside the active laser region. All these requirements can be met by using a structure with layers of different III–V and/or II–VI semiconductors.

The interface between different semiconductor materials is called a *heterojunction*.

Different materials have different energy gaps, creating energy-band discontinuities at the heterojunction, very much alike the discontinuity between Si and SiO$_2$ discussed in Section 2.5.3. These discontinuities can help achieve the condition given by Eq. (8.40). Also, different materials have different refractive indices, which can help to achieve total beam reflection at the parallel interfaces, so that the light is confined inside the active region. Molecular beam epitaxy is used to deposit one semiconductor layer over another, creating the heterojunctions. To achieve the necessary high-quality interfaces, the wafers remain inside a high-vacuum chamber, while the reacting gasses are changed as a transition from one layer to another is to be made.

The choice of the semiconductor that will emit the light (the active layer) depends on what light frequency (color) is needed. Figure 8.11 shows a laser with GaAs as the active layer,[5] as an illustration of a typical semiconductor laser. To create appropriate energy-band discontinuities, layers with wider energy gaps are needed at each side of the active layer. In this example, N-type and P-type AlGaAs are used for this purpose.[6] Let us concentrate on the N-AlGaAs/P-GaAs heterojunction. If the doping of N-AlGaAs is high enough, the electron quasi-Fermi level (E_{FN}) is close enough to the bottom of the conduction band so that the conduction band discontinuity with GaAs brings the bottom of the conduction band in the GaAs region below E_{FN}. Analogously, the valence-band discontinuity at the P-AlGaAs–P-GaAs heterojunction places E_{FP} below the top of the valence band. This provides the condition for population inversion when appropriate forward bias (V_F) is

---

[5]The energy gap of the GaAs (1.42 eV) corresponds to 900 nm light, which is in the infrared region.

[6]The energy gap of Al$_x$Ga$_{1-x}$As varies between 1.42 eV for pure GaAs ($x = 0$) and 2.9 eV for pure AlAs ($x = 1$).

applied. The emitted light is reflected backward and forward by the mirror surfaces capturing enough photons to trigger the stimulated emission. The photons that get through the partially transparent mirror (the useful laser beam) have to be replaced by new electrons and holes provided by the laser forward current (I_F).

The conduction-band discontinuities at both heterojunctions (N-AlGaAs–P-GaAs and P-GaAs–P-AlGaAs) are important. They create a potential well that traps the minority carriers (electrons) so that they cannot diffuse away from the active region, as would happen in the case of an ordinary P–N junction. This *carrier confinement* is important for maintaining the population inversion and maximizing the stimulated recombination and

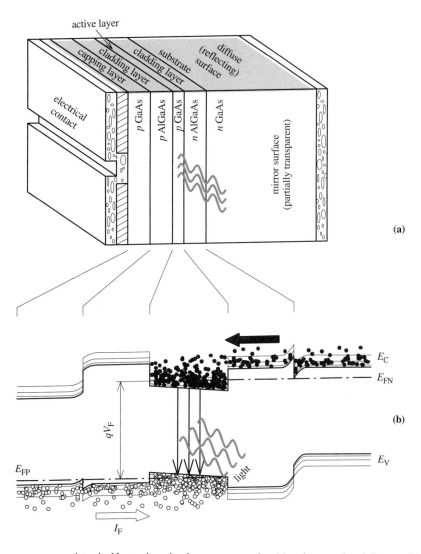

Figure 8.11 A typical heterojunction laser: cross section (a) and energy-band diagram (b).

light emission. On the other hand, the *light confinement* is achieved by the fact that the refractive index of AlGaAs is smaller than the refractive index of GaAs.

The outside GaAs layers, the P-type capping layer and the N-type substrate layer, also create heterojunctions with the adjacent P-type and N-type AlGaAs layers. These do lead to energy band discontinuities, however, the associated depletion layers are narrow because of high doping, so they do not present a practical problem for the current flow.

PROBLEMS

8.1 The top P-type layer of a P–N⁺ GaAs LED can be considered much wider than the diffusion length of the minority carriers, which is $L_n = 5 \ \mu$m. The other technological parameters and constants are as follows: $N_A = 5 \times 10^{16} \ \text{cm}^{-3}$, $D_n = 30 \ \text{cm}^2/\text{Vs}$, $n_i = 2.1 \times 10^6 \ \text{cm}^{-3}$, the diode area $A_J = 4 \ \text{mm}^2$, and the radiative recombination efficiency $\eta_Q = 0.7$. If the forward bias of the diode is $V_D = 1$ V, calculate

(a) the number of photons emitted per unit time, $P_{opt} A_J / h\nu$;

(b) the optical power of the emitted light, assuming that the photon energy is equal to the band gap ($E_g = 1.42$ eV). $\boxed{\text{A}}$

8.2 The voltage v_{IN} in the circuit of Fig. 8.1 rapidly changes from 5 to 0 V at $t = t_0$. As a result, the concentration of minority electrons starts changing as in Fig. 8.12. Solve the continuity equation (8.8) to find the change of the maximum electron concentration with time: $n_{p-max}(t)$. If $\tau_n = 10$ ns, how long does it take for the maximum of the excess electron concentration to drop to half of its original value?

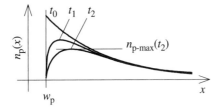

Figure 8.12 Concentration diagrams of minority electrons during LED turn-off (stored charge removal).

8.3 The continuity equation is solved in Section 8.1.3 to derive the concentration distribution of minority carriers in a long diode ($W_{anode} \gg L_n$).

(a) Modify this solution using the following more general boundary conditions for the excess electron concentrations at the beginning ($x = w_p$) and the end ($x = W_{anode} + w_p$) of the neutral region, respectively: $\delta n_p(w_p) = n_p(w_p) - n_{pe}$ and $\delta n_p(W_{anode} + w_p) = 0$. $\boxed{\text{A}}$

(b) Obtain the linear approximation of the gradient of electron concentration for a short diode ($W_{anode} \ll L_n$). *Note:* $\exp(\pm W_{anode}/L_n) \approx 1 \pm W_{anode}/L_n$ and $\exp\{\pm[W_{anode} - (x - w_p)]/L_n\} \approx 1 \pm [W_{anode} - (x - w_p)]/L_n$. Compare this result for a short diode, and Eq. (8.18) for a long diode, to the respective saturation current equations [Eqs. (3.15) and (3.16)].

8.4 (a) For the diode of Problem 8.1, calculate the transit time τ_T (SPICE parameter).

(b) Derive the dependence of τ_T on D_n for a short diode ($W_{anode} \ll L_n$) and calculate the transit time for $W_{anode} = 1\ \mu m$. $\boxed{A}$

8.5 The short circuit current of a silicon P^+–N solar cell at room temperature is $I_{sc} \approx I_{photo} = 100$ mA. The technological parameters of the cell are $A_J = 4\ cm^2$, $N_D = 5 \times 10^{16}\ cm^{-3}$, $\mu_p = 380\ cm^2/Vs$, $L_p = 10\ \mu m$, and $n \approx 1$.

(a) Derive the equation for the open-circuit voltage, V_{oc}, and calculate V_{oc}. $\boxed{A}$

(b) Calculate the maximum power that can be obtained from this cell. What load resistance R_L is needed to extract this power?

(c) How many cells, operating at the maximum power, have to be connected in series so that the voltage is 12 V? How many 12-V cells have to be connected in parallel so that the maximum power is 5.7 W? $\boxed{A}$

8.6 A PIN photodiode has junction area $A_J = 1\ mm^2$ and fully depleted "intrinsic" region $W_I = 100\ \mu m$. Calculate the photocurrent, assuming that the light of a certain intensity generates in average 10^{19} electron–hole pairs per second in a cm^3 of the photodiode material.

8.7 Assuming that the 100-μm intrinsic layer of a PIN diode is fully depleted, and neglecting any light absorption in the very thin P-type region at the top, calculate the photocurrent for a surface generation rate of $g_0 = 5.1 \times 10^{19}\ cm^{-3}\ s^{-1}$ and absorption coefficient $\alpha = 0.05\ \mu m^{-1}$. The junction area of the diode is $A_J = 1\ mm^2$.

8.8 The power density of a 0.5-μm light is $P_{opt} = 900\ W/m^2$.

(a) Calculate the number of photons per unit area and unit time that would hit the surface of a semiconductor exposed to this light. $\boxed{A}$

(b) Due to photon absorption by the semiconductor, their density decays exponentially down to zero, as does the generation rate: $g(x) = g_0 \exp(-\alpha x)$. Neglecting reflection losses, and assuming that every absorbed photon of this light generates an electron–hole pair, calculate the surface generation rate. The absorption coefficient is $\alpha = 1\ \mu m^{-1}$.

(c) What would the photocurrent be if the PIN diode described in Problem 8.7 was exposed to this light? $\boxed{A}$

8.9 The technological parameters of a solar cell are as follows: $N_A = 3 \times 10^{16}\ cm^{-3}$, $N_D = 8 \times 10^{15}\ cm^{-3}$, $D_n = 20\ cm^2/s$, $D_p = 10\ cm^2/s$, $\tau_n = 0.3\ \mu s$, $\tau_p = 0.1\ \mu s$, and $A_J = 1\ cm^2$. The generation rate of electron–hole pairs due to the absorbed light can be considered constant around the P–N junction: $g_n = g_p = 5 \times 10^{19}\ cm^{-3}$.

(a) Calculate the short circuit current.

(b) Is the drift or diffusion photocurrent dominant, and what fraction of the total current is due to the dominant mechanism?

(c) What are the maximum concentrations of the minority electrons and holes?

8.10 Due to manufacturing problems and increased defect levels, the actual lifetimes of the solar cell considered in Problem 8.9 are $\tau_n = 3$ ns and $\tau_p = 1$ ns. Repeat the calculations of Problem 8.9. [A]

REVIEW QUESTIONS

R-8.1 How can the light intensity in LEDs be varied?

R-8.2 How are different colors achieved with the LEDs.

R-8.3 What is the minority carrier lifetime? How is it related to the transit time (SPICE parameter)?

R-8.4 Does the photocurrent in P–N junctions depend on the reverse-bias voltage? Why?

R-8.5 How is the P–N junction biased (forward or reverse) in photodetector circuits and how in solar cell circuits? Why?

R-8.6 Both LEDs and semiconductor lasers are basically P–N junctions that emit light. What is the difference?

R-8.7 Why do photons stimulate electron–hole recombination? Do they stimulate emission of photons with the same or with different energy $h\nu$?

R-8.8 What is optical amplification? In principle, can it be used to amplify a weakened optical signal carrying a communication signal through an optical fiber?

R-8.9 If the P–N junction is in equilibrium or close to equilibrium ($E_{FN} \approx E_{FP}$), will the stimulated emission or absorption dominate? Would the stimulated emission and absorption rates be the same given that they are both proportional to the light intensity (photon density)? Do they depend on anything else?

R-8.10 What is needed to achieve a higher rate of stimulated emission compared to the rate of absorption? What is this condition called?

R-8.11 What is achieved by the heterojunctions in a typical semiconductor laser? Are they related to the population inversion? Minority-carrier confinement? Light confinement?

Chapter *9*

MICROWAVE FETs AND DIODES

Electromagnetic waves with frequencies in a specific range, as illustrated in Fig. 9.1, are referred to as *microwaves*. As Fig. 9.1 also shows, light, x-rays, and γ-rays are all electromagnetic waves with different frequencies. As with light, electrons can both emit and absorb microwaves, as long as they gain/lose energy equivalent to the energy of the emitted/absorbed microwave. The energy of an electron moving with a constant speed does not change, and there is no way it can emit a microwave.[1] However, if a time-varying field is applied (as a simple example, think of a sinusoidal field), the velocity of an electron varies and an electromagnetic wave is radiated. The radiation is stronger at faster velocity changes, which means at higher frequencies. The radiated electromagnetic wave (radiofrequency or microwave) travels through space, and can be collected at a very distant point by the reverse process to cause an electric current. These effects enable wireless communications.

The wireless communication systems involve semiconductor devices that help to generate microwave power and to modulate and amplify microwaves. It is important to mention that the concept of connecting a load or a source to a semiconductor device by zero-impedance wires does not work at high frequencies. Physical structures that will guide the microwaves from place to place, called *transmission lines*, are needed. This adds a number of specific, although extremely important issues, like impedance matching. However, if we limit ourselves to semiconductor-device concepts, the amplification principle illustrated in Fig. 4.4 with the generic transistor remains relevant. In addition to that, BJTs and MOSFETs, previously described in Chapters 6 and 5, respectively, are used as amplifying devices at the lower end of microwave frequencies. The aim of this chapter is to introduce the most important alternative semiconductor microwave devices, some of them being quite specific but quite necessary at higher frequencies.

The first section introduces some structural and material specifics, related to semiconductor microwave devices. In particular, material properties of Si and GaAs are compared in the context of different device structures and their suitability for microwave applications. Sections 9.2, 9.3, and 9.4 introduce the junction field-effect transistor (JFET), metal–semiconductor field-effect transistor (MESFET), and high-electron mobility tran-

---

[1]In the case of light emission, described in Chapter 8, the potential energy of the electron changes as it falls to a lower level/band.

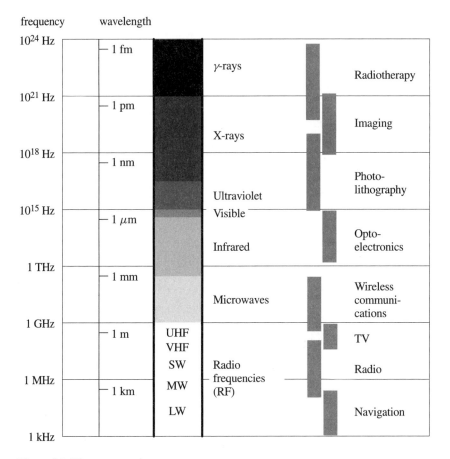

Figure 9.1 Electromagnetic waves.

sistor (HEMT) as the most common microwave FETs. However, it should be mentioned that the JFET is the simplest, the cheapest, and the most robust FET, and consequently is used in other applications not related to microwaves. Also, the MESFET as the simplest GaAs FET, as well as the sophisticated GaAs HEMT, are used for high-frequency digital applications in addition to linear microwave applications. The sections introducing these three types of FETs are relevant for any application of these devices, as the descriptions concentrate on their structures, and in particular differences from the MOSFET structure. SPICE models and parameters are provided for the JFET and the MESFET, to complete the list of SPICE semiconductor devices. As the mathematical equations modeling the JFET and the MESFET are very similar to MOSFET equations, they are not specifically derived, and no examples and problems are presented. In a sense, many complete sections can be considered as examples that follow the detailed MOSFET description, to cover the broader area of field-effect transistors. Clearly, these sections can be very helpful for a better understanding of the MOSFET.

Section 9.5 introduces negative resistance diodes. These are specific diodes that are used to generate the microwaves (oscillators) as well as microwave amplifiers.[2]

9.1 GALLIUM ARSENIDE VERSUS SILICON

Gallium arsenide, being a compound (III–V) material, is difficult to grow, especially in comparison with the well-developed and relatively cheap technology for growth of high-quality monocrystalline silicon. Although this manufacturing-related fact is not unimportant, the following will concentrate on electrical properties.

9.1.1 Dielectric–Semiconductor Interface: Enhancement versus Depletion FETs

Bulk material properties, as will be shown in the following sections, are definitely in favor of GaAs. Not surprisingly, engineers and technologists believed that GaAs would become the dominant semiconductor material. It has not happened, and there is no real indication this will ever happen. This is mostly due to the fact that no device-quality dielectric–GaAs interface could be created. Deposition of silicon dioxide (or any other good dielectric) leaves far too many traps at the interface, while the thermally grown oxide of GaAs is a poor dielectric. Consequently, the MOSFET structure could not be implemented in GaAs to industrial standards.

Figure 9.2 compares the MOSFET structure, which necessitates a gate dielectric to create a normally-off device, to a depletion-type FET created without a gate dielectric (this is a normally-on device, where the depletion layer of a reverse-biased P–N junction or Schottky contact is used to control the width of the built-in channel, and therefore the FET current). This comparison shows that the only disadvantage of the depletion-type device is the fact that a current flows at zero gate voltage. This is extremely important for complex digital circuits, as normally-off, voltage-controlled switches are necessary to minimize the power dissipation.[3] With 1 mA flowing through a normally-on FET, the power dissipated per logic cell would be >1 mW, which means an impossible >1 kW for a million-cell IC! However, this is not a limiting factor for amplifiers where circuit complexity is not an issue. As for the amplification principle, an input bias circuit can be designed so that the signal amplification can equally well be achieved by either normally-off or normally-on device. At lower frequencies (radiofrequencies), the MOSFET may still be advantageous because of its fairly linear transfer characteristic and superior output dynamic resistance (less pronounced Early effect). At higher frequencies, however, the input capacitance becomes the dominant factor (a higher gate-to-source capacitance shunts lower-frequency signals to the grounded source). As Fig. 9.2 illustrates, the depletion-type FET offers a smaller input capacitance.

---

[2]This is a conceptually different signal/power amplification, described in Section 9.5.1.

[3]This is discussed in Sections 4.2.3 and 5.2.3.

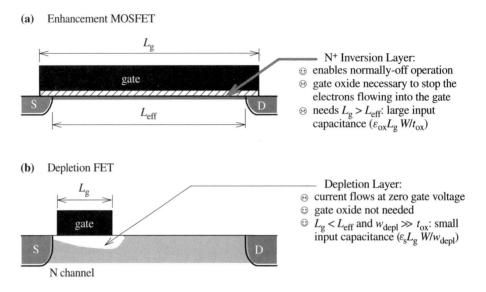

(a) Enhancement MOSFET

N+ Inversion Layer:
- ☺ enables normally-off operation
- ☹ gate oxide necessary to stop the electrons flowing into the gate
- ☹ needs $L_g > L_{eff}$: large input capacitance $(\varepsilon_{ox} L_g \, W/t_{ox})$

(b) Depletion FET

Depletion Layer:
- ☹ current flows at zero gate voltage
- ☺ gate oxide not needed
- ☺ $L_g < L_{eff}$ and $w_{depl} \gg t_{ox}$: small input capacitance $(\varepsilon_s L_g \, W/w_{depl})$

Figure 9.2 Comparison of enhancement-type MOSFET and depletion-type FET.

- The unique quality of the Si–SiO$_2$ interface, enabling the MOSFET as an implementation of a normally-off, voltage-controlled switch, is the decisive factor for the dominance of silicon technology. Power dissipation problems severely restrict the complexity of digital circuits that can be designed with any alternative device structure. The inadequate quality of GaAs–gate dielectric interfaces significantly limits digital applications of GaAs.

- Depletion-type FETs, which do not need a gate dielectric, are as good as if not a superior device structure for amplifiers. The gate-dielectric-related problems do not mask the other superior material properties of GaAs.

9.1.2 Energy Gap

The room temperature energy gap of GaAs is 1.42 eV, which is significantly wider than the silicon energy gap of 1.12 eV. As a consequence, the GaAs substrate can appear more as an insulator than a semiconductor at very low doping levels. Low-doped high-resistive GaAs is referred to as a *semiinsulator*. It could be argued that this is the most important advantage of GaAs over Si. The need for specific isolation structures (like the reverse-biased P–N junction and the field oxide in silicon technologies) is eliminated. To create an integrated circuit, the devices are placed into a single semiinsulating GaAs substrate.

The great advantage of the semiinsulating substrate is not a simplified technology process, but significantly improved high-frequency performance. The high-frequency response in silicon devices is not limited by the time the current carriers take to flow through the device, but by the time needed to charge/discharge the parasitic capacitances. For the case of a CMOS inverter, this is explained in Section 5.2.3. Parasitic capacitances are associated not only with the isolation P–N junctions, but also appear between the interconnecting metal lines and the semiconductor substrate (parasitic MOS capacitors). The use of a semiinsulating substrate eliminates the need for isolation P–N junctions, and eliminates the parasitic MOS capacitors, dramatically reducing the parasitic capacitances.

The semiinsulating GaAs substrate enables proper device operation at very high frequencies. Accordingly, GaAs devices and integrated circuits are typically used for microwave analog and high-frequency digital applications.

9.1.3 Electron Mobility and Saturation Velocity

As explained in Section 1.4, electron mobility depends on the effective mass of the electrons. The concept of effective mass is closely related to the E–k diagrams, which have systematically been developed through Figs. 1.16, 1.29, 8.5, and now Fig. 9.3. Figure 1.16 uses the model of a free electron, where the E–k diagram is a parabola given by

$$E_{\text{kin}} = \left(\frac{h}{2\pi} \right)^2 \frac{k^2}{2m^*} \tag{9.1}$$

Obviously, the "sharpness" of the parabola depends on m^*: a lighter effective mass corresponds to a "sharper" parabola. Figure 1.29 shows how the electrons gain kinetic energy when exposed to an electric field. Figure 8.5 introduces the case of indirect semiconductors (like silicon).

Parabolic E–k dependencies, associated with the free-electron model, were assumed in the simplified E–k diagrams presented in the previous sections. However, E–k dependencies are not parabolic for the entire range of k in real crystals. Figure 9.3 shows the E–k diagrams of GaAs and Si in more detail. We can see that the E–k dependencies do have parabolic shapes around the conduction-band minima and the valence-band maxima. The parabola around the absolute conduction-band minimum is much "sharper" in the case of GaAs than in Si. Consequently, the mobility and the drift velocity[4] of GaAs electrons are much higher.

Drift velocity-vs-field dependencies for Si and GaAs are shown in Fig. 9.4. It can be seen that the drift velocities of electrons are six to seven times higher in GaAs compared to Si at low electric fields (<0.5 V/μm). The maximum drift velocity in GaAs reaches a value of about 0.2 μm/ps at about 0.5 V/μm, while the drift velocity in Si is about 0.03 μm/ps at that field. Equivalently, we can say that the low-field electron mobility is six to seven times higher in GaAs compared to Si, which is observed from the higher slope of the low-field region of the v_d-vs-E curve.

High electron mobility is a very useful property. It is associated with low channel resistance and high device current, which means

---

[4]The concepts and relationship between the drift velocity and mobility are introduced in Section 1.4. Briefly, the drift velocity is directly related to the device current, as given by Eq. (1.47). The drift velocity depends on the applied electric field. At low electric fields, this relationship is linear, the mobility being the slope of this linear dependence [Eq. (1.49)].

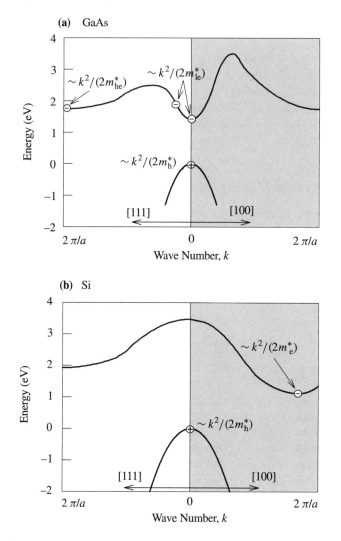

Figure 9.3 *E–k* diagrams of GaAs (a) and Si (b). Note the illustration of light (narrow conduction-band minimum) and heavy (broad conduction-band minimum) electrons for GaAs.

- fast charging/discharging of parasitic capacitances, and consequently fast device and IC response;

- large transconductance $g_m = dI_D/dV_{GS}$, as input voltage and associated channel-carrier density changes produce large current changes;

- low noise, due to the fact that the electron scattering in the channel is reduced.

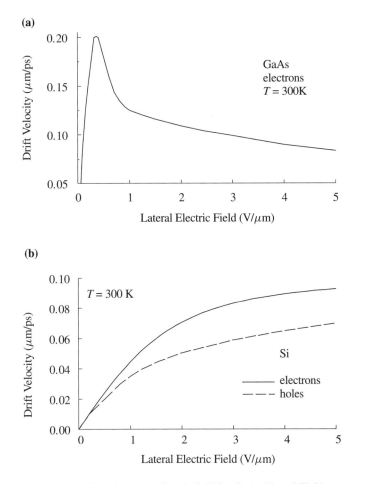

Figure 9.4 Drift velocity vs electric field for GaAs (a) and Si (b).

The mobility-related advantage of GaAs, however, is limited to the low-field region. Figure 9.4 shows that the drift velocity of electrons in GaAs is reduced at higher electric fields, and saturates at about the same level as in the case of Si. The fact that only a couple of volts across a 1-μm channel are needed to place the average field in the "high-field" region means that the modern devices are very likely to operate in velocity-saturation regime. Although this undoubtedly shows that the mobility-related advantage of GaAs is not as large as the low-field mobility ratio indicates, it does not mean that the advantage can disappear altogether. Even in the case of micron and submicron channel lengths, the device does not operate in the high-field region all the time, because the voltage across the channel changes, typically from very small to the maximum value. Also, the lateral field is not uniform at high voltages: the highest field appears at the drain end of the channel, and it reduces along the channel to a relatively low value at the source end. A part of the channel almost always operates in the low-field region, which means the resistance of this part of the channel is much smaller for GaAs.

9.1.4 Dynamic Negative Resistance

The reduction of the drift velocity, shown in Fig. 9.4a, appears as a quite strange phenomenon. However, it can be explained by the E–k diagram of Fig. 9.3a. At high electric fields, a number of electrons gain enough energy to appear at the local conduction-band minimum ($k = 2\pi/a$ in the [111] direction). As the E–k diagram shows, the parabola associated with this minimum is much broader, which means the electrons are much heavier there. The masses of light (the narrow minimum) and heavy (the broad minimum) electrons are denoted by m_{le}^* and m_{he}^*, respectively, in Fig. 9.3a. As a larger fraction of the electrons populates the broad conduction-band minimum at higher electric fields, the average drift velocity is reduced because of the increased dominance of the heavy electrons. This phenomenon can be observed as negative dynamic resistance, known as the Gunn effect. The negative dynamic resistance is a very useful effect, enabling microwave oscillators and amplifiers at very high frequencies, as described in Section 9.5.

9.2 JFET

JFET is very similar to the depletion (normally-on) MOSFET. The only difference is that a reverse-biased P–N junction is used for DC isolation of the JFET gate, as opposed to isolation by gate oxide in the case of the depletion MOSFET. The device name, junction FET (JFET), reflects this difference. This section provides a description and the energy-band diagrams of the JFET. This complements the detailed description of the enhancement (normally-off) MOSFET, given in Chapter 5. SPICE does include the JFET, and this section also provides a systematized reference to SPICE models and parameters.

9.2.1 JFET Structure

The JFET is usually used as a discrete device, although it can be implemented in integrated-circuit technology as well. Figure 9.5 shows the principal JFET structure. This is the case of *N-channel JFET*, where the device current is due to the flow of electrons from the source N^+ region, through the N-type layer called the channel, into the N^+ drain. The N channel is sandwiched between the top and bottom P^+ layers, whose separation determines the channel thickness. The top and bottom P^+ layers create P–N junctions with the N-type channel. The depletion layers associated with these junctions predominantly expand into the N layer, as its doping level is much lower than the doping of P^+ regions. The dotted line in Fig. 9.5 shows the edges of the depletion layers associated with the upper and lower P–N junctions. It is the distance between the two depletion layers, not the separation between the P–N junctions themselves, that is equal to the electrically effective channel thickness.

The upper and/or lower P–N junctions can be reverse biased, which increases the depletion layer widths, reducing the channel thickness, and therefore reducing the current

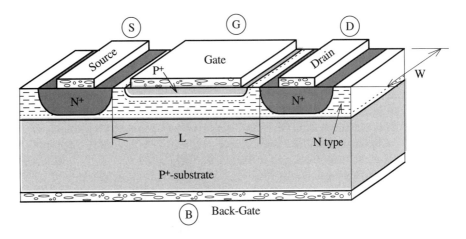

Figure 9.5 JFET structure.

that can flow through the channel. In the extreme case, the depletion layers extend over the whole thickness of the N-type layer, reducing the electrically effective channel thickness to zero, and therefore reducing the JFET current to zero. This shows that the current can be controlled by the value of negative voltage applied to the upper and/or lower P$^+$ layer. Either of the two P$^+$ layers can be used as a gate. To distinguish between them, they are labeled as *gate* and *back gate* in Fig. 9.5.

The JFET structure is similar to the depletion-type MOSFET. The N$^+$ source and drain are the same, the built-in N-type channel that connects the drain and the source is the same, and the bottom P-type substrate that creates the isolation junction is the same. The only difference is the top P–N junction, which appears in place of the metal oxide–semiconductor (MOS) structure. These different gate structures, however, control the current in a similar way: in both cases, the current flows through built-in channels at zero gate voltage (normally-on FETs), and negative voltage is needed to reduce the current to zero by removing the electrons from the channel.

Differences between the JFET and depletion-type MOSFET are important in some applications. In the case of the MOSFET, positive voltage can be applied to the gate without adverse effects, in fact it would only increase the channel current by attracting new electrons into the channel. In the case of the JFET, positive voltage at the gate is not desirable, as it sets the gate–channel P–N junction in forward mode, effectively short-circuiting the gate and the channel. This appears as a disadvantage of the JFET structure. However, the lack of the gate oxide in the JFET has a positive side as well: as no part of the gate-to-source voltage is wasted across the gate oxide, the current is more efficiently controlled, i.e., the transconductance $g_m = dI_D/dV_{GS}$ is higher. Also, the gate can be shorter than the channel, and the depletion layer larger than the oxide thickness, leading to a smaller input capacitance (Section 9.1.1).

The above comparison relates only to the relationship between the device structures and the electrical performance. Development, cost, and availability of either device are strongly influenced by a complex set of manufacturing issues.

9.2.2 JFET Characteristics

Figure 9.6a and b shows the cross section and the energy-band diagram along the channel, respectively, for the case of zero gate-to-source voltage and small drain-to-source voltage. The small discontinuities in the energy bands at the N^+-source–N-channel and N-channel–N^+-drain transitions are due to changes of the doping level, which is lower in the channel than in the source and drain regions. There are quite enough electrons in the channel to make a significant current when the energy bands are tilted by the applied V_{DS} voltage. This current depends linearly on V_{DS}, as it is basically limited by the slope of the energy bands (or the electric field in the channel in other words). The point Q in Fig. 9.6c is in the linear region of the $I_D - V_{DS}$ characteristic (the solid line) corresponding to $V_{GS} = 0$.

The dotted lines in Fig. 9.6a show the edges of the depletion layers. The widths of the depletion layers at every point along the channel depend on the actual reverse bias at that point. Although the heavy doping of P^+ layers maintains approximately the same potential inside those regions (zero in the example of Fig 9.6a), the voltage applied between the drain and the source distributes along the lower doped N-type channel. Taking the potential of the source as reference, the energy-band diagram of Fig 9.6b shows that the potential energy difference increases along the channel, and reaches the value of $q V_{DS}$ at the drain. The depletion-layer widths follow this trend: the narrowest depletion layers correspond to the zero reverse bias, and appear at the source end of the channel; the depletion-layer widths increase along the channel, and reach the maximum at the drain end of the channel. The electrically effective channel thickness follows an opposite trend: it is thickest at the source end and thinnest at the drain end of the channel.

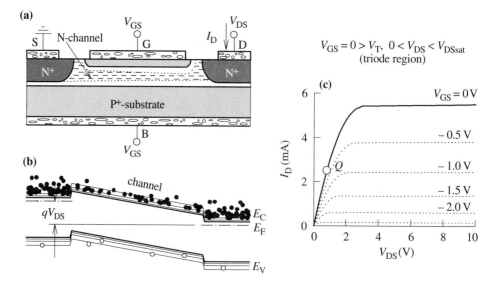

Figure 9.6 JFET in the triode region: (a) cross section, (b) energy bands, and (c) point Q showing the bias conditions on I_D–V_{DS} characteristics.

The channel thickness variation is not pronounced at small V_{DS} voltages. However, at higher V_{DS} voltages, the thickness reduction reflects in a smaller current, which is seen as a departure of the actual I_D–V_{DS} characteristic from the linear trend followed at smaller V_{DS} voltages.

It can happen that the electrically effective channel thickness becomes zero (the channel is pinched off) at sufficiently high V_{DS} voltage. The channel pinch-off would first occur at the drain end of the channel, expanding toward the source as V_{DS} is increased.

V_{DS} voltage which causes channel pinch-off at the drain end is called saturation voltage, and is labeled as V_{DSsat}.

Figure 9.7 illustrates the situation of channel pinch-off caused by high V_{DS} voltage. Analogously to the MOSFET, the voltage drop across the channel (between the source and the pinch-off point) remains constant and equal to V_{DSsat}, as any additional V_{DS} increase drops across the laterally expanding depletion layer. As the current is limited by the conditions in the channel, and not in the depletion layer (waterfall analogy with the energy-band diagram of Fig. 9.7b), this leads to saturation in the current increase with V_{DS} voltage.

It is now interesting to consider the case in which sufficient negative voltage is applied to the gate (V_{GS}) to turn the JFET off. The negative voltage applied to the P$^+$ region adds up to the positively biased channel, increasing the total reverse-bias voltage. This can lead to a situation in which the depletion layers expand over the entire N-layer thickness, even at the source end of the channel. Although Fig. 9.8a clearly illustrates this situation, it is not obvious from this cross-sectional diagram that the current through the channel is reduced to zero. The cross-sectional diagrams do not show any principal difference between the depletion layer in Fig. 9.8a and the depletion layer in Fig. 9.7a, but there is a significant difference. The depletion layer at the source end of the channel in Fig. 9.8a is controlled by the gate(s), and the electric-field lines terminate at the gates. Electrons taken from the source or any remaining part of the channel would follow the field lines only to hit the potential barriers of the reverse-biased P–N junctions. Consequently, no electron current can flow through this depletion layer. As opposed to this, the depletion layer in Fig. 9.7a is controlled by the drain, and the electric-field lines terminate at the drain. Electrons taken by the field from the end of the N-channel (the pinch-off point) are quickly transported to the drain.

Energy-band diagrams can clearly show this difference. The negative voltage at the gates increases the potential energy in the N-type layer (remember, negative electric potential corresponds to positive potential energy). Therefore, to modify the energy-band diagram of Fig. 9.7b (when $V_{GS} = 0$), to represent the case of $V_{GS} < 0$, the energy bands between the source and the drain should be lifted by an amount that corresponds to the reduced electric potential by the negative gate voltage. Figure 9.8b shows the energy-band diagram in this case. It can clearly be seen that the electrons from the source cannot flow into the drain, because of the high potential barrier created by the gate voltage.

As the JFET is a device very similar to the MOSFET, it is useful to compare the energy-band diagrams presented in this section with the corresponding MOSFET energy-band diagrams. Figures 9.6 and 5.5 on one hand, and Figs. 9.7 and 5.7 on the other hand, show similar energy-band diagrams for the triode and saturation regions, respectively. The energy-band diagrams illustrating the cutoff region are different, but that is only because

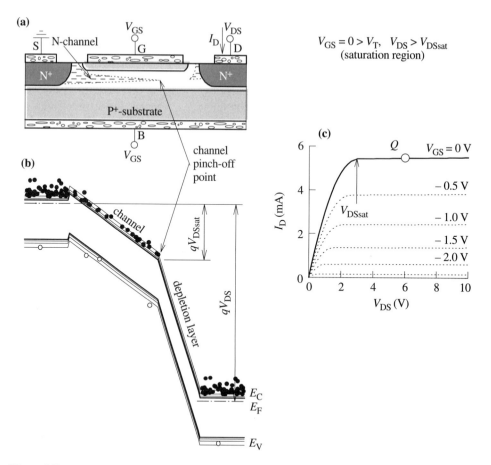

Figure 9.7 JFET in the saturation region: (a) cross section, (b) energy bands, and (c) point Q showing the bias conditions on I_D–V_{DS} characteristics.

they are shown for two different drain-to-source voltages: $V_{DS} = 0$ in Fig. 5.2 and high V_{DS} in Fig. 9.8. The MOSFET diagram of Fig. 5.2 can illustrate the JFET case at $V_{DS} = 0$, and vice versa, the JFET diagram of Fig. 9.8 applies to the MOSFET with high V_{DS} voltage applied.

9.2.3 SPICE Model and Parameters

SPICE parameters and mathematical equations modeling the dependence of the output JFET current I_D on the terminal voltages V_{GS} and V_{DS} are given in Table 9.1. The equations modeling the JFET are very similar to the SPICE LEVEL 1 model of the MOSFET. In fact, the triode-region equations become equivalent if the gain factor β of the JFET is taken to be half of the MOSFET value. It should be noted that the threshold voltage (also called pinch-off voltage in the case of JFET) is negative for N-channel devices, as the JFET and the depletion-type MOSFET are normally-on FETs. Therefore, $V_{GS} - V_T$ is positive for any

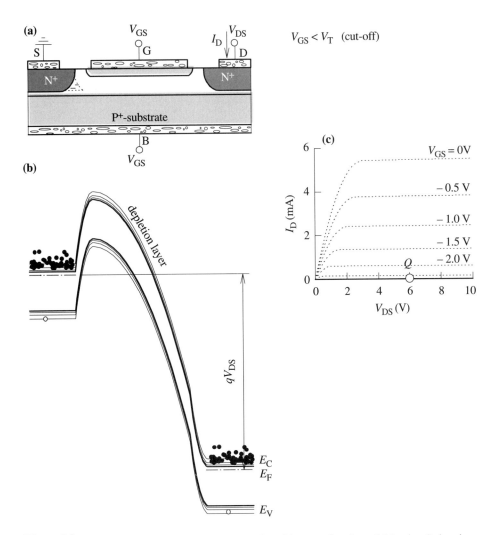

Figure 9.8 JFET in the cutoff region: (a) cross section, (b) energy bands, and (c) point Q showing the bias conditions on I_D–V_{DS} characteristics.

$V_{GS} > V_T$. P-channel devices are opposite, the threshold voltage is positive and $V_{GS} - V_T$ is negative when the FETs are on ($V_{GS} < V_T$), and V_{DS} voltage is negative as well. This gives a positive number for the current I_D in both cases, however, the current is assumed in the opposite direction, as shown in Table 9.1. As for measurement of the two key parameters, the threshold voltage V_T and the gain factor β, a procedure analogous to the one described in Section 5.4.2 (Fig. 5.25) can be used.

The multiplier $\lambda(V_{DS}) = 1 \pm \lambda V_{DS}$ is to include the effect of a small current increase with V_{DS} in the saturation region, or the finite dynamic output resistance in other words. The SPICE model for this effect is analogous to the Early model used in BJTs. Noting that the

TABLE 9.1 Summary of SPICE JFET Model: The principal Equivalent-Circuit Elements

Static Parameters

Symbol	SPICE Keyword	Parameter Name	Typical Value		Unit
			N channel	P channel	
V_T	Vto	Threshold voltage (pinch-off voltage)	-3	3	V
β	Beta	Gain factor (transconductance coefficient)	5×10^{-3}		A/V^2
λ	Lambda	Reciprocal Early voltage (channel-length modulation)	0.002		V^{-1}

Voltage-Controlled Current-Source Model

N-channel JFET

Off: $V_{GS} \leq$ Vto
Triode: $V_{GS} >$ Vto, and $0 < V_{DS} < V_{DSsat}$
Satur.: $V_{GS} >$ Vto, and $V_{DS} \geq V_{DSsat} > 0$

P-channel JFET

Off: $V_{GS} \geq$ Vto
Triode: $V_{GS} <$ Vto, and $0 > V_{DS} > V_{DSsat}$
Satur.: $V_{GS} <$ Vto, and $V_{DS} \leq V_{DSsat} < 0$

$$V_{DSsat} = V_{GS} - \text{Vto}$$

$$I_D = \begin{cases} 0, & \text{off region} \\ \text{Beta } [2(V_{GS} - V_T)V_{DS} - V_{DS}^2], & \text{triode region} \\ \text{Beta } (V_{GS} - V_T)^2 \lambda(V_{DS}), & \text{saturation region} \end{cases}$$

$\lambda(V_{DS}) = 1 + \text{Lambda } V_{DS}$ $\lambda(V_{DS}) = 1 - \text{Lambda } V_{DS}$

output-voltage-independent current is multiplied by $(1 + |\lambda V_{DS}|)$ and $(1 + |(1/V_A) V_{CE}|)$ in the cases of JFET and BJT, respectively, we can say that λ means the reciprocal Early voltage $1/V_A$. Therefore, the parameter measurement procedure described in Section 6.4 can be applied to obtain the parameter λ. Physically, the slight current increase in the saturation region is due to channel length shortening as the pinch-off point moves toward the source. Consequently, the parameter λ is frequently called the *channel-length modulation coefficient*.

Whereas the JFET can have two separate gates, the SPICE model assumes a single V_{GS} voltage. This is sufficient to cover the two most frequent application arrangements: (1) the two gates are connected together, electrically forming a single gate, and (2) one of the gates is connected to a constant voltage or grounded. These configurations are included through the JFET parameters: (1) when the gates are connected together, the gain factor is increased (doubled in the case of symmetrical gates), and the absolute value of the threshold voltage (or pinch-off voltage) is reduced, and (2) when a constant reverse-bias voltage is applied to one of the gates, the absolute value of the threshold voltage is also reduced, as it helps the other gate to pinch the channel off.

Table 9.2 summarizes the parameters that are available to include some of the parasitic elements associated with the JFET structure, and gives the complete large signal equivalent circuit used in SPICE. As always, the parasitic capacitances are of special importance, as they determine the high-frequency behavior of the JFET. In the equivalent circuit of Table 9.2, the capacitances are included through the diode model given in Table 3.5. Therefore, the parameter measurement techniques are the same as described in Sections 2.4.3 and 3.2.1. It should be mentioned, however, that the stored-charge capacitance is neglected (no τ_T parameter appears in the list), which is justified by the fact that the gate-to-channel diodes D_S and D_D are normally reverse biased.

TABLE 9.2 Summary of SPICE JFET Model: Parasitic Elements

		Parasitic-Element-Related Parameters			
Symbol	**SPICE Keyword**	**Related Parasitic Element**	**Parameter Name**	**Typical Value**	**Unit**
R_D	Rd	R_D	Drain resistance	10	Ω
R_S	Rs	R_S	Source resistance	10	Ω
I_S	IS	D_S, D_D	Saturation current	1×10^{-14}	A
n	N	D_S, D_D	Emission coefficient	1	—
$C_{GD}(0)$	Cgd	D_D	Gate-drain zero-bias capacitance	4×10^{-12}	F
$C_{GS}(0)$	Cgs	D_S	Gate-source zero-bias capacitance	4×10^{-12}	F
V_{bi}	PB	D_S, D_D	Built-in potential	0.8	V
m	M	D_S, D_D	Grading coefficient	$\frac{1}{3} - \frac{1}{2}$	—

Large-Signal Equivalent Circuit

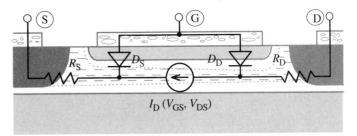

NOTE: Diodes and the current direction shown for N-channel JFET.
         Reverse for P-channel JFET.

$I_D(V_{GS}, V_{DS})$ is given in Table 9.1

$\left.\begin{matrix} D_S \\ D_D \end{matrix}\right\}$ according to the diode model of Table 3.5

The equivalent circuit of Table 9.2 can properly include all the parasitic capacitances for the case of mutually connected gates. If one of the gates is grounded, or biased with a constant voltage, the top and bottom P–N junctions are not connected in parallel, and cannot precisely be modeled by a single set of D_S and D_D diodes. Nonetheless, the parameters $C_{GS}(0)$ and $C_{GD}(0)$ can be adjusted to provide close fitting in that case. If more precise fitting of the capacitance–voltage dependencies is required, it will be necessary to add a set of diodes to the JFET, to include the effects of the second gate-to-channel junction.

9.3 MESFET

Conceptually, the metal-semiconductor field-effect transistor (MESFET) is not different from the JFET. A structural difference is that the gate is created as a Schottky diode (contact) rather than a P–N junction diode. An important practical difference is related to the fact that the MESFET is the most frequent FET implementation in GaAs, while the JFET is normally made in Si.

9.3.1 MESFET Structure

Figure 9.9 illustrates the MESFET structure. Although it is very similar to the JFET structure of Fig. 9.5, the main difference is that neither the upper nor the lower P–N junctions appear in the MESFET structure. A semiinsulating GaAs substrate defines the thickness of the N-type electron channel. The electrically effective thickness of the channel can be altered by the reverse bias of a Schottky diode, created by deposition of appropriate metal onto the N-type GaAs layer. The role of the Schottky diode is equivalent to the role of the upper P–N junction in the case of the JFET.

It should also be noted that the MESFET is surrounded by a semiinsulating GaAs. Although this does not affect the principal characteristics of the device, it is an important factor in terms of device speed. Section 9.1 discussed the benefits of parasitic capacitance reduction associated with the semiinsulating substrate.

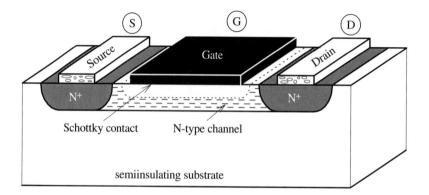

Figure 9.9 MESFET structure.

9.3.2 MESFET Characteristics

Device operation, energy-band diagrams, and consequently current–voltage characteristics are similar to those previously introduced with the MOSFET and JFET. Section 5.1.3 describes two mechanisms of current saturation: channel pinch-off and drift velocity saturation. The mechanism responsible for the current saturation in JFETs is typically the channel pinch-off. As opposed to this, the current in a typical GaAs MESFET saturates due to velocity saturation. This difference influences the device modeling, however, this problem will be considered in the next section.

In this section, we show the transfer (Fig. 9.10a) and the output (Fig. 9.10b) characteristics, with the purpose of more carefully discussing the previously mentioned limitation to V_{GS} voltage.

The characteristics shown in Fig. 9.10 are for a normally-on (depletion-type) MESFET. A significant current flows at $V_{GS} = 0$, and a negative $V_{GS} = V_T$ is needed to turn the device off. It was stated in Section 9.2.1 that positive V_{GS} was undesirable as it could turn the input diode on. Strictly speaking, some positive V_{GS} voltage can be applied before the input diode is turned on. In the example of Fig. 9.10a, the diode turn-on limit is shown to be about $V_{GS} = 0.4$ V.

The fact that some positive V_{GS} voltage can be applied makes it possible to design a normally-off MESFET. To achieve this, the N-type GaAs layer is made thinner than the depletion layer width of the Schottky diode at $V_{GS} = 0$ V. Therefore, the N-channel is pinched off by the depletion layer, and no current flows at $V_{GS} = 0$ V. The current can start flowing if positive V_{GS} voltage is applied to narrow the depletion-layer width. The V_{GS} voltage at which the current starts flowing, which is the threshold voltage V_T, is positive in this case. A practical problem with this type of normally-off FET is that the threshold voltage, no matter how close to zero, is too close to the diode turn-on limit for V_{GS}. This not only makes the manufacturing requirements very strict, but also limits the input voltage range and consequently input noise margin that circuits made of these devices

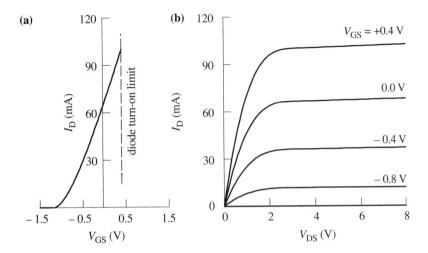

Figure 9.10 Transfer and output characteristic of a MESFET.

can handle. Nonetheless, so many circuits perform much better with normally-off FETs that these MESFETs are used, especially in high-frequency digital circuits.

9.3.3 SPICE Model and Parameters

As mentioned in the previous section, the current saturation in a typical MESFET is due to the drift velocity saturation, which occurs before the channel would be pinched off by the drain voltage. The drift velocity–electric field curve of GaAs, shown in Fig. 9.4a, is very complex. This leads to a complicated distribution of the lateral electric field in the channel, which is not possible to model by simple and yet physically based equations. Consequently, SPICE uses empirical equations, which are presented in Table 9.3.

TABLE 9.3 Summary of SPICE MESFET Model: The Principal Equivalent-Circuit Element

Static Parameters

Symbol	SPICE Keyword	Parameter Name	Typical Value	Unit
	LEVEL	Model type (1=Curtice, 2=Raytheon)		
V_T	Vto	threshold voltage (pinch-off voltage)	-2	V
β	Beta	Gain factor (transconductance coefficient)	0.1	A/V^2
λ	Lambda	Reciprocal Early voltage (channel-length modulation)	0.005	V^{-1}
α_{sat}	Alpha	Saturation coefficient	1	V^{-1}
b	B	β reduction coefficient (LEVEL 2 only)	1	V^{-1}

Voltage-Controlled Current-Source Model

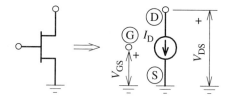

LEVEL = 1 *(Curtice model)*

$$I_D = \begin{cases} 0 & \text{for } V_{GS} \le \texttt{Vto} \\ \texttt{Beta}(V_{GS} - \texttt{Vto})^2(1 + \texttt{Lambda}\,V_{DS})\tanh(\texttt{Alpha}\,V_{DS}) & \text{for } V_{GS} > \texttt{Vto} \end{cases}$$

LEVEL = 2 *(Raytheon model)*

$$I_D = \begin{cases} 0 & \text{for } V_{GS} \le \texttt{Vto} \\ \frac{\texttt{Beta}}{1 + \texttt{B}\,(V_{GS} - \texttt{Vto})}(V_{GS} - \texttt{Vto})^2(1 + \texttt{Lambda}\,V_{DS})\,t(\alpha_{sat}\,V_{DS}) & \text{for } V_{GS} > \texttt{Vto} \end{cases}$$

$$t(\alpha_{sat}\,V_{DS}) = \begin{cases} 1 - \left(1 - \frac{\texttt{Alpha}\,V_{DS}}{3}\right)^3 & \text{for } \texttt{Alpha}\,V_{DS} \le 3 \\ 1 & \text{for } \texttt{Alpha}\,V_{DS} > 3 \end{cases}$$

[a]The parasitic elements are as in Table 8.2)

Two levels of MESFET models are available in SPICE: LEVEL 1 is an earlier developed Curtice model, and LEVEL 2 is a newer Raytheon model. The levels are selected by specifying the LEVEL as an input parameter.

Three essential parameters of the Curtice model are the threshold voltage V_T, the gain factor β, and the saturation coefficient α_{sat}. Although the meaning of V_T and β is the same as for MOSFET, the saturation effect is modeled differently [the tanh $(\alpha_{sat}V_{DS})$ term], which makes it easier to measure V_T and β in the saturation rather than the linear region. As $\tanh(\alpha_{sat}V_{DS}) \approx 1$ in the saturation region, and assuming $1 + \lambda V_{DS} \approx 1$, the drain current becomes

$$I_D \approx \beta(V_{GS} - V_T)^2 \tag{9.2}$$

Plotting the squared root of I_D current versus V_{GS} produces a line that intersects the V_{GS} axis at V_T and whose slope is equal to $\sqrt{\beta}$:

$$\sqrt{I_D} = \sqrt{\beta}(V_{GS} - V_T) \tag{9.3}$$

The measurement of α_{sat} parameter is not as easy. The best way is to assume an initial value that is close to $3/V_{DSsat}$ (V_{DSsat} being the saturation voltage), and then use nonlinear curve fitting to more precisely determine the values of all the parameters, including α_{sat}.

The term $(1 + \lambda V_{DS})$ accounts for the finite output dynamic resistance, in the same way as for the JFET. As the JFET section explains, this parameter has a meaning analogous to the Early voltage in BJTs, and is therefore measured in an analogous way.

The Curtice model predicts a parabolic increase of the drain current with gate-to-source voltage [Eq. (9.2)], which frequently fails to properly fit the experimental data. The newer Raytheon model (LEVEL 2 in SPICE) introduces an additional parameter to correct this problem. The new parameter b, called the β reduction coefficient, is analogous to the mobility modulation constant θ of MOSFETs. The measurement of this parameter is analogous to the measurement of θ described in Section 5.4.2, except that the measurements are taken in the saturation region. Noting that $t(\alpha_{sat}V_{DS}) = 1$ in the saturation region, and assuming $1 + \lambda V_{DS} \approx 1$, the Raytheon equation of Table 9.3 can be written as

$$\underbrace{\frac{I_D}{(V_{GS} - V_T)^2}}_{S} = \frac{\beta}{1 + b(V_{GS} - V_T)} \tag{9.4}$$

This equation can further be modified to the form analogous to Eq. (5.51):

$$\frac{\beta}{S} - 1 = b(V_{GS} - V_T) \tag{9.5}$$

Obviously, the parameter b is the slope of the $[(\beta/S) - 1]$-vs-$(V_{GS} - V_T)$ plot.

Another difference introduced in the Raytheon model is the replacement of tanh $(\alpha_{sat}V_{DS})$ by the computationally more effective terms $1 - [1 - (\alpha_{sat}V_{DS}/3)]^3$ in the triode, and 1 in the saturation region.

The large signal equivalent circuit of the JFET, shown in Table 9.2, can be applied to the MESFET case. Although Schottky diodes rather than P–N junction diodes should appear in the case of MESFET, there is no difference from a modeling point of view. The difference is taken into account by specifying appropriate parameter values. The same list of parameters applies, with a difference that the usual SPICE keyword for the built-in voltage is VBI rather than PB. For the sake of completeness, it should be mentioned that the SPICE equivalent circuit of the MESFET includes a resistor in series with the gate (R_G) and a capacitor in parallel with the current source (C_{DS}). The usual SPICE keywords for these parameters are RG and CDS, respectively. As MESFETs are frequently used in high-frequency applications, these parameters are helpful for more precise simulation.

9.4 HEMT

The high-electron mobility transistor (HEMT) is also called a modulation-doped field-effect transistor (MODFET), a heterostructure field-effect transistor (HFET), and a selectively doped heterojunction transistor (SDHT). It can operate at frequencies higher than 10 GHz with ultralow noise (<1 dB) and high gain. These are unique performance figures among solid-state devices, making the HEMT very attractive for satellite receivers and mobile and wireless applications.

The unique properties of this transistor are related to its specific type of carrier channel, known as two-dimensional electron gas (2DEG). This section describes the two-dimensional electron gas, the associated quantum-mechanical effects, and the way it is used to create the HEMT.

9.4.1 Two-Dimensional Electron Gas (2DEG)

Section 3.3.3 on tunneling breakdown mentions that the electrons exhibit wave properties and that the electrons are not precisely confined in minute volumes of space. Wave properties of the electrons are irrelevant as long as the device dimensions are much larger than the electron wavelength. Field-effect transistors, however, typically create a triangular potential well, like the one illustrated in Fig. 9.11. This is the case with the ordinary MOSFETs (Fig. 5.3) and also, as will be shown later, with the HEMT. As the width of the triangular potential well approaches zero at the tip, we have a situation in which the electron wavelength must become larger than the potential well width. In this case the wave properties of electrons become important, involving quantum-mechanical effects that are described in more detail in Chapter 12.

Figure 9.11 shows that no electron can be placed at energies lower than the energy level that corresponds to the half-wavelength of the electron. This lowest possible energy level for the electrons is the "first energy subband." Electrons at this level do not move freely and they are not randomly scattered in the direction of the potential well cross section (x-direction in Fig. 9.11). In this direction, the electrons appear as standing waves. In the other two directions (y and z), the potential well is much larger than the electron wavelength, and the electrons move freely and are randomly scattered. Therefore, what is seen as a single

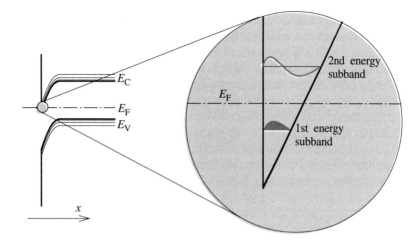

Figure 9.11 Illustration of energy level quantization, which creates two-dimensional electron gas (2DEG).

energy level in the x-direction, appears as a two-dimensional energy subband in the y and z dimensions. The concept of "electron gas" can be applied only to these two dimensions, and, consequently, the electrons in this situation are referred to as the two-dimensional electron gas (2DEG).

Figure 9.11 illustrates the second energy subband, which appears at the energy level corresponding to full electron wavelength. This illustrates the fact that the energy separation between the energy levels (subbands) depends on the electron wavelength and the slope of the potential well side. The later is directly related to the electric field that creates the triangular potential well. The effect of significant separation between particular energy levels is known as the quantization effect. A pure two-dimensional electron gas appears with a very strong quantization effect, when the lowest energy subband is populated, while all the other energy subbands are empty.

This is not the case for silicon MOSFETs, where the electrons appear in more than four subbands, as the energy separation between the subbands is not very large compared to the thermal energy. Electrons can move between different energy subbands through different mechanisms of interband scattering. This situation is referred to as a *quasi-two-dimensional electron gas*, to distinguish it from the 2DEG in which any form of carrier scattering is eliminated in the direction of the standing wave.

In the case of a silicon MOSFET, the wide energy gap of SiO_2 grown on Si creates the conduction-band discontinuity that forms the triangular energy potential well. This effect is achieved by what is known as *energy gap engineering* in the case of GaAs. Figure 9.12 illustrates the cross section and the energy-band diagram of an AlGaAs–GaAs heterojunction. The energy gap of AlGaAs is wider, and it creates a conduction-band discontinuity at the interface with GaAs. This leads to the formation of a triangular potential well. The combination of the conduction-band discontinuity, the energy band bending on the GaAs side, and the electron wavelength in GaAs leads to a quite pronounced energy level quantization. Very importantly, the position of the Fermi level is above the lowest energy subband, which means that the lowest subband is populated with electrons—no external field is needed to create the 2DEG.

The quality of the AlGaAs–GaAs interface is extraordinary. It is created by a continuous process of molecular beam epitaxy, which enables a change from GaAs to AlGaAs within a single atomic layer by changing the gas composition inside the epitaxial chamber. In addition, no doping atoms appear in the area of 2DEG. The inserted thin buffer layer of undoped AlGaAs separates the heavily doped N^+ AlGaAs from the 2DEG. Consequently, interface roughness scattering and Coulomb scattering from the doping ions are virtually eliminated. This adds to the fact that any form of scattering is eliminated in one of the three space dimensions. The reduced scattering means high electron mobility in the 2DEG, and it also means reduced noise.

9.4.2 HEMT Structure and Characteristics

The high electron mobility and the low noise associated with the 2DEG are very useful for high-frequency and low-noise applications. The 2DEG created at the AlGaAs–GaAs heterojunction makes the HEMT channel. As Fig. 9.13 illustrates, the channel is contacted

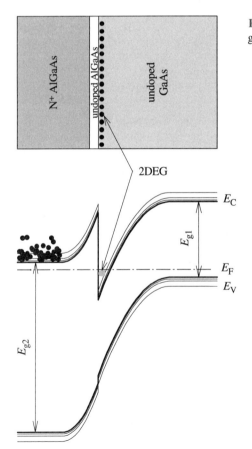

Figure 9.12 Cross section and energy-band diagram of an AlGaAs–GaAs system.

at the ends to create source and drain terminals. Also a gate in the form of a Schottky contact is created between the source and drain, to provide the means of controlling the device current. Negative gate voltages reduce the electron concentration in the channel, and can completely repel the electrons, turning the device off. Therefore, the HEMT appears as a depletion-type FET. The transfer and output characteristics are analogous to those previously described for the depletion-type MOSFET, JFET, and MESFET.

The electron path from the source to the drain is indicated by the line and the arrows in Fig. 9.13. Electrons pass through a couple of heterojunctions between the source terminal and the 2DEG, as well as between the 2DEG and the drain terminal. The energy-band diagrams illustrate the barrier shapes associated with these heterojunctions at typical drain-to-source bias V_{DS}. It can be seen that electrons have to tunnel from the N^+ GaAs source into the N^+ AlGaAs layer, and tunnel again from the drain end of the 2DEG back into the N^+ AlGaAs layer. This introduces significant source and drain resistances, which

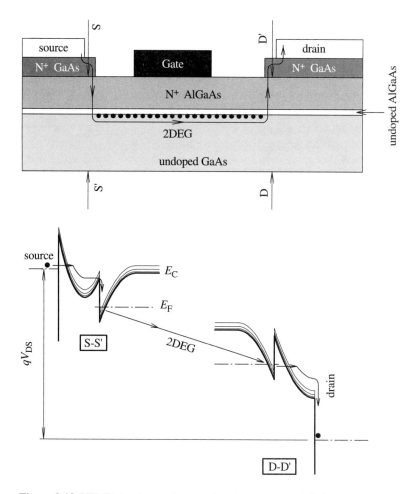

Figure 9.13 HEMT structure and energy bands at source and drain contacts.

adversely affect the high-speed and noise performance of the HEMT. Consequently, a lot of modifications to source and drain contacts have been introduced to minimize the adverse effects of the source and drain resistances. The gate resistance is also very important, especially in terms of noise performance. The gate is typically made in a mushroom shape to cut its resistance while not increasing the input capacitance, which improves the noise figure of the HEMT.

9.5 NEGATIVE RESISTANCE DIODES

Negative-resistance diodes are active microwave devices that can be used as amplifiers and oscillators at frequencies up to 100 GHz (despite the continuous improvement of FETs, their performance is still inferior or even inadequate at these frequencies). As the name suggests, a common characteristic of the negative-resistance diodes is the negative-resistance (or negative-conductance) phenomenon. More precisely, we are dealing here with *negative-dynamic (or differential) resistance (NDR)*, or, alternatively, negative-dynamic (differential) conductance. This means that a voltage increase causes a current decrease ($r = dv/di < 0$), which is different from the fact that the ratio between the instantaneous voltage and current is still positive ($R = v/i > 0$). The negative-dynamic resistance causes current and voltage to be 180° out of phase with each other. As a consequence, the signal power[5] is negative, which means that a negative-resistance diode does not dissipate but generates signal power. Again, this should not be confused with the total power, which is positive, expressing the fact that the device efficiency is <100%.

The signal amplification by two-terminal negative-resistance diodes is conceptually different from the amplification by the voltage-controlled current sources (three terminal BJTs and FETs). Section 9.5.1 of this chapter describes the principles of amplification and oscillation by negative-resistance devices. Sections 9.5.2 and 9.5.3 introduce Gunn and IMPATT diodes as representatives of *transferred electron devices* (TED) and *avalanche transit time diodes*, respectively. Finally, Section 9.5.4 briefly describes the tunnel diode.

9.5.1 Amplification and Oscillation by Dynamic Negative Resistance

When describing the amplification principle by a voltage-controlled current source in Section 4.1.3, it was explained that a load line with a *negative slope* was used (Fig. 4.4). The negative slope was achieved by a combination of load resistance and a DC power source, which basically meant that the transistor was converting the DC voltage and power into signal voltage and power. A part of the current–voltage characteristic of a negative-resistance diode exhibits a *negative slope*. This region of the current–voltage characteristic can also be used to produce signal voltage and current 180° out of phase with each other, thus negative signal power.

---

[5]Signal voltage times signal current.

Figure 9.14a shows that the load (G_L) is connected in parallel with a negative-conductance diode (g). This does not mean that the diode can deliver a signal power to the load out of nothing—it still needs a DC power source, and because of the parallel connection the power source would be effectively in the form of a current source I_{BB}, connected in parallel to the signal source i_s.[6]

Figure 9.14b shows a typical i_D–v_D characteristic of a negative conductance diode, with the negative conductance part of the curve highlighted. To use this characteristic for a graphic analysis of the circuit, the load line representing the load (G_L) and the DC bias (I_{BB}—not shown in Fig. 9.14a) is needed. When the source current $i_s = 0$ (DC bias point),

$$i_D = I_{BB} - G_L v_D \tag{9.6}$$

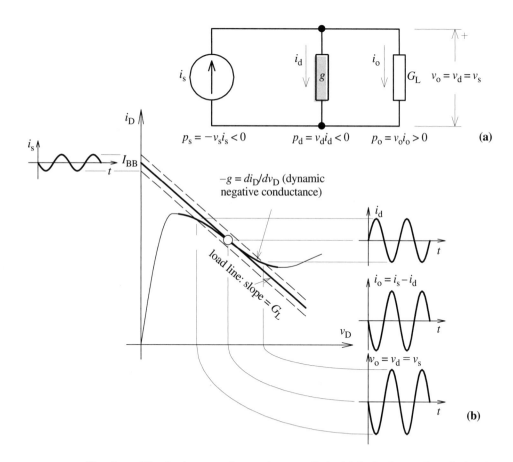

Figure 9.14 Signal amplification by a negative-conductance diode: (a) the fundamental small-signal equivalent circuit and (b) graphic load-line analysis.

---

[6]The actual implementation of the DC current source (I_{BB}) is a separate issue, but typically an ordinary voltage source separated by a lossy low-pass filter is used.

to satisfy Kirchhoff's current law at the parallel connection of g and G_L. Equation (9.6) is the needed load line, which can be drawn by observing the following two facts: (1) $i_D = I_{BB}$ for $v_D = 0$, which means the line intersects the i_D axis at I_{BB}, and (2) the slope of the line is $-G_L$. The solution is found at the intersection between the load line and the diode characteristic. The absolute values of i_D and v_D are both positive, which means that the diode dissipates (uses) instantaneous power supplied by the DC source.

When a signal current i_s is added to the DC bias current I_{BB}, the load line is shifted up or down (depending on the sign of the signal current), but the slope is not changed. This is illustrated by the dashed lines in Fig. 9.14b. It can be seen that a small decrease/increase in the incoming current, caused by the superimposed signal current (i_s) on the bias current (I_{BB}), causes a rather large increase/decrease in the diode current. Therefore, a current gain is achieved. More importantly, the increase in i_D is accompanied by a decrease in v_D, which means the signal voltage is 180° out of phase with the signal current flowing through the diode. This means negative signal power $p_d = v_d i_d$, or signal power generation in other words. On the other hand, the signal current flowing through the load (i_o) is in phase with the voltage, which means the load is using power $p_o = v_o i_o$.

The current gain ($A = i_o/i_s$) and the power gain ($A = p_o/p_s$) are identical to each other, because $v_o = v_s$. Figure 9.14b shows that the magnitude of the i_d current depends on the slopes of both the negative-conductance part of the diode curve (g) and the load line (G_L). This further means that the magnitude of the output current ($i_o = i_s - i_d$) and hence the current/power gain depends on both g and G_L. To find this dependence, let us start from Kirchhoff's current law,

$$\frac{i_o}{A} = i_d + i_o \tag{9.7}$$

where the source current i_s is expressed in terms of the output current, using the gain definition $A = i_o/i_s$. Further, express the currents in terms of the conductances and the unique voltage $v_o = v_d = v_s$ by applying Ohm's law:

$$\frac{G_L}{A} v_o = -|g| v_o + G_L v_o \tag{9.8}$$

When applying Ohm's law to the negative-conductance element, care is taken to express that the current i_d is 180° out of phase with the voltage $v_d = v_o$. This leads to the following equation for the gain:

$$A = \frac{G_L}{G_L - |g|} \tag{9.9}$$

We can conclude from this equation that

$$G_L > |g| \tag{9.10}$$

is a condition for stable amplification. With this condition, $p_s = -v_o i_s = -v_o^2 G_L/A < 0$ (power generated), $p_d = v_o i_s = -|g| v_o^2 < 0$ (power generated), and $p_o = v_o i_o = G_L v_o^2 >$

0 (power used). If $G_L < |g|$, the gain A would be negative, and v_d and i_d inverted. With this, the input signal source consumes power ($p_s > 0$) while the negative-conductance diode still generates power ($p_d < 0$). This is a situation in which the generated signal power by the negative-conductance diode is unrelated to the signal source. Is this possible?

To clarify the question of power generation unrelated to any excitation, let us remove the signal source ($i_s = 0$) but not the supply current I_{BB}. In addition to that, let G_L be adjusted so that

$$G_L = |g| \tag{9.11}$$

In this case, $A \to \infty$, which does mean that no input signal is needed to have a finite output signal current ($i_o = Ai_s = \infty \times 0 =$ finite value). An electronic system that produces an output signal without any input excitation, and thus generates the signal, is called an *oscillator*. Referring to Fig. 9.14b, the condition $G_L = |g|$ means that the load line overlaps the diode curve in the negative-conductance section. There is not a unique $i_D–v_D$ solution (intersection between the load line and the diode curve) in this case, meaning that the circuit is unstable. The operating point will oscillate, as governed by either interna' physical processes in the diode (unresonant mode) or by a resonant circuit if such a circuit is connected (resonant mode).

■ **Example 9.1 Negative-Resistance Oscillator**

The current–voltage characteristic of a negative resistance diode is shown in Fig. 9.15, where $i_A = 200\ \mu A$, $i_B = 60{,}200\ \mu A$, $v_A = 16$ V, and $v_B = 10$ V.

(a) Find the value of the load conductance G_L so that this diode is used as an oscillator.

(b) Assuming maximum signal amplitude, calculate the *power conversion efficiency* of this oscillator ($\eta = \bar{p}_o/\bar{p}_S$, where $\bar{p}_o$ is the average output signal power and $\bar{p}_S$ is the average instantaneous power supplied by the DC source).

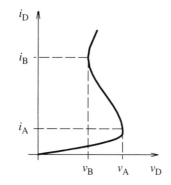

Figure 9.15 Current–voltage characteristic of a negative-resistance diode.

Solution:

(a) The oscillation condition is given by Eq. (9.11) as $G_L = |g|$. Assuming linear i_D–v_D dependence between i_A and i_B, $|g|$ can be estimated as

$$|g| \approx (i_B - i_A)/(v_A - v_B) = \frac{60{,}200 - 200}{16 - 10} = 10 \text{ mA/V} = 10 \text{ mS}$$

Therefore, the load conductance should be

$$G_L = 10 \text{ mS}$$

(b) To achieve maximum signal amplitude, the signal current and voltage should oscillate around the central (I_D, V_D) point, where I_D and V_D are $(60.2+0.2)/2 = 30.2$ mA and $(16 + 10)/2 = 13$ V, respectively. The peak amplitudes of the signal are then $I_m = i_B - I_D = (I_B - I_A)/2 = 30$ mA and $V_m = (V_A - V_B)/2 = 3$ V. Therefore, the voltage and current signals can be expressed as $v_o = v_d = V_m \sin(\omega t)$, and $i_o = i_s - i_d = -i_d = I_m \sin(\omega t)$. With this, the average output signal power is found as

$$\bar{p}_o = \frac{1}{T} \int_0^T v_o i_o \, dt = \frac{V_m I_m}{\pi} \int_0^\pi \sin^2(\omega t) d(\omega t) = V_m I_m/2$$

The instantaneous power delivered by the DC source is $p_S = I_{BB} v_D$, where I_{BB} is the current of the DC current source supplying the power, and v_D is the voltage across the current source, which is equal to the voltage across the diode, as well as the output. As $v_D = V_D - V_m \sin(\omega t)$, the average power delivered by the source can be expressed as

$$\bar{p}_S = \frac{I_{BB}}{2\pi} \int_0^{2\pi} v_D d(\omega t) = I_{BB} V_D - \frac{I_{BB} V_m}{2\pi} \underbrace{\int_0^{2\pi} \sin(\omega t) d(\omega t)}_{=0} = I_{BB} V_D$$

I_{BB} is the current at which the load line intersects the i_D axis. As the load line passes through the central (I_D, V_D) point, and has a slope of $-G_L$, the following relationship can be established:

$$\frac{I_{BB} - I_D}{V_D} = G_L$$

which further means that

$$I_{BB} = G_L V_D + I_D = 160.2 \text{ mA}$$

The power conversion efficiency is, therefore,

$$\eta = \frac{\bar{p}_o}{\bar{p}_S} = \frac{V_m I_m}{2 V_D I_{BB}} = \frac{3 \times 30}{2 \times 13 \times 160.2} = 0.0216 = 21.6\%$$

9.5.2 Gunn Diode

Gunn diodes represent a group of devices called *transferred electron devices (TED)*. TEDs are used as oscillators and amplifiers, covering the frequency range from 1 to 100 GHz, with output power capabilities greater than 1 W. Principally, these devices are made of plain N-type semiconductor pieces (no P–N junctions), with ohmic contacts at two opposite sides. The name Gunn diode is typically used for GaAs-based diodes, while the other options are InP and CdTe.

The appearance of negative-dynamic resistance in such a simple structure is due to the specific E–k dependence, which has two close minima, as shown in Fig. 9.3a for GaAs. As explained in Section 9.1.4, the electrons in the lower E–k valley have much smaller effective mass, and consequently they possess much higher mobility μ_l (note that the subscript l refers to lower E–k valley and not "low mobility"). Due to the fact that the higher E–k valley is much wider, electrons appear as much heavier there, and consequently their mobility μ_h is much lower. At small voltages, thus small electric fields inside the semiconductor, all the electrons are in the lower E–k valley. The mobility of all the electrons is high (μ_l), and the conductance $G = qAn\mu_l/L$ is large. This is illustrated by the large-slope dashed line in Fig. 9.16. At very large voltages, thus very large electric fields, most of the electrons gain sufficient energy to appear in the higher E–k valley. The mobility of the electrons is now low (μ_h), and the conductance $G = qAn\mu_h/L$ is small, as shown by the small-slope dashed line in Fig. 9.16. In the medium voltage range, a voltage increase causes a transfer of a number of electrons from the lower to the higher E–k valley, which reduces the average drift velocity of the electrons (Fig. 9.4a), and therefore reduces the current. This current reduction due to a voltage increase is the effect of negative-dynamic conductance $g = dI/dV < 0$.

The semiconductor is not stable in the negative-conductance region—it cannot establish a unique electric field when biased so that V/L exceeds the critical electric field (the critical electric field corresponds to the maximum drift velocity in Fig. 9.4a). To explain this, let us consider a packet of electrons injected by a negative cathode into GaAs biased beyond

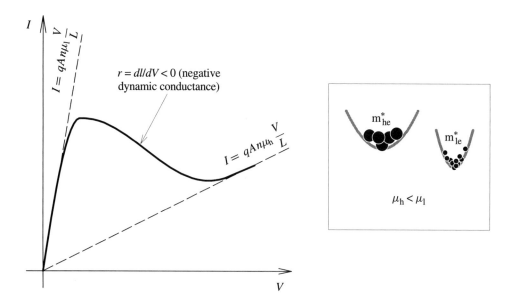

Figure 9.16 Illustration of the negative-dynamic conductance/resistance due to the Gunn effect.

the critical V/L. As the electrons move toward the anode, the electric field back toward the cathode is reduced, while the electric field toward the anode is slightly increased. This is illustrated in Fig. 9.17a. In normal circumstances, the packet of electrons would disperse itself, as the electrons leaving the packet toward the anode move faster in the region of stronger field. These electrons are not adequately replaced by electrons that come from the cathode at a slower rate. As a result the semiconductor stabilizes itself. In the case of negative v_d–E slope (negative mobility), the situation is exactly opposite: fewer electrons leave the packet moving toward the anode than the number of electrons coming from the cathode. As a result, the packet of electrons grows.

To visualize this effect by the energy bands, the region of strong electric field, such that the v_d–E slope is negative (negative mobility), is shaded in the conduction bands shown in Fig 9.17. Normally, we think of the electrons as rolling down faster when the slope of the conduction band is larger. The shaded area is not like this, and this simple model breaks down here. Still, we can gain some graphic insight if we think of the shaded area as a very "dense" medium, so that the particles move faster in the clear than the shaded area.[7] With this, we can visualize the effect of continuous electron accumulation as the accumulation layer moves down toward the anode. The increasing electron concentration in the accumulation layer increases the electric field toward the anode, which reduces the drift velocity of the electrons in this region, reducing the anode current. Therefore, *a voltage increase results with a current decrease*, as illustrated in Fig. 9.17b.

As the accumulation process continues, and the accumulation layer moves toward the anode, further increasing the electric field, the drift velocity reduction approaches saturation. At some point the increased electron concentration starts dominating the anode current, which leads to a current increase (Fig. 9.17c). The increased current causes a voltage reduction on the negative-slope load line, which only helps the current increase by causing a smaller field in the negative-mobility region than it would be at a higher voltage.

As the electron concentration in the accumulation layer drops down due to the electrons terminating at the anode, the electric field back toward the cathode is being increased, eventually above the critical level (Fig. 9.17d). By the time all the electrons from the accumulation layer are collected by the anode, a new electron "packet" is being formed at the cathode end of the semiconductor,[8] and the cycle is back to the situation represented by Fig. 9.17a.

The above explanation shows that the oscillation period is approximately equal to the time that it takes the accumulation layer to drift from the cathode to the anode. Assuming average drift velocity $\bar{v}_d$, the oscillation frequency is given by

$$f = \frac{\bar{v}_d}{L} \tag{9.12}$$

where L is the length of the sample.

---

[7]To be quite precise, we need to assume that this imaginary medium gets "denser" as the slope of the bands increases—this is to include the fact that v_d reduces with E.

[8]The creation of this electron "packet," or the initiation of the accumulation layer, is explained by the fact that the rate of electrons injected into the semiconductor is higher than their drift velocity in the high-field region.

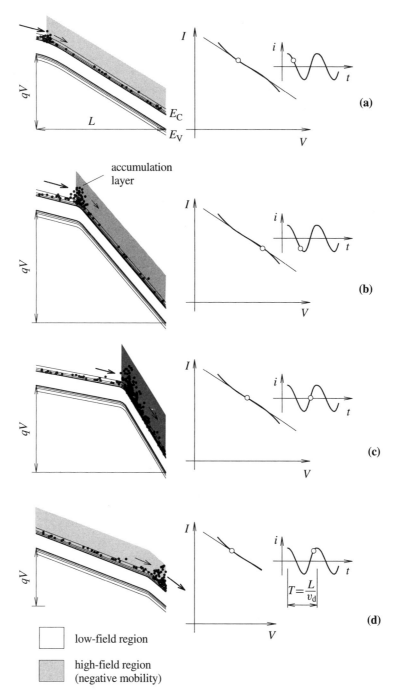

Figure 9.17 Illustration of a Gunn-effect oscillator.

This oscillation mechanism is referred to as either an *accumulation-layer mode*, *transit-time mode*, or *Gunn-oscillation mode*. It is a nonresonant mode, for the oscillation frequency is set by the physical parameters of the diode (length and electron velocity). A number of more complex modes of operation are possible. For example, at higher doping concentration, or with longer samples, dipoles of accumulation and depletion layers are formed. In this case the diode is operated at a so-called *transit-time dipole-layer mode*. Additional oscillation modes include the *quenched dipole-layer mode* and *limited-space-charge accumulation (LSA) mode*. At very low doping levels, or very short samples, the device can operate in the *stable amplification mode*, also called *uniform-field mode*. In this case, there are not enough electrons to create an accumulation layer. Consequently, the electric field is uniform so that an I–V characteristic with a negative-resistance part (Fig. 9.16) results from a simple scaling of the velocity-field characteristic (Fig. 9.4a). As the spontaneous oscillation is avoided in this mode, the device can be used as a stable amplifier of signals with frequencies near the transit-time frequency. The other types of transferred-electron devices, namely InP and CdTe diodes, are similar, but with different physical parameters.

9.5.3 IMPATT Diode

The IMPATT diode represents *avalanche transit-time devices*, which can operate at frequencies higher than 100 GHz, providing the highest continuous power of all semiconductor microwave devices. IMPATT stands for **imp**act ionization **a**valanche **t**ransit **t**ime. These devices can be implemented in Si as well as GaAs.

The IMPATT diode is a resonant device, as it requires a resonant circuit for its operation. The parallel connection of an inductor and capacitor is the resonant circuit in Fig. 9.18a. Structurally, the IMPATT diode is typically a PIN diode. In Fig. 9.18, the "intrinsic" region is fully depleted and is sandwiched by heavily doped N^+- and moderately doped P-type regions that serve as contacts. Practically, the "intrinsic" region would be a very lightly doped P-type region. The DC reverse-bias voltage V_{BB} sets the diode very close to avalanche breakdown. The resonant circuit is designed so that the positive oscillating voltage v, superimposed to the bias voltage V_B, takes the diode in avalanche mode. The electric field in the structure is the strongest at the N^+–I junction, and the field exceeds the critical breakdown value only in a narrow region around the N^+–I junction. This leads to a generation of electron and hole avalanches, as illustrated in Fig. 9.18b. The electrons are quickly neutralized by the positive charge from the nearby anode, while the holes start drifting through the depletion layer toward the cathode. Assuming drift velocity v_d and sample length L, it will take a time of $\tau = L/v_d$ for the holes to reach the cathode, causing maximum cathode current. If the resonant circuit is designed so that its frequency is

$$\frac{T}{2} = \tau \Rightarrow f = \frac{v_d}{2L} \tag{9.13}$$

the voltage is at its minimum when the current reaches its maximum. The voltage and current are 180° out of phase with each other, which means that negative-dynamic resistance is established.

The I–V characteristic of the avalanche diodes is S shaped, as in Fig. 9.15, as opposed to the N-shaped I–V characteristic shown in Fig. 9.16. This is because very small levels

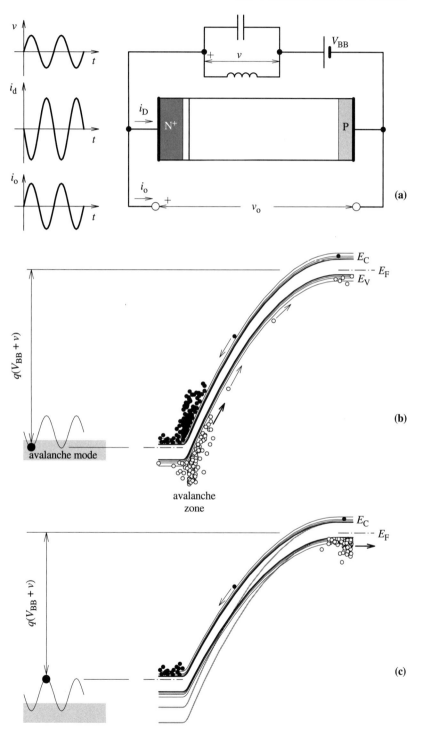

Figure 9.18 The principle of IMPATT diode operation.

of current relate to any voltage between 0 and the breakdown.[9] In a situation like the one illustrated in Fig. 9.18, the negative-dynamic resistance causes a voltage decrease as the current is increased. However, if the diode is set in a continuous-breakdown mode, the structure behaves as an ordinary small-value resistor, because of the abundance of current carriers.

9.5.4 Tunnel Diode

Tunnel diodes are small-power small-voltage devices that can be used as microwave amplifiers and oscillators, although they are being displaced by the other semiconductor devices. The tunnel diodes, also known as Esaki diodes, have an historic importance related to the Nobel-prize discovery of the tunnel effect by L. Esaki in 1958. The tunnel diode is

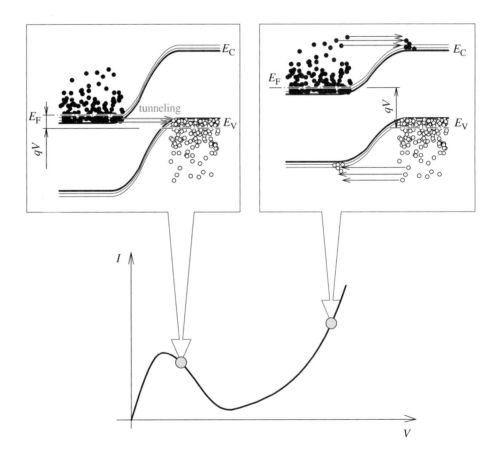

Figure 9.19 The principle of tunnel diode operation.

---

[9]This is opposite to the Gunn diode, in which a small voltage is related to a very small resistance, due to the high mobility of the conducting electrons.

a P–N junction diode with heavy doping (in the order of 10^{20} cm^{-3}) in both P-type and N-type regions. The heavy doping is related to two important facts for the operation of the tunnel diode: (1) The Fermi level is inside the conduction band in the N-type material, reflecting the extreme concentration of electrons, and inside the valence band in the P-type region, reflecting the extreme concentration of holes. (2) The depletion layer separating the N-type and P-type region is very narrow ($<$10 nm). As a result, (1) the electrons at the bottom of the conduction band in the N-type region are energetically aligned to the holes in the valence band of the P-type region (Fig. 9.19), and (2) the space separation between them is comparable to the characteristic length of the electron wave function. As described in Sections 3.3.3 and 12.3.2, this enables the electrons to tunnel through the depletion-layer barrier.

Referring to Fig. 9.19, we see that an increasing tunneling current flows as an increasing forward-bias voltage is applied. However, the forward bias splits the quasi-Fermi levels reducing the energy overlap between the N-type conduction and P-type valence bands. This leads to a tunneling current decrease as the voltage is increased. This is the negative-resistance region that is utilized by microwave oscillators and amplifiers. As the energy overlap disappears, the tunneling current drops to zero. At this point, however, the normal diode current already dominates the total current. This current is due to the electrons (and holes) being able to go over the depletion-layer barrier, and it increases as an increased voltage reduces the barrier height. This explains the N-shaped $I–V$ curve of the tunneling diode, shown in Fig. 9.19.

PROBLEMS

9.1 Power gain that can be achieved by a negative-conductance diode is given by Eq. (9.9). Derive this equation using the power balance principle.

9.2 Referring to Example 9.1:

(a) Sketch v_d, i_d, and i_s, positioning them appropriately with respect to I_D and V_D.

(b) Shade the rectangle defined by the axes and V_D and I_{BB} lines. What does the area $V_D I_{BB}$ represent?

(c) Observe the triangle defined by i_B and V_D lines and the diode characteristic. Compare it to the triangle defined by i_A, V_D and the diode characteristic. How does the area of each of these triangles relate to the signal power?

9.3 The negative-resistance region of a Gunn diode is defined by the following two voltage–current points: $(V_1, I_1) = (8 \text{ V}, 12 \text{ mA})$ and $(V_2, I_2) = (12 \text{ V}, 2 \text{ mA})$. Calculate G_L to maximize the power gain of a sinusoidal signal with peak current $I_{mi} = 0.1$ mA.

9.4 The voltage across a Gunn diode oscillates between $V_{min} = 5.5$ V and $V_{max} = 10$ V. Knowing that the effective length of the diode is $L = 3.5$ μm, and using the $v_d–E$ dependence given in Fig. 9.4a, estimate the oscillation frequency. $\boxed{A}$

9.5 Find the resonant frequency for a silicon N$^+$P$^-$P IMPATT diode with effective length $L_{P^-} = 10$ μm.

REVIEW QUESTIONS

R-9.1 Low-field electron mobility is about seven times higher in GaAs than Si. Why is GaAs not used for high-speed complex digital ICs?

R-9.2 Can the MOSFET structure be implemented in GaAs?

R-9.3 Is the power dissipation due to the current flowing through normally-on FETs a limiting factor for complex digital ICs? How about amplifiers?

R-9.4 Is the input capacitance of a depletion-type FET larger or smaller than the input capacitance of a comparable MOSFET? Is this advantageous for high-frequency analog applications?

R-9.5 Which material provides better high-frequency performance of depletion-type FETs, Si or GaAs? Why?

R-9.6 What is the main difference between the MOSFET and JFET?

R-9.7 Is the JFET a normally-on or normally-off device?

R-9.8 What happens if positive gate voltage is applied to the gate of an N-channel JFET? What if this voltage is smaller than 0.7 V?

R-9.9 What is the mechanism of current saturation in silicon JFETs?

R-9.10 The threshold voltage, also called pinch-off voltage, is the most important SPICE parameter of the JFET. Is the threshold voltage of an N-channel JFET positive or negative?

R-9.11 What are the most important parasitic elements inherently present in the JFET structure, and what are the related SPICE parameters?

R-9.12 Can the MESFET structure be implemented in silicon? If yes, what would be the difference from the JFET structure?

R-9.13 Can a GaAs MESFET be designed as a normally-on (enhancement type) device? If no, why, and if yes, are there any application constraints?

R-9.14 What is the mechanism of current saturation in GaAs MESFETs?

R-9.15 Can the concepts of particle motion be used in the case of two-dimensional electron gas (2DEG)? If yes, in how many dimensions?

R-9.16 Is the high electron mobility in the 2DEG created at the AlGaAs/GaAs heterojunction due to the fact that the gas is two dimensional, due to reduced interface scattering, or due to reduced Coulomb scattering?

R-9.17 Is the reduced carrier scattering in the HEMT channel related to its superior low-noise performance?

R-9.18 Can an analog signal be amplified by a two-terminal device? If yes, how?

R-9.19 Is the signal power $p_d = gv_d^2$ positive or negative if the dynamic conductance of the device g is negative? What does "negative signal power" mean anyway?

R-9.20 What happens if a negative-resistance diode is biased so that the load line overlaps the negative-resistance region (in other words, there is no unique intersection between the load line and the diode characteristic)?

R-9.21 Are Gunn diodes based on P–N junctions? Schottky contacts?

R-9.22 Is the Gunn effect possible in Si?

R-9.23 Is it possible to have a Gunn-diode-based oscillator without an external resonator circuit?

R-9.24 Is the electron drift velocity related to oscillation frequency that can be achieved by a Gunn diode? If yes, why and how?

R-9.25 Is the electron drift velocity related to the oscillation frequency in the case of an IMPATT diode? Is there any difference from the case of a Gunn diode?

R-9.26 Does the IMPATT diode need an external resonator?

R-9.27 Do the N-shaped I–V characteristics of Gunn and tunnel diodes mean that their physical principles are similar, while quite different from the avalanche diodes that exhibit S-shaped I–V characteristics?

R-9.28 Is the tunnel diode based on a P–N junction? If yes, is the P–N junction forward or reverse biased in the negative-resistance mode?

R-9.29 Is the operating voltage of a tunnel diode higher or lower than an IMPATT diode? Is this related to the output power capabilities?

Chapter 10

POWER DEVICES

Power electronic circuits are used to convert electrical energy from the form supplied by a source to the form required by the load. A typical example of power conversion is rectification and filtering of the AC line voltage to provide a constant DC voltage. Other types of conversion include DC–AC, DC–DC, AC–AC, and combinations of the previous types. As explained in Section 3.1.1, diodes are used to rectify an AC voltage (AC–DC conversion). Plain rectification produces "wavy" voltage, even after filtering with large-value capacitors. *Voltage regulation* is needed to obtain the needed "smooth" voltage. Such regulation is almost invariably based on *switching* techniques.[1] The other types of power conversion also involve switching techniques. Devices such as BJT, MOSFET, JFET, IGBT (insulated gate bipolar transistor), and thyristor are used as controlled switches in these circuits.

Two fundamental parameters of a semiconductor switch (either a diode or a controlled switch) are (1) the voltage that can be sustained by an open switch (*blocking voltage*), which is determined by the breakdown voltage of the device, and (2) the parasitic resistance of the closed switch (*on resistance*), which relates to the current capability of the device.

In addition to switches and capacitors, the power circuits typically involve inductors and transformers. Large inductance values are necessary at lower switching frequencies, making the inductors and transformers inconveniently large and heavy. High switching frequencies are the only solution to this problem. Consequently, the switching characteristics of power devices are almost as important as the on resistance and the blocking voltage.

The first section of this chapter briefly describes two basic DC–DC converter circuits to set a representative background for the following considerations of semiconductor power devices. Sections 10.2 and 10.3 describe power-related specifics of devices already introduced, diode and MOSFET, respectively. Sections 10.4 and 10.5 introduce IGBT and thyristor, respectively, as alternative controlled switches to the MOSFET and BJT.

---

[1]It is possible to regulate the voltage by allowing the excess voltage to drop across the controlled-variable resistance of a BJT or MOSFET. This so-called *linear regulation* is not efficient as it inevitably involves power dissipation by the regulating resistance.

10.1 POWER DEVICES IN SWITCH-MODE POWER CIRCUITS

With ideal components, including the controlled switch and the diode, the circuits of Fig. 10.1 are 100% efficient, regulated DC–DC converters. The 100% efficiency comes from the fact that the operation principles of these circuits involve no resistors. The inductors and capacitors can store electrical energy, ideally, without any loss. In many ways, the inductor is a device that is complementary to the capacitor: no DC current can flow through a capacitor and no DC voltage can drop across an inductor; fast current changes are passed by a capacitor and fast voltage changes are possible across an inductor; and the voltage across a capacitor and current through an inductor cannot change instantly. These facts and symmetry are expressed by the fundamental current–voltage relationships:

$$\text{Capacitor:} \quad i = C\frac{dv}{dt}$$

$$\text{Inductor:} \quad v = L\frac{di}{dt} \tag{10.1}$$

where t is the time, C is the capacitance, and L is the inductance.

Periodic switching of the controlled switch in these circuits results with DC–DC conversion. To show this, it should be observed that some energy is transferred from the unregulated DC source (V_{IN}) to the inductor when the switch is closed, and the same amount of energy is transferred from the inductor to the load and the capacitor when the switch is open. When the switch is closed, the capacitor provides the current and energy taken by

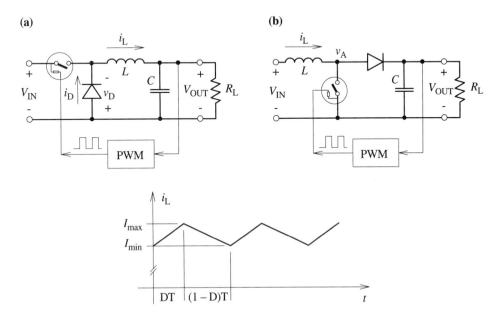

Figure 10.1 Step-down (buck) (a) and step-up (boost) (b) regulated DC–DC converters.

the load resistance R_L.[2] The inductor current increases from I_{min} to I_{max} when the switch is closed and decreases from I_{max} back to I_{min} when the switch is open.[3] Applying Eq. (10.1) to the inductor in Fig. 10.1a, and neglecting the voltage drop across the diode, the following set of equations is obtained:

$$\frac{V_{IN} - V_{OUT}}{L} = \frac{I_{max} - I_{min}}{DT}, \quad \text{switch closed}$$

$$-\frac{V_{OUT}}{L} = \frac{I_{min} - I_{max}}{(1-D)T}, \quad \text{switch open} \tag{10.2}$$

where T is the switching period, D is the fraction of time that the switch is closed (duty cycle), and $di/dt \approx (I_{max} - I_{min})/(DT)$. Eliminating I_{max} and I_{min} from the above set of equations leads to

$$V_{OUT} = D V_{IN} \tag{10.3}$$

Therefore, the load voltage depends solely on the input voltage and the duty cycle. The control circuit (PWM) adjusts the duty cycle so that any variations in the input voltage (as in the case of the "wavy" voltage obtained by AC rectification) are compensated. The circuit of Fig. 10.1a is called a *step-down* or *buck* DC–DC converter, because the output voltage is smaller than the input voltage. The circuit of Fig. 10.1b is a *step-up* or *boost* DC–DC converter. This can again be shown by expressing the inductor current for an open and a closed switch:

$$\left. \begin{array}{l} \dfrac{V_{IN}}{L} = \dfrac{I_{max} - I_{min}}{DT}, \quad \text{switch closed} \\[2mm] \dfrac{V_{IN} - V_{OUT}}{L} = \dfrac{I_{min} - I_{max}}{(1-D)T}, \quad \text{switch open} \end{array} \right\} \Rightarrow V_{OUT} = \frac{1}{1-D} V_{IN} \tag{10.4}$$

Switching power circuits cannot produce perfectly smooth voltages, the magnitude of the inevitable ripple depending on the switching frequency and the inductor/capacitor values. Higher frequencies and higher inductances and capacitances correspond to smaller ripples. As acceptable or desirable physical size of the power-electronic circuits limits the inductor and capacitor values, an increase in the switching frequency remains as the only alternative for ripple and/or circuit size reduction.

Clearly, the highly desirable size reduction of power-electronic circuits critically depends on the high-frequency switching performance of the semiconductor devices used as switches. Some devices, for example Schottky diodes and MOSFETs, exhibit apparently superior switching characteristics over the alternative P–N diodes and BJTs. The same type of device can be designed to provide better switching performance at the expense of the other fundamental characteristics, such as on resistance and blocking voltage. The switching characteristics, on resistance and blocking voltage, are so profoundly interwoven that only

---

[2]This applies to both circuits.

[3]If the current starts from I_{min}, it has to drop back to I_{min} at the end of a cycle, otherwise the steady state is not reached.

a good grasp of underlying semiconductor physics can guarantee optimum circuit design and component selection.

10.2 POWER DIODES

Both P–N junctions and Schottky contacts are used for power diodes. The principles of P–N junctions and Schottky contacts are introduced in Chapters 2 and 3. Nonetheless, there are some power-circuit aspects that warrant emphasis. The following sections discuss the most important power-related specifics of both the P–N junction and Schottky diodes.

10.2.1 PIN Diode

The power diodes are used in circuits in which relatively high voltages have to be rectified. The breakdown voltage of a P–N junction can be higher than 1000 V, provided that at least one side (P or N) is very lightly doped. Equations (2.20) and (2.33)–(2.34) and Fig. 2.8c show that the maximum electric field, which appears right at the P–N junction, is proportional to $\sqrt{N_{A,D}}$, where $N_{A,D}$ is the concentration of the lower doped region. As Eqs. (2.33)–(2.34) show, a lower $N_{A,D}$ allows a higher reverse-bias voltage V_R to be applied before the critical electric field is achieved. The critical (breakdown) electric field in silicon is in the order of several tens of volts per micrometer.

The forward voltage drop is also very important in power-electronic circuits. As an example, assume that the output voltage of the buck DC–DC converter in Fig. 10.1a is $V_{OUT} = 5$ V, while the forward voltage drop across the diode is $V_F = 1$ V. V_F is comparable to V_{OUT} and it will cause a significant power loss. To obtain an output power of $V_{OUT}I_{OUT}$, power in excess of $V_F I_{OUT}$ is dissipated by the diode. Assuming 50% duty cycle, the efficiency of the converter is less than $50 + 50 \times 5/(5 + 1) = 92\%$. The forward voltage drop is a significant barrier in the attempt to achieve efficiencies as close as possible to the theoretical limit of 100%.

The low-doped region needed for the high breakdown voltage in reverse mode inevitably increases the series resistance, and therefore the forward voltage drop. This would be especially dramatic if the power diode was created by diffusion of a P-type layer into a very low-doped N-type substrate (as in Fig. 1.10). However, there is no need to limit the doping level of the whole substrate, for its significant thickness would introduce a large series resistance. In practical power diodes, the relatively thick substrate (needed for mechanical strength only) is heavily doped. The high breakdown voltage is achieved by depositing a very low-doped epitaxial layer. A heavily doped layer at the top completes the structure. The low-doped region, sandwiched between N$^+$ and P$^+$ layers, is usually labeled by I ("insulator"). It is also commonly referred to as *drift layer*. Accordingly, this type of diode is referred to as a *PIN diode*. In reality, the "I" layer is either a P-type or N-type layer with very low doping concentration ($<10^{14}$cm^{-3}). Another difference from the P–N junction shown in Fig. 1.10 is that the curved sections of the P–N junction are avoided. This is because the field is stronger at the sharpest sections of the curve, reducing the breakdown voltage that could be achieved by a planar P–N junction. A variety of etching and surface passivation techniques are used to avoid this problem.

When used in switching power circuits, like the ones in Fig. 10.1, the switching characteristics of the diode become very important. The diode can neither be turned on nor off instantly. The times needed for both *forward recovery* (t_{fr}) and *reverse recovery* (t_{rr}) are very important, especially as higher switching frequencies are needed to reduce the size of power-electronic circuits. Figure 10.2 shows the voltage and current waveforms for a power PIN diode used in the circuit of Fig. 10.1a.

When the switch is opened at time $t = 0$, the inductor current starts flowing through the diode ($I_F = i_L$). In steady state, the forward current of minority carriers is due to diffusion. This is explained in Section 3.1.3, Eq. (3.4) and Fig. 3.6 showing that an appropriate gradient of minority-carrier concentration is needed so that the diffusion current can transport the minority carriers injected over the barrier. The concentration gradient of minority carriers is not established instantly. As a consequence, the diffusion is not the dominant mechanism of minority-carrier transport at the beginning. A significant electric field is established in the lightly doped neutral region to sustain the forced current by drifting the minority carriers, as well as injecting more majority carriers over the P–N junction barrier. This electric field adds to the electric field in the depletion layer resulting in the voltage overshoot observed during the turn-on period of the diode. The time needed to establish the steady-state profile of the minority carriers (the stored charge), and therefore to reach the steady-state forward condition, is called *forward recovery time* (t_{fr} in Fig. 10.2a). The voltage overshoot is undesirable, as it results in increased power dissipation ($> V_F I_F$) during the forward recovery

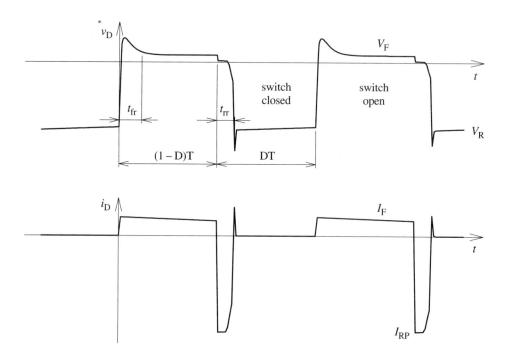

Figure 10.2 Switching response of a power diode in the buck converter of Fig. 10.1a: t_{fr}, forward recovery time, t_{rr}, reverse recovery time.

period. The SPICE model of the diode (Section 3.2) does not include this effect. A parallel $L-R$ circuit, added in series to the diode, can be used as the simplest equivalent circuit of this effect.

When the switch is closed at time $t = (1-D)T$, the diode should turn off, and a reverse-bias voltage $V_R = V_{IN}$ should appear across the diode. However, this cannot happen as long as there is a stored charge of minority carriers at the P–N junction. This effect is described in Section 3.2.2, Eq. (3.23) showing that the stored charge Q_s is given by

$$Q_s = \tau_T I_F \tag{10.5}$$

where τ_T is the transit time (SPICE parameter) and $I_F = I_D$ is the forward diode current. As mentioned in Section 3.2.2, electron–hole recombination helps the process of stored charge removal, however, the minority-carrier lifetime is too long in the low-doped PIN diodes to rely on this mechanism. In the circuit of Fig. 10.1a, the stored charge has to be drained through the switch with on resistance R_{on} and the voltage source V_{IN}. Neglecting the small voltage drop across the diode itself, the discharging current is set at

$$I_{RP} = \frac{V_{IN}}{R_{on}} \tag{10.6}$$

The discharge time can be calculated as

$$t_{rs} = \frac{Q_s}{I_{RP}} \tag{10.7}$$

In Eq. (10.7), any contribution from the recombination mechanism is, clearly, neglected. Therefore, the time given by Eq. (10.7) is a worst case estimate. The contribution of recombination can become significant if (1) the recombination mechanism is enhanced by special manufacturing techniques (some of them are discussed below) and (2) the discharging (reverse-bias) voltage is too low so that the discharging current is too small. In general, if the estimated t_{rs} is not significantly smaller than τ_T, there is enough time for the recombination mechanism to remove a significant portion of the stored charge. Therefore, if t_{rs} calculated by Eq. (10.7) is in the order of τ_T or larger, it is significantly overestimated.

After the removal of the stored charge, the charge of the depletion-layer capacitance C_d (Section 2.2.2) has to be set at a level that corresponds to the reverse-bias voltage V_R. During this process, the diode behaves more like an ordinary capacitor, with the current and voltage setting at exponential rate with time constant $C_d R_{on}$. The times t_{rs} and $C_d R_{on}$ add up to the total *reverse recovery time* t_{rr}.

The main problem with t_{rr}, especially t_{rs}, is that it limits the switching speed. As Q_s [Eq. (10.5)] and τ_T (Section 8.1.3) are related to the recombination rate and minority-carrier lifetime, a recombination rate increase would reduce these times. The recombination rate can be increased by introducing energy levels into the energy gap in a way analogous to the introduction of energy levels by the doping atoms. The best effects are achieved when the energy level to help the recombination is at the mid-gap. Suitable elements as recombination centers in silicon are gold and platinum. Consequently, diffusion of gold or platinum into the "I" region is used to improve the switching performance of PIN diodes. Another technique is high-energy electron irradiation, which introduces recombination energy levels in the energy gap by creating damage in the silicon crystalline structure. However, too high a concentration of recombination centers leads to a sharp increase in the forward voltage drop, which limits the reduction of transit time by this approach.

■ Example 10.1 Stored-Charge Removal—t_{rs} Time

In the buck DC–DC converter of Fig. 10.1a, $V_{IN} = 6$ V, $\bar{i}_L = 3.5$ A, and the transit time of the diode (SPICE parameter) is $\tau_T = 5$ μs. Calculate the peak reverse current of the diode I_{RP}, and the time it takes for the removal of the stored charge t_{rs}, for two values of the on resistance of the switch: (a) $R_{on} = 0.5$ Ω, (b) $R_{on} = 1.0$ Ω.

Solution:

(a) When the switch is opened, forward-bias current $I_F = i_L$ flows through the diode. The stored charge, caused by this current (Eq. 3.23), is

$$Q_s = \tau_T I_F = 5 \times 10^{-6} \times 3.5 = 1.75 \times 10^{-5} \text{ C}$$

When the switch is closed, the minority holes stored at the cathode side of the junction move toward the positive V_{IN} terminal, while the electrons stored at the anode side moves toward the negative V_{IN} terminal. This is the reverse peak current I_{RP} that removes the stored charge. The value of this current is set by the value of the input voltage (neglecting the small voltage drop across the diode itself) and the switch resistance:

$$I_{RP} = V_{IN}/R_{on} = 6/0.5 = 12 \text{ A}$$

The diode is not turned off for as long as there is stored charge and the discharge current flows through the circuit. The time that is needed for the stored charge removal by this current is

$$t_{rs} = Q_s/I_{RP} = 1.75 \times 10^{-5}/24 = 1.46 \text{ } \mu s$$

(b) Answer: $I_{RP} = 6$ A, $t_{rs} = 2.92$ μs.

10.2.2 Schottky Diode

Schottky diodes, introduced in Section 3.4, are single-carrier devices. Although there are minority carriers in the semiconductor, they play an insignificant role as far as the current flow through the metal–semiconductor contact is concerned. Consequently, the effects associated with the stored charge of minority carriers do not exist in Schottky diodes. This means there is no forward voltage overshoot, and $t_{rs} = 0$ (no reverse-recovery time is spent on a stored charge removal). Clearly, these devices have superior switching characteristics, and are consequently very useful in modern high-frequency power circuits.

The avalanche breakdown of a Schottky diode depends on the parameters of the semiconductor layer that creates the metal–semiconductor contact. Therefore, it is essentially the same as in the case of a PIN diode. Similar to the PIN diode, the very lightly doped layer, needed to take the high reverse-bias voltage, is deposited on to a heavily doped substrate. Typically, Pt, W, Cr, or Mo is used as the metal electrode (the anode), while an N^-–N^+ Si or GaAs structure is used on the cathode side. As before, the heavy doping of the substrate reduces the series resistance, and therefore the forward voltage drop.

Like the PIN diode, the forward current depends exponentially on the forward-bias voltage [Eq. (3.30)]. However, the current of the P–N junction diode is limited by the diffusion of minority carriers. Equation (3.16) for the saturation current I_S shows this fact. In the absence of minority carriers, the dominant and limiting current mechanism in the case of the power Schottky diodes is *thermionic emission* of electrons from the metal into the semiconductor. This process has Arrhenius-type dependence on the barrier height $q\phi_B$ and temperature:

$$I_S = A_J A^* T^2 e^{-q\phi_B/(kT)} \tag{10.8}$$

In the above equation for the saturation current of the Schottky diode, A_J is the diode area, T is absolute temperature, $q\phi_B$ is the barrier height, and A^* is the so-called effective Richardson constant. For N-type Si, $A^* \approx 120\ \mathrm{A\ cm^{-2}\ K^{-2}}$, and for N-type GaAs, $A^* \approx 140$ $\mathrm{A\ cm^{-2}\ K^{-2}}$.

Fig. 3.20 and Eq. (3.27) show that the barrier height is equal to the difference between the metal work function and the semiconductor affinity. Selecting a metal with appropriate work function $q\phi_m$, the current I_S can be set at a high enough level to achieve the needed forward current at a lower forward voltage. Obviously, this can significantly alleviate the problem of power dissipation, which is another important advantage of Schottky diodes over PIN diodes. Unfortunately, some $V_F I_F$ power loss is inevitable, as it is theoretically impossible to achieve the ideal diode characteristic (the dashed line in Fig. 3.2c). The forward voltage drop is reduced at the expense of I_S increase. It is theoretically possible to cut the barrier height to zero, which will result with zero V_F, but in this case the rectifying feature of the contact is completely lost and it appears as a perfect zero-resistance contact in either direction. Clearly, a trade-off between the good effects of V_F reduction and the bad effects of I_S increase is necessary.

The undesirable increase in I_S is exacerbated by the so-called barrier lowering effect. This effect is due to the appearance of an image force between the electrons in the semiconductor and the nearby highly conductive metal. It is shown that the barrier height reduction is[4]

$$\Delta\phi_B = \sqrt{\frac{q E_{max}}{4\pi \varepsilon_s}} \tag{10.9}$$

where E_{max} is the maximum electric field. This field is the same as for the abrupt P–N junction [Eqs. (2.20) and (2.30)] and is therefore proportional to $\sqrt{V_R}$. Clearly, a significant

---

[4]E.H. Rhoderic, *Metal-Semiconductor Contacts*, Clarendon Press, Oxford, 1978; S.M. Sze, *Physics of Semiconductor Devices*, Wiley, New York, 1981.

increase in the reverse-bias current I_S of the Schottky diode occurs with increasing reverse-bias voltage.

Example 10.2 PIN versus Schottky Diode

This example compares Si PIN and Schottky power diodes, with the same area $A_J = 0.1$ cm^2, and the same doping level in the low-doped region of $N_D = 5 \times 10^{14}$ cm^{-3}. The Schottky diode is created by depositing W onto the low-doped Si.

(a) Assuming maximum allowable field of $E_{max} = 20$ V/μ m, calculate the needed width of the drift layer. What reverse-bias voltage (maximum operating voltage) corresponds to E_{max}?

(b) Calculate the saturation currents of the PIN and the Schottky diodes and compare them. Assume the following value for the diffusion coefficient of minority holes: $D_p = 50$ cm^2/s.

(c) Neglecting the parasitic resistances, calculate and compare the forward voltage drops for an operating current $I_F = 5$ A. Discuss the results in terms of power loss. (Assume ideal emission coefficient $n = 1$.)

(d) Assuming mobility of $\mu_n = 1400$ cm^2/Vs, calculate the resistance of the drift region. What are the forward voltage drops when this resistance is included?

(e) Including the barrier lowering effect, calculate the reverse current of the Schottky diode at maximum reverse voltage. Discuss the result in terms of power dissipation.

Solution:

(a) The theory of asymmetrical abrupt P–N junction, presented in Section 2.3.2, can also be applied to the case of the Schottky diode. This is because a significant depletion layer appears only in the N-drift region in both cases. From Eq. (2.20), the maximum depletion-layer width is obtained as

$$w_{n-max} = \varepsilon_s E_{max}/(q N_D) = \frac{11.8 \times 8.85 \times 10^{-12} 20 \times 10^6}{1.6 \times \times 10^{-19} \times 5 \times 10^{20}} = 26\ \mu m$$

Therefore, the width of the drift layer should be $w_{n-epi} \approx 26\mu m$. Neglecting the small V_{bi} in comparison to V_R, the maximum operating voltage is calculated from Eq. (2.34) as

$$V_{R-max} = w_{n-max}^2 q N_D/(2\varepsilon_s) = (26 \times 10^{-6})^2 \frac{1.6 \times 10^{-19} \times 5 \times 10^{20}}{2 \times 11.8 \times 8.85 \times 10^{-12}} = 260\ V$$

(b) Noting that $N_A \gg N_D$, Eq. (3.15) can be simplified to calculate the saturation current of the PIN diode:

$$I_S = A_J q n_i^2 \frac{D_p}{w_{n-epi} N_D} = 10^{-5} \times 1.6 \times 10^{-19} \times (1.02 \times 10^{16})^2$$

$$\frac{0.005}{26 \times 10^{-6} \times 5 \times 10^{20}} = 6.4 \times 10^{-11} \text{ A}$$

The barrier height $q\phi_B$ is needed to be able to calculate the saturation current of the Schottky diode. From Eq. (3.27) and Table 2.6, we find

$$q\phi_B = q\phi_m - q\chi_s = 4.6 - 4.05 = 0.55 \text{ eV}$$

From Eq. (10.8),

$$I_S = A_J A^* T^2 e^{-q\phi_B/(kT)} = 10^{-5} \times 120 \times 10^4 \times 300^2 \times e^{-0.55/0.2585}$$

$$= 6.2 \times 10^{-4} \text{ A}$$

The saturation current of the Schottky diode is $6.2 \times 10^{-4}/6.4 \times 10^{-11} \approx 10^7$ times higher.

(c) Neglecting the parasitic resistances, the forward voltage drop can be obtained from the equation

$$I_F = I_S e^{V_F/V_t}$$

in the case of both PIN and Schottky diodes. For the PIN diode, it is

$$V_F = V_t \ln(I_F/I_S) = 0.02585 \times \ln(5/6.4 \times 10^{-11}) = 0.65 \text{ V}$$

For the Schottky diode,

$$V_F = 0.02585 \times \ln(5/6.2 \times 10^{-4}) = 0.23 \text{ V}$$

The series contact resistances, the resistance of the drift region, and the substrate resistance will add a voltage drop of the same magnitude, so the actual forward voltage drops will be significantly higher—part (d) of this Example. Nonetheless, the difference between the PIN and Schottky diodes of about 0.4 V will remain. At 5 A and assumed 50% duty cycle, this means a difference of $0.5 \times 0.4 \times 5 = 1$ W of power dissipation. Of course, this does not include the power loss due to the forward voltage overshoot in the case of the PIN diode.

(d) The conductivity of the drift region is [Eq. (1.18)]

$$\sigma = q\mu_n N_D = 1.6 \times 10^{-19} \times 0.14 \times 5 \times 10^{20} = 11.2 \ (\Omega \text{ m})^{-1}$$

The resistance is then

$$R = w_{n-epi}/(\sigma A_J) = \frac{26 \times 10^{-6}}{11.2 \times 10^{-5}} = 0.232 \ \Omega$$

With $I_F = 5$ A, the forward voltage drops of the PIN and Schottky diodes are $V_F = 0.65 + 0.232 \times 5 = 1.81$ V and $V_F = 0.23 + 0.232 \times 5 = 1.39$ V, respectively.

(e) The barrier height lowering is given by Eq. (10.9):

$$\Delta\phi_B = \sqrt{qE_{max}/(4\pi\varepsilon_s)} = \sqrt{\frac{1.6 \times 10^{-19} \times 20 \times 10^6}{4\pi \times 11.8 \times 8.85 \times 10^{-12}}} = 0.05 \text{ V}$$

The reverse-bias current with the lower barrier $q\phi_B - q\Delta\phi_B = 0.55 - 0.05 = 0.50$ is

$$I_R = I_S(V_{R-max}) = 10^{-5} \times 120 \times 10^4 \times 300^2 \times e^{-0.50/0.2585} = 4.3 \text{ mA}$$

The power loss due to this current is not insignificant at $V_{R-max} = 260$ V. Assuming again 50% duty cycle, it is $0.5 \times 260 \times 4.3 \times 10^{-3} = 0.56$ W.

10.3 POWER MOSFET

MOSFETs can be used as voltage-controlled switches in the circuits of Fig. 10.1. The fundamental characteristics of the voltage-controlled switch are shown in Fig. 4.5. In open-switch mode, the MOSFET should be able to withstand high voltages—this is referred to as *forward blocking capability*. In close-switch mode, the MOSFET should conduct high-level currents (I_D) with minimum voltage drop (V_D) to minimize the power loss ($V_D I_D$). In other words, the *on resistance* $R_{on} = V_{DS}/I_D$ should be as small as possible. To achieve these requirements, the basic MOSFET structure of Chapter 5 is significantly modified.

There are a number of different power MOSFET structures, but perhaps the most accepted one is the **v**ertical **d**ouble-diffused MOSFET structure (VDMOSFET or DMOS), shown in Fig. 10.3. There is only one way of achieving a significant current by a field-effect transistor, and this is to significantly increase the effective channel width of the MOSFET, while minimizing or at least maintaining the channel length to a small value. Tens of centimeters of effective channel width and micrometers of channel length are needed to achieve amperes of current. To facilitate a MOSFET with a tens of centimeters wide and only several micrometers long channel, a multiple cell structure is used. The top view of the VDMOSFET (Fig. 10.3a) shows the packing of hexagonal MOSFET cells to minimize the occupied area. As Fig. 10.3 indicates, all the MOSFET cells are effectively connected in parallel: (1) the silicon substrate is used as the common drain of the MOSFET cells, (2) all the N$^+$-source and P-well regions are connected by the top metalization, and (3) although windows are etched in the polysilicon layer to enable the creation of the P-type and N$^+$-type regions, the polysilicon gates remain electrically connected to each other. This means that the effective channel width of the MOSFET is equal to the perimeter of a single cell multiplied by the number of cells used in the structure.

As Fig. 10.3 (b) indicates, the P-type region is created using the polysilicon layer as the mask, and the same edge of the polysilicon layer is used to define one side of the N$^+$

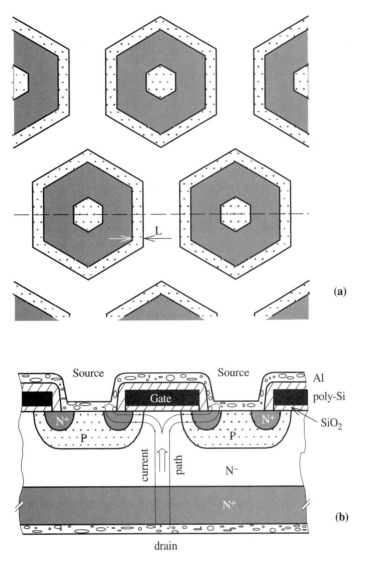

Figure 10.3 Top view (a) and cross section (b) of a VDMOSFET.

region. The difference between the edges of the P-type and N^+-type regions at the surface is due to different P-type and N^+-type lateral diffusions. This difference determines the channel length L, as illustrated in Fig. 10.3. On one side of the P-type channel region is the N^+ source and on the other is the N^- drain. The surface region of the drain is very low doped (N^-)—*drift region*. Analogously to the drift region in the case of diodes, this region becomes depleted at high V_{DS} voltages to take the voltage drop across itself, enabling the needed forward blocking capability. It is created by epitaxial deposition of low-doped silicon onto a heavily doped N^+ substrate. The heavy doping of the substrate is needed to reduce the parasitic resistance inside the drain-neutral region.

Clearly, a sufficiently wide and low-doped drift region is needed to achieve the minimum acceptable breakdown voltage. Any increase in the width or reduction in the doping concentration, however, results in an increase in the on resistance. It may not be possible to find an acceptable trade-off between the needed blocking capability and on resistance by adjusting the doping level and width of the drift region. The on resistance can be reduced, while maintaining the desired breakdown voltage, by an increase in the channel width. This is a quite simple and quite efficient approach, which can even be applied after the device has been manufactured, as discrete MOSFETs can be paralleled to achieve the needed on resistance. A related effect is *positive temperature coefficient*: if a cell (or a discrete MOSFET operating in parallel combination with others) conducts more current, its on resistance increases because of the mobility reduction, which drops the current down to the stable shared value. The importance of this effect is seen when MOSFETs are compared to BJTs. BJTs have a negative temperature coefficient, which means that a current density increase leads to on resistance reduction, which causes a further current density increase, and so on. This so-called *thermal runaway* can destroy the device.

Is there a practically important limit to the reduction of *on resistance* by paralleling MOSFETs or MOSFET cells? In the case of BJTs, even if the thermal runaway is ignored, the high input current needed to maintain the closed-switch state would be a significant problem. The input of a MOSFET is a capacitor—no DC current is needed to maintain either closed or open-switch states. But what about the current that is needed to charge or discharge the input capacitance in order to open or close the switch? Because of the large number of cells used, the input capacitance of a power MOSFET is significant. Take as an example $C_{in} = 10$ nF, to perform an estimation of the order of magnitude of the charging/discharging current. If this capacitance is to be charged from 0 to $Q = CV = 10$ nF $\times$ 10 V in $\Delta t = 0.1 \mu s$ by a constant current, the charging current has to be $I = \Delta Q / \Delta t = 1$ A. This is not a small current, and it increases proportionally with the switching frequency and the input capacitance.[5] Clearly, the performance of the circuits driving the power MOSFET used as a switch becomes extremely important. However, no driving circuit will be able to provide unlimited charging/discharging current at unlimited frequency. The input capacitance of the MOSFET poses an ultimate limit to the increase of switching frequency.

As in Fig. 10.3, the source and bulk of a power MOSFET are internally connected. Consequently, the capacitance of the drain–bulk N–P junction appears as a fairly large output capacitance. This can also be a limiting factor, although it can conveniently be utilized as a part of resonant circuits. The same junction acts as an integral power diode, connected across the output. This limits the voltage across the switch to positive values, which can also be a convienient feature. However, if a switch with reverse blocking capability is required, it would appear as a limitation.

The power capability of a power MOSFET is, clearly, another essential characteristic. If the maximum current of a power MOSFET is rated at 5 A, and the blocking voltage is at 200 V, it does not mean this MOSFET can dissipate $200 \times 5 = 1$ kW of power. The maximum current is stated for a fully switched on MOSFET ($v_{GS-HIGH}$ line in Fig. 4.5). If the on resistance is $R_{on} = 0.1\ \Omega$, the voltage drop across the MOSFET is $v_{DS} = R_{on}I_D = 0.1 \times 5 = 0.5$ V, and the dissipated power is $0.5 \times 5 = 0.25$ W. When the MOSFET

---

[5]"Proportional to input capacitance" effectively means "inversely proportional to the on resistance."

is in off mode, no power is dissipated, as $I_D = 0$ (v_{GS-LOW} line in Fig. 4.5). Ideally, the MOSFET would be in one of these two states, and the dissipated power would never exceed $R_{on}I_D^2$. However, during transitions between those two states, a significantly larger power is dissipated as neither v_{DS} nor i_D is small. To avoid overheating and failure of the MOSFET, i_D and v_{DS} should stay inside the so-called *safe operating area (SOA)*, which is defined by I_{max}, V_{DS-max}, and $i_D - v_{DS}$ points corresponding to the maximum power P_{max}[6]. The maximum power depends on how efficiently the heat is removed. MOSFETs can withstand short pulses of higher power compared to the steady-state level. Because the power MOSFETs are mainly used in switching circuits, the maximum power is typically expressed for a given pulse duration.

10.4 IGBT

The comparison of MOSFETs and BJTs, given in Section 6.1.5, shows that the main advantage of BJTs is the fact that the whole cross section is utilized for current flow. This makes the BJT structure superior in terms of achievable output currents. However, to maintain a power BJT switch in the on state, an input base current as high as one-fifth of the output current may be needed. This is a serious drawback, for a 100-A device may need 20 A of base current, which significantly complicates the input-drive circuits and reduces the power efficiency. MOSFETs as field-effect devices do not need any current to maintain the on state. Still, the input currents needed to charge/discharge the input capacitance may become quite significant if too many MOSFET cells are paralleled to achieve a high drain current (this is described in the previous section).

It can be said that the input characteristics of MOSFETs and output characteristics of BJTs are needed to create a device capable of switching high currents. It is possible to combine a MOSFET and a BJT to create such a device. Connecting the drain of an N-channel MOSFET to the collector, and the source to the base of an NPN BJT, the power MOSFET is utilized as a driver device supplying the base current to the BJT. Developing this principle further, an integrated device with field-effect input control and bipolar output action was created. Reflecting its principal features, this device is called an *insulated gate bipolar transistor (IGBT)*. A shorter name, *insulated gate transistor (IGT)*, is used as well, and it has also been called a *conductivity-modulated field-effect transistor (COMFET)*. Commercial IGBTs with blocking voltage exceeding 500 V and able to switch hundreds of amperes of currents have been developed. IGBTs are replacing power BJTs in many applications.

The cross sections of IGBT and VDMOSFET, given in Figs. 10.4a and 10.3b, respectively, show that these devices are structurally very similar. The only difference is that the N^+ substrate is replaced by the P^+ layer in the case of an IGBT. This creates the PN^-P^+ structure of the BJT. It is very easy to explain the operation of the device in the off state (zero gate voltage). In this case, the N^- layer is floating, and the IGBT appears as a BJT with unconnected base, therefore it behaves as two P–N junction diodes connected back to back. The applied voltage drops across the reverse-biased P–N junction, and no significant

---

[6]i_D-v_{DS} points corresponding to the maximum power P_{max} define a hyperbola on the i_D-v_{DS} graph.

current flows through the device. In the case of negative v_{CE}, the breakdown voltage of the lower N^-P^+ junction (V_{RB}) determines the forward blocking capability. In the case of positive v_{CE}, the breakdown voltage of the upper PN^- junction (V_{FB}) determines the reverse blocking capability. The existence of the reverse blocking capability is a significant difference from the power MOSFET. In applications in which the integrated diode of the MOSFET is utilized, the reverse blocking of the IGBT is a disadvantage. However, in applications in which reverse blocking is needed (in addition to the forward blocking), this feature of the IGBT appears as a very significant advantage.

When sufficient gate voltage is applied, a strong inversion layer connects the N^- layer to the N^+ regions, and further on to the electrode labeled as *emitter*. A significant part of the applied collector-to-emitter voltage drops across the lower N^-P^+ junction, setting it in forward-bias mode. As a result, holes are injected into the N^- base region, most of them finishing at the *emitter terminal* through the P-type region (collector of the PNP BJT).[7] These holes make most of the on-state current. For as long as the inversion layer is strong enough to neglect any voltage drop across it, the device appears as a diode with a significant current capability, as almost the whole cross section is utilized. Figure 10.4a illustrates that the MOSFET supplies electrons to maintain the forward bias of the lower N^-P^+ junction.

When the v_{CE} voltage is increased, its electric field opposes the gate field, reducing the concentration of electrons in the inversion layer. This can result in channel pinch-off, enabling a significant voltage to drop across the depleted channel area. The forward bias of the junction injecting the holes, and therefore the device current, does not increase with any further increase of v_{CE}. The device is in its *saturation region*, which is illustrated in Fig. 10.4b for the sake of completeness, although it is not useful for a device used as a switch.

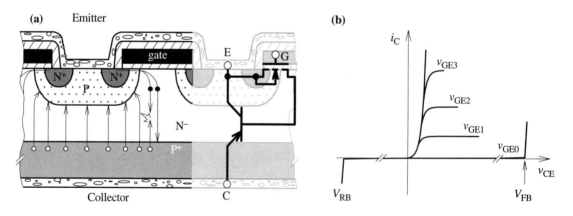

Figure 10.4 IGBT: (a) cross section of a cell with MOSFET and BJT symbols to illustrate the IGBT components and (b) current–voltage characteristics.

---

[7]The adopted labeling of the terminals as *emitter* and *collector* comes from the previously described connection of a MOSFET and NPN BJT. It can be confusing in the case of the IGBT, as the *collector terminal* is connected to the *emitter* of the inside PNP BJT, while the *emitter terminal* is connected to the *collector* of the PNP BJT.

In conclusion, it can be reiterated that the IGBT integrates the superior input and output current performances of the MOSFET and BJT, respectively. However, the "selection" of superior characteristics is limited to the current capability. Being a BJT-like device from the output, it inevitably suffers from the effects associated with storage of minority carriers. As discussed in the sections on the PIN diode and the power MOSFET, this results in an inferior switching performance compared to the one-carrier devices such as the MOSFET. As far as SPICE simulation is concerned, there is no IGBT device model with specific model parameters. The device is modeled by its equivalent MOSFET–BJT circuit, as shown in Fig. 10.4a. Consequently, the MOSFET and BJT model parameters introduced in Chapters 5 and 6 are used.

10.5 THYRISTOR

Thyristors are four-layer PNPN devices that are capable of blocking thousands of volts in the off state and conducting thousands of amperes of current in the on state. They work on the principle of an internal regenerative mechanism that leads to the so-called *latch-up* effect. Clearly, a basic understanding of the latch-up effect is necessary to understand the specifics of thyristors used as power switches.

Although thyristors utilize the latch-up effect, this effect is a potential problem in any other device that involves four layers (PNPN). An example of such a device is the IGBT described in the previous section. Also, PNPN structures are inherently present in the very common CMOS structures (Section 5.2.3). Unwanted latch-up of the parasitic thyristor structures may lead to permanent damage of these devices. This is a reliability problem, described for the case of CMOS devices in Section 11.2.6. Again, understanding the latch-up effect is necessary to understand the technological, structural, and application measures used to avoid latch-up-related problems.

The most common thyristor type is the *silicon-controlled rectifier (SCR)*. Its cross section is shown in Fig. 10.5, which also shows the two-transistor model that is used to explain the regenerative mechanism involved in the thyristor operation. It can be seen that the PNP BJT collector is the same region as the NPN BJT base, while the NPN collector is the same as the PNP base. A small gate current can trigger closed-loop amplification (the BJTs amplify each others collector current) until both transistors enter saturation, providing a low-resistance path between the anode and the cathode.

The following mathematical analysis provides the basis for a more detailed discussion. The anode current i_A, being in fact the emitter current of the PNP BJT, is given as

$$i_A = (1 + \beta_{pnp})I_{B-pnp} \tag{10.10}$$

I_{B-pnp} is the same current as the collector current of the NPN BJT, and can therefore be expressed as

$$I_{C-npn} \equiv I_{B-pnp} = \beta_{npn}(\overbrace{i_G + \underbrace{\beta_{pnp}I_{B-pnp} + I_{CB0-pnp}}_{I_{C-pnp}}}^{I_{B-npn}}) + I_{CB0-npn} \tag{10.11}$$

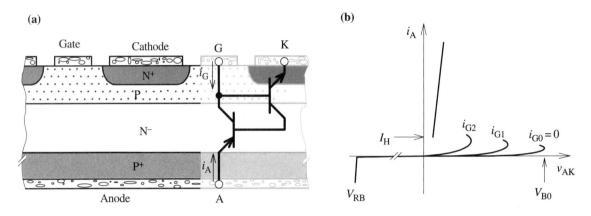

Figure 10.5 SCR: (a) cross section of a cell with a two-transistor model and (b) current–voltage characteristics.

In Eq. (10.11), $I_{CB0-pnp}$ and $I_{CB0-npn}$ are the reverse-bias collector-base leakage currents of the PNP and NPN BJTs, respectively.[8] Because of the closed loop, I_{B-pnp} appears on both the left-hand and right-hand sides of Eq. (10.11). Extracting I_{B-pnp},

$$I_{B-pnp} = \frac{\beta_{npn}(i_G + I_{CB0-pnp}) + I_{CB0-npn}}{1 - \beta_{pnp}\beta_{npn}} \qquad (10.12)$$

and putting it into Eq. (10.10), the following expression for the anode current is obtained:

$$i_A = \frac{1 + \beta_{pnp}}{1 - \beta_{pnp}\beta_{npn}}[\beta_{npn}(i_G + I_{CB0-pnp}) + I_{CB0-npn}] \qquad (10.13)$$

Assuming $i_G = 0$ and negligible leakage currents ($I_{CB0-pnp} \approx 0$ and $I_{CB0-npn} \approx 0$), the anode current is negligible, and the thyristor behaves as an open switch even for $v_{AK} > 0$. This is because the base-collector N^-P junction is reverse biased and the base currents are ≈ 0, so both BJTs are in the cutoff mode. However, if v_{AK} is increased, both $I_{CB0-npn}$ and $I_{CB0-pnp}$ are increased, leading to the appearance of some i_A current. At very small transistor currents, the current gains (βs) are small as well, but their values increase as the transistor current is increased.[9] At some voltage $v_{AK} = V_{B0}$, the transistor currents increase the current gains to the point at which $\beta_{pnp}\beta_{npn} \rightarrow 1$. According to Eq. (10.13), $i_A \rightarrow \infty$ at this condition. The regenerative mechanism has occurred, the BJTs enter saturation mode, and Eq. (10.13) no longer applies. All the P–N junctions are forward biased, which means the voltage drop between the anode and the cathode is very small, and the current flowing between the anode and the cathode is limited by the external circuit. The thyristor is latched up and it behaves as a closed switch with small parasitic resistance. The latch-up happens

---

[8]Section 6.1.2 describes the BJT currents.

[9]This effect is shown in Fig. 6.29 and described in Section 6.3.4.

quickly, and no trace connecting V_{B0} to the resistor-like characteristic of the "on" thyristor is shown in Fig. 10.5b.

In the above description, the latching occurred at voltage V_{B0} due to the internal leakage currents, in the absence of any gate current. The voltage V_{B0} is the forward blocking voltage of the thyristor. If some gate current is provided, the critical condition $\beta_{pnp}\beta_{npn}$ is reached at a smaller v_{AK} voltage, enabling latch-up with smaller anode voltages, as shown in Fig. 10.5b.

Once the SCR is latched up, it cannot be switched off by setting the gate current to zero as the regenerative process is self-sustainable. The thyristor can be turned off by reducing the current to a level at which the current gain product is $\beta_{pnp}\beta_{npn} < 1$. The minimum current at which the thyristor is still on is called the *holding current* (I_H in Fig. 10.5b). Another type of thyristor, called an *MOS controlled thyristor*, integrates a MOSFET whose gate can be used to control the switching of the thyristor. When latched up, the thyristor current creates stored charge at the forward-biased P–N junctions, which means that the switching characteristics exhibit the P–N junction-type delay, due to the need for stored-charge removal.

For negative v_{AK}, both base-emitter junctions (P^+N^- and PN^+) are reverse biased, which means no current flows as both BJTs are in cutoff. Therefore, this type of thyristor exhibits reverse blocking capability. The depletion layer is widest in the lowest-concentration N^- layer, which therefore supports most of the reverse v_{AK} voltage. The breakdown voltage of this junction determines the reverse blocking voltage V_{RB}.

This type of reverse blocking capability is unwanted in some applications, or more precisely, it is desirable that the thyristor conducts current in both directions once it is latched up. In this case, TRIAC has to be used. This is a thyristor that, in a sense, is a pair of SCRs connected in antiparallel.

PROBLEMS

10.1 In the boost DC–DC converter of Fig. 10.1b, $\bar{i}_L = 5$ A, $V_{OUT} = 5$ V, the transit time of the diode is $\tau_T = 5\ \mu s$, and the on resistance of the switch is $R_{on} = 0.5\ \Omega$. (a) Sketch the voltage v_A and the current through the diode i_D. (b) Calculate the reverse diode current I_{RP} that discharges the stored charge, and find the time t_{rs} needed for the stored charge to be removed.

10.2 To reduce the series resistance, the width of the low-doped region of the PIN and Schottky diode considered in Example 10.2 is cut down to $w_{n-epi} = 10\mu m$. (a) Calculate the achieved reduction in forward voltage drop ΔV_F. (b) Calculate the associated reduction in maximum reverse-bias voltage ΔV_{R-max}. [A]

10.3 The following are the dimensions of a hexagonal VDMOSFET (like the one in Fig. 10.3): the length of the polysilicon gate $L_g = 20\ \mu m$, the distance between two P layers (under the gate) $L_{P-P} = 10\ \mu m$, the distance between two N^+ layers (under the gate) $L_{N^+-N^+} = 18\ \mu m$, the gate-oxide thickness $t_{ox} = 80$ nm, and the thickness of the insulating oxide between the polysilicon gate and the source metalization $t_I = 1\mu m$. If the MOSFET consists of $15,000$ hexagonal cells, calculate

(a) the gate-to-drain (C_{GD}) capacitance,

(b) the gate-to-source (C_{GS}) capacitance. [A]

10.4 A power MOSFET is switched on and off by a 1-MHz pulse signal, with 0 and 10 V as the low- and high-voltage levels, respectively. Assuming a constant input capacitance of 7 nF and total series resistance in the charging/discharging circuit of 5 Ω, calculate the average dissipated power by the input-control circuit.

REVIEW QUESTIONS

R-10.1 As the channel resistance of a MOSFET varies with the gate voltage, the MOSFET can be used as a voltage-controlled resistor. Varying this controllable resistance so that any output load variations are compensated, a constant output voltage can be achieved. This simple principle is employed by so-called *linear power supplies*. Why is it beneficial to use switch-mode power supplies?

R-10.2 Is it theoretically possible to have 100% efficient switch-mode power supply?

R-10.3 Can switching power circuits be built by using a PIN or Schottky diode as the only type of switch?

R-10.4 Is the switching frequency important if ideal components are used? If the physical size of the inductors (and transformers if present) is considered, is it better to use higher or lower switching frequency?

R-10.5 Can the PIN diode impose a limit to the switching frequency?

R-10.6 Do power Schottky diodes exhibit the stored-charge-related effects (forward voltage overshoot, and constant-current reverse recovery)?

R-10.7 Are the switching characteristics of a PIN diode (e.g., forward voltage overshoot and reverse recovery) related to the power efficiency?

R-10.8 Is the reverse blocking voltage related to the power efficiency? Indirectly?

R-10.9 Low-doping concentration and sufficient width of the "I" region (drift region) are needed to achieve desired blocking capability. Does the "I" (drift) region affect the forward voltage?

R-10.10 If a PIN and a Schottky diode have identical drift regions, will the forward voltages be the same?

R-10.11 Is the forward voltage drop of a diode (either PIN or Schottky) related to the power efficiency?

R-10.12 Given better switching characteristics and smaller forward voltage, is the superiority of Schottky diodes over PIN diodes total? In other words, is there any important disadvantage of power Schottky diodes? If yes, what?

R-10.13 Is the I_S parameter in the I_D–V_D expression related to the forward voltage drop? If yes, how?

R-10.14 If two controlled switches are connected in parallel, and one of them takes a larger share of the current, it will heat more, reaching higher operating temperatures. In the case of BJTs, the temperature increases the current, while in the case of MOSFETs it decreases the current. How does this difference affect the stability of the parallel connection?

R-10.15 The channel thickness of MOSFETs is inherently limited to several nanometers (the penetration of the electric field creating the channel). Tens of centimeters of channel width are necessary to compensate for this limitation, and therefore to achieve an acceptable channel resistance. How is it possible to place a tens of centimeters-wide power MOSFET in a package not larger than 1 cm in diameter?

R-10.16 Paralleling MOSFETs or MOSFET cells increases the input capacitance. Can this represent a practical problem, given that no DC current flows through the input capacitance?

R-10.17 Charging and discharging of the input capacitance has obvious implications for the maximum switching frequency. If there is no leakage through the capacitor, do charging and discharging affect the power efficiency?

R-10.18 Is the forward blocking voltage of a MOSFET related to the on resistance? If yes, what is the relationship?

R-10.19 The drain current of a power MOSFET increases as the gate voltage is increased above the threshold voltage, but then quickly enters "saturation." Does that mean that the resistance of the drift region dominates the on resistance? If no, why does this happen?

R-10.20 IGBT integrates a MOSFET and a BJT to achieve the advantages of MOSFET input controllability and BJT output current capability. However, MOSFETs can easily be paralleled, which increases the current capability. Do IGBTs exhibit a real advantage in terms of current capability? If yes, there must be a limit/problem with paralleling the MOSFETs. What would it be?

R-10.21 For an IGBT in an on state, minority carriers are injected into the drift region. Does that mean that the on resistance is smaller than it would be if an equivalent drift region was used in a power MOSFET?

R-10.22 IGBT retains all the stored-charge-related effects found in BJTs. Does this make them inferior to MOSFETs in terms of high-frequency switching performance?

R-10.23 Thyristors are controlled switches that consist of four layers (PNPN). With either voltage polarity across the switch, there is always a reverse-biased P–N junction. How can a conductive path ever be established through the PNPN structure to set the switch in an on state?

R-10.24 What happens if the collector current of a transistor "A" is amplified by a transistor "B" and the amplified current is fed back into the base of transistor "A"? Is there a limit to this closed-loop amplification? Can a finite "loop gain" be reached if the maximum $\beta_{pnp}\beta_{npn}$ product is < 1? What if $\beta_{pnp}\beta_{npn} > 1$?

R-10.25 Consider the PNPN structure of an IGBT. If $\beta_{pnp-max}\beta_{npn-max} < 1$, does that mean that this structure can never be latched up?

R-10.26 Assume that a thyristor conducts 1000 A at 1 V of forward voltage drop, which means the on resistance is 1 mΩ. Can this thyristor be used as a small-resistance switch for low current applications (say <1 A) to achieve a negligible forward voltage drop (<1 mV)? A negative answer implies minimum-current limit. Why should there be a "minimum-current limit"?

Chapter *11*

SEMICONDUCTOR DEVICE RELIABILITY

Semiconductor devices and integrated circuits (ICs) are basically classified according to the electrical function they perform. Semiconductor devices and integrated circuits performing nominally the same function, however, do not necessarily perform it equally well. The concept of *quality* is used to express how well the required function is performed. As an example, let us consider two 741 integrated operational amplifiers. An operational amplifier is of higher quality if it has a higher gain, wider frequency bandwidth, larger common-mode rejection ratio, smaller input offset voltage, larger maximum supply voltages, larger sweep rate, etc. All these characteristics can be regarded as *conformance* figures. The conformance is, however, only one side of the quality. On the other side is the issue of how long the device or circuit will exhibit the initial performance figures. The concept of *reliability* is used to express this time dimension of the quality. The relation between the concepts of *quality, conformance*, and *reliability* is illustrated in Fig. 11.1.

Measurement and presentation of the conformance figures are straightforward; any conformance parameter can be measured directly and its value expressed. The situation is, however, different with determination andpresentation of the reliability. The reliability depends, in principle, on application conditions, which means it is not possible to establish

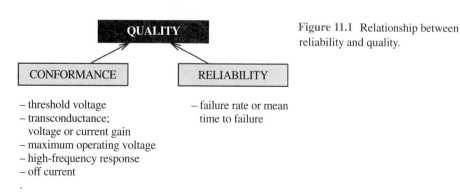

Figure 11.1 Relationship between reliability and quality.

461

an exact and unique reliability figure for a given device or IC. In addition, the reliability determination itself, regardless of the application conditions used, cannot be made by direct measurements. This is mainly because of practical constraints. Theoretically, it is possible to determine the mean time to failure directly if a corresponding number of ICs is exposed to working conditions and times to failure of each of them are recorded. This is, however, practically meaningless; such a test would last for tens of years, and by the time the data are collected nobody would be interested in them. That is why *accelerated tests* have to be applied to obtain the results in a reasonable time of 1 or 2 months. These tests are designed to accelerate particular failure mechanism(s) that are expected to dominate in given application conditions.

To understand reliability figures claimed by a manufacturer, it is obviously necessary to understand the method (or particular accelerated tests) used to obtain the data. Further, to make sure that the reliability data are relevant for particular application conditions, it is necessary to check weather all the relevant failure mechanisms are covered by the tests used. Obviously, dealing with reliability requires knowledge of failure mechanisms and reliability testing in addition to basic reliability concepts. In this chapter, reliability concepts and models are briefly reviewed first. After that, the most frequent failure mechanisms are described. Finally, procedures for reliability screening and reliability measurement are considered.

11.1 BASIC RELIABILITY CONCEPTS

11.1.1 Failure Rate

The most frequently used reliability parameter is *failure rate*. The failure rate expresses the fraction of devices that fails in a considered time interval Δt. The basic unit for the failure rate is *number of failures per hour*, or simply h^{-1}. The failure rate is, however, frequently expressed in %/h, or any multiple unit such as 1%/1000 h. When the failure rate is very low, the failure unit (FIT) is used, which is defined as 1 failure/10^9 h (1 FIT = 1 failure/10^9 h).

The failure rate changes during device operation. The general form of failure rate dependence on operation time has a bathtub shape, as illustrated in Fig. 11.2. In the initial period, the failure rate is very high, but it decreases with the time. The failures in this period are called infant mortalities or early failures. These failures are caused by gross defects, which remain inactive during the testing, but easily activate themselves when the device is exposed to exploitation conditions for some time. An example of such defects is the significantly reduced width of the interconnect metal line, illustrated in Fig. 11.3a. This line will conduct the current during the testing so the device shows no failure. However, the current density in the thinned section of the metal line is increased, giving rise to heating of the metal line at that spot. The increased current density and temperature will break the line after a short operation time. It is possible to eliminate the devices with early failure problems by reliability screening, which is discussed in detail in Section 11.3.

After the early-failure period comes a steady-state period with a small number of failures, uniformly distributed in time. The failures in the steady-state period are normally due to a combined action of a number of unrelated causes. As an example, consider the minor defect on the metal line, illustrated in Fig. 11.3b. This defect will not lead to a failure by

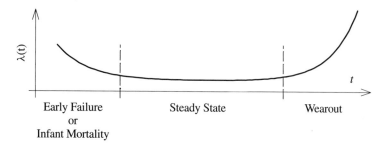

Figure 11.2 Failure rate versus time bathtub curve.

Figure 11.3 Defects on metal line: (a) gross defect causing early failure and (b) minor defect that may cause steady-state failure if combined with other factors.

itself. However, the failure may happen if some other factors are present. For example, the design is such that a large current is flowing through the line, the temperature is increased, and a transient peak voltage appears when the power is switched-on. The combination of all these factors is very likely to cause burning of the metal line.

Finally, there is a wear-out period in which the failure rate increases as a result of aging of the materials used in ICs. A metal line with no obvious defects can exhibit increased resistance due to exposure of the material to increased temperature for a long time. ICs normally do not enter this period as they are typically outdated before that.

11.1.2 Reliability and Failure Functions

Mathematically, the reliability is defined as the probability that a device will work to time t without a failure. This probability is normally denoted by $R(t)$ and is frequently called the *reliability function*. The complementary probability, i.e., the probability that the device will fail at or before time t, is normally denoted by $F(t)$. Obviously, the relationship between the functions $R(t)$ and $F(t)$ is given by

$$R(t) = 1 - F(t) \tag{11.1}$$

The function $F(t)$, called the *failure function*, is a cumulative distribution function (cdf) as it satisfies the following conditions:

$$
\begin{aligned}
F(t) &= 0 \quad \text{for } t < 0 \\
0 \le F(t) &\le F(t + \Delta t) \quad \text{for } 0 \le t \le \infty \\
F(t) &\to 1 \quad \text{for } t \to \infty
\end{aligned}
\tag{11.2}
$$

The derivative of $F(t)$ with respect to time is the corresponding probability density function (pdf) and is denoted by $f(t)$. The pdf is related to cdf as

$$f(t) = \frac{d}{dt} F(t) \tag{11.3}$$

or

$$F(t) = \int_0^t f(t) \, dt \tag{11.4}$$

As the possible time interval is from 0 to ∞, the pdf function $f(t)$ has to satisfy the following condition:

$$\int_0^\infty f(t) \, dt = 1 \tag{11.5}$$

Equation (11.5) is the mathematical representation of the fact that the device considered must fail sometimes between now and infinity.

The reliability and failure functions can be related to the failure rate $\lambda(t)$. Let us express the probability that a device will fail in the time interval Δt. This probability is

$$F(t + \Delta t) - F(t) = R(t) - R(t + \Delta t) \tag{11.6}$$

The fraction of devices that fails in the time interval Δt is then

$$\frac{R(t) - R(t + \Delta t)}{R(t)} \tag{11.7}$$

To obtain the failure rate, the above equation for the fraction of devices that fails in the time interval Δt is divided by the time interval itself:

$$\lambda = \frac{1}{\Delta t} \frac{R(t) - R(t + \Delta t)}{R(t)} \tag{11.8}$$

In the limit when $\Delta t = dt \rightarrow 0$, the failure rate is given by

$$\lambda = -\frac{1}{R(t)} \frac{dR(t)}{dt} = -\frac{d}{dt} \ln R(t) \tag{11.9}$$

Equation (11.9) gives the general relationship between the failure rate and the reliability function. The considerations given in the previous section show that the steady-state period is most important in terms of semiconductor device and IC reliability modeling and testing. In the steady-state region, $\lambda(t) = \lambda = \text{const}$. Solving Eq. (11.9) for this case, the reliability function is obtained as

$$R(t) = e^{-\lambda t} \tag{11.10}$$

The probability density function $f(t)$ is then

$$f(t) = \frac{dF(t)}{dt} = -\frac{dR(t)}{dt} = \lambda e^{-\lambda t} \tag{11.11}$$

which is the exponential distribution function, with the failure rate λ as the only parameter.

The mathematical difference between $f(t)$ and λ is obvious: $f(t)$ is a probability distribution function, while λ is its parameter. However, it is important to distinguish between the physical meanings of the failure distribution function $f(t)$ and the failure rate λ. Both quantities express how many devices are failing per unit time, or more precisely the *normalized* number of devices failing per unit time. Yet, there is a difference between them. What is obvious is that λ is a constant in the steady-state region, while $f(t)$ decreases exponentially with time according to Eq. (11.11).

Mathematically, the difference is due to the normalization factor. As $f(t)$ basically shows how the *number* of failures changes in time (dNF/dt), it is normalized by a factor that is the total number of devices available at the beginning of the consideration. On the other hand, $\lambda(t)$ expresses the failure rate at a particular instant of time, and is not related in any way to the initial number of devices. It simply shows what portion (or percentage) of the currently available devices is failing during a particular small instant of time. Therefore, the normalization factor in this case is the number of presently available devices.

In the steady-state region, the probability that a device will fail in a considered time instant is a constant, and as a consequence the failure rate is a constant. However, the actual number of devices failing per unit time (dNF/dt) decreases because the devices are failing all the time and the number of devices available for failing decreases. The exponential distribution Eq. (11.11) shows that a constant failure rate leads to an exponential decrease of the portion (or percentage) of devices failing per unit time $f(t)$.

Finally, it is very useful to relate the failure rate to the cumulative failure distribution function $F(t)$. From Eqs. (11.4) and (11.11), this relationship is obtained as

$$F(t) = 1 - e^{-\lambda t} \tag{11.12}$$

11.1.3 Mean Time to Failure

The probability density function $f(t)$ can be used to estimate the average device lifetime, more frequently called the mean time to failure ($MTTF$). For a large sample size, the mean time to failure is identical to the mathematical concept of expected time to failure, which is given by

$$MTTF = \int_0^\infty t f(t) \, dt \tag{11.13}$$

To clarify the importance and meaning of the concept of "large sample size," let us consider the extreme opposite case, which is a sample size of 1. In this case the mean time to failure is equal to the actual lifetime of the sample device, and can obviously be different from the expected time to failure due to the fact that not all the devices fail at the expected time to failure.

For the case of a constant failure rate, and therefore exponential probability density function $f(t)$:

$$MTTF = \int_0^\infty t\lambda e^{-\lambda t}\, dt \qquad (11.14)$$

$MTTF$ is obtained as

$$\boxed{MTTF = \frac{1}{\lambda}} \qquad (11.15)$$

■ **Example 11.1 The Meaning of *MTTF***

This example illustrates the importance of understanding the statistical background of reliability modeling, proving that a particular "common-sense approach" tried by many students is wrong.

What percentage of devices is expected to fail in the first year of operation if the mean time to failure is 2 years? How about when the mean time to failure is 10 years?

Solution: If you say 50%, you are wrong. Two years is the mean time to failure, which does not mean that all the devices will fail after 2 years (50% in the first year and 50% in the second year). You believe 50% will fail after the second year and 25% after the first year? Wrong again. The number of failures is not uniformly distributed in time, and this kind of approach does not help. The normalized cumulative number of failures $F(t)$ is a distribution given by Eq. (11.12). The parameter of the distribution is the failure rate λ, which can be obtained from the mean time to failure:

$$\lambda = \frac{1}{MTTF} = \frac{1}{2} = 0.5 \text{ year}^{-1}$$

Using Eq. (11.12) we calculate that

$$F(1 \text{ year}) = 1 - e^{-0.5 \times 1} = 0.393$$

This means that 39.3% of the devices is expected to fail after the first year of operation.

For the case of $MTTF = 10$ years, the percentage of devices failing after the first year is 9.5%. This is close to the "common-sense" guess of 10%, but only because there is a small percentage of failures involved and $1 - e^{-\lambda t} \approx \lambda t$. The "common-sense" guess of 70% failures after the seventh year is much further from the correct value of 50.3%, because a much larger percentage of failures is involved in this case.

■ **Example 11.2 Estimating λ and *MTTF***

To test IC reliability, 100 ICs are subject to the accelerated test for 90 days. Two ICs failed during the test. Estimate the failure rate and the mean time to failure.

Solution: The total number of failures after $90 \times 24 = 2160$ h of testing is 2. This means that the cumulative failure distribution function at $t = 2,169$ h is $F(t) = 2/100 = 0.02$. Using Eq. (11.12), we find that

$$\lambda = -\ln[(1 - F(t)])/t = -\ln(1 - 0.02)/2160 = 9.353 \times 10^{-6} \text{ h}^{-1}$$

The mean time to failure is simply $MTTF = 1/\lambda = 106{,}918$ h.

Given that a small percentage of failures is involved in this case, this problem could be solved by estimating the *MTTF* first. As ICs worked for 2160 h, and as there are 100 ICs, we have $2160 \times 100 = 216{,}000$ IC-h. With two failed ICs, $MTTF$ is estimated to as $216{,}000/2 = 108{,}000$ h. The failure rate is $\lambda = 1/MTTF = 9.26 \times 10^{-6} \text{ h}^{-1}$. Keep in mind that this approach is approximate, and can be used only in the case of a small percentage of failures.

A comment. The failure rate in this example is very high. If we have a system of 100 ICs, we can expect a failure for each of $106{,}918/100 = 1{,}069$ h, which is 44.5 days. It should be kept in mind, however, that this failure rate is for ICs subject to accelerated testing. Determination of the corresponding failure rate at normal operating conditions is explained in Section 11.4. Another point to make is related to the fact that the above estimates of $MTTF$ and λ are *point estimates*. If the test was conducted again, different values would be obtained for the $MTTF$ and λ. This problem is again considered in Section 11.4.

11.2 FAILURE MECHANISMS

The failures in ICs can be classified in at least three different ways: according to *failure modes*, according to *failure mechanisms*, and according to *failure causes*. It is important to distinguish between the concepts of the failure mode, mechanism, and cause. The failure mode is the observed result of a failure, such as an open circuit, short circuit, or parameter degradation. The failure mechanism is the physical, chemical, or other process that results in a failure. An example of a failure mechanism is corrosion of aluminum lines. The effect of a corrosion-induced failure will be observed as an open circuit, which is the failure mode. Finally, the failure cause is a circumstance during design, production, testing, or operation that initiates or contributes to a failure mechanism. The above example of corrosion-induced failure can be related to many different failure causes: defects in the package and passivation layer allowing moisture to penetrate to the chip, moisture trapped in the cavity of a ceramic package, contamination of the chip surface, an electrical potential applied to the aluminum line, or defects in the aluminum line.

Consideration of IC failures by way of failure modes provides insight into the types of degradations of IC characteristics that we may expect, but it is not helpful in terms of understanding the issues related to reliability testing. The IC failures are frequently introduced through the failure causes. This approach, however, complicates the explanations as a single failure mechanism can be initiated by a combination of failure causes. In the following text, the IC failures are considered by way of failure mechanisms. It is believed that this approach enables an efficient introduction to IC failure concepts, an understanding of which is necessary when dealing with IC reliability.

11.2.1 Formation of the Purple Plague

Gold wires are normally used for connections between IC circuit pins and aluminum bonding pads on the chip. Solid-state diffusion of gold and aluminum atoms frequently leads to creation of gold–aluminum intermetallic, so-called *purple plague*. This intermetallic compound is a conductor and it does not lead to a failure by itself. The problem is due to different diffusion rates of gold atoms into the aluminum and aluminum atoms into the gold. As a result of this difference in the diffusion rates, voids are created at the bond, which weakens the strength of the bond so that it can completely lift up causing an open circuit.

The formation of the purple plague obviously depends on the diffusion of aluminum and gold atoms. As the diffusion coefficient is proportional to $\exp(-E_A/kT)$, the temperature dependence of the formation of the purple plague is modeled by the Arrhenius reaction rate equation:

$$r = Ae^{-E_A/kT} \tag{11.16}$$

In Eq. (11.16), r is the reaction rate, E_A is the activation energy, and k is the Boltzmann constant, T is the absolute temperature, A is the frequency factor constant. The activation energy is around 1.0 eV. The failure rate due to the formation of purple plague is proportional to the reaction rate r. Exposure of the ICs to a high temperature is, obviously, a simple and efficient accelerated test, which accounts for this failure mechanism.

The formation of the purple plague is a very frequent failure mechanism, occurring due to power dissipation in the chip that heats the bonds. It can also occur, or at least be started, during packaging if a high-temperature bonding technique is used.

■ **Example 11.3** λ **and** *MTTF* **from Accelerated Testing**

Two hundred ICs are selected for a high-temperature accelerated test. Half of the ICs is subject to $T_1 = 100°C$, while the other half is subject to $T_2 = 125°C$. After 30 days of testing, 13 ICs from the 125°C group and 2 IC from the 100°C group failed. Assuming that the failures are due to the formation of purple plague, estimate the mean time to failure at operating temperature $T_o = 75°C$ and room temperature ($T_r = 25°C$).

Solution: Determine first $\lambda(T_1)$ and $\lambda(T_2)$, which is done in a way analogous to Example 11.2:

$$\lambda(T_1) = -\ln[1 - F(t)]/t = -\ln(1 - 0.02)/(30 \times 24) = 2.8 \times 10^{-5} \text{ h}^{-1}$$
$$\lambda(T_2) = -\ln[1 - F(t)]/t = -\ln(1 - 0.13)/(30 \times 24) = 1.93 \times 10^{-4} \text{ h}^{-1}$$

According to the Arrhenius rate equation, the failure rate dependence on temperature is given by

$$\lambda(T) = Ae^{-E_A/kT}$$

where A and E_A are parameters that can be determined from the accelerated test data. The activation energy can be obtained in the following way:

$$\lambda(T_2)/\lambda(T_1) = \exp\left[\frac{E_A}{k}\left(\frac{1}{T_1} - \frac{1}{T_2}\right)\right]$$

$$E_A = \frac{k \ln[\lambda(T_2)/\lambda(T_1)]}{1/T_1 - 1/T_2}$$

Using the values for $\lambda(T_1)$ and $\lambda(T_2)$, the following value for E_A is obtained:

$$E_A = \frac{8.62 \times 10^{-5} \ln(1.93 \times 10^{-4}/2.8 \times 10^{-5})}{1/(273.15 + 100) - 1/(273.15 + 125)} = 0.99 \text{ eV}$$

The coefficient A is determined next:

$$A = \lambda(T_2)e^{E_A/kT} = 6.32 \times 10^8 \text{ h}^{-1}$$

The failure rates at operating and room temperatures can now be calculated:

$$\lambda(T_o) = Ae^{-E_A/kT_o} = 3.08 \times 10^{-6} \text{ h}^{-1}$$

$$\lambda(T_r) = Ae^{-E_A/kT_r} = 1.23 \times 10^{-8} \text{ h}^{-1}$$

This means the mean time to failure at the operating temperature is $1/\lambda(T_o) = 37$ years, while the mean time to failure at room temperature is $1/\lambda(T_r) = 9{,}290$ years!

11.2.2 Electromigration

Electromigration is a phenomenon of displacement of metal atoms, caused by transfer of momentum from the electrons flowing through a conductive line. The atoms are "pushed" by the electrons, leaving behind vacancies that accumulate and as a result narrow the cross section of the conductor. This increases the current density and causes localized heating, which enhances the electromigration process, until the conductor line is completely opened. The atoms that are "pushed" toward the positive end of the conductor also accumulate, creating hillocks and whiskers. This can cause shorts between neighboring lines, or cracks in the oxide passivation.

The temperature (T) and the current density (j) are the most important factors in the process of electromigration. The temperature influences the process of electromigration through an increase in the diffusion constant of the atoms, and is therefore modeled by the Arrhenius rate equation. The current density is also an important factor, as the electromigration process directly depends on the momentum that the atoms obtain from the stream of electrons. A widely accepted equation for the $MTTF$ due to electromigration is[1]

$$MTTF = Aj^n e^{E_A/kT} \tag{11.17}$$

The parameters A and n take into account the dependence on other factors such as grain size, impurity content, metal line dimensions (length, width, and thickness), and overcoat

---

[1]*Source*: A.G. Sabnis, *VLSI Reliability* (VLSI Electronics, Microstructure Science, Vol. 22), Academic Press, San Diego, 1990.

of the metal. A typical value of the parameter n is between 1 and 2. The third parameter, the activation energy E_A, depends on factors such as grain size and dissipation efficiency. That is why the reported data on the activation energy vary over a wide range of values, from 0.3 to 1.3 eV. The described exponential temperature dependence of the electromigration process shows that high-temperature tests can efficiently be used as accelerated reliability tests.

The tendency of shrinking the IC device dimensions, and therefore aluminum and polysilicon lines, favors the electromigration process. Take the so-called "constant-voltage" scaling rule as an example. With the constant-voltage scaling rule, the current flow is increased by a factor S when the dimensions are decreased by a factor S. As the voltage is kept constant, this results in an increase of the current density (j) by a factor S^3 (S due to the increase in the current and S^2 due to the decrease in the conductor cross section). According to Eq. (11.17), the electromigration limited time to failure will be reduced by a factor S^{3n}. This illustrates the importance of monitoring electromigration in modern ICs.

11.2.3 Corrosion of Metalization

The corrosion of aluminum metalization is a frequent IC failure mechanism, which can cause open circuits. The electrochemical process of corrosion is initiated by electrolysis of water molecules (H_2O), and develops as follows:

1. H^+ ions are formed at a positively biased Al line (anode).
2. H^+ ions diffuse to a negatively biased Al line (cathode).
3. The oxygen that remains at the anode creates a layer of oxide that acts as an insulating barrier and limits any further corrosion.
4. At the cathode, H^+ ions react with Al to create Al^{3+} ions.
5. If impurity mobile ions, such as sodium and potassium, and water molecules are present, they chemically react to produce OH^- ions.
6. Created OH^- ions react with Al^{3+} ions forming $Al(OH)_3$, which is a dark gray or black corrosion product.

Phosphosilicate passivation glass (PSG) can be used to immobilize the sodium and potassium ions, which should prevent the contamination of the underlying layers. In the presence of water, however, phosphoric acid is created releasing the sodium and potassium ions.

The rate of corrosion obviously depends on many factors. The presence of water molecules is the first important factor. Not more than 1000 ppm$_V$ of moisture may be sufficient to cause a failure. In hermetically sealed packages, the moisture either remains trapped inside the cavity or is introduced through leaks in sealant glasses around the pins. The plastic packages themselves absorb and emit moisture. However, how much water will appear on the metalization itself depends on the efficiency of the passivating glass layer. Thermal expansion can lead to creation of cracks in the passivation layer. The diffusion of H^+ ions from positively biased to negatively biased aluminum lines (step 2 in the corrosion process) is the next important factor. The current of ions between the adjacent biased aluminum lines, called *corrosion current*, increases exponentially with the temperature (Arrhenius-type dependence). The grain structure of the metalization is also important, because the sodium ions tend to accumulate at the grain boundaries.

The corrosion process can obviously be accelerated by an increase of temperature and humidity. So-called 85°C/85% RH/*bias*, or the *temperature–humidity–bias* (THB) test, is an industry-wide standard for accelerating moisture-related failure mechanisms. It is important to emphasize, however, that the acceleration by temperature is limited by water evaporation from the chip surface. The frequently used 125°C/*bias* accelerated test (sometimes called *the life test*) can lead to a much smaller number of corrosion-related failures than the testing at normal operating conditions (room temperature). It is also believed that the lower power dissipation of CMOS ICs, and therefore lower chip temperature, is the cause of more pronounced corrosion problems in these ICs as compared to bipolar ICs.

11.2.4 Charge Trapping in Gate Oxide, Creation of Interface Traps, Hot-Carrier Generation

Properties of the gate oxide are extremely important for the performance of MOSFETs. There are charges in the gate oxide and traps at the oxide–semiconductor interface that significantly influence the characteristics of MOSFETs. The origin of interface traps and oxide charge is discussed in Section 2.5.2 and illustrated in Fig. 2.14. Creation of these charges and traps during device operation leads to instabilities in device characteristics influencing reliability.

The *gate-oxide charge* affects the channel carriers by its electric field, consequently affecting device characteristics. The two most important MOSFET parameters are the threshold voltage and the effective channel carrier mobility. The positive gate-oxide charge mainly influences the threshold voltage, decreasing its value in N-channel MOSFETs and increasing it in P-channel MOSFETs. As for the *interface traps*, they easily capture and release carriers from the channel when the gate voltage changes the surface potential, moving the interface trap levels above or below the Fermi level. As a fraction of the gate-induced channel carriers is captured by the interface traps, the presence of interface traps increases the threshold voltage of the MOSFET and reduces the effective channel-carrier mobility. Some of the oxide-network defects can be close enough to the silicon substrate to exchange the charge with the silicon through tunneling. Because of the tunneling barrier, charging and discharging of these defects are relatively slow processes (anything between 30 ms and days or even months). This kind of oxide defects appears as *slow traps*. They are also known as *border traps*, to indicate that they appear at the border between the silicon and the bulk of silicon dioxide.

As mentioned above, the gate-oxide charge and interface traps present before the exploitation of the device started are not important in view of reliability. However, the oxide charge and border and interface trap densities can be increased during device operation, leading to a threshold voltage shift and/or mobility reduction. This drift of the MOSFET parameters during exploitation may lead to a *functional failure* of the circuit, for example, causing switching errors in logic circuits as a result of shifted threshold voltage. Additionally, the MOSFET parameter drift may cause a change in the circuit parameters, and once the circuit parameters change beyond specified limits, the circuit is classified as failed. This is so-called *parametric failure*.

A number of failure causes can initiate the mechanisms leading to changes of oxide charge and trap densities during device operation. Figure 11.4 illustrates the relationship between the most frequent causes and the corresponding failure mechanisms. As the diagram of Fig. 11.4 illustrates, the failure mechanisms are linked to each other, in the sense that hole trapping leads to border trap, and finally interface trap creation. This is because the defects at the interface are energetically the most stable defects, and the defects away from the interface tend to be transformed into interface defects through a variety of mechanisms. A typical manifestation of this link is the frequently observed "turn-around" of the threshold voltage of N-channel MOSFETs, illustrated in Fig. 11.5. The initial threshold voltage decrease is due to hole trapping, which increases the density of positive oxide charge qN_{oc}. With time, a number of holes are transferred from the oxide defects to the interface, neutralizing the corresponding oxide traps, and creating interface traps. The interface traps increase the threshold voltage of MOSFETs, and once their effect becomes dominant, the threshold voltage starts increasing with time. As the interface traps accumulate at the interface, the threshold voltage can increase above the initial value.

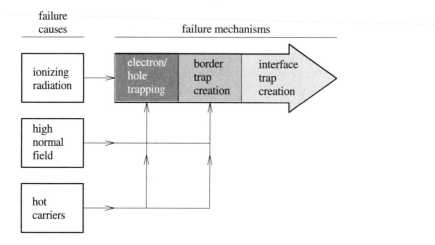

Figure 11.4 Failure causes and mechanisms leading to changes of oxide charge, and border and interface trap densities.

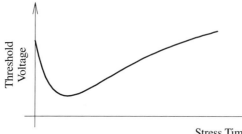

Figure 11.5 The threshold voltage "turn-around" effect is a manifestation of the link between hole trapping and the creation of interface traps.

The three most important failure causes, related to oxide charge and border and interface traps are (1) high normal electric field, (2) hot carriers, and (3) ionizing radiation. These failure causes and the associated mechanisms are described in more details in the following.

High Normal Electric Field

Even moderate gate voltages can produce high electric fields when the oxide thickness is reduced to enable reductions in MOSFET lateral size. A high oxide field is related to bent energy bands and bent oxide-defect levels, which create triangular potential barrier with the energy bands in the silicon, as illustrated in Fig. 11.6. A higher electric field means a larger slope of the oxide-defect levels, and consequently a narrower energy barrier and a more pronounced electron tunneling through the barrier.

Figure 11.6a and b illustrates the cases of negative and positive gate bias, respectively. In the case of negative gate bias, electrons tunnel from the neutral defects into the silicon valence band (or equivalently, holes tunnel from the silicon valence band onto the oxide defects), creating positive oxide charge. In the case of positive gate bias, the electrons can tunnel from a defect level into the oxide conduction band, and roll down into the gate (not shown in Fig. 11.6), leaving behind positively charged defects. As can be seen, a positive oxide charge is created with either gate bias. Mechanisms analogous to those illustrated in Fig. 11.6 can lead to creation of a negative oxide charge, however, it is the positive charge creation that is typically observed in practice. The reason for this is the type of defects and the position of energy levels they introduce into the oxide energy gap.

The rate of formation of the gate oxide charge, and consequently interface traps, is proportional to the tunneling probability, and is therefore expressed as

$$r = Ae^{-E_0/E_{ox}} \tag{11.18}$$

In Eq. (11.18), E_{ox} is the electric field in the oxide and A and E_0 are model parameters. As can be seen, the failure rate due to the tunneling of holes and electrons into and from the oxide traps depends exponentially on the applied gate-oxide field. On the other hand, this mechanism shows only a weak dependence on temperature. That is why the best way of accelerating this mechanism is to increase the operating voltage during testing. This mechanism is especially important for devices that normally work with high fields in the oxide such as EEPROM memory devices. Scaling down of device dimensions without a proportional scaling of the supply voltage increases the gate-oxide field and, consequently, favors this mechanism even in ordinary MOSFETs.

Hot Carriers

Scaling down of MOSFETs also increases the lateral field in the channel, which has proved a greater reliability problem than the normal field increase. The increased lateral electric field is responsible for injection of so-called hot carriers into the oxide. The hot carriers are electrons and holes in the channel of MOSFETs accelerated by the lateral field to the extent

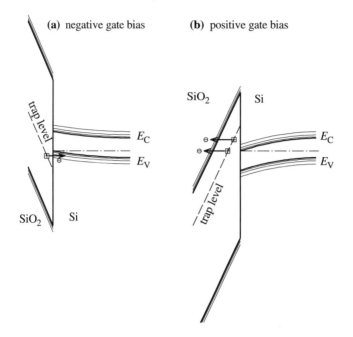

Figure 11.6 Mechanisms of positive gate-oxide charge creation by a high gate-oxide field: (a) negative gate bias and (b) positive gate bias.

that they have kinetic energies much larger than the thermal kinetic energy ($\propto kT$, where T is the actual silicon-crystal temperature). To account for this increase in the kinetic energy of the channel carriers, the concept of temperature of the electron or hole gas is introduced. When the lateral electric field in the channel is high, the electron or hole gas is not in thermal equilibrium with the silicon lattice, and can exhibit much higher temperature ($>1000°C$, while the silicon lattice is at room temperature). Electrons with such a high temperature, or kinetic energy, are able to overcome the potential barrier between the silicon substrate and the gate oxide (≈ 3.2 eV) if elastically scattered toward the gate. Such a mechanism of electron injection into the gate oxide is called channel hot-electron injection, and is illustrated in Fig. 11.7. Generally, it is not observed that the holes are injected in a similar way, which is due to the higher potential barrier between the silicon and the oxide for the case of holes (≈ 4.6 eV).

In addition to the above described process, the highly accelerated channel carriers can initiate the avalanche process, and therefore generation of highly energetic electron–hole pairs. The generated hot electrons and holes can be attracted by the normal field and consequently injected into the oxide, as illustrated in Fig. 11.8. This mechanism appears when the MOSFET is working in the saturation mode, which means the channel is pinched-off somewhere between the source and drain. In the case of N-channel MOSFETs, the direction of the normal field between the source and the pinch-off point is toward the substrate, so the injection of electrons is favored. The normal field in the region between

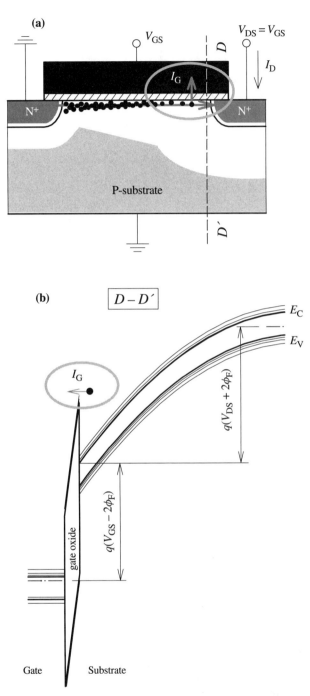

Figure 11.7 Mechanism of channel hot-electron injection into the gate oxide: (a) illustration at the device cross section and (b) energy bands at the point of hot-electron injection.

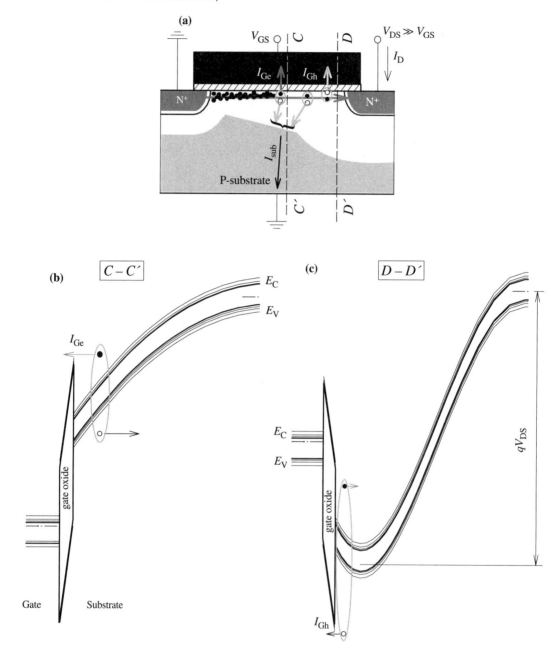

Figure 11.8 Mechanism of avalanche hot-carrier injection into the gate oxide: (a) illustration at the device cross section, (b) energy bands in the channel region (injection of electrons), and (c) energy bands in the pinch-off region (injection of holes).

the pinch-off point and the drain is, however, toward the gate, which favors the injection of avalanche-generated holes into the oxide.

Although both electrons and holes can be injected into the oxide, the trapping efficiency is not the same for the two types of carriers. There are not many defects that can efficiently trap electrons, and the electrons normally pass through the oxide being attracted by the positively biased gate electrode. The probability that an electron appearing in the oxide is trapped can be as low as 10^{-5}, which means only one in 100,000 injected electrons is trapped. On the other hand, there is a number of oxide defects (refer to Fig. 2.14) that can very efficiently trap the holes. Almost every hole appearing in the oxide is trapped by oxide defects. Consequently, the most frequently observed degradations related to hot carriers are due to the trapping of holes injected into the gate oxide. Due to the high efficiency of hole trapping, even a low level of avalanching in the channel may lead to a significant shift of device parameters, leading to a failure of the IC.

The number of holes injected into the oxide is obviously proportional to the number of holes generated in the avalanche process. It can be seen from Fig. 11.8a that only a part of the generated holes is injected into the oxide, the other part being collected by the substrate-contact electrode. The holes flowing into the substrate contact make the substrate current I_{sub} that can be measured. Consequently, the substrate current can be related to the hot-carrier limited device lifetime t. It has been found that this relation is as follows[2]:

$$t = A(I_{sub})^{-m} \tag{11.19}$$

where t is the lifetime corresponding to a predetermined shift of threshold voltage/channel-carrier mobility/transconductance, while A and m are parameters of the model. The lifetime is used here instead of the mean time to failure because the device characteristic degradation does not exhibit a pronounced randomness, which enables direct modeling of the lifetime, without involving the statistical considerations from Section 11.1.

The substrate current increases exponentially with the increase of the maximum lateral electric field in the channel E_{max}[3]:

$$I_{sub} \approx C e^{-B/E_{max}} \tag{11.20}$$

where B and C are constants. The maximum electric field increases with increase of the drain voltage and decrease of the channel length. This means that the hot-carrier degradation is exponentially accelerated by an increase in the drain voltage. This also means that a decrease in the channel length exponentially accelerates the hot-carrier degradation as well. That is why the hot-carrier degradation problems appear as one of the most important limiting factors for further scaling down of MOSFET dimensions. A lightly doped drain

---

[2]*Source*: E. Takeda and N. Suzuki, "An empirical model for device degradation due to hot-carrier injection," *IEEE Electron Device Lett.*, **EDL-4**, 111–113, 1983.

[3]*Source*: J.R. Brews, "The submicron MOSFET," in *High-Speed Semiconductor Devices*, S.M. Sze, ed., Wiley, New York, 1990 (p. 139).

(LDD) structure, like the one presented in Fig. 7.1, has successfully been used to enable the development of submicrometer MOSFETs. A number of other special MOSFET structures are being investigated to further decrease hot-carrier problems.

The accelerated tests for detection of hot-carrier-induced degradation are generally based on increased bias voltage, which consequently increases the lateral field in the MOSFET channel and the substrate current. Monitoring the substrate current is necessary as the extrapolation of the lifetime is based on use of Eq. (11.19). It is important to clarify the influence of temperature, because the high-temperature–bias tests are frequently considered as general reliability tests. It is important to realize that increase in the ambient, and therefore the crystal temperature decreases the temperature (or kinetic energy) of the hot carriers. This is because of the increased phonon scattering of the carriers, which means reduced mean scattering length. The kinetic energy gained by a carrier between two scattering events is proportional to $q E_{lat} l_{sc}$, where E_{lat} is the lateral electric field and l_{sc} is the scattering length, or distance passed by the carrier between the two scattering events. It is obvious that high-temperature accelerated tests can mask the hot-carrier-related reliability problems. Reducing the ambient temperature to a certain value helps accelerate the failures.

Ionizing Radiation

In some specific applications, in particular space and the military, semiconductor devices can be exposed to ionizing radiation. In this case, the radiation is breaking silicon–oxygen bonds in the oxide generating free electron–hole pairs. As can be expected, electrons leave the oxide while the holes remain trapped at the oxide defects, giving rise to an accumulation of the positive gate-oxide charge. With time, a part of the positive oxide charge is converted into interface traps, according to the sequence of mechanisms shown previously in Fig. 11.4. The amount of created positive charge, and therefore accumulated interface traps, proportionally depends on the dose of radiation. Once the dose is high enough, the corresponding shift of device parameters is recorded as a parametric failure.

■ **Example 11.4 Hot-Carrier Limited Lifetime**

MOSFETs with two different channel lengths ($L = 0.8\ \mu m$ and $L = 0.5\ \mu m$) are subject to hot-carrier accelerated testing. The test conditions and the results obtained are given in Table 11.1. Determine the lifetimes at operating voltage $V_D = 5.5$ V for the both types of devices. The substrate currents measured at the operating voltage are 0.248 and 0.490 μA for devices with the longer and shorter channel, respectively.

TABLE 11.1 Results of Hot-Carrier Accelerated Testing

	$L = 0.8\ \mu m$		$L = 0.5\ \mu m$	
	Test 1	Test 2	Test 1	Test 2
$I_{sub-max}(\mu A)$	7.65	13.5	6.10	12.6
Lifetime (min)	4,467	813	4,266	490

Solution: The lifetime at the operating voltage can be calculated from Eq. (11.19) when the values of the parameters m and A are known. These parameters can be determined from the accelerated test data. To determine the parameter m, Eq. (11.19) is applied to the two given test conditions and the two obtained equation divided:

$$\frac{t_1}{t_2} = \left(\frac{I_{sub1}}{I_{sub2}}\right)^{-m}$$

From the above equation we have

$$m = -\frac{\log(t_1/t_2)}{\log(I_{sub1}/I_{sub2})} = -\frac{\log(4,467/813)}{\log(7.65/13.50)} = 3.00$$

The parameter A is then calculated as

$$A = t_1(I_{sub1})^m = 4,467\,(7.65)^{3.00} = 1,999,863$$

Having the parameters A and m, the lifetime at the operating conditions is calculated as

$$t = A(I_{sub-op})^{-m} = 1,999,863\,(0.248)^{-3.00} = 131,112,837\,\text{min} = 249.3\,\text{years}$$

Similarly, the following results are obtained for the device with channel length $L = 0.5\ \mu$m: $m = 2.98$, $A = 933,907$, $t = 7,825,630$ min $= 14.9$ years.

■ Example 11.5 EEPROM Reliability

N-channel MOSFETs with gate-oxide thickness $t_{ox} = 5$ nm are to be used as bit-select transistors in an EEPROM array. During write/erase operations the gate of the MOSFET has to be biased with $V_{G-op} = 3.2$ V. To test the device, accelerated tests with $V_{G1} = 3.8$ V and $V_{G2} = 4.0$ V are performed. It is shown that the devices fail after $t_1 = 190$ s and $t_2 = 70$ s with the stress voltages $V_{G1} = 3.8$ V and $V_{G2} = 4.0$ V, respectively. If the duration of a write/erase operation is 8 ms, find how many write/erase operations these devices are expected to perform successfully.

Solution: High voltage (V_G) at the gates of these MOSFETs, or the corresponding high gate-oxide field $E_{ox} = V_G/t_{ox}$, produces charges in the gate oxide, the rate of which is given by Eq. (11.18). Time to failure, or lifetime, is then proportional to the reciprocal value of this rate:

$$t = \frac{1}{A_t}e^{E_0/E_{ox}}$$

This equation can be used to determine the lifetime at the operating voltage once values of the parameters E_0 and A_t aredetermined. Determine first the stress and operating gate-oxide fields: $E_{ox-op} = V_{G-op}/t_{ox} = 6.4$ MV/cm, $E_{ox1} = V_{G1}/t_{ox} = 7.6$ MV/cm, and $E_{ox2} = V_{G2}/t_{ox} = 8.0$ MV/cm. Now, the parameters E_0 and A_t can be determined from the accelerated test data:

$$\frac{t_2}{t_1} = \exp[E_0(1/E_{ox2} - 1/E_{ox1})]$$

$$E_0 = \frac{\ln(t_2/t_1)}{(1/E_{ox2}) - (1/E_{ox1})} = 151.8 \text{MV/cm}$$

$$A = \frac{\exp(E_0/E_{ox1})}{t_1} = 2,487,180 \ s^{-1}$$

$$t_{op} = \frac{1}{A_t} \exp(E_0/E_{ox-op}) = 8039 \ s$$

Finally, it is easy to find that the operating lifetime of 8039 s corresponds to 8, 039/0.008 ≈ 10^6 write/erase operations.

11.2.5 Gate-Oxide Breakdown

MOSFET efficiency requires the gate oxide to be as thin as possible. Gate-oxide breakdown limits the minimum oxide thickness at a given operating voltage. MOSFETs are designed to operate with gate-oxide fields significantly lower than the critical (breakdown) field. Nonetheless, it is possible that the gate-oxide field becomes higher than the critical (breakdown) field during device operation, which leads to gate-oxide breakdown and therefore IC failure.

A typical distribution of the breakdown field is shown in Fig. 11.9. It illustrates that there are three different breakdown modes: low, medium, and high breakdown-field modes. The low breakdown-field mode ($E_{BD} < 0.1$ V/nm) is attributed to the existence of pinholes in the gate oxide. The medium breakdown-field mode (0.1 V/nm < E_{BD} < 0.8 V/nm) is caused by weak spots in the gate oxide. This type of breakdown is typically catastrophic—any subsequent measurement shows a low breakdown field. The high breakdown-field mode is associated with defect-free oxides. This breakdown is not catastrophic, and is observed as an increase in the current flowing through the oxide due to tunneling of electrons through the oxide barrier (Fowler–Nordheim current).

Exposure of the gate oxide to high fields during the operation may lead to the creation of a charge in the gate oxide, as explained in the previous section. Charge can also be trapped in the gate oxide after injection of hot carriers into the oxide. This charge, which is mainly positive, decreases the oxide field on the anode side but increases the oxide field on the cathode side. The increased field on the cathode side increases the current flowing through the oxide, and eventually a breakdown happens. This breakdown, which takes some time to occur, is called a *time-dependent dielectric breakdown* (TDDB). It is the TDDB that affects the reliability because it happens during device exploitation.

It is not difficult to imagine that the time to breakdown t_B, or lifetime, is proportional to the reciprocal value of the Fowler–Nordheim current, or in other words to the probability of electron tunneling through the silicon–oxide barrier [refer to Eq. (11.18)]. Therefore,

$$t_B = \tau_0(T)e^{E_{0B}/E_{ox}} \tag{11.21}$$

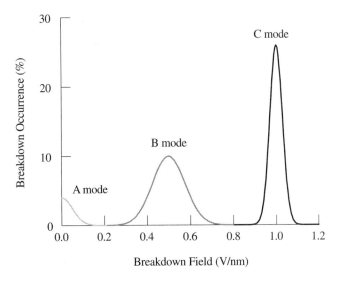

Figure 11.9 Typical breakdown-field distribution.

where $\tau_0(T)$ and E_{0B} are parameters and E_{ox} is the gate-oxide field. This equation can be applied to any of the above-described breakdown types, even to those from low and medium breakdown-field modes, but the time to breakdown can be extremely short so the device could fail during the testing. As expected, short times to breakdown are associated with the existence of pinholes or weak spots in the gate oxide. To account for the breakdowns due to the pinholes and weak spots, the effective (thinned) gate-oxide thickness should be used when determining the gate-oxide field E_{ox}. This gives increased value of the gate-oxide field ($E_{ox} = V_{ox}/t_{ox-eff}$) and therefore significantly reduced time to breakdown t_B. Because of that, it is more convenient to express the time to breakdown in the following way:

$$t_B = \tau_0(T)e^{E_{0B}t_{ox-eff}/V_{ox}} \tag{11.22}$$

where V_{ox} is the voltage applied across the gate oxide and t_{ox-eff} is the effective gate-oxide thickness.

It is obvious that gate-oxide breakdown is easily accelerated by an increase in the voltage applied to the gate of MOSFETs. The increase in temperature also accelerates the breakdown mechanism. In fact, both parameters in Eq. (11.22), τ_0 and E_{0B}, are temperature dependent. The temperature dependence of τ_0 is especially important because of its Arrhenius form:

$$\tau_0(T) = \tau_0 e^{-E_A/kT} \tag{11.23}$$

Therefore, increased temperature can equally well be used to accelerate the breakdown-related failures. The activation energy E_A, which appears as an additional parameter, depends on the applied gate-oxide field and gate-oxide quality.

A carefully designed IC could be expected to be quite immune to the oxide-breakdown-related failures. The situation is, however, not as simple as specifying the maximum gate voltage. The human body can be charged to very high electrostatic potentials (several kilovolts). If an electrostatically charged body comes in contact with the IC pins, *electrostatic discharge* (ESD) occurs and the gate oxide can be destroyed. This can happen while handling ICs, during assembling or repairing of systems containing the ICs. The presence of an electrostatically charged body can induce breakdown even when the device is not touched. The high field of a body, electrostatically charged to several kilovolts, and approaching MOSFET devices induces a very large voltage drop across the highly resistive gate oxide. Exposure of devices to high voltages induced in this way does not last long, but the voltages induced can be so high that the breakdown happens almost instantly. To minimize occurrence of ESD-induced gate-oxide breakdown, the input of any MOSFET IC is protected by an input circuit that limits the amount of voltage that can be applied to the gates of the input MOSFETs. Input protection circuits are normally diode–resistor circuits placed between the input pin and the gates of the input MOSFETs, as illustrated in Fig. 11.10. The role of input protection circuits is to short circuit the input pin to the V_+ rail if input voltage is higher than V_+ or to the V_- rail if the input voltage is lower than V_- (refer to Section 3.1.1). The presence of input protection circuits improves IC reliability significantly, but if ICs are frequently exposed to high electrostatic fields, the input protection diodes may fail. Normal high-field and high-temperature accelerating tests will not discover sensitivity of the IC to ESD-related failures. Special tests are needed, in which ESD pulses, obtained from specially designed circuits simulating ESD from the human body, are applied to the IC.

Transients represent additional sources of high-voltage spikes appearing at the gate of a MOSFET. Again, the gate oxide is not exposed to increased gate voltage for a long time, however, the voltage can be high enough to cause a sudden breakdown of the gate oxide. This breakdown mechanism may also elude the standard high-voltage–high-temperature tests, especially if a static bias is applied during the testing.

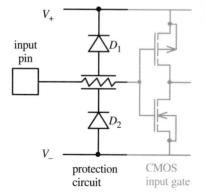

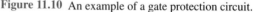

Figure 11.10 An example of a gate protection circuit.

Gate-oxide breakdown-related failures are quite frequent in MOSFET ICs. In future MOSFET ICs, gate-oxide thickness will be further decreased to enable proper functioning of scaled-down MOSFETs. This trend will most likely make the failures due to gate-oxide breakdown even more frequent.

11.2.6 Activation of Parasitic Structures

There are several parasitic structures inherently present in ICs that are not active in normal conditions, but if activated can lead to failures. These structures include parasitic MOSFETs, parasitic BJTs (lateral and vertical), and parasitic thyristor structure.

The parasitic MOSFETs appear in the field region, between the active MOSFETs: the field oxide is the gate oxide of the parasitic MOSFET, the metalization running over the field oxide is the gate, and the source and drain regions of the neighboring active MOSFETs are also the source and drain of the parasitic MOSFET. In normal conditions, the parasitic MOSFET is off as its threshold voltage is designed to be larger than the supply voltage. However, if a large amount of charge is created in the field oxide, due to the exposure of the IC to radiation or to the high electric field appearing across the oxide, the threshold voltage of the parasitic MOSFET can be decreased and the MOSFET activated. This would cause leakage current between two neighboring devices, and therefore failure of the IC.

The most important parasitic structure, from the reliability point of view, seems to be the thyristor structure. It is inherently present in the structure of both standard bipolar and CMOS ICs. Activation of the thyristor structure (the so called *latch-up* effect) is, however, much more frequent in CMOS ICs. The parasitic thyristor structure appearing in an N-well CMOS inverter is illustrated in Fig. 11.11. As can be seen, it is a PNPN structure that can be represented by an appropriate connection of a PNP and an NPN BJTs (Fig. 11.11b). No current flows through the parasitic thyristor structure in normal conditions (the thyristor is in the *off* mode), and its presence does not disturb the normal operation of the IC. However, if the structure is activated, its resistance drops to virtually zero (Fig. 11.11c), which short circuits V_+ and V_- supply rails causing permanent damage of the IC due to the flow of unlimited current.

It is important to consider the conditions under which the thyristor structure can be activated. In normal working conditions the PNP and NPN transistors involved in the structure (Fig. 11.11b) are in the cutoff mode: N-well–P-substrate junction (collector–base junction of both BJTs) is always reverse biased, while base–emitter junctions of both BJTs are short-circuited by the chip metalization. Nonetheless, if a significant leakage current flows through resistors R_S and R_W, the voltage drop across the internal parts of the base–emitter junctions can become sufficient to turn the BJTs into the active mode. A significant leakage current through R_S and R_W can appear if the structure experiences a large transient voltage because of noise or electrostatic discharge. Being in the active mode, the emitter current of the PNP BJT can be expressed as

$$I_{Epnp} = (1 + \beta_{pnp})I_{Bpnp} \tag{11.24}$$

The base of the PNP BJT is in fact the same region as the collector of the NPN BJT. Then the base current of the PNP BJT is in fact the collector current of the NPN BJT, which means it can be expressed as

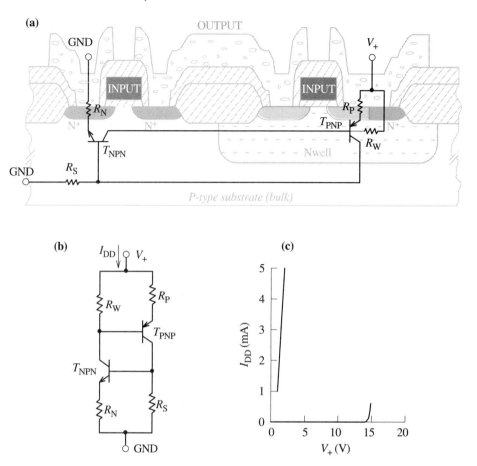

Figure 11.11 The parasitic thyristor structure in an N-well CMOS inverter: (a) cross section of the CMOS inverter illustrating the components of the structure, (b) the equivalent circuit, and (c) the I–V characteristic.

$$I_{Bpnp} \equiv I_{Cnpn} = \frac{I_{Enpn}}{1 + (1/\beta_{npn})} \tag{11.25}$$

Replacing I_{Bpnp} from Eq. (11.25) into Eq. (11.24) we obtain

$$I_{Epnp} = \frac{1 + \beta_{pnp}}{1 + (1/\beta_{npn})} I_{Enpn} \tag{11.26}$$

From the circuit in Fig. 11.11b, it is obvious that the current flowing through the structure is

$$I_{DD} = I_{Epnp} + I_{R_W} = I_{Enpn} + I_{R_S} \tag{11.27}$$

If I_{Enpn} is eliminated from Eqs. (11.26) and (11.27), the current flowing through the structure can be expressed as

$$I_{DD} = I_{Epnp} + I_{R_W} = \frac{\beta_{pnp}(1 + \beta_{pnp})(I_{R_W} - I_{R_S})}{1 - \beta_{npn}\beta_{pnp}} + I_{R_W} \qquad (11.28)$$

At the onset of the active mode the current gains β_{npn} and β_{pnp} are small, so that $\beta_{npn} \times \beta_{pnp} <$ 1. A limited current given by Eq. (11.28) flows through the structure in that case. If the transient voltage is sufficiently large, however, the leakage current will further increase the emitter currents of the transistors. As emitter currents increase, the current gains β_{npn} and β_{pnp} increase, since the BJTs enter more efficient regions of active-mode operations. As $\beta_{npn} \times \beta_{pnp}$ approaches unity, the denominator in Eq. (11.28) approaches zero causing a sudden increase in the current flowing through the thyristor structure; switching occurs as shown in Fig. 11.11c.

From the above consideration, it is obvious that latch-up can be prevented by ensuring that $\beta_{npn-max} \times \beta_{pnp-max} < 1$. Also, careful design of the circuit layout can reduce the probability of the latch-up event. Use of SOI technology eliminates the problem by eliminating the thyristor four-layer structure altogether.

11.2.7 Soft-Error Mechanisms

Materials used for IC packages frequently contain traces of elements such as uranium, thorium, and americium. These materials emit α particles in the energy range of 4 to 9 MeV. An average α particle, entering the semiconductor substrate with the energy of 5 MeV, penetrates to a depth of about 25 μm generating on its way up to 2.5×10^6 electron–hole pairs. Electrons and holes generated in P–N junction depletion layers are immediately separated by the built-in electric field, electrons being pushed into the N-type region and holes into the P-type region of the semiconductor. Electrons and holes generated in the neutral regions diffuse before they are recombined or enter a depletion layer, in which case they are again collected by the electric field.

Electron and holes generated in this way can change the state of a digital cell, causing an error in the circuit function. They, however, do not cause permanent damage (unless the thyristor structure is activated), and the circuit continues normal operation once the mistaken information is corrected. To illustrate soft-error mechanisms, take the example of the MOS dynamic RAM cell, previously shown in Fig. 7.5. The part of the memory cell that stores the information is the MOS capacitor, shown again in Fig. 11.12a. When logic "1" state is stored at the cell, the capacitor is charged, which means excess positive hole charge is stored at the semiconductor plate, more exactly in the potential well created by the P$^+$ region. Logic "0" state is represented by no charge at the capacitor plates, which means no extra holes appearing in the P$^+$ region. If an α particle passes through the capacitor, it generates electron–hole pairs in the depletion layer of the P–N junction, the holes being collected in the P$^+$ region, as illustrated by the energy-band diagram of Fig. 11.12b. If the memory cell was in logic "0" state before the appearance of the α particle, a soft error occurs as the cell is filled by holes that correspond to the logic "1" state.

The probability of soft error occurring in a logic cell is normally very small, but when multiplied by the large number of logic cells existing in a modern IC, the probability of

(a)   **(b)**

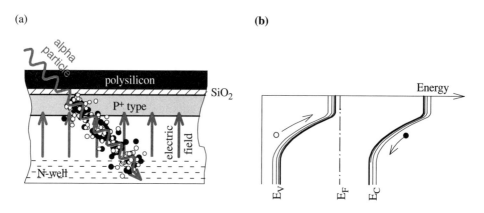

Figure 11.12 Illustration of a soft error in an MOS dynamic RAM cell.

soft error occurring in an IC becomes significant. Of course, the rate of soft-error failures significantly depends on the materials used for the encapsulation of the IC.

11.3 RELIABILITY SCREENING

As explained in Section 11.1, weak devices fail at the beginning of their operating life (early failure region in Fig. 11.2). It is extremely beneficial to remove these devices before the exploitation starts, and in that way significantly reduce the maintenance cost. The procedure of weak-device removal is called *reliability screening*.

The reliability screening is achieved by aging the devices up to the point at which the steady-state region begins (refer to the bathtub curve shown in Fig. 11.2). In other words, the early-failure region of the bathtub curve is cut off by the reliability screening, so that the real exploitation of devices starts at the beginning of the steady-state region. The devices that fail during the aging process are the devices that exhibit early-failure problem. The devices that do not fail during this aging period are the devices free of gross defects (they can still fail during the operation but only due to a random combination of several factors, as explained in Section 11.1).

The most straightforward aging procedure would be to expose the devices to working conditions for the time period equivalent to the early-failure period. However, at least two factors make such a procedure impractical. The first factor is duration of the aging process, and therefore the screening procedure. The early-failure period can be longer than a year, which is short in view of the mean time to failure, but is extremely long in view of the duration of the screening procedure. Screening procedures longer than a few weeks cannot be practically accepted. That is why appropriate accelerating tests should be used for reliability screening. As explained in the previous section, the rate of failure mechanisms strongly depend on the operation conditions, such as temperature, bias, humidity, appearance of electrostatic discharge events, etc. Therefore, a failure mechanism can be accelerated by making use of harsher screening conditions compared to the operation conditions, for example, increase in the temperature or voltage applied. The screening conditions, however, should be carefully designed, as neither *understress* nor *overstress* is desirable. The understress, of course, means that not all the weak devices are removed, and therefore the reliability screening is not efficient. The overstress, on the other hand, leads to overaging of "healthy" devices, which reduces their mean time to failure. The overaging can be quite significant as, for example, 20 hours of accelerated testing may correspond to years of normal operating life.

The second factor that causes the aging conditions to be different from the operating conditions is the diversity of operating conditions. A device in an IC can work in very different bias conditions. For example, the gate of a MOSFET in a digital circuit used as a detector, which rarely changes the logic state, could be biased at high voltage for years. In contrast, the gate voltage of a MOSFET changes hundreds of millions of times a second if the IC is used in a computer system. Also, an IC can work in moderate temperature and environment conditions, or can be exposed to high operating temperature accompanied by high relative humidity and salinity. Some ICs are required to work in the environment of increased γ- or x-radiation, or increased electrostatic field, which may induce an ESD event. Different application conditions favor different failure mechanisms. Aging ICs with a high-temperature–bias stress test, most likely will not eliminate the devices prone to radiation or ESD-related failures. A 100% effective screening would require elimination of any weak IC, regardless of the intended application. It does not matter that an IC is not expected to be exposed to frequent ESD events, it might still be better not to use it if it is so weak that it would fail after a single ESD event. However, it should be kept in mind that a poorly designed screening procedure could damage healthy ICs. If an IC is not expected to operate in conditions of increased γ- or x-radiation, it is better not to apply any radiation-hardness test, because it would require exposure of all ICs to a certain dose of radiation, which can easily make ICs sensitive to some other failure mechanisms, such as oxide breakdown.

Design of the screening procedure is, obviously, not a trivial problem. There is no commonly accepted screening procedure, although U.S. military standards strictly define tests and test conditions that are to be used in screening the ICs for military applications. The following list illustrates the typical reliability screening tests.

1. *Internal Visual.* This is an internal visual inspection of IC chips, in which devices with visible defects are removed. It should be mentioned that the IC chips with defects that cause immediate failures are already removed during the final electrical testing. This inspection is to remove ICs with visible defects expected to lead to an early failure.
2. *Stabilization Bake.* During this step, the ICs are exposed to 125°C with no bias applied for 24 h. It is expected that this high-temperature treatment will ensure that physical and chemical processes initiated during the IC manufacture and assembly are completed.
3. *Thermal Shock.* In this test, the ICs are suddenly cooled (−65°C) and heated (150°C) by dipping in cold and hot liquid baths. The cold and hot thermal shock is repeated 15 times.
4. *Temperature Cycle.* In this test, the ICs are transferred from a cold chamber (−65°C) to a hot chamber (150°C) and back to the cold chamber, this cycle being repeated 10 times. This test, together with the above-mentioned thermal shock, is

expected to eliminate ICs with assembly-related problems, which cannot therefore withstand the induced expansions and contractions of the materials used.

5. *Mechanical Shock.* The ICs are accelerated to 20,000 *g* and then suddenly stopped.

6. *Centrifuge.* The ICs are subject to acceleration of 30,000 *g*. Mechanical shock and centrifuge tests are used to detect mechanical and structural weaknesses in the packages.

7. *Hermeticity.* To test hermeticity of the packages, ICs are exposed to a He atmosphere with increased pressure. If there are cracks or pinholes in the package He enters into the IC package. When sunk into a liquid immediately after that, the He makes bubbles coming out of the package having cracks or pinholes.

8. *Critical Electrical Parameters.* The critical electrical parameters are measured before burn-in (next step).

9. *Burn-in.* This is the most important screening test. The ICs are operated at 125°C for 168 h or longer under normal operating bias conditions. As described in the previous section, most of the failure mechanisms are accelerated by temperature. The high temperature (125°C) is expected to accelerate the early failures and enable their elimination after only 168 h of operation. As "normal operating bias conditions" cannot generally be established, many electronic-system manufacturers repeat the burn-in test once the ICs are installed onto the printed circuit boards.

10. *Final Electrical.* This test is used to detect failed ICs during the screening testing.

11. *X-Ray Radiograph.* Inspection by *x*-ray radiography may detect defects in bonding not detected by previously applied tests, but its main purpose is checking for particles encapsulated in the cavity of ceramic packages. The presence of conductive free particles in the cavity of a ceramic package can cause a short circuit.

12. *External Visual.* Defects on the IC pins and package are checked.

Finally, let us emphasize that all the ICs produced should pass through the tests of a screening procedure. There is no sampling, this is a 100% testing.

11.4 RELIABILITY MEASUREMENT

When the early-failure period (Fig. 11.2) is cut off by the reliability screening procedure, devices start their real operating life at the beginning of the steady-state region. In the steady-state region, the devices still fail, due to a random combination of several factors. It is important to predict how frequent these failures are going to be, or in other words what the failure rate is. As shown in Section 11.1, the reciprocal value of the failure rate gives the mean time to failure ($MTTF$) for the case of constant failure rate. Determination of the failure rate, or mean time to failure, is referred to as the *reliability measurement*. The reliability measurement is not as straightforward as the measurement of conformance parameters. Three fundamental problems are associated with the reliability measurement. Solutions to these problems requires use of specific techniques. These problems and the techniques used for reliability measurement are explained in the following sections.

11.4.1 Problem 1: Destructive Nature of the Measurement. Solution: Sampling Technique

Determination of the mean time to failure is destructive—one can know how long a device has worked once the device has failed. To overcome this problem, a sampling technique is applied. A certain number (n) of devices is randomly taken from the batch of processed devices for the purpose of reliability testing. The test devices are then operated until enough of them fail, to enable calculation of the mean time to failure.

> The mean time to failure calculated from sample data is denoted by $mttf$, to distinguish it from the mean time to failure of the whole batch, which is denoted by $MTTF$.

Naturally, the mean-time-to-failure value obtained from the sample data ($mttf$) is taken as the best estimate of the mean time to failure of the whole batch of devices ($MTTF$).

The application of the sampling technique to solve the problem of the destructive nature of the reliability measurement appears very simple. The only question is how large a sample size should be taken. Too large a sample size is obviously costly, as too many devices would have to be destroyed. **How small a sample size is acceptable?** Can we take one only device to represent tho whole batch of devices?

To answer this important question, let us imagine the following situation. A technician selects 30 devices and determines the mean time to failure as 17.1 years. However, the boss is suspicious, and takes an additional 30 devices, which give a mean time to failure of 14.6 years. Who is correct? Try to find out a third time, and the result is 15.2 years! Try hundred of times and plot the results as shown in Fig. 11.13.

> The mean time to failure obtained using a finite sample size n is a random variable. Its distribution is centered around the mean time to failure of the main batch ($\overline{mttf} = MTTF$), however, it has a finite standard deviation (σ_{mttf}). The standard deviation σ_{mttf} expresses the error of the mean-time-to-failure measurement, introduced by the sampling technique.

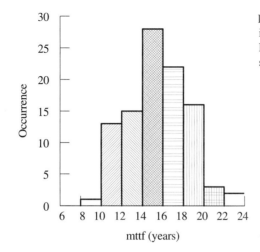

Figure 11.13 Histogram obtained by a hundred independent estimates of mean time to failure. Each time, 30 different devices are used as a sample.

For the case of a large sample size ($n \geq 30$), the mean time to failure determined by sampling can be expressed by the normal (Gaussian) distribution:

$$f(mttf) = \frac{1}{\sigma_{mttf}\sqrt{2\pi}} \exp\left[-\frac{(mttf - \overline{mttf})^2}{2\sigma_{mttf}^2}\right] \tag{11.29}$$

where the distribution parameters, the mean $\overline{mttf}$, and the standard deviation σ_{mttf} are given by

$$\overline{mttf} = \frac{1}{\lambda} = MTTF \tag{11.30}$$

$$\sigma_{mttf} = \frac{1}{\sqrt{n}}\frac{1}{\lambda} = \frac{MTTF}{\sqrt{n}} \tag{11.31}$$

Equation 11.31 shows that σ_{mttf} depends on the sample size n. This effect is illustrated in Fig. 11.14.

A larger sample size corresponds to a smaller standard deviation σ_{mttf} of the estimated mean time to failure $mttf$. This means that a larger sample size corresponds to a smaller measurement error.

In the extreme case of $n \rightarrow \infty$ (the whole batch is tested), the standard deviation is $\sigma_{mttf} = 0$, which means the *mean time to failure* is calculated without any error. Note that this does not mean that the *lifetime* of a particular device is precisely determined. The lifetime of a particular device is a random variable obeying the exponential distribution given by Eq. (11.11).

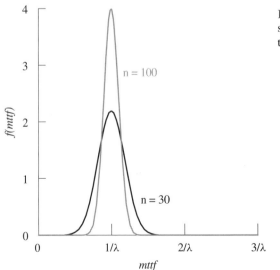

Figure 11.14 Illustration of the effect of sample size n on standard deviation of the mean-time-to-failure estimation.

The standard deviation is frequently defined as the *absolute measurement error*. The standard deviation normalized by the mean value then becomes the corresponding *relative error*. In our case, the absolute error is σ_{mttf}, while the relative error is $\sigma_{\text{mttf}}/\overline{mttf}$. Consequently, the reliability measurement result can be expressed as $\overline{mttf} \pm \sigma_{\text{mttf}}$. It should be mentioned that this does not mean the batch mean time to failure $MTTF$ will definitely be between $\alpha_l = \overline{mttf} - \sigma_{\text{mttf}}$ and $\alpha_u = \overline{mttf} + \sigma_{\text{mttf}}$. In fact, the probability that the real value will be in the limits of $\overline{mttf} \pm \sigma_{\text{mttf}}$ is 0.68, as

$$\int_{\overline{mttf}-\sigma_{\text{mttf}}}^{\overline{mttf}+\sigma_{\text{mttf}}} f(mttf)d(mttf) = 0.68 \tag{11.32}$$

This probability is frequently called the *confidence level*, in which case the corresponding lower and upper limits are called *confidence limits*. The general relationship between the confidence limits α_l and α_u and the confidence level is given by

$$\int_{\alpha_l}^{\alpha_u} f(mttf)d(mttf) = \text{Confidence Level} \tag{11.33}$$

Table 11.2 gives the confidence limits corresponding to the most frequently used confidence levels.

TABLE 11.2 Confidence Limits for Different Confidence Levels

Confidence Level	Lower Confidence Limit	Upper Confidence Limit
0.68	$\overline{mttf} - \sigma_{\text{mttf}}$	$\overline{mttf} + \sigma_{\text{mttf}}$
0.90	$\overline{mttf} - 1.65\sigma_{\text{mttf}}$	$\overline{mttf} + 1.65\sigma_{\text{mttf}}$
0.95	$\overline{mttf} - 2\sigma_{\text{mttf}}$	$\overline{mttf} + 2\sigma_{\text{mttf}}$
0.99	$\overline{mttf} - 2.58\sigma_{\text{mttf}}$	$\overline{mttf} + 2.58\sigma_{\text{mttf}}$
0.997	$\overline{mttf} - 3\sigma_{\text{mttf}}$	$\overline{mttf} + 3\sigma_{\text{mttf}}$

▨ **Example 11.6 Error in *MTTF* Estimation**

The standard deviation of $MTTF$ estimation needs to be smaller than 10% of the lifetime standard deviation. Find the needed sample size.

Solution: The lifetime standard deviation (σ_t) is the standard deviation of the exponential distribution that describes the distribution of the lifetime [Eq. (11.11)]. The standard deviation of exponential distribution (11.11) is

$$\sigma_t = \frac{1}{\lambda}$$

Using Eq. (11.31), the sample size can be expressed as

$$n = \frac{\sigma_t^2}{\sigma_{mttf}^2}$$

As σ_{mttf} should be smaller than 10% of the lifetime standard deviation, which means $\sigma_{mttf} < 0.1\sigma_t$, it is obvious that

$$n > \left(\frac{1}{0.1}\right)^2 = 100$$

11.4.2 Problem 2: Duration of the Measurement. Solution: Accelerated Testing

The problem of test duration is solved in a similar way to the case of reliability screening, namely by accelerated testing. The accelerated testing shortens the device lifetime, and therefore enables the determination of $MTTF$ or λ with much shorter tests. The factor by which the test time is shortened is called the *acceleration factor*, which can be expressed in terms of the mean times to failure, or failure rates, at stress and operating conditions in the following way:

$$AF = \frac{MTTF_{oper}}{MTTF_{stress}} = \frac{\lambda_{stress}}{\lambda_{oper}} \tag{11.34}$$

Obviously, the mean time to failure obtained with the accelerated testing is simply multiplied by the acceleration factor to obtain the mean time to failure at the operating conditions. Example 11.1 provides an illustration of how the failure rate and the mean time to failure can be determined from accelerated test data. Examples 11.3, 11.4, and 11.5 illustrate how the accelerated test data are extrapolated to operating conditions for the cases of failure mechanisms accelerated by temperature, current density, and electric field, respectively.

11.4.3 Problem 3: Different Failure Mechanisms and Operating Conditions. Solution: Combination of Reliability Tests

The accelerated testing should be carefully designed so as not to induce new failures by overstressing the devices. However, it is equally important that the testing used covers all the failure mechanisms that are likely to develop during exploitation. Therefore, the testing conditions should properly reflect the operating conditions. System manufacturers can know that ICs will work in a particular computer or control system, which enables them to perform accelerated testing by operating the ICs assembled onto the printed circuit boards. However, this approach cannot be applied by IC manufacturers. They use a set of static and dynamic bias arrangements to ensure that all the possible failure mechanisms are covered. The use of both static and dynamic bias is necessary as the static bias does not affect the failures induced by transients, while the dynamic bias may, for example, mask the failures due to creation of oxide charges.

Even when the bias conditions are established properly, not all the possible problems are overcome. It may not be enough to simply operate the ICs at increased temperature and extrapolate the data. For example, testing at 125°C can mask the failures due to aluminum corrosion and, especially, the failures related to hot-carrier effects. If it turns out that some of the masked failure mechanisms are pronounced, the reliability measurement becomes irrelevant. It is necessary to apply a set of reliability tests to ensure that all the failure mechanisms that are likely to influence the reliability are taken into account. There is no general set of tests applicable to all the intended application conditions. For example, if any abnormal environment condition is expected, such as a significant level of ionizing radiation, an appropriate test should be used to account for this condition. Nevertheless, it is possible to list the most frequently used reliability tests.[4] A brief description of these tests is given in the following. Table 11.3 summarizes the coverage of the failure mechanisms considered in Section 11.2 by the tests described below.

1. *High-Temperature Operating-Bias (HTOB) Test.* This is the most frequently used test; sometimes it is called *the life test.* The ICs are operated at 125 and 150°C for 1000 hours or more under normal operating conditions. This test accelerates mainly diffusion-related failure mechanisms (formation of the purple plague, electromigration, instabilities due to ionic contamination, which is classified under "charge trapping" in Table 11.3), and also other thermally activated mechanisms such as the dielectric breakdown.

2. *High-Temperature High-Bias (HTHB) Test.* In this test, the devices are operated at both increased voltages and temperature. This test can be more efficient than the *HTOB* test in accelerating the dielectric breakdown and charge trapping related mechanisms.

3. *Temperature–Humidity-Bias (THB) Test.* The ICs are operated in an environment of 85% relative humidity and 85°C temperature under a static reverse bias for 1000 h or more. The test accelerates the corrosion of metalization.

4. *Room Temperature High-Bias (RTHB) Test.* The devices are operated at increased voltages and room temperature for about 10 days or longer, and drifts in specified parameters are monitored. The test is targeted at accelerating the hot-carrier-related

TABLE 11.3 Coverage of Failure Mechanisms by Reliability Tests

Failure Mechanism	HTOB	HTHB	THB	RTHB	α	ESD
1. Formation of the purple plague	✓	✓				
2. Electromigration	✓	✓				
3. Corrosion of metalization			✓			
4. Charge trapping,	✓	✓		✓		
interface trap creation,	✓	✓		✓		
hot-carrier generation				✓		
5. Dielectric breakdown	✓	✓				✓
6. Activation of parasitic structures	✓	✓		✓		
7. Soft-error mechanisms					✓	

[4]*Source*: A.G. Sabnis, *VLSI Reliability* (VLSI Electronics, Microstructure Science, Vol. 22), Academic Press, San Diego, 1990.

mechanisms. It is even better when the test is conducted at temperatures lower than room temperature (for example, -10 to $-20°C$), because the decrease in the temperature accelerates the failure mechanism.

5. *α-Particle Test.* The devices are bombarded by α particles from a concentrated laboratory source such as ^{241}Am. The aim of the test is to determine the soft-error rate.

6. *Electrostatic Discharge (ESD) Test.* Special testing equipment is used to generate electrostatic discharge that models electrostatic discharge events from both a human body and a charged device. The test is used to determine the ESD threshold of devices.

When a set of reliability tests is performed, and the mean times to failure or the failure rates determined for each of them, the overall data on $MTTF$ and λ should be expressed. Frequently, there is no practical need to combine the test data, because one of the tests dominates with the number of failures. In that case, the overall $MTTF$ and λ are simply taken to be equal to the values obtained from that test. If this is not the case, however, then the test data have to be combined. This is not difficult if it is assumed that different tests induce different and, therefore, independent failures. In that case, the overall failure rate is given by

$$\lambda_{\text{oper}} = \sum_{i=1}^{nt} \lambda_{\text{oper}-i} \tag{11.35}$$

where $\lambda_{\text{oper}-i}$ corresponds to the ith failure mechanism, and nt is the number of independent failure mechanisms involved.

Understanding the tests and failure mechanisms can help establish a conclusion about the (in)dependence of the failures, although in some cases a complete failure analysis may be required. For example, if the failure rates from the tests *HTOB, α,* and *ESD* are to be combined, Eq. (11.35) can directly be applied, as the failure mechanisms activated by these tests are most likely independent. A different example is when the failures from *HTOB* and *HTHB* tests are pronounced. In that case, especially if the number of failures induced by the *HTHB* test is higher than the number of failures induced by the *HTOB* test, it is better not to take into consideration the failures from *HTOB* test. This is because the failures induced by the *HTOB* are most likely induced by the *HTHB* as well.

■ **Example 11.7** *MTTF* **Due to Combined Failure Mechanisms**

The measured mean times to failure, associated with two independent failure mechanisms, are 15 and 21 years, respectively. Express the overall mean time to failure.

Solution: From Eq. (11.35) we find that

$$\lambda_{\text{oper}} = \frac{1}{MTTF_{\text{oper}-1}} + \frac{1}{MTTF_{\text{oper}-2}} = \frac{1}{15} + \frac{1}{21} = 0.1143 \frac{1}{\text{years}}$$

Therefore, $MTTF_{\text{oper}} = 1/\lambda_{\text{oper}} = 8.75$ years.

11.1 The results of an accelerated reliability test are given in Table 11.4. Estimate the failure rate and the mean time to failure.   [A]

11.2 Consider a telephone system consisting of 1000 blocks, each block having in average 10 ICs. Estimate the number of system failures per month ($NSFM$) and the percentage of system blocks failing in 10 years ($PSBF$) if the failure rates of the ICs are 10, 100, and 1000 FIT, respectively.   [A]

11.3 Using a sample size of $n = 100$, the mean time to failure of a batch of ICs is estimated to be $\overline{mttf} = 15$ years. Find the standard deviation of the estimated mean time to failure (σ_{mttf}). Find the upper and lower mean-time-to-failure values ($mttf_{max}$ and $mttf_{min}$) corresponding the confidence level of 95% ($\overline{mttf} \pm 2\sigma_{mttf}$).

11.4 An accelerated test is performed on 10 ICs and time to failures are recorded. The results are given in Table 11.5. Determine λ_{stress} and $MTTF_{stress}$ after (a) the third failure ($t = 299$ h) and (b) the ninth failure ($t = 1,951$ h). Compare and comment on the results.

11.5 The lifetime of a group of MOSFETs, subject to a hot-carrier accelerated test, is 4266 min. Knowing that the acceleration factor for the hot-carrier stressing is given by

$$AF = \left(\frac{I_{sub-oper}}{I_{sub-stress}} \right)^{-3}$$

and knowing that the substrate currents under operating and stress conditions are $I_{sub-oper} = 0.248 \ \mu A$ and $I_{sub-stress} = 6.10 \ \mu A$, respectively, determine the lifetime under operating conditions.

11.6 Two groups of 100 ICs in each group are selected for reliability testing. The ICs from the first group are subjected to the high-temperature operating bias (HTOB) test, while the ICs from the second group are subjected to temperature–humidity-bias (THB) tests. It is

TABLE 11.4 Results of Accelerated Reliability Test

	Unit									
	1	2	3	4	5	6	7	8	9	10
Operating time (h)	7,000	6,000	5,500	6,800	7,200	9,000	7,800	6,100	9,100	6,400
Number of failures	0	0	0	1	0	0	1	0	1	0

TABLE 11.5 Accelerated Test Results

	Failure Number								
	1	2	3	4	5	6	7	8	9
Time (t) (h)	89	251	299	384	545	770	1,019	1,420	1,951
$1 - F(t)$	0.9	0.8	0.7	0.6	0.5	0.4	0.3	0.2	0.1

found that 50% of the ICs from the first group failed after 545 h, while 50% of the ICs from the second group failed after 1010 h. Assuming that the accelerated tests enforce independent failure mechanisms, and using an acceleration factor of 260 for both tests, find the best estimate for the mean time to failure ($MTTF$).

11.7 Table 11.6 shows the results of three accelerated tests related to three independent failure mechanisms appearing in a batch of ICs. What percentage of the ICs from this batch is expected to fail after the first year of operation? [A]

REVIEW QUESTIONS

R-11.1 How is reliability related to quality?

R-11.2 Can reliability be expressed quantitatively (using numbers) or can it be expressed only descriptively (high, medium, low)?

R-11.3 If the answer to the above question is yes, does it mean that we can say precisely when a given device is going to fail? If no, what does the reliability measurements give?

R-11.4 Take the situation of a batch of devices that fails at a constant rate λ. Sketch the change of total number of failed devices with time. What is this function called?

R-11.5 Is the fact that the failures in the steady-state region are due to a number of mutually independent causes related to the fact that the failure rate λ is constant? Give an example of mutually independent causes leading to a single failure.

R-11.6 How is the mean time to failure related to the failure rate?

R-11.7 What is the most effective test for the purple plague-related failures?

R-11.8 It is discovered under a microscope that the left-hand side of a metal line is open, while hillocks appear on the right-hand side of the line. Which side of the line was positively biased?

R-11.9 Does the temperature accelerate the electromigration process? If yes, why?

R-11.10 Is the 125°C test under operating bias effective for checking ICs against metal corrosion?

TABLE 11.6 Results of Three Different Accelerated Tests

Failure Mechanism	AF	MTTF
Al corrosion	1,000	256 h 7 min
Hot carriers	100	1928 h 50 min
Electromigration	10,000	42 h 19 min

R-11.11 Is sodium contamination important for the process of aluminum corrosion?

R-11.12 What MOSFET parameters are affected by hole trapping in the gate oxide?

R-11.13 Can trapped holes generate border and interface traps?

R-11.14 What MOSFET parameters are affected by the interface traps?

R-11.15 What is a hot carrier?

R-11.16 Do the hot carriers themselves damage MOSFETs? If not, how do they affect the reliability?

R-11.17 Which are the two most important mechanisms of hot-carrier injection into the gate oxide?

R-11.18 Are the injected holes or the injected electrons into the gate oxide more damaging? Why?

R-11.19 Does increased temperature accelerate hot-carrier-related failure mechanisms?

R-11.20 What is time-dependent dielectric breakdown (TDDB)? Is it related to oxide charge creation due to a high oxide field? If yes, how?

R-11.21 Does activation of the parasitic thyristor structure (latch-up) in CMOS ICs damage the ICs permanently?

R-11.22 Do α particles and the related soft errors damage the ICs permanently?

R-11.23 Should all the manufactured ICs be subject to the tests of a prescribed screening procedure? If yes, is the cost of this testing justified?

R-11.24 Should all the manufactured ICs be subject to testing for the purpose of reliability measurement? If not, is the high testing cost the only problem?

R-11.25 Using small sample sizes for reliability measurement is cheaper. Is the measurement process affected by the sample size? How?

R-11.26 Would you agree to pay a higher cost for high reliability ICs if the IC manufacturer refuses to provide the list of reliability tests applied (a company secret)? If no, why would you not believe the claimed mean time to failure?

Chapter *12*

QUANTUM MECHANICS

A number of device effects have been explained in terms of seemingly odd phenomena and concepts, such as tunneling, electron wavelength, etc. This type of phenomena can be observed only when we deal with small things, however, the semiconductor devices are small enough and many of these phenomena appear as the main reason for a particular device behavior. The theory that is used to explain and predict the behavior of small things is known as *quantum mechanics*.[1]

The descriptions of the device effects, given in the main text of the book, provide the basic insight into the underlying phenomena of quantum mechanics. However, as these phenomena and concepts appear strange and abstract, there is a need for a deeper insight into the principles of quantum mechanics. This chapter addresses this need.

12.1 WAVE FUNCTION

The central concept of quantum mechanics is the *wave–particle duality*, mentioned in Section 1.5.1. A complex mathematical function, consisting of real and imaginary parts, is used to model the dual wave–particle properties. This function is called the *wave function*, and is labeled by ψ.

Figure 12.1 illustrates the simplest form of the wave function. Imagine a point circulating in the complex plane. The *rate of change of phase with time* (radians per second) is called the *angular frequency* ω. As the whole circle has 2π radians, the angular frequency is

$$\omega = \frac{2\pi}{T} \tag{12.1}$$

the *period T* being the time that it takes to complete a circle.

The position of any point in the complex plane can be expressed by a complex number $a + jb$, where a and b are the real and imaginary parts, respectively, while $j = \sqrt{-1}$.

---

[1]Quantum mechanics is a general theory, involving the laws of *classical physics*.

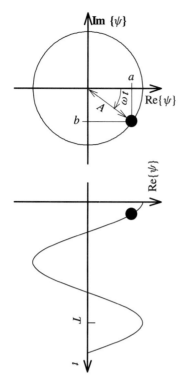

Figure 12.1 Illustration of the $\psi(t) = A\ \exp(-j\omega t)$ function and its relationship to waves.

Alternatively, it can be expressed as $A \exp(j\varphi)$, where the distance from the origin (A) is called the *amplitude* and the angle with respect to the real axis (φ) is called the *phase*. In our case, $\varphi = -\omega t$. Therefore, the position of the circulating point can be expressed by the following complex function:

$$\psi(t) = Ae^{-j\omega t} \tag{12.2}$$

or, alternatively:

$$\psi(t) = \underbrace{A\ \cos(\omega t)}_{Re\{\psi\}} - j\ \underbrace{A\ \sin(\omega t)}_{Im\{\psi\}} \tag{12.3}$$

Obviously, the real part of ψ is a cosine function of time.

Now, imagine that the point is attached to a membrane causing it to follow the cosine $Re\{\psi\}$ oscillation. Imagine further that the membrane oscillation is transferred to air particles, or any other set of particles. The push and pull of the membrane causes peaks and valleys of particle concentration that travel in a direction perpendicular to the membrane, say the x-direction. If λ is the distance that a peak (or a valley) travels as the oscillating point completes a whole circle (2π radians), then the *rate of change of phase with distance* (radians per meter) is

$$k = \frac{2\pi}{\lambda} \tag{12.4}$$

More frequently, k is referred to as the *wave number*, or the *wave vector* in the case of three-dimensional presentation (**k**).

In general, a traveled distance x is related to the change of phase as $\varphi = kx$ (or $\varphi = \mathbf{kr}$ in the three-dimensional case). Therefore, we have obtained the following wave function of time t and distance x:

$$\psi(x, t) = A e^{-j(\omega t - kx)} \tag{12.5}$$

The real part of this wave function,

$$Re\{\psi(x, t)\} = A \, \cos(\omega t - kx) \tag{12.6}$$

is plotted in Fig. 12.2. Obviously, λ is the wavelength, and is related to the period T through the wave speed v:

$$\lambda = vT \tag{12.7}$$

Equations (12.1), (12.4), and (12.7) show that the wave number and the angular frequency are also related to each other through the wave speed:

$$\omega = vk \tag{12.8}$$

Equation (12.5) describes a wave that is not localized in space (x-direction).

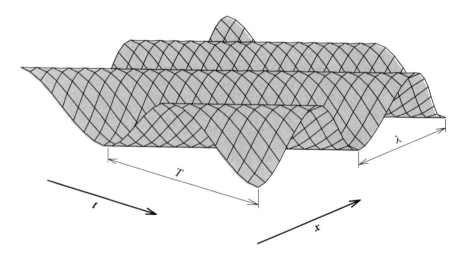

Figure 12.2 Plot of $Re\{\psi(x, t)\} = A \, \cos(\omega t - kx)$—illustration of a plane wave.

To illustrate this important point, let us say that the considered wave is *sound*, produced by the oscillating membrane. As the wave is not decaying in the x-direction, we will hear a constant sound level at any point x. An analogous example can be made with light, which will appear with constant intensity at any point x. Mathematically, the wave intensity is expressed as $\psi(x, t)\psi^*(x, t)$, where ψ^* is the complex conjugate (the signs of js reversed). It is easy to show that the intensity of the wave given by Eq. (12.5) is indeed time and space independent: $\psi(x, t)\psi^*(x, t) = A^2$.

The above result looks perfectly normal when applied to sound and light. However, how about electrons? As mentioned in Section 1.5.1, electron diffraction and interference demonstrate beyond any doubt that the electrons can behave as waves as much as the light does. However, if Eq. (12.5) is applied to an electron, the fact that this electron would not be localized in space appears really confusing. This may be mathematically possible, but we know for a fact that there are electrons that are very much localized inside our devices. Obviously, Eq. (12.5) is too simple to account for this case.

There are many ways of making Eq. (12.5) more sophisticated in an attempt to come up with a wave function that could describe our localized electron. But what is the chance of guessing the right function? In addition, there may be different wave functions to describe electrons in different conditions! Should we try the magic rule from mathematics that says that any function can be approximated by a polynomial, or another superposition-type function such as the Taylor series? Looking at Eq. (12.5), we notice that it looks like a single term (harmonic) of what is known as the Fourier series. Therefore, our needed function $\psi(x, t)$ can be approximated as

$$\psi(x, t) \approx \sum_{n=-N}^{N} \underbrace{A\Delta k f(k_n)}_{A_n} e^{-j(\omega_n t - k_n x)} \tag{12.9}$$

What is happening here is a superposition of cosine waves of different frequencies (wave numbers) and amplitudes. Physically, this means we need a large number $(2N)$ of "cosine wave machines" described earlier (Figs. 12.1 and 12.2), operating at different angular frequencies ω_n and maximum amplitudes A_n. In addition to that, we need to place all these "machines" close to each other so that they have a perfectly blended effect on the air, or any other set of particles.

Mathematically, we can push things even further and take the largest possible N (which is ∞), and the smallest possible Δk (which is dk). In this case, the sum in Eq. (12.9) becomes an integral:

$$\psi(x, t) = A \int_{-\infty}^{\infty} f(k)e^{-j(\omega t - kx)} \, dk \tag{12.10}$$

The wave function $\psi(x, t)$ expressed in this way is called a *wave packet*, where $f(k)$ is a *spectral function* specifying the amplitude of the harmonic wave with wave number k and angular frequency $\omega = vk$.

There are so many things naturally distributed according to the normal, or Gaussian distribution, that it seems practically the most relevant to take the normal distribution as an example of the spectral function $f(k)$:

$$f(k) = \frac{1}{\sqrt{2\pi}\,\sigma_k} \exp\left[-\frac{(k-k_0)^2}{2\sigma_k^2}\right] \tag{12.11}$$

This is a bell-shaped function centered at $k = k_0$ with width σ_k. In terms of our "cosine wave machines," this means that most of the "machines" would be producing waves with frequencies close to $\omega_0 = vk_0$, with the number of "machines" decaying as the frequency difference from ω_0 increases.

The integral in Eq. (12.10), with the normal distribution as the spectral function $f(k)$, can be explicitly performed, leading to the following form of the wave function:

$$\psi(x,t) = Ae^{-(\sigma_k^2/2)(vt-x)^2}e^{-j(\omega_0 t - k_0 x)} \tag{12.12}$$

The real part of this function,

$$Re\{\psi(x,t)\} = Ae^{-(\sigma_k^2/2)(vt-x)^2}\cos(\omega_0 t - k_0 x) \tag{12.13}$$

is plotted in Fig. 12.3. We can see that our normally distributed "cosine wave machines" produced a kind of localized wave packet that travels in the x-direction. If our membranes were generating sound, this time we would not hear a constant sound level at any x point. If the wave packet was light, we would not see the same light intensity at any x point. Standing at a single x point, we would hear or see the wave packet passing by us as a lump of sound or light, traveling with speed v.

Obviously, the concept of wave packet is a lot more general than the single harmonic wave, and appears as a tool that can model the wave–particle duality. Let us clarify this point further by assuming that the wave packet of Fig. 12.3 represents an electron. Imagine first that the electron is to pass through a slit the width of which is equal to 200 wavelengths λ. As Fig. 12.4b illustrates, the electron would get through this slit not even noticing it. This electron appears as a genuine particle. However, if we make the slit narrow, as in Fig. 12.4c, the electron would appear as a pure wave.

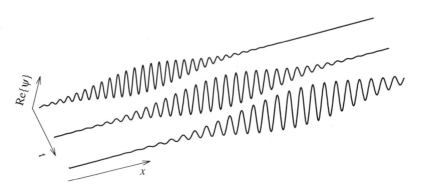

Figure 12.3 Illustration of a wave packet, traveling in the x-direction.

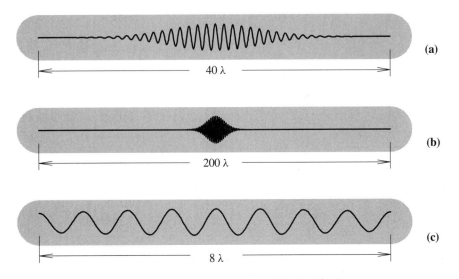

Figure 12.4 A wave packet (a) can model the wave–particle duality: it appears as a particle in an environment characterized by dimensions much larger than its width (b), while it appears as a wave when the dimensions shrink down (c).

12.2 HEISENBERG UNCERTAINTY PRINCIPLE

Playing with wave packet and slit widths becomes really interesting when we notice that the space dimension of the wave packet depends on the width of our spectral function, σ_k. In Eq. (12.12), it is the term $A \exp[-(\sigma_k^2/2)(vt - x)^2]$ that determines the shape and the width of the wave packet in space. This is again the familiar Gaussian bell-shaped profile, centered at $x = vt$, with a width of $\Delta x = 1/\sigma_k$. Figure 12.5 correlates the space width of the wave packet, Δx, with the width of the spectral function, σ_k.

We observe that we can better concentrate the wave packet in space if we use a wider spectral function. Labeling the spectral function width as $\Delta k = \sigma_k$, the following simple relationship can be written:

$$\Delta x \, \Delta k = 1 \tag{12.14}$$

Obviously, Δx can be made narrow enough to fit as comfortably in the 8λ slit of Fig. 12.4c as the wave packet of Fig. 12.4b fits in the 200λ slit. **Does this mean the increase in spectral function width (Δk) makes the wave packet look more like a particle?**

To answer this important question, we need to specify first the characteristic properties of a particle. In other words, the answer to the above question would depend on the exact definition of a particle. If all that we want from a particle is to be as localized in space as possible, then the answer is affirmative. However, from the position of classical physics we expect a specific momentum $p = m_0 v$ to be associated with any particle. The momentum appears as fundamental a property of a particle as the wavelength is for

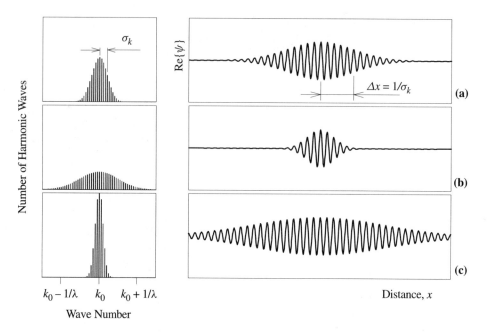

Figure 12.5 Relationship between the wave packet width in space, Δx, and the width of the corresponding spectral function, σ_k.

a wave. The acceptance of light as particle meant that there was a momentum associated with waves:

$$p = \frac{h}{\lambda} = \frac{h\nu}{c} \tag{12.15}$$

where $\nu = 1/T = \omega/2\pi$ is the light frequency, c is the light speed, and h is Planck's constant. De Broglie generalized this link between the momentum and wavelength to associate wavelengths with what was perceived as classical particles [Eq. (1.56)]. Eliminating λ from Eqs. (12.4) and (12.15), the following link between the momentum p and the wave number k is established:

$$p = \frac{h}{2\pi}k = \hbar k \tag{12.16}$$

where $\hbar = h/2\pi$.

The direct link between the momentum and the wave number has important implications in our attempt to make the wave packet look more like a particle by widening the wave number distribution. According to Eqs. (12.14) and (12.16), we can achieve a better space localization of the wave packet (a smaller Δx) only at the expense of a wider momentum distribution. We can achieve a perfect space localization of our particle ($\Delta x = 0$), but we would completely lose the information on its momentum ($\Delta p = \infty$). This is hardly a perfect particle. On the other hand, we can fix the momentum ($\Delta p = 0$), which means fixing the wave number and the wavelength [$p = (h/2\pi)k = h/\lambda$]. This is the case of the single harmonic wave introduced at the beginning of this chapter [Eq. (12.5) and Fig. 12.2], which lacked any localization in space ($\Delta x = \infty$). It definitely looks more like a perfect wave than a perfect particle.

There is no a perfect particle with a precisely determined momentum and a precisely determined space position.

This general fact, applicable to any object, is known as the Heisenberg uncertainty principle. It is more frequently expressed in the following way:

The uncertainties in the position and momentum of a particle at any instant must have their product greater than Planck's constant:

$$\Delta x \Delta p \geq \frac{\hbar}{2} \tag{12.17}$$

There may be some reservation related to the choice of the Gaussian spectral function. In other words, you may think there could be another spectral function that would give a smaller $\Delta x \Delta p$ product. However, it is exactly the Gaussian wave packet that has the smallest spread in space and momentum; any other function leads to even more pronounced uncertainty.

You may have a problem accepting that all these, wave packets, wave functions, and the uncertainty principle in particular, are so general that they apply to any object (particle or wave). In this case you will probably like the following problem: let us take a ball with mass $m_0 = 1$ kg, and let us put it in the center of a football field. The ball is stationary, there is no momentum $p = m_0 v$, therefore there is no momentum uncertainty Δp. The problem with the uncertainty principle, you may think, is that the ball remains stationary in the center, so there is no Δx uncertainty either. This may seem like sound reasoning, but it is too descriptive; let us therefore take some numbers. How well is our center defined? Is it within $\Delta x = 1$ μm, or better? For a reference, let me remind you that the size of a silicon atom is about 0.2 nm $= 2 \times 10^{-10}$ m. Let us assume a better precision than we could ever achieve: $\Delta x = 10^{-10}$ m. According to the uncertainty principle, this means $\Delta p > \hbar/\Delta x = 1.05 \times 10^{-34}/(2 \times 10^{-10}) = 5.25 \times 10^{-25}$ kg m/s. As the mass of the ball is 1 kg, its speed uncertainty is $\Delta v > 5.25 \times 10^{-25}$ m/s. To declare that the uncertainty principle does not apply, you will have to ensure that your ball is not moving with this speed. Let us say you could invent a measurement system that could record when it moved for 1 nm. You would still need to wait for 1 nm/$(5.25 \times 10^{-25}$ m/s$) = 1.9 \times 10^{15}$ s, which is more than 60 million years!

The aim of the above example is to demonstrate that no "common sense" concept, developed from everyday observations of macroobjects and particles, can demonstrate that the uncertainty principle is not generally valid. However, we made it clear at the beginning of this chapter that quantum mechanics was needed to explain and predict the behavior only of small things. It is certainly true that quantum mechanics effects are not important when we deal with large and heavy objects.

Going back to small things, let us take a reasonable number, 5 nm, as the width of a device channel in which some electrons are confined. The position uncertainty is obviously not worse than $\Delta x = 5$ nm. The corresponding momentum uncertainty is $\Delta p > 1.05 \times 10^{-34}/(2 \times 5 \times 10^{-9}) = 1.05 \times 10^{-26}$ kg m/s, which corresponds to an approximate speed uncertainty of $\Delta v = 1.05 \times 10^{-26}/9.1 \times 10^{-31} = 1.15 \times 10^5$ m/s. This is by no means negligible; actually, it is comparable to the thermal velocity of electrons in a silicon crystal.

We have used oscillating membranes ("wave packet machines") and sound to illustrate the wave packet. If applied to sound, everything looks clear: (1) Eq. (12.12) is the complex wave function of the sound packet, (2) Eq. (12.13) is the real instantaneous amplitude of the sound (plotted in Figs. 12.3–12.5), and (3) the intensity of the sound is

$$|\psi(x,t)|^2 = \psi(x,t)\psi^*(x,t) = A^2 e^{-\sigma_k^2(vt-x)^2} \tag{12.18}$$

The intensity of this sound packet, plotted in Fig. 12.6, is not too hard to understand: there are air particles that vibrate within a localized domain, producing a sound packet with sound intensity as in Fig. 12.6. The sound packet may appear as either a particle or a wave, depending on its relative size. In this example, the wave function is applied to an ensemble of particles. However, **what is the meaning of the intensity $|\psi(x,t)|^2$ when the wave function is applied to a single particle, such as a single electron?**

The answer to this question is in the relationship between the concepts of probability and statistics. It appears perfectly meaningful to say there are 2×10^{19} air particles per cm^3. Let us assume that an extremely good vacuum can be achieved, so good that there is only one particle left in a room of $200 \times 300 \times 167$ cm $= 10^7$ cm^3. In this case, we can say only that there are $1/10^7 = 0.0000001$ particles per cm^3. What does this mean? The answer is: *probability* to find this particle in a specified volume of 1 cm^3. Let us assume that our membranes ("wave packet machines") are installed in this room with the single particle inside. This will affect the probability of finding the particle in our specified 1 cm^3, depending on the specific wave function intensity $|\psi(x,t)|^2$.

If the wave function $\psi(x,t)$ is to be a representation of a single particle (electron), then the intensity $\psi(x,t)\psi^*(x,t)$ relates to the probability of finding this particle (electron) at point x and time t.

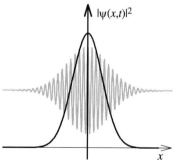

Figure 12.6 Illustration of wave packet intensity.

12.3 SCHRÖDINGER EQUATION

We have used a specific wave function $\psi(x, t)$ to introduce the concept of a wave packet and to demonstrate that it can model the particle–wave duality. Although this specifically selected wave function [Eq. (12.12)] provides the smallest spread in space and momentum, it can by no means be used for any object in any situation. Obviously, we need to be able to find somehow the wave function that would specifically model a particular object (say an electron) in specific conditions (say a MOSFET with a certain electrical bias).

In 1926, Schrödinger postulated a differential equation that results in the needed wave function when solved with appropriate initial and/or boundary conditions. To obtain as good an understanding of the Schrödinger equation as possible, we will relate it to the continuity equation. The continuity equation was introduced and applied to the case of particle flow in Section 1.3.4 [Eq. (1.32)]. Subsequently it was applied to the electric flux density in order to introduce the Poisson equation in Section 2.3.1. Let us write a general form of the continuity equation (1.32) by replacing the particle concentration N with *something* abstract, labeled by ψ. In this case the particle flow (current) J has to be replaced by the flow of this *something*, which we label by ι:

$$\frac{\partial \iota}{\partial x} = -\frac{\partial \psi}{\partial t} \tag{12.19}$$

This equation means that if more of *something* flows out then into a considered point ($\partial \iota / \partial x > 0$), then this *something* decreases at that point ($-\partial \psi / \partial t$). This is a kind of conservation principle that applies to the wave function $\psi(x, t)$. This conservation principle can be related to the fact that particle concentration at point x can be decreased only if some of the particles flow somewhere else.

Let us again take the example of our bell-shaped wave packet, this time appearing somewhere in space free of any external influence. Can there be any flow ι in this case? If it can, then it can happen only by a kind of *diffusion* process in which the sharpness of the bell-shaped profile is being reduced until it becomes flat. The diffusion current is modeled by Eq. (1.25), which in the generalized form indicates that the flow of *something* ι is proportional to the gradient of this *something*:

$$\iota \propto \frac{\partial \psi}{\partial x} \tag{12.20}$$

Eliminating the flow ι from Eqs. (12.19) and (12.20), a generalized Fick's equation [Eq. (1.34) in the case of particle concentration] is obtained:

$$\frac{\partial \psi}{\partial t} \propto \frac{\partial^2 \psi}{\partial x^2} \tag{12.21}$$

This equation is analogous to the Schrödinger equation applied to a free particle represented by an arbitrary wave packet $\psi(x, t)$:

$$i\hbar \frac{\partial \psi(x, t)}{\partial t} = -\frac{\hbar^2}{2m} \frac{\partial^2 \psi(x, t)}{\partial x^2} \tag{12.22}$$

where m is the particle mass.

Equation (12.22) can be generalized to describe the motion of a particle in a force field represented by a potential energy $E_{pot}(x)$[2]:

$$j\hbar \frac{\partial \psi(x, t)}{\partial t} = -\frac{\hbar^2}{2m} \frac{\partial^2 \psi(x, t)}{\partial x^2} + E_{pot}(x)\psi(x, t) \tag{12.23}$$

This is the one-dimensional time-dependent Schrödinger equation. The three-dimensional form can be written as

$$j\hbar \frac{\partial \psi}{\partial t} = -\frac{\hbar^2}{2m} \nabla^2 \psi + E_{pot}\psi \tag{12.24}$$

where $\nabla^2 = \partial^2/\partial x^2 + \partial^2/\partial y^2 + \partial^2/\partial z^2$, $\psi = \psi(x, y, z, t)$, and $E_{pot} = E_{pot}(x, y, z)$. However, in the following we will limit ourselves to the one-dimensional case.

The variables of the time-dependent Schrödinger equation [x and t in Eq. 12.23)] can be separated if the following form of the wave function is used: $\psi(x, t) = \psi(x)\chi(t)$. In this case, Eq. (12.23) can be transformed into

$$j \frac{\hbar}{\chi(t)} \frac{\partial \chi(t)}{\partial t} = -\frac{\hbar^2}{2m\psi(x)} \frac{\partial^2 \psi(x)}{\partial x^2} + E_{pot}(x) \tag{12.25}$$

The left-hand side of this equation is a function of time alone, while the right-hand side is a function of position alone. This is possible only when the two sides are equal to a constant. The constant is in the units of energy, and it actually represents the total energy E_{tot}:

$$j \frac{\hbar}{\chi(t)} \frac{\partial \chi(t)}{\partial t} = E_{tot}$$

$$-\frac{\hbar^2}{2m\psi(x)} \frac{\partial^2 \psi(x)}{\partial x^2} + E_{pot}(x) = E_{tot} \tag{12.26}$$

Obviously, the time-independent wave function $\psi(x)$ has to satisfy the following *time-independent Schrödinger equation*:

$$-\frac{\hbar^2}{2m} \frac{d^2 \psi(x)}{dx^2} + E_{pot}(x)\psi(x) = E_{tot}\psi(x) \tag{12.27}$$

In the following, we will use Eq. (12.27) to find the wave function of electrons in some characteristic situations.

12.3.1 Electron in a Potential Well

The behavior of electrons in a potential well is of great practical importance. As mentioned in Section 1.5.1, the atom electrons are confined inside a funnel-shaped potential well. Also,

---

[2]This is analogous to using current with drift, in addition to the diffusion term, in the continuity equation [Eq. (8.27)].

the electrons in a MOSFET or HEMT channel appear in a triangular potential well. We will use here the simplest case, which is the potential well with infinite walls, to illustrate the important aspects of electrons in a potential well.

The infinite potential well, illustrated in Fig. 12.7a, is mathematically defined as

$$E_{pot}(x) = \begin{cases} 0 & \text{for } 0 < x < W \\ \infty & \text{for } x \leq 0 \text{ and } x \geq W \end{cases} \quad (12.28)$$

Consequently,

$$\frac{d^2\psi(x)}{dx^2} + \frac{2m}{\hbar^2}E_{tot}\psi(x) = 0 \quad \text{for } 0 < x < W$$

$$\psi(x) = 0 \quad \text{for } x \leq 0 \text{ and } x \geq W \quad (12.29)$$

Mathematically, this problem is reduced to solving the Schrödinger equation for a free particle ($E_{pot} = 0$) with the following boundary conditions: $\psi(0) = 0$ and $\psi(W) = 0$. The solution of this type of differential equation can be expressed as

$$\psi(x) = A_+ e^{s_1 x} + A_- e^{s_2 x} \quad (12.30)$$

where $s_{1,2}$ are the roots of its characteristic equation[3]:

$$s^2 + \underbrace{\frac{2m}{\hbar^2}E_{tot}}_{k^2} = 0 \quad (12.31)$$

As the solution of the characteristic equation $s^2 = -k^2$ is $s_{1,2} = \pm jk$, the general solution is expressed as

$$\psi(x) = A_+ e^{jkx} + A_- e^{-jkx} \quad (12.32)$$

Comparing this wave function to the time-independent part of Eq. (12.5), we can see that it consists of two plane waves traveling in the opposite directions: (1) $A_+ \exp(jkx)$, traveling in the positive x-direction, and (2) $A_- \exp(-jkx)$, traveling in the negative x direction. Equation (12.32) represents the general solution for the electron wave function in free space ($E_{pot} = 0$), as the superposition of the two plane waves can account for any possible situation in terms of boundary conditions. When the electron is trapped between two walls, both plane waves are relevant as the electron is reflected backward and forward by the walls. Steady states are still possible, but only in the form of standing waves, as illustrated in Fig. 12.7b. Obviously, the standing waves are formed only for specific values of $k = 2\pi/\lambda$, which correspond to integer multiples of half-wavelengths $\lambda/2$. Noting that $k = 2\pi/\lambda$ is determined by the total energy E_{tot}, according to Eq. (12.31), the following important conclusion is reached:

---

[3]s^2 represents the second derivative, $d^2\psi(x)/dx^2$, while $\psi(x)$ itself is represented by $s^0 = 1$.

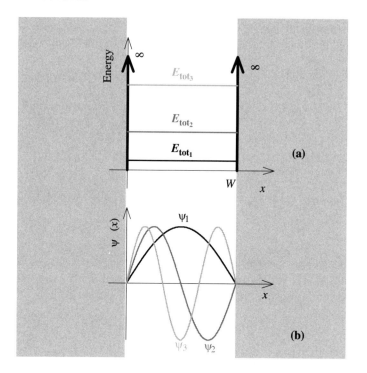

Figure 12.7 A particle in an infinite potential well: the possible energy levels (a) correspond to standing waves with integer half-wavelength multiples (b).

The electrons in a potential well cannot have any value of total energy. Only specific energy values are possible (Fig. 12.7a). This effect is called *energy quantization*.

The two constants in the general solution (12.32), A_+ and A_-, can be determined so to obtain the particular wave function $\psi(x)$ representing our specifically defined case (electrons inside a potential well of width W and infinitely high walls). Applying the boundary condition $\psi(0) = 0$, we find the following relationship between the constants A_+ and A_-:

$$A_+ e^0 + A_- e^0 = 0 \Rightarrow A_+ = -A_- \tag{12.33}$$

After this, Eq. (12.32) can be transformed as

$$\psi(x) = A_+ \cos(kx) + jA_+ \sin(kx) - A_+ \underbrace{\cos(-kx)}_{=\cos(kx)} - jA_+ \underbrace{\sin(-kx)}_{=-sin(kx)}$$

$$= \underbrace{2jA_+}_{A} \sin(kx) \tag{12.34}$$

where A is a new constant, replacing the constant A_+.

The second boundary condition is $\psi(W) = 0$. With the exception of the trivial case $A = 0$, no other value for A in Eq. (12.34) can satisfy this boundary condition. However, $\sin(kW)$ can be zero for a number of kW values, which means that the possible solutions are determined by specific (and discrete) k values. As k is related to the total energy [Eq. (12.31)], this means that the possible solutions are determined by specific (and discrete) energy values. This is the same conclusion related to the *energy quantization* effect, only it is reached in a different way. Given that $\sin(kW) = 0$ for $kW = n\pi$, where $n = 1, 2, 3, \ldots$, the possible energy levels are found as

$$E_{\text{tot}n} = n^2 \frac{\pi^2 \hbar^2}{2m W^2} \qquad (n = 1, 2, 3, \ldots) \tag{12.35}$$

These energy levels are illustrated in Fig. 12.7a.

There is an additional condition that has to be satisfied by the wave function that is to represent electrons in the potential well. It was explained earlier that $|\psi(x)|^2 = \psi(x)\psi^*(x)$ is the probability of finding an electron at x. If an electron is trapped inside the potential well, then it has to be somewhere between 0 and W, which means

$$\int_0^W \psi(x)\psi^*(x)\, dx = 1 \tag{12.36}$$

This is called the normalization condition. The normalization condition determines a specific value of the constant A, as

$$\int_0^W A^2 \left(\sin \frac{n\pi}{W}x \right)^2 dx = 1 \Rightarrow A = \sqrt{\frac{2}{W}} \tag{12.37}$$

Therefore, the final solution is

$$\psi_n(x) = \sqrt{\frac{2}{W}} \sin \frac{n\pi}{W}x \qquad (n = 1, 2, 3, \ldots) \tag{12.38}$$

where $\psi_1(x), \psi_2(x), \psi_3(x), \ldots (n = 1, 2, 3, \ldots)$ represent different electron states (electrons at different energy levels $E_{\text{tot}n}$). The wave functions $\psi_1(x)$, $\psi_2(x)$, and $\psi_3(x)$ are illustrated in Fig. 12.7b.

The example of an infinite potential well qualitatively explains the appearance of a two-dimensional electron gas as a result of energy quantization in the HEMT channel (Section 9.4.1). However, Eq. (12.35) cannot be used to calculate the energy levels, because the potential well of Fig. 12.7a is not a precise approximation of the triangular potential well appearing in the HEMT channel (Fig. 9.11). The triangular potential well can be defined as

$$E_{\text{pot}} = \infty \qquad \text{for } x \le 0$$
$$E_{\text{pot}} = q E_s x \qquad \text{for } x > 0 \tag{12.39}$$

where E_s is the electric field at the surface of the semiconductor. In this case, the following equation has to be solved:

$$-\frac{\hbar^2}{2m^*}\frac{d^2\psi(x)}{dx^2} + qE_sx\psi(x) = E_{tot}\psi(x) \qquad \text{for } x > 0 \qquad (12.40)$$

where m^* is the effective mass of electrons. Wave functions in the form of Airy functions[4] satisfy Eq. (12.40). The numerical evaluation of the Airy functions yields a complicated series, however, approximate values for the zeros lead to the following approximate equation for the energy levels[5]:

$$E_{tot\,n} \approx \left(\frac{\hbar^2}{2m^*}\right)^{1/3}\left[\frac{3\pi q}{2}\left(n - \frac{1}{4}\right)E_s\right]^{2/3} \qquad (n = 1, 2, 3, \ldots) \qquad (12.41)$$

12.3.2 Tunneling

Another important quantum-mechanical effect, observed in semiconductor devices, is tunneling (Section 3.3.3). To illustrate the effect of tunneling, let us find the wave function $\psi(x)$ for the case when electrons are approaching a potential-energy barrier, as in Fig. 12.8. This potential barrier is mathematically defined as

$$E_{pot}(x) = \begin{cases} 0 & \text{for } x < 0 \\ E_{pot} & \text{for } 0 \le x \le W \\ 0 & \text{for } x > W \end{cases} \qquad (12.42)$$

Writing the Schrödinger equation in the following forms:

$$\frac{d^2\psi(x)}{dx^2} + \underbrace{\frac{2m}{\hbar^2}E_{tot}}_{k^2}\,\psi(x) = 0 \qquad \text{for } x < 0 \text{ and } x > W \qquad (12.43)$$

and

$$\frac{d^2\psi(x)}{dx^2} - \underbrace{\frac{2m}{\hbar^2}(E_{pot} - E_{tot})}_{\kappa^2}\,\psi(x) = 0 \qquad \text{for } 0 \le x \le W \qquad (12.44)$$

the general solution can be expressed as

$$\psi(x) = \begin{cases} A_+e^{jkx} + A_-e^{-jkx} & \text{for } x < 0 \\ B_+e^{\kappa x} + B_-e^{-\kappa x} & \text{for } 0 \le x \le W \\ C_+e^{jkx} + C_-e^{-jkx} & \text{for } x > W \end{cases} \qquad (12.45)$$

Again, the electron wave function appears as a superposition of two plane waves traveling in the opposite directions when $E_{pot} = 0$ [the first and third rows in Eq. (12.45)]. In the region where $E_{pot} \neq 0$, two different cases have to be considered: $E_{tot} > E_{pot}$ and $E_{tot} < E_{pot}$.

---

[4]Special functions that are combinations of Bessel and modified Bessel functions

[5]*Source*: T. Ando, A.B. Fowler, and F. Stern, "Electronic properties of two-dimensional systems," *Rev. Modern Phys.*, **54**, 437–672 (1982).

Figure 12.8 Illustration of tunneling.

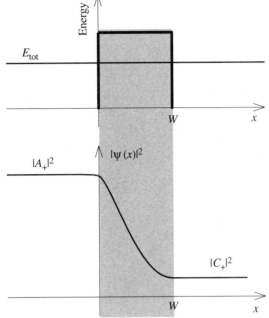

CASE 1: $E_{tot} > E_{pot}$. In this case, classical mechanics predicts that a particle with energy E_{tot} should go over the lower barrier, E_{pot}, without any interference. However, the wave function inside the barrier region ($B_+e^{\kappa x}$ if we limit ourselves to the wave traveling in the positive x-direction) is different from the incident wave (A_+e^{jkx}). As $\kappa^2 = (2m/\hbar^2)(E_{pot} - E_{tot})$ is negative, κ can be expressed as $\kappa = jk_E$, where $k_E = \sqrt{(2m/\hbar^2)(E_{tot} - E_{pot})}$ is the wave number inside the barrier region. This means that the incident plane wave continues to travel as a plane wave through the barrier region ($B_+e^{jk_Ex}$), however with an increased wavelength $\lambda_E = 2\pi/k_E$ ($k_E < k$). In addition to that, the intensity of the wave ($|B_+|^2$) is reduced as there is a finite probability ($|A_-|^2$) that the particle is reflected by the barrier.

CASE 2: $E_{tot} < E_{pot}$ *(Tunneling)*. Classical mechanics predicts that a particle cannot go over a potential barrier that is higher than the total energy of the particle. However, the wave function of the particle does not suddenly vanish[6] when it hits the barrier as $B_+e^{\kappa x} \neq 0$. This time, $\kappa = -\sqrt{(2m/\hbar^2)(E_{pot} - E_{tot})}$ is a real number, and $\exp(\kappa x)$ causes an exponential decay of the wave function intensity in the barrier region. This is illustrated in Fig. 12.8. If the barrier is wide, so that $W \gg 1/|\kappa|$, the wave function would practically drop to zero inside the potential barrier. This means that $C_+ \approx C_- \approx 0$. However, if the potential barrier is not as wide, the wave function does not drop to zero at the end of the

---

[6]In the specific case of $E_{pot} = \infty$ (infinite barrier height), the wave function is equal to zero inside the barrier region as $\kappa = -\sqrt{(2m/\hbar^2)(E_{pot} - E_{tot})} = -\infty$ and $\exp(-\infty x) = 0$.

potential barrier ($x = W$), which means that $C_+ \neq 0$. There is a finite probability ($|C_+|^2$) of finding the particle beyond the barrier ($x > W$), which is the tunneling effect. Probability $|C_+|^2$, normalized by the probability of a particle hitting the barrier ($|A_+|^2$), defines the *tunneling probability*, also called the *tunneling coefficient*.

Applying appropriate boundary conditions,[7] it is possible to express the tunneling probability $T = |C_+|^2/|A_+|^2$ in terms of the barrier height and width, and the mass and energy of the particle. However, many practical barriers are not well represented by this simple rectangular tunneling barrier. There is an approximation method, known as the WKB[8] method, which can be used to determine the tunneling probability for the case of an arbitrary barrier. An arbitrary energy barrier $E_{pot}(x)$ is represented by

$$\kappa(x) = \sqrt{\frac{2m}{\hbar^2}[E_{pot}(x) - E_{tot}]} \qquad \text{for} \quad \begin{array}{c} E_{tot} < E_{pot} \\ a \leq x \leq b \end{array} \qquad (12.46)$$

where a and b are the points at which the particle energy E_{tot} intersects the tunneling barrier $E_{pot}(x)$. According to the WKB approximation, the tunneling probability is

$$T \approx \exp\left[-2\int_a^b \kappa(x)\,dx\right] \qquad (12.47)$$

Obviously, the tunneling probability exponentially decays with $b - a$ (which is the barrier width) and $\sqrt{E_{pot}}$.

REVIEW QUESTIONS

R-12.1 Is the theory of quantum mechanics applicable only to small things?

R-12.2 Do large particles possess wave-like properties? Can they be observed (like "diffraction of a soccer ball")?

R-12.3 A sum of single-harmonic wave functions with distributed wave vectors creates a wave packet, "localized" in space. Are the two "parameters" of the wave packet, the width of the wave vector distribution Δk and the space localization Δx, independent?

R-12.4 One way of expressing the Heisenberg uncertainty principle is that we cannot measure the position of an electron with better accuracy than $\Delta x = \hbar/2\Delta p$, if the momentum is determined with precision Δp. Is this a fundamental limitation of our measurement techniques, or is it simply the way the small things are?

R-12.5 Is the Heisenberg principle at all related to the "dual" wave–particle behavior of electrons? If yes, how?

---

[7] The wave function and its derivative have to be continuous at each interface

[8] Wentzel, Kramers, and Brillouin

R-12.6 Solving the Schrödinger equation one obtains the electron wave function. Is this the function that describes the "shape" of the wave packet?

R-12.7 What is the wave function good for in semiconductor-device electronics: can it help to determine the electron concentration and/or the electron mobility?

R-12.8 Does solving the Schrödinger equation provide any other information about the electrons, such as the electron energy?

R-12.9 The probability that an electron wave will be reflected by an energy barrier is 0.9. What is the remaining probability of 0.1?

R-12.10 Is the electron wave function related to the tunneling probability?

R-12.11 Is there any relationship between the Schrödinger equation and the WKB method for determining of the tunneling probability? If yes, which method (WKB or solving the Schrödinger equation) would you use to find the tunneling probability for the triangular barrier created by the semiconductor–oxide interface and the gate-oxide field in a MOSFET?

Appendix *A*

BASIC INTEGRATED-CIRCUIT CONCEPTS AND ECONOMICS

Fabrication of modern integrated circuits (ICs) requires more than 10 photolithography sequences and more than 200 processing steps in total. The fabrication involves very sophisticated and expensive equipment, the purest materials ever developed, precisely defined process conditions, and the most rigorous process control procedures. Yet a single IC rarely costs more than $1000, and some ICs can cost less than a dollar to manufacture. This is possible because thousands of ICs are processed together. Although the size of an individual IC is rarely larger than 1×1 cm^2, the processing is performed on *wafers*, which in the case of silicon can be as large as 30 cm in diameter. Some of the processing steps, like mask aligning during photolithography, are performed on single wafers. However, most of the processing (cleaning, etching, diffusion, etc.) is performed on batches of 25 or more wafers.

Figure A.1 illustrates a wafer containing hundreds of *chips*. The complete pattern (layout) of the IC being manufactured is replicated on every chip, excluding the test chip sites. The test chips contain specifically designed test devices (test structures), which are used for in-line process controls.

Regardless of the extreme cleanliness of the fabrication environment and materials used, and regardless of the strict processing conditions, defects still occur during processing.

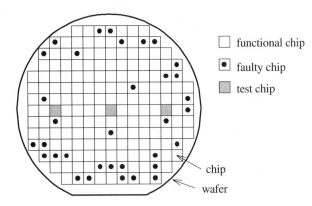

□ functional chip

⊡ faulty chip

▨ test chip

chip

wafer

Figure A.1 Wafers containing many IC chips are used for semiconductor processing to enable mass production.

517

A single defect, leading to an open circuit, short circuit, or excessive leakage, is enough to render the whole chip (IC) faulty. The ratio between the number of chips functioning properly at the end of chip processing and the total number of chips is called *chip yield*. Obviously, a low yield processing results in increased production cost per functional chip, as the functional chips share the cost of processing the chips that will turn out to be faulty. At the end of chip processing, every chip is electrically tested on the wafer by a process known as *electrical probing*. The chips that do not function properly are marked, usually by a red ink dot.

After electrical probing, the wafers are cut into individual chips to enable assembly into individual IC packages. The following are the main steps of the assembly process:

- Chip attachment to IC carriers (also known as die attachment).
- Connection of externally accessible *IC pins* to circuit terminals on the chip, known as *bonding pads*, by a thin golden wire. This process step is called *bonding*.
- Chips encapsulation, usually in a plastic package. A layer of glass, surrounding the externally accessible IC pins, is used to join the IC carrier and a cap in the case of ceramic packages. Ceramic packaging is more expensive, and is used for high-reliability applications.
- Labeling.
- Final testing, to detect and remove the ICs not functioning properly at the end of the manufacturing process. *Assembly yield* expresses the fraction of properly functioning ICs at the end of the assembly process.

The assembly process is expensive, as the ICs have to be processed individually. The assembly cost can dominate the final IC cost. This explains why electrical probing at the end of wafer processing is necessary. Of course, the chips marked as nonfunctional at that stage do not enter the assembly phase.

Integrated circuits can be classified into two general categories: *fixed-function* (sometimes called *off-the-shelf*) and *application-specific ICs* (ASICs).

Fixed-function ICs are designed by their manufacturers to meet the needs of a wide variety of applications. Examples of digital fixed-function ICs are 7400-series and 4000-series of logic circuits, while an example of a linear fixed-function IC is the operational amplifier. ASICs are designed by the end user to meet the specific requirements of a circuit and are usually produced by an IC manufacturer as specified by the end user.

Design of electronic circuits using fixed-function ICs together with discrete devices can be referred to as the classical approach to electronics circuit design. This approach divides the electronics engineers into *integrated-circuit makers* and *integrated-circuit users*. The application-specific ICs frequently appear as a better alternative, especially in terms of ever increasing demand for miniaturized electronic circuits, reduced cost, reduced power consumption, increased reliability, etc.

There are several types of ASICs: programmable logic devices (PLD), gate and analog arrays, standard cell ICs, and full-custom ICs. PLDs consist of a range of logic gates that can automatically be interconnected to perform a desired function. In the case of gate arrays, prefabricated logic gates are interconnected by one or more customized masks. Similar to the digital gate arrays, analog arrays contain either a range of prefabricated basic analog circuits (such as amplifiers, current sources, etc.) or discrete devices, which are interconnected into an analog circuit by a customized mask. With the standard cell approach, the IC designer extracts circuit cells from a software library and composes them on the masks to form the final layout of the IC. In full custom ICs, the customer designs all details of the integrated circuit.

The choice of a particular type of ASIC will mainly depend on the estimated cost per integrated circuit, which further depends on the number of ICs being produced. It is obvious that in the case of high design cost, as in full-custom ICs, a large number of ICs is required to share this cost. With a large number of produced units, the cost of full-custom designed ICs can become the lowest available as full-custom design enables a fully optimized circuit. If a smaller number of units is needed, however, the design cost has to be reduced, and that is where ICs designed using standard cells or partially prefabricated gate arrays become the cheapest option. Finally, in the case of PLDs the design cost becomes almost negligible compared to the cost of the chip itself.

ASIC is the technology that gives the *users* what they need—the power to design their own ICs. ASIC technology, however, does not simply convert a traditional *IC user* into an *IC maker*. A sound understanding of integrated-circuit technologies is necessary in order to be able to design a customized or semicustomized integrated circuit.

■ Example A.1 IC Production Cost and Number of Wafers

An estimate of different costs associated with an IC production is given in Table A.1. Each wafer contains 500 chips. Determine the dependence of the integrated-circuit production cost on the number of ICs being produced (use the following numbers of wafers: 10, 100, 1000, and 10,000). Plot the dependence and comment on the results.

TABLE A.1 Different Costs Associated with IC Production

Design cost (DC)	$100,000
Wafer production cost (WPC)	$250
Assembly and final testing cost per chip ($AFTC$)	$0.40

Solution:

The integrated-circuit production cost ($ICPC$) can be expressed as follows:

$$ICPC = \underbrace{\frac{DC}{NW \times NCPW}}_{\text{design cost per chip}} + \underbrace{\frac{WPC}{NCPW}}_{\text{WPC per chip}} + AFTC$$

where NW is the number of wafers produced and $NCPW$ is the number of chips per wafer. Using the data from Table A.1, $ICPC$ dependence on the number of wafers produced is obtained as

$$ICPC = \frac{200}{NW} + 0.50 + 0.40$$

The results obtained for different numbers of wafers are plotted in Fig. A.2. It can be seen that the integrated-circuit production cost is dramatically reduced as the number of wafers produced is increased. This illustrates the need for mass production of ICs, in order to achieve low production cost.

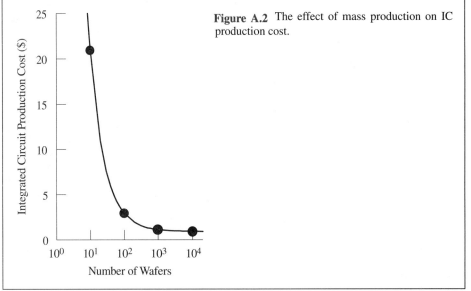

Figure A.2 The effect of mass production on IC production cost.

■ **Example A.2 Effects of Yield**

(a) It is implicitly assumed in the previous example that the yield is 100%. If the chip yield is $Y_c = 0.65$ and the assembly yield is $Y_a = 0.90$, repeat the calculations from Example A.1.

(b) A corresponding investment can increase the yield, as illustrated in Table A.2. Calculate the dependence of the IC production cost on the investments in yield increase given in Table A.2, and comment on the results. Take the example of 500,000 produced chips.

TABLE A.2 The Effects of Investment on Chip Yield

Investment	Chip Yield
$20,000	0.825
$80,000	0.93
$150,000	0.96

Solution:

(a) The yield loss increases the integrated-circuit production cost, as the functional chips have to share the cost of processing the nonfunctional chips. The yield factor can be included in the cost calculations if the $ICPC$ terms are divided by the corresponding yield data:

$$ICPC = \frac{\frac{DC}{NW \times NCPW} + \frac{WPC}{NCPW}}{Y_c} + \frac{AFTC}{Y_a}$$

The following results are obtained: \$32.0 for $NW = 10$, \$4.3 for $NW = 100$, \$1.5 for $NW = 1000$, and \$1.2 for $NW = 10,000$.

(b) The investments made (INV) add an additional term to the cost expression:

$$ICPC = \frac{\frac{DC}{NW \times NCPW} + \frac{WPC}{NCPW} + \frac{INV}{NW \times NCPW}}{Y_c} + \frac{AFTC}{Y_a}$$

For the case of $NW \times NCPW = 500{,}000$, the above expression becomes

$$ICPC = \frac{0.70 + \frac{INV}{500{,}000}}{Y_c} + 0.44$$

which gives the following results: \$1.52 before any investment is made, \$1.34 for a \$20,000 investment, \$1.36 for an \$80,000 investment, and \$1.48 for a \$150,000 investment. These results are plotted in Fig. A.3. It can be seen that there is an optimum investment in the chip yield increase, which gives the lowest cost.

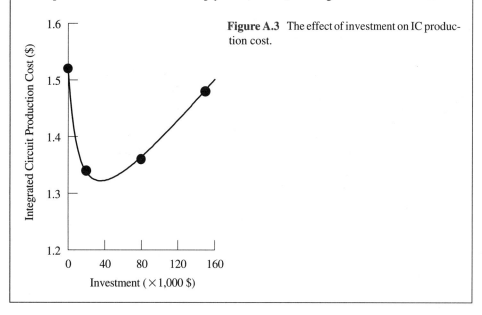

Figure A.3 The effect of investment on IC production cost.

■ Example A.3 Cost of Gate Arrays

Using gate arrays, the IC design cost can be significantly reduced. In this case an array of devices and logic gates is already designed and produced, and only the final mask defining the device interconnection has to be designed. However, this approach is not as efficient as the full-custom design in terms of the area that is occupied by the integrated circuit, and as a result the number of chips per wafer is smaller. In addition, the yield is decreased.

Repeat the calculations from Example A.2a for a gate-array-based design, using the data from Table A.3. Compare the results with the results from Example A.2a.

TABLE A.3 Different Costs Associated with an ASIC Production

Design cost (DC)	Reduced	$10,000
Wafer production cost (WPC)	Same	$250
Assembly and final testing cost per chip ($AFTC$)	Same	$0.40
Number of chips per wafer ($NCPW$)	Decreased	400
Chip yield (Y_c)	Decreased	0.55
Assembly yield (Y_a)	Same	0.90

Solution:

Using the data from Table A.3, $ICPC$ dependence on the number of wafers produced is obtained as

$$ICPC = \frac{\frac{DC}{NW \times NCPW} + \frac{WPC}{NCPW}}{Y_c} + \frac{AFTC}{Y_a} = \frac{\frac{25}{NW} + 0.625}{0.55} + 0.44$$

which gives the following results: $6.1 for $NW = 10$, $2.0 for $NW = 100$, $1.6 for $NW = 1,000$, and $1.6 for $NW = 10,000$. These results are plotted in Fig. A.4, along with the results obtained in Example A.2a for the full-custom approach. It can be seen that the gate-array approach dramatically reduces production cost for the cases of smaller number of wafers, however, it becomes more expensive when a large number of wafers are produced.

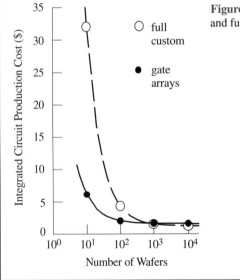

Figure A.4 ASIC production cost for gate arrays and full custom design.

Appendix *B*

CRYSTAL LATTICES, PLANES, AND DIRECTIONS

The atoms of a crystalline material are regularly placed in points that define a particular *crystal lattice*. The regularity of the atom placement means that the pattern of a *unit cell* is replicated to build the entire crystal. Figure B.1 illustrates the unit cell of a *cubic lattice*, which is the simplest three-dimensional crystal lattice. Important parameters of any crystal lattice are the lengths of the unit cell edges. Cubic crystals have unit-cell edges of the same length (a), called a *crystal lattice constant*. In general, unit cell edges can be different ($a \neq b \neq c$).

It is convenient to express the positions of atoms and different crystallographic planes

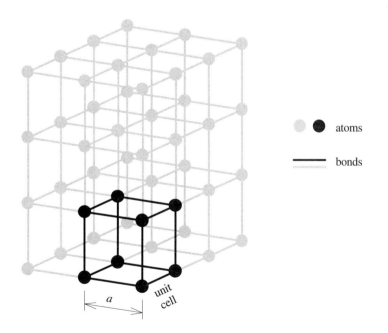

Figure B.1 Simple cubic lattice and its unit cell (a is the lattice constant).

in terms of the lengths of the unit cell edges, a, b, and c. Any plane in space may be described by the equation

$$h\frac{x}{a} + k\frac{y}{b} + l\frac{z}{c} = 1 \qquad (B.1)$$

where a/h, b/k, and c/l are intercepts of the x-, y- and z-axes respectively. In general, h, k, and l can take any value. For example, $h = k = l = 1/2$ defines the plane that intersects the x-, y-, and z-axes at $2a$, $2b$, and $2c$, respectively. However, this plane is crystallographically identical to the plane that intersects the axes at a, b, and c, in which case $h = k = l = 1$. For that reason, h, k, and l are defined as integers in the case of crystals.

Characteristic crystallographic planes are defined by a set of integers, (hkl), known as Miller indices.

In the case of negative intercepts, the minus sign is placed above the corresponding Miller index, for example $(\bar{h}kl)$.

Figure B.2 illustrates three important planes in cubic crystals. The shaded plane labeled as {100} intersects the x-axis at a, and it is parallel to the y- and z-axes (mathematically, it intersects the y- and z-axes at infinity, as $a/0 = \infty$). Analogously, the plane (010) intersects the y-axis at a and is parallel to the x- and z-axes. As there is no crystallographic difference between the (100), (010), and (001) planes, they are uniquely labeled as {100}, where the braces {} indicate that the notation is for all the planes of equivalent symmetry.

The second plane shown in Fig. B.2, labeled as {110}, intersects two axes at a and is parallel to the third axis. Finally, the {111} plane intersects all the axes at a.

In addition to planes, it is necessary to describe directions in crystals. By convention, the direction perpendicular to the (hkl) plane is labeled $[hkl]$. A set of equivalent directions is labeled as $< hkl >$, for example, $< 100 >$ represents $[100]$, $[010]$, $[001]$, $[\bar{1}00]$, etc. Figure B.3 illustrates the most important directions in cubic crystals.

In addition to the simple cubic lattice (Fig. B.1), a number of more complex lattice structures have cubic unit cells. Figure B.4 illustrates two additional cases: body-centered cubic (Fig. B.4b) and face-centered cubic (Fig. B.4c) unit cells. In the case of

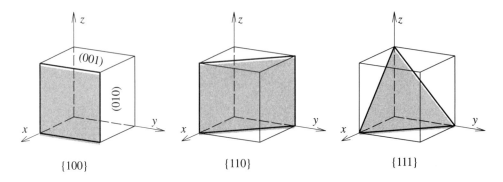

Figure B.2 Miller indices for the three most important planes in cubic crystals.

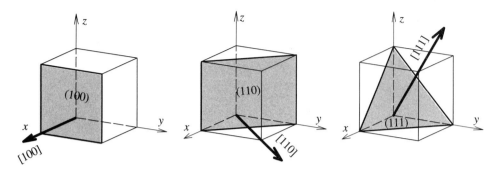

Figure B.3 Important directions in cubic crystals.

a body-centered cubic cell, there is an additional atom in the center of the cube, while in the case of a face-centered cubic cell, six additional atoms are centered on the six faces.

The most important semiconductor materials, Si and GaAs, also have cubic unit cells, although with a lot more complex structure. Figure B.5 shows the unit cell of silicon, which is the same as the unit cell of diamond, germanium, and some other crystals. Because of the fascination with diamond, this type of structure is usually referred to as a *diamond unit cell* or *diamond lattice structure*. It is important to note the tetrahedral relationship of the atoms in the small cube, where the atom in the center of the tetrahedron creates identical bonds with the four atoms in the four tetrahedron corners. This is because of the fact that these elements have four outer-shell electrons, creating four covalent bonds with the neighboring atoms. In the case of a silicon crystal, the distance between the nearest neighbors is 0.118 nm, while the crystal lattice constant (the cube edge a) is 0.5428 nm.

The unit cell of GaAs is very similar, where gallium and arsenic atoms occupy alternating sites, so that each gallium atom is linked to four arsenic atoms and vice versa. This type of lattice is referred to as a *zincblade lattice structure*.

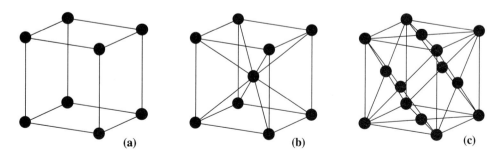

Figure B.4 Three different types of cubic unit cells: (a) simple cubic, (b) body-centered cubic, and (c) face-centered cubic.

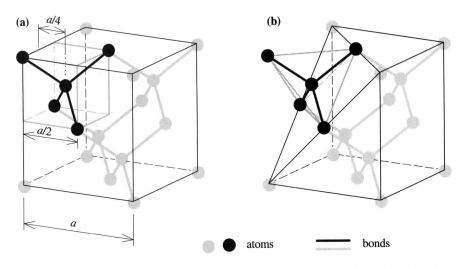

Figure B.5 Diamond unit cell, illustrated in two ways to show (a) the cubic unit cell and (b) the inherent tetrahedral structure.

Appendix *C*

HALL EFFECT AND SUMMARY OF KINETIC PHENOMENA

The motion of charge carriers under the action of internal or external fields, concentration gradients, temperature gradients, or any other internal or external conditions, leads to a number of different kinetic phenomena. Table C.1 provides a summary of the kinetic phenomena, important in terms of different applications of semiconductor devices. As can be seen, the most important phenomena are described in the main text of the book. The following text describes the Hall effect in more detail.

The Hall effect is related to the force that acts on a charged particle that moves in a magnetic field. In the general case of arbitrary velocity and magnetic field directions, the force is expressed as

$$\mathbf{F} = q\mathbf{v} \times \mathbf{B} \qquad (C.1)$$

where $\mathbf{v}$ is the particle velocity, $\mathbf{B}$ is the magnetic flux density, and the vector cross product is $\mathbf{v} \times \mathbf{B} = |\mathbf{v}||\mathbf{B}| \sin[\angle(\mathbf{v}, \mathbf{B})]$. Figure C.1 illustrates the Hall effect. Let us assume that a current of holes (I) is established through a P-type semiconductor bar so that the holes move in the y-direction with the velocity equal to drift velocity $v_d = v_y$. At time $t = 0$, a magnetic field perpendicular to the current flow, B_z, is established. As a hole enters the semiconductor bar with velocity v_y, its trajectory is deviated by the force $F_x = -qv_yB_z$, causing it to hit the side of the semiconductor bar (Fig. C.1a). The holes accumulating at one side of the bar create an electric field $-E_x$ that counteracts the magnetic-field force:

$$qE_x = qv_yB_z \sin[\underbrace{\angle(v_y, B_z)}_{-90°}] \Rightarrow -E_x = v_yB_z \qquad (C.2)$$

This establishes a steady state where the holes continue to flow through the semiconductor bar in the y-direction (Fig. C.1b). If the width of the semiconductor bar (the dimension in the x-direction) is W, a voltage V_H equal to $-E_xW$ can be measured between the opposite sides. This voltage is referred to as the Hall voltage, while the corresponding field is referred to as the Hall field ($E_H = -E_x$ in Fig. C.1b).

Assuming that all the holes move with the drift velocity $v_y = v_d$, and relating the drift

TABLE C.1 A Summary of Kinetic Phenomena in Semiconductors

Name	Description	Cause	Effect
Drift	Sections 1.1.4, 1.4.1, 1.4.2, 1.4.4	Electric field (electric-potential gradient)	Electric current
Diffusion	Sections 1.3.1, 1.3.4, 1.4.4, 8.1.3	Concentration gradient	Particle current
Photoelectric effect	Section 8.2	Light	Photocurrent and/or voltage
Light emission	Section 8.1	Current	Light
Hall effect	Fig. C.1 (this appendix)	Magnetic field	Hall voltage
Seebeck effect	Temperature gradient causes a carrier concentration gradient; this causes either current flow in a circuit consisting of at least two junctions (P–N, metal–semiconductor, or two different metals), or voltage if the circuit is cut at an arbitrary point; the thermal current and thermal voltage are proportional to the temperature difference, the proportionality coefficient depending on the material properties	Temperature gradient	Thermal current and/or voltage
Peltier effect	A current flow against (along) a built-in junction field causes heat release (absorption) (junction heating or cooling)	Electric current	Temperature increase or decrease
Piezo resistive effect	Variation of the resistivity (conductivity) of a semiconductor caused by strain; the physical reason is the variation of the energy-band pattern	Mechanical strain	Changed resistivity

velocity to current density j_y [$j_y = qpv_d$ according to Eq. (1.47)], the Hall field can be expressed as

$$E_H = \frac{j_y}{qp} B_z \qquad (C.3)$$

which further leads to the following expression for the Hall voltage:

$$V_H = \frac{1}{qp} \frac{I}{t_s} B_z \qquad (C.4)$$

where t_s is the sample thickness. When the effect was discovered, Hall observed that the field E_H is directly proportional to the magnetic flux density and the current density, and inversely proportional to the sample thickness:

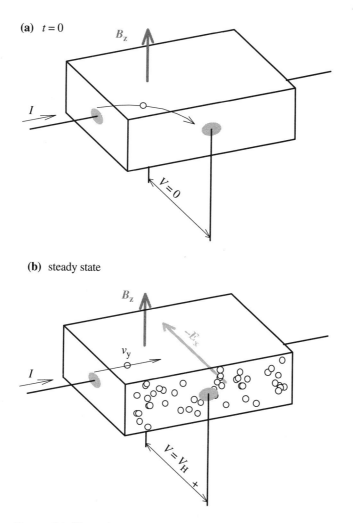

(a) $t = 0$

B_z

I

$V = 0$

(b) steady state

B_z

$-E_x$

v_y

I

$V = V_H$

Figure C.1 Illustration of the Hall effect: (a) the force due to the magnetic field B_z deviates the hole trajectory, and (b) accumulated holes create the Hall field $E_H = -E_x$, which counteracts the force from the magnetic field B_z.

$$V_H = R_H \frac{I B_z}{t_s} \tag{C.5}$$

where the proportionality constant R_H is now known as the Hall coefficient. Comparing the empirical Eq. (C.5) to Eq. (C.4), the Hall coefficient is obtained as

$$R_H = \frac{1}{qp} \tag{C.6}$$

Alternatively, R_H can be expressed as

$$R_H = \frac{\mu_p}{\sigma_p} \tag{C.7}$$

given that $\sigma_p = q\mu_p p$.

In reality, the drift velocity v_d represents only the average carrier velocity, and it cannot be said that $v_y = v_d$ for any carrier. This means that the above theory can strictly be applied only to the carriers moving with average velocity. Consequently, the mobility in Eq. (C.7) is called the Hall mobility, to indicate that it generally has a value different from the drift mobility. This inaccuracy can be compensated for by introducing a factor r in the expression relating the empirical Hall coefficient to the semiconductor properties:

$$R_H = \begin{cases} \frac{r}{qp} & \text{for holes} \\ -\frac{r}{qn} & \text{for electrons} \end{cases} \tag{C.8}$$

The factor r is typically between 1 and 2.

Using Eq. (C.5), the Hall coefficient can experimentally be determined by measuring the Hall voltage V_H for a given current I and magnetic flux density B_z, and knowing the sample thickness t_s. The measured value of the Hall coefficient can be used to calculate the carrier concentration, p or n, using Eq. (C.8).

It is interesting to note that the polarity of the Hall voltage V_H depends on the type of semiconductor used. If an N-type semiconductor is used in the example of Fig. C.1, with the same current direction, the electrons would move in the negative y-direction, however, the force due to the magnetic field B_z is in the same direction because of the negative charge of the electrons: $(-q)(-v_y)B_z = qv_yB_z$. This means the electrons would accumulate at the same side as the holes in Fig. C.1b, leading to an E_x field in the opposite direction, and therefore a V_H voltage with the opposite polarity. Therefore, the polarity of the Hall voltage can be used to determine the semiconductor type.

Practical applications of the Hall effect include semiconductor material characterization techniques and magnetic sensors.

Appendix *D*

SUMMARY OF EQUATIONS AND KEY POINTS

1. RESISTORS: INTRODUCTION TO SEMICONDUCTORS

1.1 *Sheet resistance* $R_S = 1/(\bar{\sigma} x_j)$ involves all the technological parameters that influence the resistance of a semiconductor layer. It is expressed in $\Omega/\square$ to indicate that it means resistance per square, so that the total resistance is

$$R = R_S \frac{L}{W}$$

where the geometric parameters determine the number of squares L/W.

1.2 The electric field E_y causes *drift current* of charged particles, as expressed by the differential form of Ohm's law

$$j_y = \sigma E_y = -\sigma \frac{d\varphi}{dy} \qquad \boxed{\text{drift current}}$$

1.3 There are two types of current carriers in semiconductors, negatively charged *electrons* and positively charged *holes*, so that conductivity is

$$\sigma = qn\mu_n + qp\mu_p$$

where n and p are the *concentrations*, and μ_n and μ_p are the *mobilities* of electrons and holes, respectively.

1.4 By N-type doping, the concentration of electrons can technologically be set at $n \approx N_D$, where N_D is the concentration of *donors*. By P-type doping, the concentration of holes can be set at $p \approx N_A$, where N_A is the concentration of *acceptor* atoms. In thermal equilibrium,

$$np = n_i^2$$

531

where n_i is the *intrinsic carrier concentration* (a temperature-dependent constant). An increase in n leads to a decrease in p because of more pronounced electron–hole *recombination*.

1.5 A random thermal motion results in an effective *diffusion current* when there is a gradient of the particle concentration:

$$J_{\text{diff}} = -D\frac{\partial N}{\partial x}$$

1.6 A current density gradient (a difference in the number of particles flowing *in* and *out* per unit time) leads to a change in the particle concentration in time, as expressed by the *continuity equation*:

$$\frac{\partial J}{\partial x} = -\frac{\partial N}{\partial t}$$

1.7 Diffusion of atoms in a solid material is insignificant at room temperature, but it is pronounced at significantly high temperatures. This is expressed by the Arrhenius-type dependence of the diffusion coefficient:

$$D = D_0 \exp\left(-\frac{E_A}{kT}\right)$$

1.8 Diffusion through selectively etched masking windows is used to create doped layers. Doping by diffusion is typically performed in two steps: (1) *constant-source diffusion* where the surface concentration is constant, while the number of atoms per unit area (*dose*) is continuously increasing, and (2) *drive-in* diffusion, where no new atoms are added (the dose is constant) but the existing ones are diffused deeper. The final diffusion profile has a Gaussian shape:

$$N(x, t) = N_s \exp\left(-\frac{x^2}{4Dt}\right)$$

1.9 The concept of *current density* relates to the *velocity* of a certain *concentration* of particles. In the case of drift current of electrons, $j = -qnv_d$, where v_d is the *drift velocity*. The drift velocity is directly related to the electric field causing the drift:

$$v_d = \begin{cases} -\mu_n E & \text{for electrons} \\ \mu_p E & \text{for holes} \end{cases}$$

where the proportionality coefficient $\mu_{n,p}$ is called the *mobility*. By introducing the concept of mobility, the fundamental current equation ($j = -qnv_d$) is converted into the differential Ohm's law ($j = qn\mu_n E = \sigma E$). The mobility is constant at low electric fields, but it drops at high fields to express the *drift-velocity saturation* effect.

1.10 The mobility value is basically determined by *scattering* and the *effective mass* of the carriers. Two dominant scattering mechanisms are phonon scattering (the mobility decreases with temperature) and Coulomb scattering by the ionized doping atoms (the mobility increases with temperature).

1.11 Independently of any drift current, a gradient of the carrier concentration, causes a diffusion current that is significant even at room temperature:

$$j_{\text{diff}} = \begin{cases} q D_n \frac{dn}{dx} & \text{for electrons} \\ -q D_p \frac{dp}{dx} & \text{for holes} \end{cases} \qquad \boxed{\text{diffusion current}}$$

While diffusion and drift are independent current mechanisms, there is a relationship between their coefficients:

$$D_{n,p} = \frac{kT}{q} \mu_{n,p}$$

1.12 The energy distribution of electrons is given by the Fermi–Dirac distribution function, which can be approximated by the Maxwell–Boltzmann distribution for *nondegenerate semiconductors*:

$$f = \frac{1}{1 + e^{(E-E_F)/kT}} \rightarrow \boxed{\text{for } E - E_F \gg kT} \rightarrow f \approx e^{-(E-E_F)/kT}$$

where E_F is the reference energy level called the *Fermi level*.

1.13 A finite probability f of finding an electron with energy E does not mean such an electron could exist. In semiconductors, there are *bands* of possible energy levels (*electron states*), separated by *energy gaps*. The two energy bands of interest are the *conduction band*, in which the mobile electrons appear, and the *valence band*, in which the mobile holes appear. They are separated by an energy gap E_g, which is the most important characteristic of a semiconductor. The probability f ($f_h = 1 - f$) has to be multiplied by the concentration of the effective density of states to obtain the concentration of mobile electrons (holes):

$$n = N_C e^{-(E_C - E_F)/kT}, \qquad p = N_V e^{-(E_F - E_V)/kT}$$

1.14 The energy-band model is a powerful tool for understanding the phenomena exploited by different semiconductor devices. A band diagram in its essential form indicates the top of the valence band E_V, the bottom of the conduction band E_C (they are separated by the energy gap E_g), and the Fermi level E_F, whose position expresses the doping type and level. A slope in the energy–space diagrams indicates an electric field: an energy difference due to band bending in space, expressed in eV, is numerically equal to the negative electric-potential difference between the two points ($E = -q\varphi$). To visualize the drift current, the electrons in the conduction band can be thought of as balls on a solid surface; an analogy for the holes in the valence band is bubbles in water.

2. CAPACITORS: REVERSE-BIASED P–N JUNCTION AND MOS STRUCTURE

2.1 A *built-in voltage* appears at any junction of materials having mobile electrons with different potential energies, that is, different positions of the Fermi level with respect to the reference vacuum level, or expressed in yet another way, different *work functions*. The built-in voltage is numerically equal to the work-function difference, expressed in eV. This way, the energy-band bending of qV_{bi} levels E_F throughout the system, indicating that the system is in thermal equilibrium.

2.2 In the case of P–N junction, E_F leveling results in removal of higher energy electrons from the N-type side of the junction and their recombination with holes from the P-type side. This creates a *depletion layer* at the P–N junction. The width of the depletion layer, w_d, defines the depletion-layer capacitance of the P–N junction:

$$C_d = \frac{\varepsilon_s}{w_d} \text{ F/m}^2$$

The net charge in the depletion layer is not zero, as there are uncompensated donor and acceptor ions. This charge is the source of the built-in electric field (energy-band slope) and built-in voltage (energy barrier qV_{bi}), where the built-in voltage is given by

$$V_{bi} = \frac{kT}{q} \ln \frac{N_D N_A}{n_i^2}$$

2.3 A reverse-biased P–N junction is not in thermal equilibrium: the Fermi level is split into electron and hole quasi-Fermi levels, so that the barrier height at the P–N junction is increased ($qV_{bi} + qV_R$). This is a barrier for the majority carriers; minority carriers flow easily through the junction, resulting in a small "leakage" current. The reverse bias, V_R, increases the depletion-layer width, reducing the depletion-layer capacitance.

2.4 The *Poisson equation*,

$$\frac{d^2\varphi}{dx^2} = -\frac{\rho_{charge}}{\varepsilon_s}, \qquad \text{where } \rho_{charge} = q(N_D - N_A + p - n)$$

has to be solved to establish the relationship between the depletion-layer width and the voltage drop across the depletion layer. In the case of abrupt and asymmetrical P–N junctions, the result is

$$w_d \approx w_p = \underbrace{\sqrt{\frac{2\varepsilon_s(V_{bi} + V_R)}{qN_A}}}_{\text{for } N_A \ll N_D}; \qquad w_d \approx w_n = \underbrace{\sqrt{\frac{2\varepsilon_s(V_{bi} + V_R)}{qN_D}}}_{\text{for } N_D \ll N_A}$$

The maximum electric field, appearing right at the P–N junction, is also found:

$$E_{max} = \frac{q}{\varepsilon_s} N_D w_n = \frac{q}{\varepsilon_s} N_A w_p$$

2.5 The SPICE equation for $C_d(V_R)$ is

$$C_d = C_d(0) \left(1 + \frac{V_R}{V_{bi}}\right)^{-m}$$

which generalizes the analytical solution for the two extreme cases ($m = 1/2$ for an abrupt and $m = 1/3$ for a linear P–N junction). The parameters $C_{d(0)}$, m, and V_{bi} can be estimated using a graphic or linear-regression method (the logarithm function "linearizes" this equation), which can further be tuned by nonlinear fitting to measured data ($C_{meas} = C_d + C_p$).

2.6 Oxidizing silicon in strictly controlled conditions creates a dielectric–semiconductor interface of unique quality in terms of electronic properties. When a metal or heavily doped polysilicon is deposited on to the thermal oxide, a metal–oxide–semiconductor (MOS) capacitor is created. The metal (polysilicon) electrode is referred to as a *gate*. Depending on the gate voltage applied, an MOS capacitor can be in one of the three modes defined in Table D.1.

2.7 At flat-band conditions ($V_G = V_{FB}$), the net charges at the capacitor plates, the electric field, and the semiconductor *surface potential* (φ_s) are all zero. A nonzero *flat-band voltage* appears due to the metal–semiconductor *work-function difference* ($q\phi_{ms}$) and the effects of the oxide charge (N_{oc}):

$$V_{FB} = \underbrace{\phi_{ms}}_{\phi_m - \phi_s} - \frac{q N_{oc}}{C_{ox}}$$

The work function is the difference between the energy of a free electron (vacuum level) and the Fermi level. The Fermi level of a heavily doped N$^+$ silicon (polysilicon) is approximately at the bottom of the conduction band, so $q\phi_m = q\chi_s$, where the electron affinity $q\chi_s$ is a material constant. The Fermi level of a heavily doped P$^+$ silicon (polysilicon) is approximately at the top of the valence band, therefore, $q\phi_m = q\chi_s + E_g$. The Fermi level in the moderately doped semiconductor substrate depends on the doping level, the Fermi potential $q\phi_F$ expressing the difference between the mid-gap and the position of the Fermi level:

TABLE D.1 Potential–Capacitance–Charge Equations for an MOS Capacitor

	Accumulation	**Depletion**	**Strong Inversion**
P-type	$V_G < V_{FB}$	$V_{FB} \leq V_G \leq V_T$	$V_T < V_G$
N-type	$V_{FB} < V_G$	$V_T \leq V_G \leq V_{FB}$	$V_G < V_T$
φ_s	Small	$|\varphi_s| \leq |2\phi_F|$	$\varphi_s \approx 2\phi_F$
C	$C = C_{ox} = \varepsilon_{ox}/t_{ox}$	$\frac{1}{C} = t_{ox}/\varepsilon_{ox} + w_d/\varepsilon_s$	LF: $C = C_{ox} = \varepsilon_{ox}/t_{ox}$
		$w_d = \sqrt{2\varepsilon_s|\varphi_s|/(qN_{A,D})}$	HF: $\frac{1}{C} = t_{ox}/\varepsilon_{ox} + w_d/\varepsilon_s$
			$w_d = \sqrt{2\varepsilon_s|2\phi_F|/(qN_{A,D})}$
$|Q|$	$Q_{ACC} = |V_{FB} - V_G|C_{ox}$	$Q_d = qN_{A,D}w_d$	$Q_I = |V_G - V_T|C_{ox}$
		$= \sqrt{2\varepsilon_s qN_{A,D}|\varphi_s|}$	$Q_d = \sqrt{2\varepsilon_s qN_{A,D}|2\phi_F|}$
			$= \gamma C_{ox}\sqrt{|2\phi_F|}$

$$q\phi_s = q\left(\chi_s + \frac{E_g}{2q} + \phi_F\right)$$

$$\phi_F = \begin{cases} +\frac{kT}{q}\ln\frac{N_A}{n_i} & \text{for P type} \\ -\frac{kT}{q}\ln\frac{N_D}{n_i} & \text{for N type} \end{cases}$$

2.8 At the onset of strong inversion ($V_G = V_T$), the surface potential φ_s is approximately at its maximum value of $|2\phi_F|$. In strong inversion, the depletion-layer width and the depletion-layer charge do not change significantly, because the voltage drop across the depletion layer (φ_s) remains approximately at the constant level of $2\phi_F$. Any gate-voltage change in the strong inversion results in a proportional change of the inversion-layer charge: $Q_I = C_{ox}(V_G - V_T)$. The threshold voltage is given by

$$V_T = V_{FB} \pm 2|\phi_F| \pm \gamma\sqrt{2|\phi_F|}$$

where the upper and lower signs are for P-type and N-type substrates, respectively, and the body factor γ is given by

$$\gamma = \frac{\sqrt{2\varepsilon_s q N_{A,D}}}{C_{ox}}$$

3. DIODES: FORWARD-BIASED P–N JUNCTION AND METAL–SEMICONDUCTOR CONTACT

3.1 When a forward-bias voltage $V_D = V_F$ is applied across a P–N junction, the barrier height (due to the built-in voltage) is reduced, and a number of majority electrons and holes are able to go over the barrier, appearing as minority carriers on the other side. Because of their exponential distribution with energy, the number of electrons and holes with energies higher than the reduced barrier height $V_{bi} - V_D$ depends exponentially on V_D. Consequently, the concentration of minority carriers at the edges of the depletion layer depends exponentially on V_D:

$$n_p(|w_p|) = n_{pe}e^{V_D/V_t}, \qquad p_n(|w_n|) = p_{ne}e^{V_D/V_t}$$

where n_{pe} and p_{ne} are the equilibrium minority-carrier concentrations ($p_{ne} = n_i^2/N_D$, $n_{pe} = n_i^2/N_A$). The injected minority carriers move through the neutral regions by diffusion, so that the diffusion coefficients of minority carriers (D_n and D_p), in addition to the applied voltage, the geometric parameters, and the doping levels, determine the overall diode current:

$$I_S = A_J q n_i^2 \left(\frac{D_n}{W_{\text{anode}}N_A} + \frac{D_p}{W_{\text{cathode}}N_D}\right)$$

$$I_D = I_S\left(e^{V_D/V_t} - 1\right)$$

3.2 In the absence of any second-order effects, the slope of $\ln I_D$–V_D plot is $1/V_t$. However, electron–hole recombination increases the low-level current, and observable voltage drops in the neutral region decrease the high level current. This effectively reduces the $\ln I_D$–V_D slope to $1/nV_t$, where $1 \leq n \leq 2$. Another important second-order effect is due to parasitic resistance r_S, appearing in series with the P–N junction. As a result, the P–N junction voltage V_{D0} is different from the voltage applied across the terminals of the diode: $V_D = V_{D0} + r_S I_D$. The static diode model in SPICE is a resistance r_S in series with a voltage-controlled current source:

$$I_D = I_S \left(e^{\frac{V_{D0}}{n V_t}} - 1 \right)$$

where the saturation current I_S and the emission coefficient n are model parameters, while V_t is the thermal voltage ($V_t \approx 26$ mV at room temperature).

3.3 The excess minority carriers, due to forward diode current, create *stored charge* at the edges of the depletion layer. The stored charge is directly proportional to the current:

$$Q_s = \tau_T I_D$$

and it changes as the applied voltage changes: $C_s = dQ_s/dV_D = \tau_T dI_D/dV_D$. C_s is effectively a capacitance, which appears in parallel with the depletion-layer capacitance. The forward diode voltage cannot be instantly changed to zero or reverse-bias voltage (the diode cannot be instantly turned off), not before the stored charge is removed. The stored charge is being removed by (1) an approximately constant discharging current [$\approx (V_R + 0.7 \text{ V})/R$], flowing through the series equivalent resistance R and the reverse-bias voltage source, in the opposite direction of the forward current, and (2) electron–hole recombination.

3.4 The electric field in the depletion layer, especially due to a reverse bias, accelerates the electrons and holes, which lose the gained kinetic energy as they collide with the crystal atoms. If the critical electric field is reached, this energy is sufficient to generate electron–hole pairs, the new electrons being accelerated to further generate electron–hole pairs, leading to *avalanche breakdown*. Diodes operating in the avalanche mode can be used as voltage references, because large current variations are supported by a very small voltage change. Another type of breakdown is due to *tunneling*: heavily doped P- and N-type regions create a very narrow depletion layer (barrier), which is further narrowed by the reverse bias to enable the electrons and holes to tunnel through the barrier. The avalanche-breakdown voltage has a positive temperature coefficient (V_{BR} increases with the temperature), while the tunneling-breakdown voltage has a negative temperature coefficient. The forward diode voltage also has a negative temperature coefficient ($\Delta V_D/dT \approx -2$ mV/$^\circ$C).

3.5 If the Fermi levels of a semiconductor and a metal are at different positions with respect to the vacuum level (different work functions $q\phi_s$ and $q\phi_m$), they create a built-in voltage at the metal–semiconductor interface:

$$V_{bi} = \phi_m - \phi_s$$

This built-in voltage is associated with a depletion layer at the semiconductor surface, and represents a barrier for the electrons in the semiconductor. The electrons in the metal also face a barrier, which is

$$q\phi_B = q\phi_m - q\chi_s$$

$q\phi_B$ is voltage independent, however, the barrier from the semiconductor side can be changed by applied voltage ($V_{bi} - V_F$; $V_{bi} + V_R$), enabling the metal–semiconductor contact to operate as a rectifying diode—a *Schottky diode*.

3.6 The current–voltage characteristic of the Schottky diode has the same form as for the P–N junction diode:

$$I_D = I_S \left(e^{V_D/(nV_t)} - 1 \right)$$

By proper selection of the metal electrode, Schottky diodes with smaller built-in voltages can be created, resulting in a smaller forward voltage drop for the same current. In the model, this is accounted for by a larger I_S, which shows that there is a direct link to an increase in the reverse-bias current.

3.7 Schottky diodes are single-carrier devices (the minority holes in metal–N-type Schottky diodes do not play a significant role in the current flow). There is no stored charge of minority carriers and there is no stored-charge capacitance and the associated switch-off delay.

3.8 A contact between a metal and a *heavily doped* semiconductor leads to a very narrow barrier (narrow depletion layer), which is further narrowed down by the applied voltage, enabling the carriers to tunnel in either direction. This type of metal–semiconductor contact acts as a small-resistance (ohmic) contact.

4. BASICS OF TRANSISTOR APPLICATIONS

4.1 Every transistor can operate in two principal modes: *voltage-controlled current source* and *voltage-controlled switch*. When the controlling voltage is smaller than a threshold value, the output of the transistor appears as an open circuit (open switch). When the controlling voltage is larger than the threshold value, the output of the transistor appears as either a small resistance at smaller output voltages (closed switch) or a current source due to current saturation with increase of the output voltage.

4.2 As the input (controlling) voltage can change the state of the transistor from "off" to "on," the *transfer characteristic* for large input signals is nonlinear. However, with a proper *input-bias circuit*, the operating point can be placed inside a linear-like region of the characteristic, as far as small signals are concerned. A voltage source and a resistor, added in series with the transistor output, create the simplest *output-bias* circuit that is needed to (1) properly bias the transistor for operation as a current source (higher output voltages) and (2) convert the current of the current source, controlled by the small input signal, into a large-amplitude output voltage (voltage amplification).

4.3 If the output-bias circuit is used in conjuction with large input signals, driving the transistor from an "off" to an "on" state, this transistor circuit acts as an inverter: for *low*-input voltage, the transistor is "off" and the *high*-voltage level of the power supply appears at the output; for *high*-input voltage, the transistor is "on" connecting the output to the *low* ground level. A serious disadvantage of this circuit is the need to have a current flowing through the resistor (internal power dissipation) in order to create a voltage drop that compensates the voltage of the power supply. A much better solution is when a *complementary transistor* is used as a dynamic resistance (instead of the ordinary resistor), creating an *inverter with complementary transistors*. This way, the resistance is small when needed (to connect the output to the power supply), and appears as almost an open circuit when the output is connected to ground and needs to be isolated from the power supply. As no power is needed to maintain either logic state, this principle enables integration of very complex digital circuits.

5. MOSFET

5.1 The *inversion layer* of an MOS capacitor, with *source* and *drain* contacts at two ends, forms a switch in the "on" mode. This is a voltage-controlled switch, for the inversion layer (that is, the *channel* connecting the source and the drain) exists for as long as there is appropriate voltage at the *gate* electrode. In other words, the electric field (the built-in plus any external due to the gate voltage) can attract carriers of the same type as the source and drain regions (the switch is in the "on" mode), and can either deplete the channel region or attract opposite-type carriers, in which case the switch is in the "off" mode (no conducting channel exists between the source and the drain terminals). This is known as the *field effect*, hence the name of the device: metal–oxide–semiconductor field-effect transistor (MOSFET).

5.2 There are N-channel MOSFETs (electrons in the inversion layer and N-type source/drain regions in a P-type substrate) and P-channel MOSFETs (holes in the channel and P-type source/drain regions in an N-type substrate). The N-channel MOSFETs are "on" when the gate voltage is larger, while the P-channel MOSFETs are "on" when the gate voltage is smaller than the *threshold voltage*:

$$V_\mathrm{T} = V_\mathrm{FB} \pm 2|\phi_\mathrm{F}| \pm \gamma\sqrt{2|\phi_F| + |V_{SB}|}$$

where the upper signs are for N-channel and the lower signs are for P-channel MOSFETs. The reverse-bias voltage across the source-to-substrate junction, $|V_{SB}|$, increases the threshold voltage (*body effect*). N-channel and P-channel MOSFETs that are "off" for $V_{GS} = 0$ are *normally-off* MOSFETs, also called *enhancement-type* MOSFETs. MOSFETs with built-in channels (they can conduct current at $V_{GS} = 0$ V) are called *normally-on*, or *depletion-type* MOSFETs.

5.3 The resistance of the channel (closed switch) is inversely proportional to $Q_I = C_{ox}(V_{GS} - V_T)$. Therefore, $R_{on} = 1/[\beta(V_{GS} - V_T)]$, which means the current is

$$I_D = \beta(V_{GS} - V_T)V_{DS}$$

At higher V_{DS} voltages, this linear dependence of I_D on V_{DS} is not maintained, for the drain starts depleting the drain end of the channel, reducing the current, which reaches its maximum at $V_{DS} = V_{DSsat}$. The region $0 \leq |V_{DS}| \leq |V_{DSsat}|$ is referred to as the *triode region* and $|V_{DSsat}| < |V_{DS}| < |V_{BR}|$ is referred to as the *saturation region*. The saturation is reached due either to (1) full depletion of the drain end of the channel (channel "pinch off") or (2) drift velocity saturation at high lateral electric fields in the channel. The saturation region enables the device to work as a current source whose current is controlled by the gate. Therefore, the MOSFET provides one possible implementation of the voltage-controlled current source used in analog circuits.

5.4 Enhancement-type N-channel and P-channel MOSFETs provide the best available voltage-controlled switches for the inverter with complementary transistors. The technology used to fabricate complex digital integrated circuits, based on the principle of the inverter with complementary transistors, is known as *CMOS* ("complementary MOS") technology. The older *P-well* technology uses N-type substrate for the P-channel MOSFETs, placing the N-channel MOSFETs in P-wells. The newer CMOS technologies use P-substrate for the N-channel and N-wells for the P-channel MOSFETs, an approach that requires *threshold-voltage adjustment*. The threshold-voltage adjustment is typically achieved by *ion implantation* of doping atoms, a process that provides excellent control of the doping dose and profile. It is also possible to create integrated circuits with N-channel MOSFETs only, using enhancement MOSFETs as drivers and depletion MOSFETs as load resistances— *NMOS technology*.

5.5 Modeling of the MOSFET current is based on Ohm's law, which accounts only for the linear region when applied literally (summary point 5.3. The rudimentary MOSFET model (SPICE LEVEL 1 model) modifies the basic Ohm's relationship in order to include the reduction of channel-carrier density (Q_I) by the drain field. A more precise model is obtained when the lateral nonuniformity of the depletion layer, which widens toward the reverse-biased drain-to-bulk junction, is included. This leads to a computationally inefficient model (SPICE LEVEL 2 model), involving the $(V_{DS} + \cdots)^{3/2}$ term. Simplifying this model by the linear and parabolic term of Taylor series leads to what is referred to as the SPICE LEVEL 3 model. This model is structurally the same as the rudimentary model, with an added factor F_B to account for the effects of the depletion-layer variations. The LEVEL 3 model for the above-threshold drain current is summarized in Table D.2

5.6 The MOSFET structure has a number of inherent parasitic elements: source-to-bulk and drain-to-bulk diodes, resistors in series with all four terminals, gate–source overlap capacitor, gate–drain overlap capacitor, and the central MOS capacitor, which is modeled by two parts: gate–source ($C_{ox}/2$ in strong inversion) and gate–drain ($C_{ox}/2$ in series with C_d, its value becomes smaller as the drain voltage expands the depletion layer). The parasitic capacitances are especially important, as they determine the high-frequency behavior.

5.7 The first estimation of SPICE MOSFET parameters can be obtained by applying mathematical transforms that "linearize" the model equations. A particularly useful technique employs measurements of V_T for different V_{SB} voltages (body effect), in order to obtain the body factor γ (it involves the substrate-doping concentration and the gate-oxide thickness).

6. BJT

6.1 A BJT in *normal active mode* acts as a voltage-controlled current source. In the case of an NPN BJT, the *forward-biased base–emitter* junction emits electrons into

TABLE D.2 MOSFET Equations[a]

$$I_D = \begin{cases} \beta\left[(V_{GS} - V_T)V_{DS} - (1 + F_B)\frac{V_{DS}^2}{2}\right], & \text{for } 0 \leq |V_{DS}| < |V_{DSsat}| \\ \frac{\beta}{2(1+F_B)}(V_{GS} - V_T)^2, & \text{for } |V_{DS}| \geq |V_{DSsat}| \end{cases}$$

$$V_{DSsat} = \frac{V_{GS} - V_T}{1 + F_B} \qquad \beta = \underbrace{\mu_{eff}\frac{\varepsilon_{ox}}{t_{ox}}}_{KP}\frac{W}{L_{eff} - L_{pinch}}$$

$$F_B = \frac{\gamma F_s}{2\sqrt{|2\phi_F| \pm V_{SB}}} + F_n \qquad \mu_{eff} = \frac{\mu_s}{1 + \mu_s|V_{DS}|/(v_{max}L_{eff})}$$

$$\mu_s = \frac{\mu_0}{1 + \theta|V_{GS} - V_T|}$$

$$V_T = V_{T0} \pm \gamma F_s\left(\sqrt{|2\phi_F| \pm V_{SB}} - \sqrt{|2\phi_F|}\right) - \sigma_D V_{DS} + F_n V_{SB}$$

Second Order Effects	Principal Model
• Mobility reduction due to vertical field (V_{GS} voltage)	$\theta = 0$
• Mobility reduction due to lateral field (V_{DS} voltage)	$v_{max} = \infty$
• Channel-length modulation in saturation region	$L_{pinch} = 0$
• Short-channel charge sharing in the depletion layer	$F_s = 1$
• Fringing-field widening of a narrow-channel depletion layer	$F_n = 0$
• Drain current increase in saturation due to DIBL	$\sigma_D = 0$

[a] Upper signs, N channel; lower signs, P channel.

the base, which diffuse very efficiently through the very narrow base region to be collected by the electric field of the *reverse-biased collector–base* junction. Even the smallest positive collector voltage collects the electrons efficiently enough so that I_C does not increase with $V_{CB} \approx V_{CE}$ (a current source). However, I_C strongly depends on the bias of the controlling B–E junction. A good analogy is a waterfall current, which does not depend on the height of the fall but on the amount of the incoming water.

6.2 In addition to the emitter electron current I_{nE}, there is also I_{pE} current due to the holes emitted from the base back into the emitter, which reduces the emitter efficiency: $\gamma_E = I_{nE}/(I_{nE} + I_{pE})$. Some of the electrons are recombined in the base, leading to a nonideal base transport factor: $\alpha_T = I_{nC}/I_{nE}$. The reverse-bias current of the base–collector junction, I_{CB0}, can usually be neglected, so that emitter-to-collector and base-to-collector current gains can be related to the technological parameters as follows:

$$\alpha = \frac{I_C}{I_E} = \alpha_T \gamma_E, \qquad \beta = \frac{\alpha}{1 - \alpha}$$

A good NPN BJT will have a much heavier doped emitter than base ($I_{nE} \gg I_{pE}$) and a very narrow base ($\alpha_T \approx 1$). When the collector and the emitter are swapped (*inverse-active mode*), α and β are significantly smaller.

6.3 If C–B is reverse biased and the forward B–E bias is very low, or the B–E junction is also reverse biased, no significant current flows through the BJT (*cutoff mode*). When both the B–E and the C–B are forward biased, they both push current into each other (*saturation mode*). With typical bias circuits, $V_{BE} > V_{BC}$, which means there is a small positive output voltage $V_{CE} = V_{BE} - V_{BC}$. Also, the net collector current maintains its direction, although it becomes smaller as the BJT is driven deeper into saturation. The cutoff and saturation modes enable a BJT to be used as a voltage-controlled switch.

6.4 There is a complementary PNP BJT. It is a mirror image of the NPN: opposite doping types, opposite voltage polarities, and opposite current directions.

6.5 There are strong similarities between the energy-band diagrams and therefore between operation principles of BJTs and MOSFETs. However, there are also very important differences. The concentration of carriers in the MOSFET channel is controlled by an electric field (no input current is needed and no input power is wasted); a significant input current, and consequently input power, is needed to keep a BJT in the saturation mode (a closed switch). On the other hand, a very shallow field penetration through the channel charge severely limits the cross-sectional area of the MOSFET channel; in the case of BJTs, the input control is over the entire P–N junction area, hence, superior current capability.

6.6 The standard bipolar IC technology is designed so to optimize the performance of the NPN BJT as the main device. A low-doped N-epitaxial layer is deposited onto P substrate, and "cut" into N-epi islands by deep P$^+$ isolation-diffusion stripes. The boundary N-epi–P junction is reverse biased to provide electrical isolation of the individual components, placed into the N-epi islands. The N-epi islands become collectors, with P-type base and N^+-emitter regions of increasing doping concentrations sequentially diffused into one another. PNP BJTs can be implemented as either substrate (P-base–N-epi–P-substrate as emitter–base–collector regions) or lateral (P-base–N-epi–P-base regions). The neutral regions of base and emitter diffusion regions, as well as the N-epi, are used as resistors. The base–collector, base–emitter, and collector–substrate P–N junctions can be used as diodes and/or capacitors. The breakdown voltage of the B–E junction is low (≈ 7 V) due to the very heavy emitter doping—this junction can be used as a reference diode.

6.7 The general version (accounting for all the modes of operation) of the Ebers–Moll model for an NPN BJT is

$$I_C = I_S \, (e^{V_{BE}/V_t} - 1) - \left(1 + \frac{1}{\beta_R}\right) I_S \, (e^{V_{BC}/V_t} - 1)$$

$$I_E = -\left(1 + \frac{1}{\beta_F}\right) I_S \, (e^{V_{BE}/V_t} - 1) + I_S \, (e^{V_{BC}/V_t} - 1)$$

$$I_B = \frac{1}{\beta_F} \, I_S \, (e^{V_{BE}/V_t} - 1) + \frac{1}{\beta_R} \, I_S \, (e^{V_{BC}/V_t} - 1)$$

The assumed current directions are toward the transistor, hence the reverse signs of the emitter current. The only difference in the PNP model is that the polarities of the voltages are opposite ($V_{BE} \rightarrow V_{EB}$, $V_{BC} \rightarrow V_{CB}$), and the current directions are assumed to be out of the transistor. In the normal active mode, $\exp(V_{BE}/V_t) \gg 1$ and $\exp(V_{BC}/V_t) - 1 \approx 1$, which simplifies the principal Ebers–Moll equations to

$$I_C = I_S e^{V_{BE}/V_t}$$

$$I_E = -\left(1 + \frac{1}{\beta_F}\right) I_S e^{V_{BE}/V_t} = -\frac{I_C}{\alpha_F}$$

$$I_B = \frac{1}{\beta_F} \, I_S e^{V_{BE}/V_t} = \frac{I_C}{\beta_F}$$

6.8 The principal model assumes a perfect current source (the collector current is fully independent of the collector voltage). In reality, the expansion of the C–B depletion layer due to an increased reverse bias leads to narrowing of the effective base width (*base width modulation*), resulting in an increase of the collector current (a finite output dynamic resistance). This is known as the Early effect, and is modeled through I_S:

$$I_S = I_{S0} \left(1 + \frac{|V_{BC}|}{|V_A|} + \frac{|V_{BE}|}{|V_B|}\right)$$

where I_{S0} is the SPICE parameter, and not I_S itself. The forward and reverse Early voltages $|V_A|$ and $|V_B|$ are also parameters.

6.9 The Gummel–Poon level in SPICE extends the Ebers–Moll equations to include additional second-order effects. A manifestation of these effects is β dependence on the collector current: it has a maximum at medium currents, being smaller at small collector currents, but with a lot more dramatic reduction at high-level injection. The reduction at small currents is modeled by adding "leakage" components to the base-current equation:

$$I_B = \frac{I_{S0}}{\beta_{FM}}(e^{V_{BE}/V_t} - 1) + C_2 I_{S0}(e^{V_{BE}/(n_{EL}V_t)} - 1)$$

$$+ \frac{I_{S0}}{\beta_{RM}}(e^{V_{BC}/V_t} - 1) + C_4 I_{S0}(e^{V_{BC}/(n_{CL}V_t)} - 1)$$

where the maximum current gains β_{FM} and β_{RM} are parameters in the model, the other parameters being C_2, C_4, n_{EL}, and n_{CL}. The reduction at high-level injection is modeled by equations that add the injected (stored) charge in the base, including the depletion-layer charge (Early effect). In the normal-active mode, and without the Early effect, the collector current is

$$I_C \approx \begin{cases} I_{S0}e^{V_{BE}/V_t}, & \text{for } I_C < I_{KF} \\ \sqrt{I_{S0}I_{KF}}e^{V_{BE}/(2V_t)} & \text{for } I_C > I_{KF} \end{cases}$$

where the high-level knee current I_{KF} is a parameter. An analogous parameter I_{KR} relates to the inverse-active mode.

6.10 The B–E and B–C P–N junctions introduce the associated parasitic capacitances, both depletion layer and stored charge. Also, there are parasitic resistances in series with all the BJT terminals. It is these parasitic elements that determine the high-frequency linear and switching performance of BJTs. In standard bipolar IC technology, the individual components are isolated by a reverse-biased P-substrate–N-epi junction. This junction also introduces a parasitic depletion-layer capacitance. In addition to that, with the N-epi–P-base junction, it creates a parasitic substrate PNP BJT (not included in SPICE). The parasitic elements associated with the IC diodes, resistors, and capacitors may need to be externally added to the SPICE models for a proper simulation of a bipolar IC.

7. ADVANCED AND SPECIFIC IC DEVICES AND TECHNOLOGIES

7.1 A reduction of MOSFET dimensions (L, W, and t_{ox}), accompanied by an appropriate substrate-doping increase, results in faster CMOS cells that occupy a smaller area. Lower doped drain and source extensions are used in small-dimension MOSFETs to reduce the electric field at the drain end of the channel. Deep-submicron MOSFETs also feature silicide gate layers and silicide source/drain contacts to decrease the gate and contact resistances. A better control over the channel region by the gate (a stronger field effect) is needed to enable further scaling down of MOSFET dimensions, so metal oxides with their high permittivity are considered as gate dielectrics.

7.2 An MOS capacitor in a "charged" or "discharged" state plays the role of a memory bit that stores the binary "1" or "0" digit. With the adjacent MOSFETs, needed to select and read/write the information, it creates a *random-access memory* (RAM) cell. The charge at the MOS capacitor recombines, so a fairly frequent refreshing is necessary. Polysilicon completely surrounded by oxide (floating gate) is a "proper" trap for the electrons, so once charged it remains in this state practically forever. The *read-only memory* (ROM) MOSFETs utilize the floating gate. The process of charging/discharging is slow and damages the separating oxide after a limited number of charge transfers (millions!), so this type of memory cell cannot be used instead of the RAMs.

7.3 Integration of bipolar and CMOS technologies (BiCMOS) is an advanced technology that makes available the superior characteristics of both MOSFETs and BJTs to IC designers. Silicon-on-insulator (SOI) technology is another advanced technology that may prove to be an important "enabling" technology, in particular for high-reliability and high-speed CMOS ICs.

8. PHOTONIC DEVICES

8.1 In direct semiconductors, the energy released due to *recombination* of an electron–hole pair may be in the form of a photon. The opposite process, photon absorption by an electron, results in *generation* of an electron–hole pair. During these processes the total energy is conserved: the electron must lose energy $h\nu$ when a photon is emitted, and the electron's energy is increased by $h\nu$ when a photon is absorbed.

8.2 The electron–hole recombination is utilized in *light-emitting diodes* (LED). LEDs are operated in a forward-bias mode, so that the forward-bias current injects minority carriers into the neutral regions, which are then recombined by the majority carriers to emit light.

8.3 The recombination process is characterized by *minority-carrier lifetimes* $\tau_{n,p}$, which together with the diffusion constants $D_{n,p}$ determine the diffusion lengths:

$$L_{n,p} = \sqrt{D_{n,p}\tau_{n,p}}$$

Also, the minority-carrier lifetimes are directly related to the *transit time* τ_T (SPICE parameter):

$$\tau_T = \begin{cases} \tau_n/2 & \text{for N}^+\text{–P junctions} \\ \tau_p/2 & \text{for P}^+\text{–N junctions} \end{cases}$$

8.4 The electron–hole generation due to absorbed light is utilized in *photodetectors* and *solar cells*. The reverse-bias current of a photodetector diode or a solar cell is increased from the normal saturation current I_S to $I_S + I_{photo}$, where for the case of uniform generation rate(s) $g_{n,p}$

$$I_{photo} = \underbrace{q A_J g_n w_d}_{\text{drift}} + \underbrace{q A_J g_n L_n + q A_J g_p L_p}_{\text{diffusion}}$$

The photodetector diodes are operated in the reverse-bias mode, as light-controlled current sources. A solar cell is directly connected to a load resistance, which results in a positive voltage drop across the cell while the current remains negative—the *negative power* means that the solar cell is delivering power.

8.5 As the particle concentration changes due to the recombination or external generation, recombination and external-generation terms appear in the complete form of *continuity equation*:

$$\frac{\partial n_p(x, t)}{\partial t} = \frac{1}{q} \frac{\partial j_n(x, t)}{\partial x} + g_n(x, t) - \frac{n_p(x, t) - n_{pe}}{\tau_n} \quad \text{for electrons}$$

$$\frac{\partial p_n(x, t)}{\partial t} = -\frac{1}{q} \frac{\partial j_p(x, t)}{\partial x} + g_p(x, t) - \frac{p_n(x, t) - p_{ne}}{\tau_p} \quad \text{for holes}$$

8.6 As opposed to electrons that obey the Pauli exclusion principle, the probability that a photon with energy $h\nu$ will be emitted is increased by a factor $(n + 1)$ if there are already n photons with this energy. This "behavior" leads to *stimulated light emission*. If an *inversion population* is reached, that is, so many electrons and holes are injected into a semiconductor material that both electron and hole quasi-Fermi levels enter the conduction and valence bands, respectively:

$$E_{FN} - E_{FP} > h\nu = E_2 - E_1$$

the stimulated emission exceeds the recombination rate. The emitted light is reflected by parallel mirrors at the ends of the semiconductor so that the intensity of photons, and therefore the probability of stimulated emission, is enhanced. These principles are utilized by *lasers* to generate highly directional beams of monochromatic light. The light that gets out of the laser is compensated by maintaining the forward-bias current of the laser above the needed *threshold current*.

8.7 To practically achieve the inversion-population condition, a narrower band-gap semiconductor is sandwiched by N^+ and P^+ semiconductors with wider band gaps. The energy-band discontinuities at the *heterojunctions* "force" the quasi-Fermi levels inside the conduction/valence band of the active semiconductor. The energy-band discontinuities also help to confine the carriers inside the active region so that they recombine rather than diffuse away. Also, the active layer in the middle has a higher refraction index, which helps to confine the generated light inside the active region.

9. MICROWAVE FETS AND DIODES

9.1 The input capacitance of Si JFETs is smaller than in a comparative MOSFET, making the JFET a superior device for high-frequency analog applications.

9.2 GaAs MESFET can be used at higher frequencies than Si MOSFETs, JFETs, and BJTs due to
  (a) higher low-field mobility of GaAs electrons, which relates to a higher transconductance;
  (b) wider band gap, which enables semiinsulating substrates, eliminating the capacitor-based isolation structures used in Si.

9.3 HEMTs utilize 2DEG channels in which the electrons are free to move in two dimensions only, while they appear as standing waves in the third dimension. 2DEG is created at the heterojunction of two undoped materials (GaAs and AlGaAs). These facts relate to reduced scattering, hence the name high-electron mobility transistor. The reduced scattering associated with the other advantages of GaAs technology makes the HEMTs most suitable solid-state devices for low-noise high-frequency applications.

9.4 The maximum operating frequency of any transistor is not limited by the time it takes for the electrons to travel through the transistor (transit time), but by the input parasitic capacitance. Negative-resistance diodes operate at or near the transit-time frequencies. The negative dynamic resistance can be employed to convert a DC supply power into signal power ($p_d = gv_d^2 < 0$ as $g < 0$). This enables amplifiers and oscillators operating at frequencies higher than any other solid-state system.

9.5 The negative dynamic resistance in a Gunn diode appears when an increasing voltage (electric field) converts an increasing number of high-mobility electrons into low-mobility electrons. This happens due to a transfer of electrons from an energetically lower E–k valley into a higher E–k valley, associated with a heavier effective mass of the electrons.

9.6 The negative dynamic resistance of an IMPATT diode is achieved by synchronizing the transit time of carriers created by avalanche breakdown to the half-period of a resonant voltage. The positive peak of the resonant voltage adds to a DC bias voltage to set the diode in avalanche mode. The carriers created take time τ (transit time) to come to the opposite electrode (maximum signal current); the half-period of the voltage is set to τ, which means it is the most negative when the current is at the positive peak.

9.7 The negative resistance of a tunnel diode is due to diminishing overlap (in energy terms) of N-type conduction and the nearby P-type valence bands, caused by Fermi level splitting by an increasing forward-bias voltage. The diminishing "energy alignment" of N-type electrons to P-type holes means fewer electrons can tunnel through the very narrow depletion layer at the P–N junction.

10. POWER DEVICES

10.1 Power electronic circuits use switches to convert electrical energy from one form to another. P–N junction (including PIN) and Schottky diodes are used as "uncontrollable" switches (rectifiers). BJTs, MOSFETs, IGBTs, and thyristors are most frequently used as controlled switches.

10.2 PIN diodes are, in principle, P–N junction diodes with an inserted low-doped ("insulating"—*I*) region, also called a *drift region*, which takes the reverse-bias voltage drop across its depletion layer. PIN diodes exhibit good reverse-blocking (off state) and current (on state) capabilities. A relatively high barrier height and the voltage drop across the drift region contribute to relatively high forward voltage drop (>1 V). In a 5-V power supply, this alone can lead to 25% power-efficiency loss. Minority carriers are involved in current conduction, causing two undesirable switching effects: (1) forward voltage overshoot due to limited rate of minority carriers accumulation to the forward-bias level of stored charge, and (2) delay and current flow needed to remove the stored charge so that the diode is switched off again.

10.3 Schottky diodes exhibit superior switching characteristics, as the current conduction involves majority carriers only, and no stored-charge effects appear. The choice of metal work function can adjust the barrier height so that the forward voltage drop is reduced, compared to PIN diodes. This inevitably leads to an increase in reverse-bias current, which is a limiting factor for the forward-voltage drop reduction. There is a significant increase in reverse-bias current with the reverse-bias voltage, due to a barrier-lowering effect. Analogously to PIN diodes, power Schottky diodes involve a drift (low-doped) region to achieve the desired reverse-blocking capability.

10.4 Power BJTs, as controlled switches, share the positive (reverse-blocking and good current capabilities) and negative (stored-charge enforced switching delays) characteristics of PIN diodes. In addition, a large input base current is needed to maintain the BJT in the on state.

10.5 Tens of thousands of MOSFET cells can be paralleled to compensate for the inherently inferior current capability and to enable a small on resistance. Being a majority-carrier field-effect device, it exhibits superior switching performance and no steady-state input current is needed to maintain either on or off states. Still, high transient currents are needed to charge/discharge the input capacitance in order to switch the power MOSFET on/off, imposing a practical limit to the increase in input capacitance due to paralleling MOSFET cells. Analogously to diodes and BJTs, a drift region is used to provide good forward-blocking capability ("forward" refers to positive drain-to-source voltage). In the on state, the drift region contributes to, and even dominates, the on resistance. Bulk and source are short-circuited in power MOSFETs, which connects the internal bulk-to-drain power diode across the output—no reverse-blocking capability.

10.6 IGBTs combine an input MOSFET and an output BJT to create a device with superior input control and output current capabilities. Its BJT-like output renders it inferior to the MOSFET in terms of high-frequency switching performance.

10.7 Thyristors are four-layer PNPN structures that can be latched into the conduction (on) state by a regenerative mechanism: the PNPN layers form PNP and NPN BJTs, with collectors connected to each others' base, amplifying and pushing the current in a closed loop until both BJTs enter the saturation mode. This mechanism works when

$\beta_{pnp}\beta_{npn} \to 1$, which does not happen if the collector currents are below a certain level. In this state, the thyristor is in the blocking (off) mode, either reverse or forward. A gate electrode is used to provide the triggering level of current—the gate is used to turn the thyristor on, but it cannot be used to turn the basic thyristor (SCR) off. The SCR turns off when the current is dropped below the *holding current* so that $\beta_{pnp}\beta_{npn} < 1$. There is an accumulation of minority carriers, and consequently the switching performance is limited by stored-charge effects. However, thyristors have superior blocking and current capabilities: they can conduct thousands of amperes of current in the on state and block thousands of volts in the off state.

11. SEMICONDUCTOR DEVICE RELIABILITY

11.1 Reliability is an important dimension of *quality*. It indicates how long the initial performance parameters of a device (or system) would remain within specified limits. Reliability is a stochastic concept, so that the probability parameters are practically meaningful when applied to a large number of units. The most important reliability parameter is the *failure rate*, expressing the fraction of units that fails in a considered time interval Δt. Generally, the time dependence of the failure rate, $\lambda(t)$, has a bathtub shape with three distinct regions: (1) early failures, (2) steady state, where λ is constant, and (3) wear out.

11.2 In the steady-state region, the failure distribution function is given by the exponential probability density function,

$$f(t) = \lambda e^{-\lambda t}, \qquad \int_0^\infty f(t)\,dt = 1$$

so that the cumulative failure function is

$$F(t) = 1 - e^{-\lambda t}$$

The *mean time to failure* for a batch of units is equal to the expected unit lifetime: $MTTF = \bar{t} = \int_0^\infty t f(t)\,dt$. In the steady-state region,

$$MTTF = \frac{1}{\lambda}$$

11.3 The most common failure mechanisms in semiconductor devices are formation of the purple plague at the chip-to-wire bonds; electromigration of atoms in the metalization lines; corrosion of the metalization; charge trapping in the oxide and interface-trap creation due either to high oxide field, hot-carrier generation, or exposure to ionizing radiation; gate-oxide breakdown; activation of parasitic structures, in particular, the inherent thyristor in CMOS ICs (latch-up effect); and soft errors. The first three mechanisms have an Arrhenius-type temperature dependence:

$$\lambda = A_0 e^{-E_A/kT}$$

In addition, the electromigration depends critically on the current density ($\lambda \propto j^{-n}$) and the corrosion on the relative humidity. The charge trapping and interface-trap creation depend exponentially on the oxide field E_{ox}:

$$\lambda = A_0 e^{-E_0/E_{\mathrm{ox}}}$$

while in the case of hot-carrier generation the lifetime is modeled through the substrate current I_{sub} of the MOSFET:

$$\bar{t} = A I_{\mathrm{sub}}^{-m}$$

11.4 Semiconductor devices are normally "aged" to a point at which the early-failure period ends and the steady-state period starts. This procedure, referred to as *reliability screening*, helps increase the reliability of the electronic systems, cutting maintenance cost. It is practically very important to shorten the "aging" time, so specifically designed procedures that accelerate the early failures are used. The most frequent screening technique is the "burn-in," where the devices are operated at increased temperature ($T = 125°C$) for a certain period of time (usually 168 h). Every device from the batch goes through the screening procedure, and it is important not to "overage" them, which would undesirably shorten the lifetime of the electronic systems that would use these devices.

11.5 Determination of the reliability parameters λ or $MTTF$, referred to as *reliability measurement*, is conceptually different from the straightforward measurement of the conformance parameters. The reliability measurement has to deal with three specific problems: (1) the destructive nature of the measurement, (2) the duration of the measurement, and (3) the existence of different failure mechanisms and their dependence on different operating conditions. The solution to the first problem is *sampling*—a sample of n devices is "destroyed" to determine its mean-time-to-failure ($mttf$), and this measurement is used as the best estimate of the mean-time-to-failure for all devices from the same batch. The sample selection is a random event, and $mttf$ is a random variable with its own distribution. For larger sample sizes ($n > 30$), this distribution is very close to the normal distribution with the standard deviation

$$\sigma_{\mathrm{mttf}} = \frac{\overline{mttf}}{\sqrt{n}} = \frac{MTTF}{\sqrt{n}} = \frac{1}{\lambda\sqrt{n}}$$

The second problem is solved by *accelerated testing*, so that the measurement time is reduced by the acceleration factor AF:

$$AF = \frac{MTTF_{\mathrm{oper}}}{MTTF_{\mathrm{stress}}} = \frac{\lambda_{\mathrm{stress}}}{\lambda_{\mathrm{oper}}}$$

The accelerated tests are based on the factors that influence the particular failure mechanisms, and they include operation at increased temperature, humidity, operating voltage, etc. The third problem is solved by a *combination of accelerated tests*. In the case of independent failures, the overall λ is the sum of the results obtained by the individual tests:

$$\lambda_{\text{oper}} = \sum_{i=1}^{nt} \lambda_{\text{oper}-i}$$

12. QUANTUM MECHANICS

12.1 Quantum mechanics is a theory that explains and predicts the behavior of small things, so small that both *particle* and *wave* properties are pronounced. The models of classical physics relate to "pure" waves and particles. The *waves* are represented by the wave function:

$$\psi(x, t) = A \, e^{-j(\omega t - kx)}$$

where $\psi(x,t)\psi^*(x,t) = |A|^2$ is the wave intensity, the *angular frequency* ω is the rate of change of phase with time, and the *wave vector/number* k is the rate of change of phase with distance. The motion of *particles* is represented by their momentum $p = mv$, where m is the mass and v is the velocity of the particle. The concepts of "particle" and "wave" cannot be distinguished in very small things, the momentum and the wave vector being fundamentally related to each other:

$$\boldsymbol{p} = \hbar k = \frac{h}{2\pi} \boldsymbol{k}$$

To model this wave–particle behavior, the concept of a *wave packet* is used. A wave packet is a superposition of a large number of single-harmonic waves (plane waves) with distributed wave numbers, so that the wave packet is to an extent localized in space. The width of the wave-number distribution (Δk) and the space localization (Δx) are related—a better space localization is possible at the expense of a larger wave vector/momentum uncertainty:

$$\Delta x \, \Delta p \geq \frac{\hbar}{2}$$

This fact is one way of expressing one of the fundamental principles of quantum mechanics, known as the *Heisenberg uncertainty principle*.

12.2 Clearly, the wave function contains the information on both the wave and the particle behavior of electrons. It is not a "universal function"—it depends on the forces imposed on the electrons, which is reflected in the potential $E_{\text{pot}}(x)$ and total E_{tot} energy of the electrons. The wave function in given circumstances can be obtained by solving the Schrödinger equation, whose one-dimensional time-independent form is

$$-\frac{\hbar^2}{2m} \frac{d^2\psi(x)}{dx^2} + E_{\text{pot}}(x)\psi(x) = E_{\text{tot}}\psi(x)$$

12.3 The simplest form of the Schrödinger equation is for an electron in a potential well with infinitely high walls and width W. This is a simple model for many real situations, like the electrons in the potential "funnel" of the atoms. The solution of the Schrödinger equation shows that electrons can take only discrete energy values (*quantization effect*): $E_{\text{tot n}} = n^2\pi^2\hbar^2/(2mW^2)$, $(n = 1, 2, 3, \dots)$. A more realistic case for the electrons in semiconductor devices (MOSFET, HEMT, etc.) is a triangular potential well, where the discrete energy levels are given by

$$E_{\text{tot n}} \approx \left(\frac{\hbar^2}{2m^*}\right)^{1/3} \left[\frac{3\pi q}{2}\left(n - \frac{1}{4}\right)E_{\text{s}}\right]^{2/3}$$

where E_{s} is the electric field creating the slope of the triangular potential well.

12.4 The Schrödinger equation can also be used to predict and describe *tunneling*—the finite probability $\psi(x)\psi^*(x)$ of finding an electron on the other side of a potential barrier. There is an approximate WKB method that provides an equation for the tunneling probability for an arbitrary energy barrier $E_{\text{pot}}(x)$:

$$T \approx \exp\left[-2\int_a^b \kappa(x)\,dx\right], \qquad \kappa(x) = \sqrt{\frac{2m}{\hbar^2}[E_{\text{pot}}(x) - E_{\text{tot}}]} \qquad \text{for} \quad \begin{matrix} E_{\text{tot}} < E_{\text{pot}} \\ a \leq x \leq b \end{matrix}$$

Appendix *E*

CONTENTS OF *COMPUTER EXERCISES MANUAL*[1]

---

[1]This manual is provided on the CD, enclosed with the book.

LIST OF SELECTED SYMBOLS

Symbol	Description	SI Unit
A	Area	m^2
A_C	Base–collector area	m^2
A_E	Base–emitter area	m^2
A_J	P–N junction area	m^2
b	β reduction coefficient, **SPICE parameter**	
BV	P–N junction breakdown voltage, **SPICE parameter**	V
C	Capacitance	F
c	The speed of light, *physical constant*, $c = 3 \times 10^8$ m/s	
C_d	Depletion-layer capacitance	F
$C_d(0)$	Zero-bias depletion-layer capacitance, **SPICE parameter**	F
C_{meas}	Measured capacitance	F
C_p	Parasitic capacitance	F
C'_{ox}	Oxide capacitance	F
C_{ox}	Oxide capacitance per unit area	F/m^2
C_2	Base–emitter leakage current coefficient, **SPICE parameter**	
C_4	Base–collector leakage current coefficient, **SPICE parameter**	
D	*Section 1.3:* Diffusion coefficient	m^2/s
	Section 2.3: Electric flux density	C/m^2
D_n	Electron diffusion coefficient	m^2/s
D_p	Hole diffusion coefficient	m^2/s
D_0	Frequency factor (diffusion-coefficient parameter)	m^2/s
$E, E(x)$	Electric field	V/m
E_A	Activation energy	J (eV)
E_C	Bottom of the conduction band	J (eV)
E_F	Fermi level	J (eV)
E_g	Energy gap, **SPICE parameter**	J (eV)

556

E_i	Mid-gap energy	J (eV)
E_{kin}	Kinetic energy	J (eV)
E_{max}	Maximum electric field	V/m
E_{ox}	Gate-oxide field	V/m
E_{pot}	Potential energy	J (eV)
E_V	Top of the valence band	J (eV)
f	Electron occupancy probability	
F_B	A factor in LEVEL 3 MOSFET model	
f_h	Hole occupancy probability	
F_s	Charge sharing factor	
$F(t)$	Cumulative failure distribution function	
$f(t)$	Density failure distribution function	
g_m	Transconductance	A/V
g_n	External generation rate of electrons	$cm^{-3}\,s^{-1}$
g_p	External generation rate of holes	$cm^{-3}\,s^{-1}$
h	Planck's constant, *physical constant*, $h = 6.626 \times 10^{-34}$ Js	
$\hbar$	$\frac{h}{2\pi}$	
I	DC current	A
I_B	Base current	A
I_C	Collector current	A
I_{CB0}	Collector–base reverse-bias current	A
I_{CS}	Base–collector saturation current	A
I_D	*Chapter 3* Diode current	A
	Chapter 4: Output transistor DC current	A
	Chapter 5: Drain DC current	A
i_D	*Chapter 4:* Instantaneous transistor current	A
	Chapter 5: Instantaneous drain current	A
i_d	*Chapter 4:* Small-signal transistor current	A
	Chapter 5: Small-signal drain current	A
I_{Dsat}	Drain saturation current	A
I_E	Emitter current	A
I_{ES}	Base–emitter saturation current	A
I_{KF}	Normal knee current, **SPICE parameter**	A
I_{KR}	Inverse knee current, **SPICE parameter**	A
I_{nC}	Electron component of the collector current	A
I_{nE}	Electron component of the emitter current	A
I_{pC}	Hole component of the collector current	A
I_{pE}	Hole component of the emitter current	A
I_S	Saturation current, **SPICE parameter**	A
I_{S0}	Saturation current for $V_{CB} = 0$, **SPICE parameter**	A
I_{sub}	Substrate current	A
J_{diff}	Diffusion current of particles	s^{-1}
j	Current density	A/m^2
j_{dr}	Drift current density	A/m^2
j_{diff}	Diffusion current density	A/m^2

j_n	Electron current density	A/m^2
j_p	Hole current density	A/m^2
k	Boltzmann's constant, *physical constant*,	
	$k = 8.62 \times 10^{-5}$ eV/K $= 1.38 \times 10^{-23}$ J/K	
$\boldsymbol{k}; k_x, k$	Wave vector; wave number	m^{-1}
KP	Transconductance parameter, **SPICE parameter**	A/V
L_{eff}	Effective MOSFET channel length	m
L_g	MOSFET gate length, **SPICE parameter**	m
L_n	Diffusion length of electrons	m
L_p	Diffusion length of holes	m
l_{sc}	Scattering length	m
m	Grading coefficient, **SPICE parameter**	
	Chapter 12: Mass	kg
m_0	Mass of electron in vacuum	kg
m^*	Effective mass	kg
$MTTF$	Mean time to failure	s
$mttf$	Mean time to failure estimated by sampling	s
$\overline{mttf}$	Mean (average) $mttf$	s
N	Particle concentration	m^{-3}
n	*Chapter 1:* Electron concentration;	m^{-3}
	also, the first quantum number	
	Chapters 3 and 9: Emission coefficient, **SPICE parameter**	
	Chapter 11: Sample size	
N_A	Acceptor concentration, **SPICE parameter**	m^{-3}
N_B	Bulk doping concentration	m^{-3}
N_C	Effective concentration of states at E_C	m^{-3}
n_{CL}	Base–collector leakage emission coefficient,	
	SPICE parameter	
N_D	Donor concentration, **SPICE parameter**	m^{-3}
n_{EL}	Base–emitter leakage emission coefficient,	
	SPICE parameter	
N_{oc}	Oxide charge density, **SPICE parameter**	m^{-2}
N_V	Effective concentration of states at E_V	m^{-3}
n_i	Intrinsic carrier concentration	m^{-3}
n_n	Concentration of electrons in N-type regions	m^{-3}
n_p	Concentration of electrons in P-type regions	m^{-3}
n_{pe}	Equilibrium concentration of electrons in P-type regions	m^{-3}
p	Hole concentration	m^{-3}
$\boldsymbol{p}, p_x, p$	Momentum	kg m/s
p_n	Concentration of holes in N-type regions	m^{-3}
p_{ne}	Equilibrium concentration of holes in N-type regions	m^{-3}
p_p	Concentration of electrons in P-type regions	m^{-3}
q	Unit of charge, *physical constant*, $q = 1.6 \times 10^{-19}$ C	

Symbol	Description	Units
Q_d	Depletion-layer charge density	C/m^2
Q_I	Inversion-layer charge density	C/m^2
R	Resistance	Ω
R_D	Drain resistance, **SPICE parameter**	Ω
R_G	Gate resistance, **SPICE parameter**	Ω
r_n	Effective recombination rate of electrons	$m^{-3}\ s^{-1}$
r_o	Dynamic output resistance	Ω
r_p	Effective recombination rate of holes	$m^{-3}\ s^{-1}$
R_S	*Chapter 1:* Sheet resistance	$\Omega/\square$
	Chapter 5 and 9: Source resistance, **SPICE parameter**	Ω
r_S	Parasitic resistance, **SPICE parameter**	Ω
T	Absolute temperature	K
	Chapter 12: Period	s
t	Time	s
t_d	Dielectric thickness	m
t_{ox}	Oxide thickness, **SPICE parameter**	m
$\boldsymbol{v}, v_y, v$	Velocity	m/s
V	DC voltage	V
V_A	Normal Early voltage, **SPICE parameter**	V^{-1}
V_B	Inverse Early voltage, **SPICE parameter**	V^{-1}
V_{bi}	Built-in voltage, **SPICE parameters**	V
V_D	DC voltage across a diode	V
v_D	Instantaneous voltage across a diode	V
v_d	Drift velocity	m/s
V_{D0}	DC voltage across a P–N junction	V
V_{DS}	*Chapter 4:* Output transistor voltage—DC; (v_{DS}, instantaneous; v_{ds}, small-signal)	V
	Chapter 5: Drain-to-source voltage	V
V_{DSsat}	Drain-to-source saturation voltage	V
V_F	Forward-bias voltage	V
V_{FB}	Flat-band voltage	V
V_G	Gate voltage	V
V_{GS}	*Chapter 4:* Input transistor voltage—DC (v_{GS}, instantaneous; v_{gs}, small-signal)	V
	Chapter 5: Gate-to-source voltage	V
v_{max}	Maximum drift velocity, **SPICE parameter**	m/s
V_R	Reverse-biased voltage	V
V_T	Threshold voltage	V
V_{T0}	Zero-bias threshold voltage, **SPICE parameter**	
V_t	Thermal voltage	V
v_{th}	Thermal velocity	m/s
V_+	Positive power-supply voltage	V
V_-	Negative power-supply voltage	V
W	*Chapter 1:* Resistor width	m
	Chapter 5: MOSFET channel width, **SPICE parameter**	m

w_{base}	Base width	m
w_d	Depletion-layer width	m
w_n	Depletion-layer width in N-type region	m
w_p	Depletion-layer width in P-type region	m
x	Space coordinate	m
x_{ch}	Channel thickness	m
x_j	P–N junction depth, **SPICE parameter**	m
x_{j-lat}	Lateral diffusion, **SPICE parameter**	m
y	Space coordinate	m
z	Space coordinate	m
α	*Chapter 2:* Optical absorption coefficient	m^{-1}
	Chapter 6: Common-emitter current gain	
α_F	Common-base current gain in normal active mode	
α_R	Common-base current gain in inverse mode	
α_{sat}	Saturation coefficient, **SPICE parameter**	V^{-1}
β	*Chapter 5:* Gain factor	A/V^2
	Chapter 6: Common-base current gain	
	Chapter 9: Gain factor, **SPICE parameter**	A/V^2
β_F	Normal common-emitter current gain, **SPICE parameter**	
β_{FM}	Maximum normal current gain, **SPICE parameter**	
β_R	Inverse common-emitter current gain, **SPICE parameter**	
β_{RM}	Maximum inverse current gain, **SPICE parameter**	
β_0	Low-field gain factor	A/V^2
γ	Body factor, **SPICE parameter**	$V^{1/2}$
δ	Width effect on V_T, **SPICE parameter**	
$\delta n_p(x)$	Excess minority electron concentration	m^{-3}
$\delta p_n(x)$	Excess minority hole concentration	m^{-3}
ε_d	Dielectric permittivity	F/m
ε_{ox}	Oxide permittivity	F/m
ε_s	Silicon permittivity	F/m
ε_0	Vacuum dielectric constant, *physical constant*,	
	$\varepsilon_0 = 8.85 \times 10^{-12}$ F/m	
η	Static feedback, **SPICE parameter**	
θ	Mobility modulation coefficient, **SPICE parameter**	V^{-1}
λ	*Chapters 1, 2, and 12:* Wavelength	m
	Chapter 9: Reciprocal Early voltage, **SPICE parameter**	V^{-1}
	Chapter 11: Failure rate	s^{-1}
μ	Carrier mobility	$V/(m^2s)$
μ_0	Low-field channel-carrier mobility, **SPICE parameter**	$V/(m^2s)$
μ_{eff}	Effective carrier mobility	$V/(m^2s)$
μ_n	Electron mobility	$V/(m^2s)$
μ_p	Hole mobility	$V/(m^2s)$
μ_s	Surface mobility	$V/(m^2s)$
ν	Light frequency	Hz

ρ	Resistivity	Ω m
ρ_{charge}	Charge density	C/m^3
σ	Conductivity	$(\Omega\ m)^{-1}$
σ_D	$V_T\!-\!V_{DS}$ dependence coefficient	
τ_F	Normal transit time, **SPICE parameter**	s
τ_R	Inverse transit time, **SPICE parameter**	s
τ_n	Electron lifetime	s
τ_p	Hole lifetime	s
τ_T	Transit time, **SPICE parameter**	s
Φ	Dose of implanted/diffused doping atoms	m^{-2}
ϕ_B	Electric-potential equivalent of the energy barrier height	V
ϕ_F	Fermi potential	V
$\lvert 2\phi_F \rvert$	Surface potential in strong inversion, **SPICE parameter**	V
ϕ_m	Electric-potential equivalent of the metal work function	V
ϕ_{ms}	Electric potential counterpart of the metal-to-semiconductor work-function difference	V
$\varphi, \varphi(x)$	Variable electric potential	V
φ_s	Potential at the semiconductor surface	V
χ_s	Electron affinity	V
ω	Angular frequency	rad/s

BIBLIOGRAPHY

Chapters 1–6

D. Foty, *MOSFET Modeling with SPICE: Principles and Practice*, Prentice-Hall, Upper Saddle River, NJ, 1997.

G. Massobrio and P. Antognetti, *Semiconductor Device Modeling with SPICE*, 2nd ed., McGraw-Hill, New York, 1993.

R.S. Muller and T.I. Kamins, *Device Electronics for Integrated Circuits*, 2nd ed., Wiley, New York, 1986.

R.F. Pierret, *Semiconductor Device Fundamentals*, Addison-Wesley, Reading, MA, 1996.

D.K. Reinhard, *Introduction to Integrated Circuit Engineering*, Houghton Mifflin, Boston, 1987.

W.R. Runyan and K.E. Bean, *Semiconductor Integrated Circuit Processing Technology*, Addison-Wesley, Reading, MA, 1990.

M. Shur, *Introduction to Electronic Devices*, Wiley, New York, 1996.

B.G. Steetman, *Solid State Electronic Devices*, 4th ed., Prentice-Hall, Englewood Cliffs, NJ, 1995.

S.M. Sze, *Physics of Semiconductor Devices*, 2nd ed., Wiley, New York, 1981.

S.M. Sze, ed., *VLSI Technology,* McGraw-Hill, New York, 1983.

W.C. Till and J.T. Luxon, *Integrated Circuits: Materials, Devices, and Fabrication*, Prentice-Hall, Englewood Cliffs, NJ, 1982.

P. Tuinenga, *SPICE: A Guide to Circuit Simulation Using PSPICE*, Prentice-Hall, Englewood Cliffs, NJ, 1988.

C.T. Wang, *Introduction to Semiconductor Technology: GaAs and Related Compounds*, Wiley, New York, 1990.

R.M. Warner and B.L. Grung, *Semiconductor-Device Electronics*, Oxford University Press, New York, 1991.

Chapter 7

J.R. Brews, "The Submicron MOSFET," in *High-Speed Semiconductor Devices*, S.M. Sze, ed., Wiley, New York, 1990.

Chapter 8

B. Mroziewicz, M. Bugajski, and W. Nakwaski, *Physics of Semiconductor Devices*, North-Holland, Amsterdam, 1991.

J. Singh, *Semiconductor Devices—An Introduction*, McGraw-Hill, New York, 1994.

S.M. Sze , *Physics of Semiconductor Devices*, 2nd ed., Wiley, New York, 1981.

Chapter 9

J.J. Carr, *Microwave & Wireless Communications Technology*, Newnes, Boston, 1997.

I.M. Gottlieb, *Practical RF Power Design Techniques*, TAB Books, Blue Ridge Summit, 1993.

T. Koryu Ishii, *Practical Microwave Electron Devices*, Academic Press, San Diego, 1990.

The Microwave Engineering Handbook, Vol. 2, *Microwave Circuits, Antennas and Propagation,* B.L. Smith and M.-H. Carpentier, eds., Chapman & Hall, London, 1993.

S.M. Sze, *Physics of Semiconductor Devices*, 2nd ed., Wiley, New York, 1981.

S. Yngvesson, *Microwave Semiconductor Devices*, Kluwer Academic, Boston, 1991.

Chapter 10

B.J. Baliga, *Modern Power Devices*, Krieger, Malabar, 1992.

M.J. Fisher, *Power Electronics*, PWS-Kent, Boston, 1991.

J.G. Kassakian, M.F. Schlecht, and G.C. Verghese, *Principles of Power Electronics*, Addison-Wesley, Reading, MA, 1992.

N. Mohan, T.M. Undeland, and W.P. Robbins, *Power Electronics: Converters, Applications and Design*, 2nd ed., Wiley, New York, 1995.

Chapter 11

W.J. Bertram, "Yield and Reliability," in *VLSI Technology*, S.M. Sze, ed., Mc-Graw Hill, New York, 1983.

J.R. Davis, *Instabilities in MOS Devices*, Gordon and Breach Science Publishers, London, 1981.

L.J. Gallace, "Reliability," in *VLSI Handbook: Silicon, Gallium Arsenide, and Superconducting Circuits*, J. Di Giacomo, ed., Mc-Graw Hill, New York, 1989.

MIL-STD-883C, *Military Standard, Test Methods and Procedures for Microelectronics*, Department of Defense, Washington, DC, 1983.

A.G. Sabnis, *VLSI Reliability* (VLSI Electronics, Microstructure Science, Vol. 22), Academic Press, San Diego, 1990.

G.L. Schnable, "Failure Mechanisms in Microelectronic Devices," in *Microelectronic Reliability, Vol. 1: Reliability, Test and Diagnostics*, E.B. Hakim, ed., Artech House, Norwood, 1989.

Chapter 12

S. Brandt and H.D. Dahmen, *The Picture Book of Quantum Mechanics*, Springer-Verlag, New York, 1995.

D.K. Ferry, *Quantum Mechanics—An Introduction for Device Physicists and Electrical Engineers*, Institute of Physics Publishing, Bristol, 1995.

J.-M. Lévy-Leblond and F. Balibar, *Quantics—Rudiments of Quantum Physics*, 2nd ed., North-Holland, Amsterdam, 1990.

Appendix A

A.V. Ferris-Prabhu, *Introduction to Semiconductor Device Yield Modeling*, Artech House, Boston, 1992.

M.R. Haskard, *An Introduction to Application Specific Integrated Circuits*, Prentice-Hall, Sydney, 1990.

D. Pellerin and M. Holley, *Practical Design Using Programmable Logic*, Prentice-Hall, Englewood Cliffs, NJ, 1991.

ANSWERS TO SELECTED PROBLEMS

CHAPTER 1

1.5 $L = 56.4\ \mu$m, $W = 28.2\ \mu$m.

1.6 $R_S = 180\ \Omega/\square$, $R_c = 5\ \Omega$.

1.12 $n = 5.9 \times 10^{12}\ \mathrm{cm}^{-3}$.

1.13 $(\Delta R/R) \times 100 = -24.5\%$.

1.18 $R_S = 1116\ \Omega/\square$.

1.19 (b) $(dx_j/x_j) \times 100 = 1.70\%$.

1.23 (c) $R = 53.3\ \Omega$.

1.27 $E = 2.59$ V/mm.

1.35 (a) $R_S(27°\mathrm{C}) = 0.625\ \Omega/\square$, $R_S(700°\mathrm{C}) = 0.172\ \Omega/\square$.

CHAPTER 2

2.2 (c) $Q_C = 0.58\ \mathrm{mC/m}^2$.

2.6 (b) $V_{bi} = 0.30$ V.

2.8 (b) $V_{bi} = \frac{kT}{q} \ln \frac{N_D N_A}{N_C N_V} + \frac{E_g}{q}$.

2.13 (a) $V_{bi} = 0.588$ V. (d) Abrupt junction: $E_{max} = 1.04\ \mathrm{V}/\mu$m;

linear junction: $E_{max} = 0.53\ \mathrm{V}/\mu$m.

2.17 $C_d(0) = 3.9$ pF, $m = 0.43$.

2.24 (a) $q\phi_{ms} = 0.203$ eV.

2.25 (c) $V_{FB} = 0.87$ V.

2.28 N-type; $Q_I = 0$ C/m^2.

2.33 (c) $N_A = 1.77 \times 10^{17}$ cm^{-3}.

2.34 $\varphi_s = 0.80$ V, $V_{ox} = 5.16$ V.

2.36 (a) $E_{ox} = E_s = 0$; (b) $E_{ox} = 2.32$ V/μm, $E_s = 0$.

2.37 (b) $V_{BRinv} = -78.20$ V.

CHAPTER 3

3.4 (b) $n_p(|w_p|) = 1.26 \times 10^{-7}$ cm^{-3}, $p_n(|w_n|) = 1.26 \times 10^{-11}$ cm^{-3}

3.5 (b) $j_n = 3.20 \times 10^{-2}$ A/m^2, $j_p = 3.00 \times 10^{-7}$ A/m^2.

3.7 (b) $V_D = 2.54$ V.

3.14 (b) $n = 1.575$, $I_S = 8.10 \times 10^{-12}$ A.

3.17 $Q_s = 2.49 \times 10^{-11}$ C.

3.18 (b) $Q_{s-rec} = 2.9 \times 10^{-10}$ C.

3.20 $p_t = 2.5$.

3.23 (a) $V_{BR} = 1067$ V; (b) $V_{BR} = 12.7$ V; (c) $W_N \geq 7.13$ μm.

3.28 (b) 5.4 eV; 0.65 eV.

3.31 (a) $C_d(0) = 0.145$ mF/m^2.

CHAPTER 5

5.4 (b) $\varphi_s = 0.172V$, $Q_d = 5.36 \times 10^{-4}$ C/m^2 for $V_{GS} = -0.5$ V; $\varphi_s = 0.605$ V, $Q_d = 1.01 \times 10^{-3}$ C/m^2 for $V_{GS} = 0$ V; (d) $\varphi_s = 0.797$ V, $Q_d = 1.15 \times 10^{-3}$ C/m^2.

5.7 1.05 times.

5.8 (b) $Q_I = 2.17$ mC/m^2 for $V_{GS} = -0.75$ V;

$$Q_I = 0 \text{ for } V_{GS} = 0 \text{ and } 0.75 \text{ V.}$$

5.12 (b) $P_{diss} = 0$ W.

5.13 (b) $P_{diss} = 2.25$ μW.

5.18 $E = 2.5$ V/μm, $j = 9.0 \times 10^{10}$ A/m^2, $\mu_{eff} = 320$ cm^2/Vs.

5.19 (b) $V_{DSsat} = -0.943$ V, $I_{Dsat} = 2.22$ mA, $I_D = 0.80$ mA.

5.20 (b) $\Delta I_D = 1.49$ mA.

5.23 (a) $r_{o2} = 100$ kΩ.

5.27 $I_D = 1.81$ mA.

5.29 $C_{GS0} = C_{GD0} = 345$ pF/m, $C_{GB0} = 172.6$ pF/m.

CHAPTER 6

6.6 (e) BJT damaged (impossible).

6.8 (f) $|I_{B-max}| = 14.33$ μA.

6.9 46.7%.

6.10 (b) $R_S = 13.95$ kΩ/□.

6.14 (a) $\gamma_E = 1/\left(1 + \frac{D_{emitter}}{D_{base}} \frac{N_{base}}{N_{emitter}} \frac{w_{base}}{w_{emitter}}\right)$.

6.15 $\beta_R = 0.65$.

6.16 $I_C = -164.88$ mA, $I_E = 82.44$ mA, $I_B = 82.44$ mA.

6.20 $V_{BR} = 618$ V.

CHAPTER 8

8.1 $P = 2.1$ mW.

8.3 (a) $\delta n_p(x) = \delta n_p(w_p) \left(e^{[W_{anode} - (x - w_p)]/L_n} - \right.$

$$\left. e^{-[W_{anode} - (x - w_p)]/L_n}\right) \Big/ \left(e^{(W_{anode}/L_n)} - e^{-(W_{anode}/L_n)}\right).$$

8.4 (b) $\tau_T = W_{anode}^2/(2D_n)$; $\tau_T = 0.167$ ns.

8.5 (a) $V_{oc} = V_t \ln(1 + I_{photo}/I_S)$; $V_{oc} = 0.565$ V. (c) $N_{series} = 25$, $N_{parallel} = 5$.

8.8 (a) $\Phi_{opt} = 2.26 \times 10^{21}$ s^{-1}m^{-2}; (c) $I_{photo} = 361.6 \, \mu$A.

8.10 (a) $I_{photo} = 3.08$ mA, (b) 89.8% is diffusion photocurrent, (c) $n_p = 1.5 \times 10^{11}$ cm^{-3}, $p_n = 5 \times 10^{10}$ cm^{-3}.

CHAPTER 9

9.4 $f = 28.6$ GHz.

CHAPTER 10

10.2 (a) $\Delta V_F = 0.715$ V, (b) $\Delta V_{R-max} = 260 - 161.7 \approx 100$ V.

10.3 (b) $C_{GS} = 1.86$ nF.

CHAPTER 11

11.1 $\lambda = 5.03 \times 10^{-5}$ h^{-1}, $MTTF = 19,878 \, h$.

11.2

λ (FIT)	NSFM	PSBF (%)
10	0.072	0.86
100	0.72	8.6
1000	7.2	86

11.7 9.55%.

INDEX